RADIO FREQUENCY POWER IN PLASMAS

Previous Proceedings in the Series of Topical Conferences on Radio Frequency Power in Plasmas

Year	Conference	Publisher	ISBN
2003	15th	AIP Conf. Proceedings Vol. 694	0-7354-0158-6
2001	14th	AIP Conf. Proceedings Vol. 595	0-7354-0038-5
1999	13th	AIP Conf. Proceedings Vol. 485	1-56396-861-4
1997	12th	AIP Conf. Proceedings Vol. 403	1-56396-709-X
1995	11th	AIP Conf. Proceedings Vol. 355	1-56396-536-4
1993	10th	AIP Conf. Proceedings Vol. 289	1-56396-264-0
1991	9th	AIP Conf. Proceedings Vol. 244	0-88318-937-2

To learn more about these titles, or the AIP Conference Proceedings Series, please visit the webpage http://proceedings.aip.org/proceedings

RADIO FREQUENCY POWER IN PLASMAS

16th Topical Conference on
Radio Frequency Power in Plasmas
Park City, Utah 11 – 13 April 2005

EDITORS
Stephen J. Wukitch
Paul T. Bonoli
MIT Plasma Science and Fusion Center

SPONSORING ORGANIZATION
American Physical Society

Melville, New York, 2005
AIP CONFERENCE PROCEEDINGS ■ 787

Editors:

Stephen J. Wukitch
MIT Plasma Science and Fusion Center
190 Albany St. NW21-103
Cambridge, MA 02139
USA

E-mail: wukitch@psfc.mit.edu

Paul T. Bonoli
MIT Plasma Science and Fusion Center
175 Albany St. NW17-121
Cambridge, MA 02139
USA

E-mail: bonoli@psfc.mit.edu

Copyright in the following articles remains the property of the authors: pp. 50-53, 54-57, 90-97, 110-113, and 273-276.

The articles on pp. 122-129 and 315-318 are covered by British Crown Copyright, EURATOM/UKAEA Fusion Association, Oxfordshire.

Authorization to photocopy items for internal or personal use, beyond the free copying permitted under the 1978 U.S. Copyright Law (see statement below), is granted by the American Institute of Physics for users registered with the Copyright Clearance Center (CCC) Transactional Reporting Service, provided that the base fee of $22.50 per copy is paid directly to CCC, 222 Rosewood Drive, Danvers, MA 01923, USA. For those organizations that have been granted a photocopy license by CCC, a separate system of payment has been arranged. The fee code for users of the Transactional Reporting Services is: ISBN/0-7354-0276-0/05/$22.50.

© 2005 American Institute of Physics

Permission is granted to quote from the AIP Conference Proceedings with the customary acknowledgment of the source. Republication of an article or portions thereof (e.g., extensive excerpts, figures, tables, etc.) in original form or in translation, as well as other types of reuse (e.g., in course packs) require formal permission from AIP and may be subject to fees. As a courtesy, the author of the original proceedings article should be informed of any request for republication/reuse. Permission may be obtained online using Rightslink. Locate the article online at http://proceedings.aip.org, then simply click on the Rightslink icon/"Permission for Reuse" link found in the article abstract. You may also address requests to: AIP Office of Rights and Permissions, Suite 1NO1, 2 Huntington Quadrangle, Melville, NY 11747-4502, USA; Fax: 516-576-2450; Tel.: 516-576-2268; E-mail: rights@aip.org.

L.C. Catalog Card No. 2005932081
ISBN   0-7354-0276-0
ISSN   0094-243X
Printed in the United States of America

CONTENTS

Preface xiii

ION CYCLOTRON RANGE OF FREQUENCIES

Nonlinear ICRF-Plasma Interactions *(Review)* 3
    J. R. Myra, D. A. D'Ippolito, D. A. Russell, L. A. Berry, E. F. Jaeger, and
    M. D. Carter

**Global and Self-consistent Simulation of ICRF Heating in Toroidal
Plasmas** *(Invited)* 15
    S. Murakami, A. Fukuyama, T. Akutsu, N. Nakajima, V. Chan, M. Choi,
    S. C. Chiu, L. Lao, V. Kasilov, T. Mutoh, R. Kumazawa, T. Seki, K. Saito,
    T. Watari, M. Isobe, T. Saida, M. Osakabe, M. Sasao and LHD
    Experimental Group

**Self-consistent, Full-wave/Fokker-Planck Calculations for Ion
Cyclotron Heating in Non-Maxwellian Plasmas** *(Invited)* 23
    E. F. Jaeger, L. A. Berry, R. W. Harvey, J. R. Myra, R. J. Dumont,
    C. K. Phillips, D. N. Smithe, D. B. Batchelor, P. T. Bonoli, M. D. Carter,
    E. D'Azevedo, D. A. D'Ippolito, and J. C. Wright

**Monte-Carlo Orbit/Full Wave Simulation of Fast Alfvén Wave (FW)
Damping on Resonant Ions in Tokamaks** *(Invited)* 31
    M. Choi, V. S. Chan, V. Tang, P. Bonoli, R. I. Pinsker, and J. Wright

**Numerical Simulation of Ion Cyclotron Heating Experiments with
Coupled Maxwell and Quasilinear Fokker-Planck Solvers** 38
    M. Brambilla, R. Bilato, C. Maggi, H.-U. Fahrbach, W. Suttrop, and the
    ASDEX Upgrade Team

Positive Quasi Linear Operator Formulation 42
    L. A. Berry, E. F. Jaeger, and the SciDAC RF-Plasma Interactions Team

**Velocity-Space Diffusion Coefficients Due to Full-Wave ICRF Fields
in Toroidal Geometry** 46
    R. W. Harvey, F. Jaeger, L. A. Berry, N. M. Ershov, A. P. Smirnov,
    P. Bonoli, J. C. Wright, D. B. Batchelor, E. D'Azevedo, M. D. Carter,
    and D. N. Smithe

**Effects of Finite Orbit Width and RF-induced Spatial Diffusion on
Ion Cyclotron Emission** 50
    T. Hellsten, T. Bergkvist, T. Johnson, and M. Laxåback

**Analysis of a Quasi-Linear Model for Ion-Cyclotron Interactions
in Tokamaks** 54
    T. Johnson, T. Hellsten, and L.-G. Eriksson

**A Simple Method to Account for Drift Orbit Effects when Modeling
Radio Frequency Heating in Tokamaks** 58
    D. Van Eester

A Full-Wave Solution of the Maxwell's Equations in 3D Plasmas 62
    P. Popovich, N. Mellet, L. Villard, and W. A. Cooper

Parametric Decay During HHFW on NSTX *(Invited)* 66
    J. R. Wilson, S. Bernabei, T. Biewer, S. Diem, J. Hosea, B. LeBlanc,
    C. K. Phillips, P. Ryan, and D. W. Swain

Absorption of Fast Waves at Moderate to High Ion Cyclotron Harmonics on DIII-D *(Invited)* 74
    R. I. Pinsker, M. Porkolab, W. W. Heidbrink, Y. Luo, C. C. Petty, R. Prater,
    M. Choi, F. W. Baity, E. Fredd, J. C. Hosea, R. W. Harvey, A. P. Smirnov,
    M. Murakami, and M. A. Van Zeeland

Electron Energy Confinement For HHFW Heating and Current Drive Phasing on NSTX 82
    J. C. Hosea, S. Bernabei, T. Biewer, B. LeBlanc, C. K. Phillips,
    J. R. Wilson, D. Stutman, P. Ryan, and D. W. Swain

Investigation of HHFW and NBI Combined Heating in NSTX 86
    B. P. LeBlanc, R. E. Bell, S. Bernabei, T. M. Biewer, J. C. Hosea, and
    J. R. Wilson

Development of RF Tools and Scenarios for ITER on JET *(Invited)* 90
    J.-M. Noterdaeme, M. Mantsinen, V. Bobkov, A. Ekedahl, L.-G. Eriksson,
    P. U. Lamalle, A. Lyssoivan, J. Mailloux, M.-L. Mayoral, F. Meo,
    I. Monakhov, K. Rantamaki, A. Salmi, M. Santala, S. Sharapov, D. Van
    Eester, JET-EFDA Task Force H and JET-EFDA Contributors

Long Pulse Plasma Heating Experiment by Ion Cyclotron Heating in LHD *(Invited)* 98
    T. Seki, T. Mutoh, R. Kumazawa, K. Saito, T. Watari, Y. Nakamura,
    M. Sakamoto, T. Watanabe, S. Kubo, T. Shimozuma, Y. Yoshimura,
    H. Igami, K. Ohkubo, Y. Takeiri, Y. Oka, K. Tsumori, M. Osakabe,
    K. Ikeda, K. Nagaoka, O. Kaneko, J. Miyazawa, S. Morita, K. Narihara,
    M. Shoji, S. Masuzaki, M. Goto, T. Morisaki, B. J. Peterson, K. Sato,
    T. Tokuzawa, N. Ashikawa, K. Nishimura, H. Funaba, H. Chikaraishi,
    N. Takeuchi, T. Notake, H. Ogawa, Y. Torii, F. Shimpo, G, Nomura,
    M. Yokota, C. Takahashi, A. Kato, Y. Takase, H. Kasahara, M. Ichimura,
    H. Higaki, Y. P. Zhao, J. G. Kwak, H. Yamada, K. Kawahata, N. Ohyabu,
    K. Ida, Y. Nagayama, N. Noda, A. Komori, S. Sudo, O. Motojima, and
    LHD Experimental Group

Modeling of Discharges with Fast Wave Power in DIII-D 106
    R. Prater, C. Choi, R. W. Harvey, J. C. Hosea, C. C. Petty, R. I. Pinsker,
    M. Porkolab, and A. P. Smirnov

Toroidal Rotation and ICRF Heating in NBI-driven Discharges in JET 110
    J. S. deGrassie, L.-G. Eriksson, J.-M. Noterdaeme, and JET EFDA
    Contributors

ICRH Current Ramp Discharges and Alfven Cascades in Alcator C-Mod 114
    M. Porkolab, E. Edlund, J. Snipes, S. Wukitch, N. Basse, P. Bonoli,
    C. Boswell, C. Fiore, N. Gorelenkov, A. Hubbard, G. J. Kramer, L. Lin,
    Y. Lin, E. Marmar, and G. Schilling

Edge Minority Heating Experiment in Alcator C-Mod 118
    S. J. Zweben, J. L. Terry, P. Bonoli, R. Budny, C. S. Chang, C. Fiore,
    J. Hughes, Y. Lin, R. Perkins, M. Porkolab, G. Schilling, S. Wukitch and
    the Alcator C-Mod Team

ICRF Heating for the Non-Activated Phase of ITER: From Inverted Minority to Mode Conversion Regime *(Invited)* 122
    M.-L. Mayoral, P. U. Lamalle, D. Van Eester, P. Beaumont, E. De La Luna,
    P. De Vries, C. Gowers, R. Felton, J. Harling, V. Kiptily, K. Lawson,
    M. Laxåback, E. Lerche, P. Lomas, M. J. Mantsinen, F. Meo,
    J.-M. Noterdaeme, I. Nunes, G. Piazza, M. Santala, and JET-EFDA
    Contributors

Current and Flow Driven in Thin Layers 130
    H. Weitzner

Investigation of Mode-Transformed Ion Cyclotron Waves at the Ion-Ion Hybrid Layer 134
    R. Bilato and M. Brambilla

Numerical Studies of Poloidal Field Effects on ICRF Mode Conversion 138
    A. Parisot, S. J. Wukitch, P. Bonoli, Y. Lin, R. Parker, M. Porkolab,
    A. K. Ram, and J. C. Wright

Modeling of Ray Splitting in a Tokamak 142
    E. R. Tracy, A. Jaun, and A. N. Kaufman

Triplicate Budden Resonance in the Presence of Sheared Flow 146
    A. N. Kaufman, A. J. Brizard, and E. R. Tracy

Theory and Practice in ICRF Antennas for Long Pulse Operation *(Invited)* 150
    L. Colas, E. Faudot, S. Brémond, S. Heuraux, R. Mitteau, M. Chantant,
    M. Goniche, V. Basiuk, G. Bosia, J. P. Gunn, and the Tore Supra Team

Investigation of "Conjugate T" Load-resilient ICRF Antenna Systems—Application to the JET ITER-Like and to a Possible ITER ICRF System *(Invited)* 158
    P. U. Lamalle, A. M. Messiaen, P. Dumortier, F. Durodié, M. Evrard,
    F. Louche, M. Vervier, and R. Weynants

Validation of a 3D/1D Simulation Tool for ICRF Antennas *(Invited)* 166
    R. Maggiora, V. Lancellotti, D. Milanesio, G. Vecchi, V. Kyrytsya,
    A. Parisot, and S. J. Wukitch

Automatic Control of ITER-like Structures 174
    G. Bosia and S. Brémond

Effects of Coupling and Asymmetries on Load Resilience of IC ITER-like Structures 178
    G. Bosia, S. Brémond, and L. Colas

Proposals for ITER Ion Cyclotron Reference Design Upgrades 182
    G. Bosia, B. Beaumont, S. Brémond, and L. Colas

RF Circuit Simulation of the JET ITER-like ICRH Antenna 186
    M. Evrard, F. Durodié, and P. U. Lamalle

Three-Dimensional Electromagnetic Modeling of the ITER ICRF Antenna (External Matching Design) 190
    F. Louche, P. U. Lamalle, P. Dumortier, and A. M. Messiaen

Status of the JET ITER-like Antenna High-Power Prototype
Test Program 194
    R. H. Goulding, F. W. Baity, F. Durodié, A. Fadnek, K. D. Freudenberg,
    J. C. Hosea, G. D. Loesser, B. E. Nelson, M. Nightingale,
    D. A. Rasmussen, D. O. Sparks, and R. Walton

Study of the ITER ICRH System with External Matching by Means
of a Mock-up Loaded by a Variable Water Load 198
    A. Messiaen, M. Vervier, P. Dumortier, P. Lamalle, and F. Louche

Experimental Proof of a Load Resilient External Matching Solution
for the ITER ICRH System 202
    M. Vervier, A. Messiaen, P. Dumortier, and P. Lamalle

Initial Operation of the Alcator C-Mod ICRF Antennas with High-Z
Metal Antenna Guards 206
    G. Schilling, S. J. Wukitch, Y. Lin, A. Parisot, M. Porkolab, and the
    Alcator C-Mod Team

Heat Loads on Tore Supra ICRF Launchers Plasma
Facing Components 210
    S. Brémond, L. Colas, M. Chantant, B. Beaumont, A. Ekedahl,
    M. Goniche, P. Moreau, and R. Mitteau

2D Modeling of DC Potential Structures Induced by RF Sheaths with
Transverse Currents in Front of ICRF Antenna 214
    E. Faudot, S. Heuraux, and L. Colas

Fusion Antenna Analysis Using the Modular Oak Ridge RF
Integration Code (MORRFIC) 218
    M. D. Carter, D. A. D'Ippolito, J. R. Myra, and D. A. Russell

Integrated Codes for ICRF-Edge Plasma Interactions 222
    D. A. D'Ippolito, J. R. Myra, D. A. Russell, and M. D. Carter

A Particle-in-Cell Approach to Time Domain Simulations of the ICRF
Edge Regions 226
    D. N. Smithe and C. K. Phillips

Prediction of Plasma-facing ICRH Antenna Behavior via a
Finite-Element Solution of Coupled Integral Equations 230
    V. Lancellotti, D. Milanesio, R. Maggiora, G. Vecchi, and V. Kyrytsya

An Alternative Method for Calculating the RF Plasma Dielectric
Response in ICRH Simulations 234
    E. A. Lerche and P. U. Lamalle

Antenna Optimization by Using Finite Element Programs 238
    F. Braun and ICRF Group

Measurements and Calculations of CW Transmission Lines for ICRF
in W7-X 242
    D. Birus, D. A. Hartmann, W. Becker and W7-X Team

The ICRH System Planned for Wendelstein 7-X 246
    D. A. Hartmann, D. Birus, J. Wendorf, and F. Wesner

Development of 2 MW Dummy Load for KSTAR ICH System 250
    J.-G. Kwak, S. J. Wang, Y. D. Bae, J. S. Yoon, and B. G. Hong

CURRENT DRIVE

Synergy in RF Current Drive *(Invited)* 257
    R. J. Dumont and G. Giruzzi
Modulated Current Drive Measurements 265
    C. C. Petty, W. A. Cox, C. B. Forest, R. J. Jayakumar, J. Lohr, T. C. Luce,
    M. A. Makowski, and R. Prater
RF Current Drive in Internal Transport Barrier 269
    Y. Peysson, J. Decker, V. Basiuk, A. Bers, G. Huysmans, and A. K. Ram
Fast Wave Current Drive in JET ITB-Plasmas 273
    T. Hellsten, M. Låxaback, T. Bergkvist, T. Johnson, M. Mantsinen,
    G. Matthews, F. Meo, F. Nguyen, J.-M. Noterdaeme, C. C. Petty,
    T. Tala, D. Van Eester, P. Andrew, P. Beaumont, V. Bobkov, M. Brix,
    J. Brzozowski, L.-G. Eriksson, C. Giroud, E. Joffrin, V. Kiptily,
    J. Mailloux, M.-L. Mayoral, I. Monakhov, R. Sartori, A. Staebler,
    E. Rachlew, E. Tennfors, A. Tuccillo, W. Walden, K.-D. Zastrow,
    and JET-EFDA Contributors

LOWER HYBRID RANGE OF FREQUENCIES

**Recent RF Experiments and Application of RF Waves to Real-Time
Control of Safety Factor Profile in JT-60U** *(Invited)* 279
    T. Suzuki, A. Isayama, S. Ide, T. Fujita, T. Oikawa, S. Sakata, M. Sueoka,
    H. Hosoyama, M. Seki, and the JT-60 Team
**Full-wave Electromagnetic Field Simulations of Lower Hybrid Waves
in Tokamaks** *(Invited)* 287
    J. C. Wright, P. T. Bonoli, M. Brambilla, E. D'Azevedo, L. A. Berry,
    D. B. Batchelor, E. F. Jaeger, M. D. Carter, C. K. Phillips, H. Okuda,
    R. W. Harvey, J. R. Myra, D. A. D'Ippolito, and D. N. Smithe
**Bridging the Spectral Gap in Lower Hybrid Current Drive by
Parametric Instability** *(Invited)* 295
    R. Cesario, C. Castaldo, A. Cardinali, and F. Paoletti
Long Pulse, Multi-MW Operation in Tore Supra 303
    G. T. Hoang on behalf of the Tore Supra Team
Lower Hybrid Current Drive Efficiency on Tore Supre and JET 307
    M. Goniche, J. F. Artaud, V. Basiuk, Y. Peysson, T. Aniel, A. Ekedahl,
    G. Giruzzi, F. Imbeaux, J. Mailloux, D. Mazon, W. Zwingman and JET
    EFDA Contributors
**Statistical Analysis of Lower Hybrid Current Drive Efficiency
on FTU** 311
    G. Calabrò, V. Pericoli Ridolfini, FTU Team, and ECRH Team
**Impurity Radiation from the LHCD Launcher During Operation in
JET and Investigation of Launcher Damage** 315
    K. K. Kirov, J. Mailloux, A. Ekedahl, LHCD Team and JET-EFDA
    Contributors

Lower Hybrid Experiments on MST 319
    M. C. Kaufman, J. A. Goetz, M. A. Thomas, D. R. Burke, and
    D. J. Clayton

Lower Hybrid Antenna Design for MST 323
    J. A. Goetz, M. A. Thomas, M. C. Kaufman, and S. P. Oliva

Implementation of LHCD Experiments on Alcator C-Mod 327
    R. Parker, N. Basse, W. Beck, S. Bernabei, R. Childs, R. Ellis,
    E. Fredd, N. Greenough, M. Grimes, D. Gwinn, J. Hosea, J. Irby,
    P. Koert, C. C. Kung, B. Labombard, J. Liptac, G. D. Loesser,
    E. Marmar, G. Schilling, D. Terry, J. Terry, R. Vieira, G. Wallace,
    J. R. Wilson, and J. Zaks

**Microstrip Directional Coupler Design for a Reduced
Height Waveguide** 331
    G. Wallace, P. Koert, R. Parker, D. Terry, and S. J. Wukitch

ELECTRON BERNSTEIN RANGE OF FREQUENCIES

**Electron Bernstein Wave Research on the National Spherical
Torus Experiment** 337
    G. Taylor, A. Bers, T. S. Bigelow, M. D. Carter, J. B. Caughman,
    J. Decker, S. Diem, P. C. Efthimion, N. M. Ershov, E. Fredd, R. W. Harvey,
    J. Hosea, F. Jaeger, J. Preinhaelter, A. K. Ram, D. A. Rasmussen,
    A. P. Smirnov, J. B. Wilgen, and J. R. Wilson

EBW Experments in the Madison Symmetric Torus 341
    J. K. Anderson, M. Cengher, W. A. Cox, C. B. Forest, S. M. McMahon,
    R. I. Pinsker, and V. Svidzinski

**Survey of EBW Mode-conversion Characteristics for Various
Boundary Conditions** 345
    H. Tanaka, H. Igami, and T. Maekawa

EBW Simulation for MAST and NSTX Experiments 349
    J. Preinhaelter, G. Taylor, V. Shevchenko, J. Urban, M. Valovic, P. Pavlo,
    L. Vahala, and G. Vahala

Self-consistent Formulation of EBW Excitation by Mode Conversion 353
    A. Bers and J. Decker

**Fully Relativistic Ray-tracing and Dispersion Relations of Electron
Bernstein Waves** 357
    E. Nelson-Melby, R. W. Harvey, A. P. Smirnov, A. K. Ram, and S. Coda

Relativistic Modifications to Electron Bernstein Waves 361
    A. K. Ram and J. Decker

Plasma Current Start-up by Outboard PF Coils in JT-60U and TST-2 365
    Y. Takase, A. Ejiri, K. Hanada, S. Ide, O. Mitarai, S. Shiraiwa,
    M. Ushigome, JT-60 Team and TST-2@K Team

ELECTRON CYCLOTRON RANGE OF FREQUENCIES

**The 10 MW, CW ECRH System for W7-X: Status and First
Integrated Tests** *(Invited)* 371
    V. Erckmann, P. Brand, H. Braune, G. Dammertz, G. Gantenbein,
    W. Kasparek, H. P. Laqua, G. Michel, M. Thumm, and the W7-X ECRH
    Teams at IPP, FZK and IPF

**Performance Optimization in ASDEX Upgrade with ECRH and
ECCD** *(Invited)* 379
    F. Leuterer, C. Angioni, R. Dux, G. Gantenbein, S. Günter, A. Manini,
    M. Maraschek, A. Mück, M. Münich, R. Neu, A. G. Peeters, F. Ryter,
    D. Wagner, H. Zohm and ASDEX Upgrade Team

**Third Harmonic X-mode ECH Top-launch on TCV
Tokamak** *(Invited)* 387
    G. Arnoux, S. Alberti, L. Porte, S. Nowak, B. Marlétaz, Ph. Marmillod,
    Y. Martin and TCV Team

**Initial Results of Multi-frequency Electron Cyclotron Heating in the
Levitated Dipole Experiment** 395
    A. K. Hansen, S. Mahar, A. C. Boxer, J. L. Ellsworth, D. T. Garnier,
    I. Karim, J. Kesner, M. Mauel, and E. E. Ortiz

**Control of the eITB Formation and Performance in Fully
Non-inductively Sustained ECCD Discharges in TCV** 399
    M. A. Henderson, R. Behn, A. Bottino, Y. Camenen, S. Coda, E. Fable,
    T. P. Goodman, An. Martynov, P. Nikkola, A. Pochelon, O. Sauter,
    C. Zucca and the TCV Team

**The Physics Performance of the Front Steering Launcher for the
ITER ECRH Upper Port** 403
    M. Henderson, R. Chavan, P. Nikkola, G. Ramponi, G. Saibene,
    F. Sanchez, O. Sauter, H. Shidara, and H. Zohm

**Performance of the ECH Transmission Lines and Launchers
in DIII-D** 407
    K. Kajiwara, A. Muir, C. B. Baxi, J. Lohr, I. A. Gorelov, M. T. Green,
    D. Ponce, and R. W. Callis

Long Pulse ECH Plasma in LHD 411
    S. Kubo, Y. Yoshimura, T. Shimozuma, H. Igami, T. Notake, R. Kumazawa,
    T. Seki, K. Saito, Y. Nakamura, T. Mutoh, K. Ohkubo and LHD
    Experimental Group

**Formation of Spherical Tokamak by ECH without Center Solenoid in
the LATE Device** 415
    H. Tanaka, M. Uchida, T. Yoshinaga, J. Yamada, S. Yamaguchi, and
    T. Maekawa

RF PLASMA APPLICATIONS

**Acceleration of Dense Flowing Plasmas using ICRF Power in the
VASIMR Experiment** *(Invited)* 421
    J. P. Squire for the VASIMR Team

Capacitive Systems for Dielectric Plasma Etch *(Invited)* **429**
    D. Hoffman
RF Power Coupling and Plasma Transport in Magnetized
Capacitive Discharges **437**
    P. M. Ryan, M. D. Carter, and D. J. Hoffman
A Simulation Approach for ICRF Plasma Thruster Antennas **441**
    G. Vecchi, L. Valitutti, V. Lancellotti, R. Maggiora, and D. Milanesio
Studies of ICRF Discharge Conditioning (ICRF-DC) on ASDEX
Upgrade, JET and TEXTOR **445**
    A. Lyssoivan, R. Koch, D. Van Eester, G. Van Wassenhove, M. Vervier,
    R. Weynants, H. G. Esser, V. Philipps, G. Sergienko, E. Gauthier,
    V. Bobkov, H.-U. Fahrbach, D. A. Hartmann, J.-M. Noterdaeme, V. Rohde,
    W. Suttrop, I. Monakhov, A. Walden, TEXTOR Team, ASDEX Upgrade
    Team and JET-EFDA Contributors
Experimental Results of the Coaxial Multipactor Experiment (CMX) **449**
    T. P. Graves, B. LaBombard, S. J. Wukitch, and I. H. Hutchinson

Author Index **453**

PREFACE

The Sixteenth Topical Conference on Radio Frequency Power in Plasmas was held on April 11-13, 2005 at the Park City Marriott in Park City, Utah. The meeting was sponsored by the Massachusetts Institute of Technology Plasma Science and Fusion Center and the American Institute of Physics and endorsed by the Division of Plasma Physics of the American Physical Society. The Conference was co-chaired by P.T. Bonoli and S.J. Wukitch of MIT PSFC. The Program Committee included Drs. A. Becoulet, D. Van Eester, A. Fukuyama, D. Hartmann, C.C. Petty, P. Ryan, and J.R. Wilson.

A total of 102 papers were presented at the Conference: two reviews, twenty-three invited papers, and 78 poster presentations. Reflecting the international nature of the meeting, 44 presentations were from the European Union, 7 from Japan, and 2 from Korea. About 90% of the papers were from plasma fusion science with a few papers from space propulsion and material processing. Of the 108 registered attendees, half were from the United States and the remainder Europe, Japan, and Korea.

The Conference program reflected recent trends in funding and scientific emphasis. An advanced scientific computation initiative has spurred full wave solutions in tokamak geometries where the phenomena to be modeled ranged from mm to tens of centimeters and self-consistent models of energetic particles and waves. Another area of emphasis is ICRF antenna design for present and next step devices. In total about 50% of the papers could be categorized as ICRF papers perhaps reflecting the perceived importance of ICRF power for future burning plasma experiments.

We would like to acknowledge the Program Committee for its diligent effort in crafting an excellent Conference program. The Conference co-chairs would like to commend the Conference site selection committee Chairman and Prof. M. Porkolab for selecting Park City and coordinating conference activities. We would like to acknowledge the superb efforts by Valerie Censabella and Carol Arlington in making this a very successful conference. A special thanks is due to Corrine Fogg for her assistance in organizing this conference. Finally, we would like to thank the many participants for their effort and traveling to Park City making the Conference a success.

The Seventeenth Topical Conference will be hosted by Oak Ridge National Laboratory in 2007.

ION CYCLOTRON RANGE OF FREQUENCIES

Nonlinear ICRF-Plasma Interactions

J. R. Myra, D. A. D'Ippolito, and D.A. Russell

Lodestar Research Corporation
Boulder, Colorado, USA

L. A. Berry, E. F. Jaeger, and M. D. Carter

Oak Ridge National Laboratory
Oak Ridge, Tennessee, USA

Abstract. The well developed linear theory of ICRF (including FW, HHFW and IBW) interactions with plasma has enjoyed considerable success in describing antenna coupling and wave propagation, and provides a well-known framework for calculating power absorption, current drive, etc. In some situations, less well studied nonlinear effects are of interest, such as flow drive, ponderomotive forces, rf sheaths, parametric decay and related interactions with the edge plasma. Standard ICRF codes have begun to integrate this physics to achieve improved modeling capabilities. This paper concentrates on basic rf-plasma-interaction physics with illustrative applications to tokamaks. For FW antennas, the parallel electric field near launching structures is known to drive rf-sheaths which can give rise to convective cells, interaction with plasma "blobs", impurity production, and edge power dissipation. In addition to sheaths, IBW waves in the edge plasma are subject to strong ponderomotive effects and parametric decay. In the core plasma, slow waves can sometimes induce nonlinear effects. Mechanisms by which these waves can influence the radial electric field and its shear are summarized, and related to the general (reactive-ponderomotive and dissipative) force on a plasma from rf waves. It is argued that there are significant opportunities now for new predictive capabilities by advances in integrated simulation.

Keywords: nonlinear, ion cyclotron range, sheath, ponderomotive, parametric, flow, shear.
PACS: 52.40.Kh, 52.35.Mw, 52.40.Fd

INTRODUCTION

RF physics is arguable one of the richest areas of plasma physics in which linear theory is both widely applicable to experiment and exceedingly rich in the complexity and subtlety of physical phenomena which it can describe. This fact underlies many decades of successful theoretical research in ion cyclotron range of frequencies (ICRF) wave physics. Nevertheless, there are a few instances when linear physics fails, and nonlinear effects can become important. The most obvious case is near the antenna where the rf fields are large, and typical rf voltages, and/or ponderomotive potentials easily exceed the local plasma temperature T_e. Wave-induced rf sheath voltages can also exceed T_e near walls and limiters where the plasma temperature is low. In addition to these situations in the edge plasma, nonlinear effects can also be important further in towards the core for slow waves, which have a small group velocity and therefore require a large electric field to carry power. In these cases, nonlinear plasma interactions can drive plasma flows and radial electric field shear, of interest for turbulence suppression and transport barrier formation.

The goal of this paper is to summarize physics concepts in these areas where nonlinear effects enter, indicate available modeling and analysis tools, and point out opportunities for new predictive capabilities. Although experimental motivation will be given, this paper will not attempt a comprehensive experimental review. Many of the relevant older experimental references have been given elsewhere.[1-3]

The outline of our paper is as follows. In the next section we consider antenna-edge interactions arising in fast wave (FW) launch from rf sheaths. This is followed by a discussion of additional effects present in ion Bernstein wave (IBW) direct launch experiments. Then IBW and ion cyclotron wave (ICW) core interactions and driven flows are reviewed. The final section presents some thoughts on the opportunities and prospects for future theoretical work involving integrated computer modeling.

FW LAUNCH ANTENNA-EDGE INTERACTIONS: RF SHEATHS

A variety of rf edge interactions have been seen on ICRF experiments for several decades.[1,2] These rf specific effects include: impurities (RF-enhanced sputtering), density rise, arcs, antenna damage, and anomalous edge power dissipation. RF antenna voltage controls the near field rf-specific effects, which are generally least severe in anti-symmetric (e.g. dipole) phasing.[4] Experience and intuition developed over the years have partly, if not mostly mitigated deleterious effects in many experiments, under most circumstances. But, it is likely that they will have important implications for present and future long pulse operation, where even small effects can have large consequences.[5]

The primary culprit for many of the observed phenomena is the rf sheath which exists at "end-plates" where the field line contacts a conductor. Important places where rf sheaths occur will be identified later. The basic physics underlying a sheath is that both species, electrons and ions, initially try to leave at their respective thermal velocities. In response to the growing charge imbalance, the plasma develops a potential to confine the electrons. This potential, which must be higher than the applied voltage at either of the two ends, reflects almost all of the electrons at the sheath entrance. The sheath width $\Delta \sim \lambda_d (eV/T_e)^{3/4}$ at each end is determined by requiring that the un-neutralized ion space charge in the sheath layer is sufficient to give rise to the requited potential drop. Here λ_d is the Debye length, e the proton charge, V the applied voltage and T_e the electron temperature. In addition to reflecting electrons, this large sheath accelerates ions into the plates causing sputtering. The energy for this accelerations comes from the circuit, and appears as lost power to the sheath. Thus, the whole process is driven by the need for charge ambipolarity.

This basic sheath physics extends immediately into an ICRF sheath,[6-9] where an oscillating voltage is applied to each plate. Electrons leave alternately out one end, then the other, escaping from the end where applied voltage is highest (and hence the reflecting barrier seen by the electrons lowest). This give rise to an oscillating parallel electron current. The central voltage oscillates up and down at twice the applied frequency, but always remains higher that the applied voltage at either end. The net effect is that there is both rectification of the applied voltage and a large second harmonic. The net sheath power dissipation (for the case of Maxwell-Boltzmann electrons) is given by[10,11]

$$P_{sh} = n_e c_s A T_e \xi I_1(\xi)/I_0(\xi) \rightarrow n_e c_s A e V_{rf} \qquad (1)$$

where $n_e c_s$ is the plasma flux, A is the projected area normal to the magnetic field, $\xi = eV_{rf}/T_e$, I_0 and I_1 are Bessel functions, and the final form is the high voltage limit. The two most important parameters are the plasma density (and flux) into the antenna, and the rf voltage.

Several other considerations are important, including the angle at which the field line strikes the plate. This influences the ion orbits in the sheath, and impacts surface physics calculations such as sputtering. Simulations have been performed to quantify these and other effects.[7,8] In particular, for shallow incidence of the field lines onto the surface, there is a significant magnetic presheath in which the ion flow transitions from being sonic along the field lines to sonic normal to the plate.[8] These types of calculations confirm that the sheath voltage drop available for ion acceleration and power dissipation is normally an order unity fraction of the applied rf voltage, because the potential drop in the sheaths is largely controlled by electron physics and simple ambipolarity considerations.

One of the goals that is being pursued in contemporary research is that of including plasma and rf-sheath effects in antenna coupling codes. Since it is still impractical to do full wave particle simulations for the rf fields and sheaths, it is important to be able to characterize the main effect of the sheaths in a simple way. A useful model[12] is to regard the electrons as an oscillating charge layer, which leaves a vacuum gap in the rf sheath. As far as the rf is concerned, this vacuum gap provides an extra capacitance in the rf circuit. This type of model was investigated[13] for plasma processing and is currently being tested in an rf antenna coupling code for fusion applications.[14] The presence of rf sheaths can modify the rf-field distribution between antenna bumper limiters.

Utilizing the fact that the sheath is a thin layer, it is also possible to analytically derive a "sheath boundary condition" that can replace the usual perfect conductor boundary conditions on the tangential component of E and the normal component of B on the surface of a conductor.[15,16] The resulting boundary condition involves the sheath width and its effective dielectric properties which can model both sheath capacitance and dissipation (resistance).

The interesting and important question concerning self-consistent rf coupling codes and sheath interactions is the competition associated with the plasma density near the antenna. A large density is favorable for good coupling (because the launched FW is usually evanescence at low densities) but this also increases the level of sheath interaction. These considerations are partly addressable in antenna design, for example by using septa and bumper limiters.[17]

RF sheaths occur where field lines containing plasma contact conducting surfaces. On the antenna itself, the geometry of these connections implies a phasing and field line angle dependence.[18] For a two-strap antenna in 0-π (dipole) phasing, and symmetrical sheath connections on the front face of the Faraday screen, there is no net rf voltage induced between contact points. However in 0-0 (monopole) phasing a large voltage can result. The voltage in the case is essentially the fraction of the end-to-end voltage along the current strap that is subtended by the contacts.

This kind of Faraday screen rf sheath is exacerbated by large misalignment of the B-field with the Faraday Screen and/or a large component of B along the current strap, and by low-$k_\parallel$ non-symmetric phasings. As a result, experiments have shown that ICRF heating in dipole phasing is much easier than ICRF current drive.

In practice, not all important sheaths are this simple: capacitive, corner and feeder effects also drive rf sheaths.[19] While the sheath voltages in 0-π phasing are smaller, they are not zero. One reason is that the field lines that cut across the corner of the antenna don't see anti-symmetric, canceling rf magnetic fluxes. A related effect is that current forced onto the current straps by the feeders leads to charges where the conductors take a sharp corner, and this can be thought of as a capacitive coupling to the rf sheaths.

Due to the grazing nature of the field lines contacting the complicated three-dimensional structure of an antenna, the contact points are very sensitive to field line location, and adjacent field lines can end up having very different induced sheath voltages because their induction circuits trap a different amount of rf magnetic flux. When adjacent field lines charge to different voltages, there is a perpendicular electric field between them. This gives rise to **E**×**B** drifts and the important concept of rf-sheath-induced convection.[20-22]

The effects of rf-induced convection have been seen indirectly in experiments. On JET, reduced particle confinement and increased SOL density scale length during monopole H-modes were attributed to rf-induced convection.[20] In Tore Supra the up/down heat flux asymmetry on the antenna was interpreted as arising from a large-scale rf-sheath driven convection roll pattern in front of the antenna.[23-25] This convection occurs because the antenna acts like a giant biased probe, charging positive all the field lines in front of it. The tokamak magnetic field gives a preferred direction to the **E**×**B** drift pattern and is responsible for convecting plasma preferentially into the bottom of the antenna. Recently,[23] it has been demonstrated that this heat flux asymmetry reverses with reversal of the tokamak B-field, consistent with the rf-driven convection mechanism (although power flow asymmetries due to the Hall term may also play a role[26]). Convective physics modifies fluxes into the antenna, affects sputtering, electron sheath heating (not discussed here) and, importantly, modifies the electron density profile in front of the antenna. Reflectometers were used to measure this effect on TFTR,[27] and show that the antenna effectively pumps on the edge plasma.

Thus, the rf antennas modify the edge density profile that they have to couple to. To treat this interaction theoretically, we consider the time-averaged vorticity or charge-balance equation,

$$\frac{c^2}{4\pi v_a^2}\frac{d}{dt}\nabla_\perp^2 \Phi = \nabla_\parallel J_\parallel + \frac{2c}{B}\mathbf{b}\times\mathbf{\kappa}\cdot\nabla p \qquad (2)$$

where v_a is the Alfvén velocity and κ is the curvature. The currents which contribute to the dynamics in Eq. (2) are ion polarization currents across the magnetic field, parallel currents which terminate on sheaths, and magnetic field line curvature. The perpendicular polarization current, which appears as the charge advection term on the left of Eq. (2), couples flux tubes in the perpendicular direction. The parallel current term describes the 1-D sheath dynamics considered earlier. The curvature term, usually neglected in rf physics, is what gives rise to low frequency edge turbulence. The edge instabilities driven by this term eject filaments of plasma called blobs into the scrape-off-layer (SOL). These blobs convect towards the antenna by a simple mechanism.[28,29] Curvature drift creates a charge separation. This gives rise to an internal electric field inside the plasma blob. The blob then convects radially as a whole due to the **E**×**B** drift. Consequently, the subject of antenna-plasma interaction is entwined with that of blobs and edge turbulence, and this interaction is fundamental to calculating the self-consistent SOL density profile in front of the antenna.[16] This *self-consistent* density is required for studies of rf coupling, impurities, antenna damage and other antenna interaction effects.

Although the total power going into rf sheaths is most problematic at high power, at low power, this same effect can be used to diagnose rf sheaths. Since, the sheath power dissipation is linear with the voltage for $eV/T_e > 1$ [see Eq. (1)], its contribution to the loading resistance in this (low but not too low) power regime scales like $1/V$. Thus, sheath power can dominate the loading – a useful result for diagnosing the existence and properties (area, local density and voltage) of rf sheaths experimentally and potentially for validation of antenna-sheath codes. This effect has been observed[30] and successfully modeled[10] to show that sheath area, voltage and ambient density are the most important parameters.

The ubiquity of antenna sheaths has motivated work into sheath mitigation by the use of insulating materials.[31] On Phaedrus it was shown that the plasma potential rise due to rf sheath rectification could be almost completely eliminated by employing insulating limiters to intercept the field lines before they contact the metal, and complete the sheath circuit.[32] Effectively, the insulator adds an additional series impedance to the plasma sheath and absorbs most of the voltage drop that would otherwise appear across the sheaths. The main challenge here will be to come up with insulating materials that can withstand a reactor environment. Boron compounds are often used in present day experiments, but novel ceramic materials have also been investigated.[33,34]

When field lines are sufficiently long (so that the plasma resistance supports a significant voltage drop along the field line) the sheaths at the two ends become "disconnected". When these sheaths are also asymmetric (different voltages), they can drive a net dc parallel current. This effect was studied on TEXTOR,[35] and more recently on JET[36] where it was found that the sheath driven currents can trigger arcs at the high voltage end in some situations. This occurred in mixed phasing experiments where there was a current path between powered monopole and dipole antennas. In this case the cross-field polarization current driven by rf convection was postulated to be part of the current path.

So far the discussion has been confined to sheath losses local to the antennas. But edge parasitic power losses are often observed in low single-pass and low-$k_\|$ phasing situations where near field sheaths do not appear to explain the whole story. One concept which can relate very well to this type of observation is that of the far field sheath, which gives a general mechanism for dissipation of wave energy in the SOL.

Edge rf fields appear on walls and limiters due to poor single pass absorption, or direct coupling to edge and surface modes.[8] Because the flux surfaces are not generally aligned with conducting boundaries, the FW polarization alone cannot satisfy the proper boundary conditions, and of necessity a slow wave with $E_\|$ is generated.[37,38] This slow wave is often evanescent. The presence of this $E_\|$ in the boundary plasma brings into play all of the sheath effects that have been discussed so far in the near field antenna context. In particular far field sheaths give a mechanism for edge power loss and impurity generation. Other dissipation mechanisms for waves at the edge are also possible, for example collisional dissipation of wave energy[8] by neutral collisions. The low-$k_\|$ modes, being less evanescent in their propagation from the core towards the walls, are most susceptible to these dissipation mechanisms.

IBW EDGE INTERACTIONS

Sheaths can be just as important for IBW edge interactions as in the FW case of the preceding section. However, the IBW case also allows a rich variety of other nonlinear physics, primarily because Bernstein waves have a small group velocity and consequently require large electric fields to carry a significant power flux. The linear theory of IBW coupling is rather well developed and has been reviewed by Ono[3] where interesting experimental results and nonlinear mechanisms are also reported. IBW coupling has met with mixed success and linear theory alone fails to describe many experiments. Coupling of power to the core has generally been better on small machines, and experiments benefit from good conditioning. In a number of cases, the application of IBW power has failed to heat the core plasma at all. Here we review some nonlinear effects which bear on the issue of getting IBW power through the SOL into the core, noting that in some cases the same physics can also be relevant to nominal FW and high harmonic fast wave (HHFW) experiments where the large pitch of B relative to the antenna current strap results in substantial $E_\|$ (slow wave) coupling.

Ponderomotive expulsion of plasma is one of the expected nonlinear mechanisms. For the slow wave, the (repulsive) ponderomotive potential is usually approximated by the jitter energy of electrons in the parallel rf field,

$$\Psi \sim \frac{1}{2} m_e \tilde{u}_\|^2 = \frac{e^2 E_\|^2}{4 m_e \omega^2} \qquad (3)$$

For representative electric fields $E_\| \sim 300$ V/cm and frequencies in the range of tens to hundred of MHz, the condition of strong nonlinear interactions, $\Psi > T$ (e.g. in the SOL or at the separatrix) is easily met in all but the highest frequency experiments.[3] Measurements from the DIII-D tokamak[39] showed that as the power is raised the effect on the *reactive* loading is the same as moving the plasma away from the antenna, and is consistent with a ponderomotive expulsion interpretation.

Many large-tokamak IBW experiments have shown that the loading *resistance* is large and insensitive to the frequency (i.e. to the location of cyclotron resonance with respect to the antenna). This feature, was not expected from traditional direct launch IBW theory, but could be explained by a linear theory model[40] which assumed ponderomotive depletion of density in front of the antenna, and allowed the energy to be absorbed at the ensuing lower hybrid resonance (LHR). A related 1-D nonlinear model,[41] which explicitly included ponderomotive profile steepening, showed enhanced wave reflection near the LHR that effectively channeled energy into a coaxial mode propagating in the halo plasma. Furthermore, the phasing properties of this mode were consistent with loading and heating efficiency measurements on TFTR.[42-44] The basic idea is that the longer *poloidal* wavelength in 0-0 poloidal phasing enables a shorter radial wavelength parasitic coaxial mode to fit in the halo plasma between the LHR and the wall. These observations may underlie some of the different IBW behaviors in large and small machines.

In IBW experiments, parametric decay instability (PDI) is often observed.[45] Observations of PDI are often correlated with edge ion heating, as noted recently in the HHFW context on NSTX.[46] Parametric decay may be important in deciding what happens to wave energy that is trapped in the edge; however, it has been difficult to measure the power going into the PDI daughter waves.

In the parametric decay interaction, there is a large amplitude pump wave at frequency and wavenumber (ω_0, $\mathbf{k}_0$), and it is presumed that there are two other modes in the plasma at frequencies and wavenumbers which add up to (ω_0, $\mathbf{k}_0$), called the daughter waves, denoted by (ω, $\mathbf{k}$), (ω_-, $\mathbf{k}_-$). The wave equation for each daughter mode is driven by a nonlinear beat current of the other daughter with the pump,

$$[(c/\omega)^2 \nabla \times \nabla \times - \varepsilon \cdot] \mathbf{E}(\omega) \propto \mathbf{E}(\omega_0)\mathbf{E}(\omega_-) \qquad (4)$$

The initially small daughter waves can be linearly unstable above a certain threshold pump wave amplitude, which typically depends on the damping rates of the daughter modes. This instability is the PDI.

In the so-called dipole approximation (i.e. that of a long wavelength pump), the theoretical analysis can be done linearly by transforming to an oscillating frame, viz. the frame of the jitter in the pump wave field. In this case the species-dependent jitter is what provides the mode coupling that make the PDI process unstable. This linear theory of PDI (that is with a fixed pump wave) is rather well developed for the FW and IBW cases of interest for fusion plasmas.[47] These types of calculations can also take into account convection of wave energy out of the interaction region in inhomogeneous plasmas. In FTU it was shown[48] that there are competing constraints on the optimal SOL density: high to reduce PDI, but low to reduce reflected power for good linear coupling.

Theoretically, fully nonlinear calculations (including pump depletion) are very difficult. For IBW, the theory must include kinetic, hot plasma dynamics, and two or three spatial dimensions for realistic results. This problem presents an opportunity for future theory and simulation.

SLOW WAVE CORE INTERACTIONS: FLOW DRIVE

In many experiments spanning several decades, it was found that directly launched IBW power could trigger improved confinement regimes in a tokamak.[49-55] In other experiments, there were observations of IBW-induced flows.[56,57] Collectively, experiments show that the IBW can drive flows, and that the IBW can sometimes enhance confinement, however the mechanisms have not been fully established experimentally.

Plasma turbulence research has shown that sheared flows can suppress turbulence.[58-60] This knowledge has stimulated theoretical work on the calculation of rf driven flows, beginning with the pioneering work of Craddock and Diamond[61] and continuing up to the present.[48,62-69] A number of 1-D and ray tracing calculations established that local absorption of IBW power at a cyclotron resonance was accompanied by redistribution of momentum that resulted in sheared poloidal (bipolar) flows.

Theoretically and experimentally, the direct launch IBW scheme for flow drive and turbulence suppression appears to be plausible, but practically it can be difficult to get the IBW power into the plasma, as noted in the previous section. This raises the question of whether mode-converted slow waves such as the IBW or ICW could be used to drive flows while avoiding the problems associated with direct IBW launch.

When the FW encounters a mode conversion (MC) layer in a multi-ion-species plasma, both the IBW and the ICW can result as mode conversion products.[68,70,71] New diagnostics, such as phase contrast imaging have allowed these waves to be observed directly,[72] and have helped to stimulate new theoretical work on flow drive, generalizing the previous work to handle MC, hot plasmas, and general electromagnetic waves.

There are basically three mechanisms by which an RF wave can induce forces on a plasma. The first one can be thought of as photon absorption, in which the rf wave energy is absorbed and imparts a proportional momentum, k/ω, to the plasma. This process is most effective for slow waves, with their relatively large k. Note that this is fundamentally a dissipative force.

The second mechanism can be described as photon reflection. In the extreme case of total reflection, the force is $2k/\omega$ times the one-way power flow. However, this mechanism is better thought of in terms of reactive ponderomotive forces, driven by the gradient of the electric field amplitude rather than as related to circulating power. It is fundamentally non-dissipative.

The third mechanism is a momentum redistribution mechanism related to the Reynolds stress. No net force can be supplied by this mechanism, but adjacent flux surfaces can acquire equal and opposite forces and thereby create sheared flows. The nonlinear stress tensor which describes this process contains both the mechanical Reynolds stress component **vv** and the electromagnetic stress **BB**. For an electromagnetic wave there can sometimes be cancellations between the two pieces.[61,66]

There are several elegant formalisms for calculating nonlinear plasma effects due to rf waves, including guiding center[73] and quiver kinetics[74] formulations. Recently, flow drive work[64,69] has been developed using a different formulation, that of the W matrix[75,76] developed to describe energy flow and absorption in the presence of nonlocality introduced by finite gyroradius effects. This formalism uses a global Fourier representation of the rf fields, and is well suited to implementation in Fourier based codes.[77] The matrix $W(\mathbf{k}, \mathbf{k}')$ is the generalization of the usual hot plasma conductivity matrix $\sigma(\mathbf{k})$ to the nonlocal case. Thus the familiar $\mathbf{J}\cdot\mathbf{E}$ expression for absorbed power is generalized to

$$P_{rf} = \frac{1}{4}\sum_{\mathbf{kk'}} e^{i(\mathbf{k'}-\mathbf{k})\cdot\mathbf{r}} \mathbf{E}_\mathbf{k}^* \cdot W(\mathbf{k},\mathbf{k}') \cdot \mathbf{E}_{\mathbf{k}'} + cc \qquad (5)$$

Analogous to the energy moment of the Vlasov equation from which P_{rf} arises, one can take the momentum moment. In this case the nonlinear driving terms are the forces, which include the Lorentz force, and the divergence of a nonlinear stress tensor which involves the second order distribution function. It is shown in Ref. 69 that the total force on a fluid element can be expressed in terms of the three basic mechanisms, direct absorption, reactive ponderomotive force, and momentum redistribution. The reactive ponderomotive term reduces exactly to well known expressions in the fluid limit. Furthermore, flux-surface-averaged plasma flows in a tokamak can be driven only by the dissipative forces. Remarkably, these may be expressed simply in terms of the W matrix. The direct absorption term can drive net flows, depends on the momentum in the waves, and is effective with either electron or ion dissipation. The dissipative stress (momentum redistribution) term drives bipolar sheared flows but no net flows. It depends on the power absorbed in the perpendicular direction and scales inversely with the cyclotron frequency, so it is only significant for ions. In general flow drive is largest for short wavelengths and narrow dissipation layers, the narrow layers implying stronger shear in the flow.

This flow drive theory was implemented in the AORSA code[68] and applied to a C-Mod mode-conversion case which generates both IBW and ICW products. The toroidal flow can be obtained by balancing the rf force with an empirical diffusion of toroidal angular momentum. For 1 MW of power, the flow is in the range of a few km/s and the peak shearing rate is about 10^4 s^{-1}, which is somewhat small for effective turbulent suppression. To date, a careful survey of parameter space for more optimal cases has not been done. Also, there are some subtleties in the converting forces to flows, that bring in both neoclassical and turbulent transport theory. More theoretical work is needed in this area, particularly including time transients and anomalous diffusion which couple the poloidal and toroidal flows in the theory. Experiments that exhibit rf-induced confinement improvement and have the diagnostic capability to make measurements of poloidal and toroidal velocity shear are also needed, as well as experimental validation of flows from mode-converted waves.

Turbulence suppression is approximated governed by the shear in the radial electric field. There are different mechanisms for modifying E_r shear by applied rf waves that can be seen from the steady state ion radial force balance equation.

$$\frac{v_\zeta B_\theta - v_\theta B_\zeta}{RB_\theta} \equiv G(\psi) = -c\left(\frac{\partial\Phi}{\partial\psi} + \frac{1}{Zen_i}\frac{\partial p_i}{\partial\psi}\right) + \left\langle\frac{c}{Zen_i}\frac{F_{i\psi}}{RB_\theta}\right\rangle_\psi \qquad (6)$$

The flux function $G(\psi)$, representing the $\mathbf{v}\times\mathbf{B}$ flow term, balances the radial electric field, the ion pressure gradient and any external radial forces. Nonlinear wave momentum processes drives flows, as discussed in the preceding paragraphs; the rf can modify the ion pressure profile locally, changing ∇p_i; and finally, in principle, the waves can exert a direct radial ponderomotive force, although in practice this is almost always negligible in large tokamaks.

INTEGRATED MODELING: THE NEW FOREFRONT

The preceding issues have implications for integrated modeling, which is an exciting new forefront for rf physics. Integrated modeling can play an important role in hardware design, scenario development (including not only core rf physics, but now also nonlinear edge rf physics) and in the interpretation of experimental results.

One promising area is the incorporation of more edge physics into antenna coupling codes, such as plasma (blobs and turbulence) in the antenna region, wave scattering from blobs and fluctuations, sheath and ponderomotive effects and surface physics (e.g. sputtering and neutral gas desorption).

Inclusion of this physics would provide a predictive capability for plasma loading with a self-consistent density profile. At present, there is no robust way of predicting in advance the antenna loading for future experiments, mainly because the edge and SOL density profile is not known. Additionally, this type of integrated modeling could predict some operational constraints on the antenna such as local power dissipation, hot spot damage, and possibly certain types of arcs. An exciting goal would be a complete self-consistent description of the effects of rf on the edge (e.g. turbulence), and visa versa. Some work in this direction is in progress.[14,16,23]

For such a computational project to succeed, validation of codes with experiments at the most fundamental level is necessary. Low power loading measurements[30] would provide a very useful tool in this regard, as well as yielding a direct experimental diagnosis of sheaths, local plasma density, and antenna-plasma interactions.

A second promising area for integrated modeling is that of more realistic edge conditions for global full wave rf codes. Typically in these codes, all the launched power is absorbed in the core no matter how weak the core absorption is. It is known from experiments that edge physics is especially important for low $k_\parallel$ cases. More realistic models of edge dissipation are needed, for example employing boundary conditions to model sheaths.[13-16] Edge collisions and neutrals may also be important in some cases. Incorporation of the missing edge physics will allow a new predictive capability for lost power and heating efficiency.

In conclusion, nonlinear effects are generally important for ICRF waves at the edge and nonlinear effects can also be important in the core for short wavelength, slow waves such as the IBW and ICW. Many important individual pieces of nonlinear RF interactions are at least partially understood as isolated phenomena. These include rf-sheaths and their role on impurities, convection, SOL currents, ponderomotive effects, far field sheaths and edge dissipation, parametric decay, and rf effects on plasma flows and E_r. In order to make this knowledge really useful in a practical way, more integration of these pieces is needed. Integrated rf-edge modeling holds out the exciting possibility of a predictive capability that has so far been elusive.

As the fusion community stands at the threshold of a burning plasma experiment, our motivation for this new predictive capability is strong. Furthermore, we are acquiring the means for such computations through grand challenge computing resources that can make these computations feasible, both from a hardware and software perspective. These circumstances provide a significant opportunity for the ICRF theory and simulation community.

ACKNOWLEDGMENTS

This work was supported by U.S. DOE grant DE-FG02-97ER54392. Discussions with R. I. Pinsker, L. Colas and B. P. LeBlanc are acknowledged as well as the involvement and support of the RF SciDAC Team.

REFERENCES

1. J.-M. Noterdaeme, AIP Conf. Proc. **244**, 71 (1992).
2. J.-M. Noterdaeme and G. Van Oost, Plasma Phys. Control. Fusion **35**, 1481 (1993).
3. M. Ono, Phys. Fluids B **5**, 241 (1993).
4. M. Bures, J. Jacquinot, K. Lawson, M. Stamp, et al., Plasma Phys. Control. Fusion **33**, 937 (1991).
5. L. Colas, E. Faudot, S. Bremond, S. Heuraux. et al., this conference.
6. H. S. Butler and G. S. Kino, Phys. Fluids **6**, 1346 (1963).
7. F. W. Perkins, Nucl. Fusion **29**, 583 (1989).
8. M. Brambilla, R. Chodura, J. Hoffmann, J. Neuhauser, Plasma Phys. Control. Nucl. Fusion Res. 1990 (IAEA, Vienna, 1991), Vol. 1, p. 723.
9. J. R. Myra, D. A. D'Ippolito and M. J. Gerver, Nuclear Fusion **30**, 845 (1990).
10. D.A. D'Ippolito and J. R. Myra, Phys. Plasmas **3**, 420 (1996).
11. G.J. Greene, Ph. D. dissertation, California Institute of Technology, Pasedena, CA, 1984.
12. M.A. Lieberman, IEEE Trans. Plasma Sci. **PS-16**, 638 (1988).
13. E.F. Jaeger, L.A. Berry, J.S. Tolliver and D.B. Batchelor, Phys. Plasmas **2**, 2597 (1995).
14. M. D. Carter, D. A. D'Ippolito, J. R. Myra and D. A. Russell, this conference.
15. J.R. Myra, D. A. D'Ippolito and M. Bures, Phys. Plasmas **1**, 2890 (1994).
16. D. A. D'Ippolito, J. R. Myra, D. A. Russell and M. D. Carter, this conference.
17. S.J. Wukitch., R.L. Boivin, P.T. Bonoli, et al., Plasma Phys. Control. Fusion **46**, 1479 (2004).
18. D.A. D'Ippolito, J.R. Myra, M. Bures, and J. Jacquinot, Plasma Phys. Cont. Fusion **33**, 607 (1991).
19. J.R. Myra, D.A. D'Ippolito and Y.L. Ho, Fusion Eng. Design **31**, 291 (1996).
20. D.A. D'Ippolito, J.R. Myra, J. Jacquinot and M. Bures, Phys. Fluids B **5**, 3603 (1993).
21. R.A. Moyer, R. Van Niewenhove, G. Van Oost, et al., J. Nucl. Mater **176-177**, 293 (1991).
22. D. Diebold, R. Majeski, T. Tanaka, et al., Nucl. Fusion **32**, 2040 (1992).
23. L. Colas, L. Costanzo, C. Desgranges. et al., Nucl. Fusion **43**, 1 (2003); L. Colas, E. Faudot, S. Brémond, S. Heuraux, this conference.
24. M. Bécouldet, L. Colas, S. Pécoul et al., Phys. Plasmas **9**, 2619 (2002).
25. E. Faudot, S. Heuraux and L. Colas, this conference.
26. E.F. Jaeger, M.D. Carter, L.A. Berry, D.B. Batchelor, et al., Nucl. Fusion **38**, 1 (1998).
27. D. A. D'Ippolito, J. R. Myra, J. H. Rogers, K. W. Hill, et al., Nucl. Fusion **38**, 1543 (1998).
28. S. I. Krasheninnikov, Phys. Lett. A **283**, 368 (2001).
29. D. A. D'Ippolito, J. R. Myra, and S. I. Krasheninnikov, Phys. Plasmas **9**, 222 (2002).
30. D. W. Swain, R. I. Pinsker, F. W. Baity, M. D. Carter, et al., Nucl. Fusion **37**, 1 (1997).
31. R. Majeski, P. H. Probert, T. Tanaka, D. Diebold, et al., Fusion Eng. Design **24**, 159 (1994).

32. J. Sorensen, D. A. Diebold, R. Majeski, N. Hershkowitz, Nucl. Fusion **36**, 173 (1996).
33. J. R. Myra, D. A. D'Ippolito, J. A. Rice and C. S. Hazelton, J. Nucl. Mater **249**, 190 (1997).
34. D. A. D'Ippolito, J. R. Myra, J.A. Rice and C.S. Hazelton, AIP Conf. Proc. **403**, 463 (1997).
35. R. Van Nieuwenhove and G. Van Oost, Plasma Phys. Control. Fusion **34**, 525 (1992).
36. D. A. D'Ippolito, J. R. Myra, P. M. Ryan, E. Righi, et al., Nucl. Fusion **42**, 1357 (2002).
37. F.W. Perkins, Bull. Am. Phys. Soc. **34**, 2093 (1989), paper 6S6.
38. J.R. Myra and D. A. D'Ippolito, Phys. Plasmas **1**, 2890 (1994).
39. M. J. Mayberry, R. I. Pinsker, C. C. Petty, M. Porkolab, et al., Nucl. Fusion **33**, 627 (1993).
40. S.C. Chiu, M.J. Mayberry, R.I. Pinsker, C.C. Petty, M. Porkolab, AIP Conf. Proc. **244**, 169 (1992).
41. D.A. Russell, J.R. Myra and D.A. D'Ippolito, Phys. Plasmas **5**, 743 (1998).
42. J.R. Wilson, R.E. Bell, S. Bernabei, et al., Phys. Plasmas **5**, 1721 (1998).
43. J.R. Myra, D. A. D'Ippolito, D.A. Russell, J.H. Rogers, T. Intrator, Phys. Plasmas **7**, 283 (2000).
44. T. Intrator, J.R. Myra and D. A. D'Ippolito, Nucl. Fusion **43**, 531 (2003).
45. R.I. Pinsker, C.C. Petty, M.J. Mayberry, M. Porkolab, et al., Nucl. Fusion **33**, 777 (1993).
46. J.R. Wilson, this conference.
47. M. Porkolab, Fusion Eng. Design **12**, 93 (1990).
48. A. Cardinali, C. Castaldo, R. Cesario, et al., Nucl. Fusion **42**, 427 (2002).
49. M. Ono, P. Beiersdorfer, R. Bell, S. Bernabei, et al., Phys. Rev. Lett. **60**, 294 (1988).
50. T. Seki, K. Kawahata, M. Ono, K. Ida, et al., AIP Conf. Proc. **244**, 138 (1991).
51. J. D. Moody, M. Porkolab, C. L. Fiore, F. S. McDermott,et al., Phys. Rev. Lett. **60**, 298 (1988).
52. B. LeBlanc, S. Batha, R. Bell, S. Bernabei, et al. Phys. Plasmas **2**, 741 (1995).
53. C.K. Phillips, M.G. Bell, R.E. Bell, S. Bernabei, et al., Nucl. Fusion **40**, 461 (2000).
54. R. Cesario, A. Cardinali, C. Castaldo, M. Leigheb, et al., Phys. Plasmas **8**, 4721 (2001).
55. B Wan, Y. Zhao, J. Li, et al., Phys. Plasmas **10**, 3703 (2003).
56. B.P. LeBlanc, R.E. Bell, S. Bernabei, J.C. Hosea, et al., Phys. Rev. Lett. **82**, 331 (1999).
57. C. Riccardi, F. De Colle, M. Fontanesi, C. C. Petty, et al., AIP Conf. Proc. **595**, 83 (2001).
58. H. Biglari, P.H. Diamond and P.W. Terry, Phys. Fluids B **2**, 1 (1990).
59. K.H. Burrell, T.N. Carlstrom, E.J. Doyle, et al., Plasma Phys. Control. Fusion **34**, 1859 (1992)
60. P.W. Terry, Rev. Mod. Phys. **72**, 109 (2000).
61. G.G. Craddock and P.H. Diamond, Phys. Rev. Lett. **67**, 1535 (1991).
62. M. Ono et al., Plasma Phys. Control. Nucl. Fusion Res. 1994 (IAEA, Vienna, 1995), Vol. 1, p. 469.
63. L.A. Berry, E.F. Jaeger and D.B. Batchelor, Phys. Rev. Lett. **82**, 1871 (1999).
64. E. F. Jaeger, L. A. Berry, and D. B. Batchelor Phys. Plasmas **7**, 3319 (2000).
65. A. G. Elfimov, G. Amarante-Segundo, et al. Phys. Rev. Lett. **84**, 1200 (2000).
66. J.R. Myra and D.A. D'Ippolito, Phys. Plasmas **7**, 3600 (2000).
67. H. Weitzner, L.A. Berry, E. F. Jaeger and D. B. Batchelor, Phys. Plasmas **7**, 564 (2000).
68. E.F. Jaeger, L.A. Berry, J.R. Myra, D.B. Batchelor, et al., Phys. Rev. Lett. **90**, 195001 (2003).
69. J. R. Myra, L. A. Berry, D. A. D'Ippolito, and E. F. Jaeger, Phys. Plasmas **11**, 1786 (2004).
70. F.W. Perkins, Nucl. Fusion **17**, 1197 (1977).
71. E. Nelson-Melby et al., Phys. Rev. Lett. **90**, 155004 (2003).
72. Y. Lin, this conference; and refs. therein.
73. J.R. Cary and A.N. Kaufman, Phys. Fluids **24**, 1238 (1981).
74. P. J. Catto et al., Phys. Fluids B **2**, 2395 (1990).
75. D.N. Smithe, Plasma Phys. Control. Fusion **31**, 1105 (1989).
76. M. Brambilla, *Kinetic Theory of Plasma Waves*, (Clarendon Press, Oxford, 1998).
77. E.F. Jaeger, L.A. Berry, E.D. D'Azevedo, et al., Phys. Plasmas **8**, 1573 (2001).

Global and Self-consistent Simulation of ICRF Heating in Toroidal Plasmas

S. Murakami[1], A. Fukuyama[1], T. Akutsu[1], N. Nakajima[2], V. Chan[3], M. Choi[3], S.C. Chiu[4], L. Lao[3], V. Kasilov[5], T. Mutoh[2], R. Kumazawa[2], T. Seki[2], K. Saito[2], T. Watari[2], M. Isobe[2], T. Saida[2], M. Osakabe[2], M. Sasao[6] and LHD Experimental Group

[1]*Department of Nuclear Engineering, Kyoto University, Kyoto 606-8501, Japan*
[2]*National Institute for Fusion Science, Oroshi, Toki, Gifu 509-5292, Japan*
[3]*General Atomics, P.O. Box 85608, San Diego, California 92186-5608, USA*
[4]*Sunrise R&M, Inc., 8585 Hopseed Lane, San Diego, California 92129, USA*
[5]*Institute of Plasma Physics, National Science Center, KIPT, Kharkov, Ukraine*
[6]*Graduate School of Engineering, Tohoku University, Sendai 980-8579, Japan*

Abstract. The ICRF heating in toroidal plasmas is studied using two global simulation codes: a full wave field solver TASK/WM and a drift kinetic equation solver GNET. The codes are applied to both tokamaks and helical systems. The full wave code TASK/WM evaluates the realistic wave electric field, in which the effect of the self-consistent non-Maxwellian velocity distribution on the wave propagation is taken into account. GNET solves a linearized drift kinetic equation (5D phase-space) for energetic ions including complicated behavior of trapped particles in helical systemsCharacteristics of energetic ion distributions in the phase space are investigatedSelf-consistent analysis including the effect of energetic ion distribution on the fast wave propagation is also reported for tokamak plasmas.

Keywords: ICRF hating, global simulation, tokamak, helical, LHD
PACS: 52.25.Dg, 52.35.Mw, 52.50.Qt, 52.55.Fa, 52.55.Hc

INTRODUCTION

ICRF heating generates highly energetic trapped ions, which drift around the torus for a long time (typically on a collisional time scale) interacting with the RF wave field. Thus, the behavior of these energetic ions is strongly affected by the characteristics of the drift motions, that depend on the magnetic field configuration. In particular, in a 3D magnetic configuration, complicated drift motions of trapped particles would play an important role in the confinement of the energetic ions and the ICRF heating process.

Additionally, since the wavelength of the ICRF heating is typically comparable to the plasma scale length and the 3D geometry effect on the RF wave field would be also important in a 3D magnetic configuration. Therefore a global simulation of ICRF heating is necessary for the accurate modeling of the plasma heating process in a 3D magnetic configuration.

Also, the tail ion distribution changes the affects the fast wave propagation and the absorption rate. Thus, the self-consistent treatment of the wave field and the energetic ion tail distribution is necessary for the accurate evaluation of the ICRF heating.

In this paper we, first, study the ICRF heating in the LHD using two global simulation codes; a drift kinetic equation solver GNET [1, 2] and a wave field solver TASK/WM[3]. Both codes are taken into account the 3D geometry using the numerically obtained 3D MHD equilibrium. Then, we describe the development of the self-consistent ICRF heating simulation in a tokamak plasma with TASK code.

GLOBAL SIMULATION IN A 3D MAGNETIC CONFIGURATION

In order to study the ICRF heating in a 3D magnetic field configuration we have been developing a global simulation code combining two global codes; GNET and TASK/WM.

GNET solves a linearized drift kinetic equation for energetic ions including complicated behavior of trapped particles in 5-D phase space

$$\frac{\partial f}{\partial t} + (\mathbf{v}_{//} + \mathbf{v}_D) \cdot \nabla f + \mathbf{a} \cdot \nabla_{\mathbf{v}} f - C(f) - Q_{ICRF}(f) - L_{particle} = S_{particle} \qquad (1)$$

where $C(f)$ and Q_{ICRF} are the linear Coulomb Collision operator and the ICF heating term. $S_{particle}$ is the particle source term by ionization of neutral particle and the radial profile of the source is evaluated using AURORA code

. The particle sink (loss) term, $L_{particle}$, consists of two parts; one is the loss by the charge exchange loss assuming the same neutral particle profile as the source term calculation and the other is the loss by the orbit loss escaping outside of outermost flux surface.

The Q_{ICRF} term is modeled by the Monte Carlo method. When the test particle pass through the resonance layer where $\omega - k_{//} v_{//} = n\omega_c$, the perpendicular velocity of this particle, $v_{\perp 0}$, is changed by the following amount

$$\Delta v_{\perp} = \sqrt{\left(v_{\perp 0} + \frac{q}{2m} I |E_+| J_{n-1}(k_{\perp}\rho) \cos\phi_r \right)^2 + \frac{q^2}{4m^2} \{I |E_+| J_{n-1}(k_{\perp}\rho)\}^2 \sin^2\phi_r} - v_{\perp 0}$$

$$\approx \frac{q}{2m} I |E_+| J_{n-1}(k_{\perp}\rho) \cos\phi_r + \frac{q^2}{8m^2 v_{\perp 0}} \{I |E_+| J_{n-1}(k_{\perp}\rho)\}^2 \sin^2\phi_r \qquad (2)$$

where E_+ and ϕ_r are the RF wave electric fields and random phase, respectively. Also, q, m, ρ, J_n are the charge, mass and the Larmor radius of the particle, and n-th Bessel function, respectively. The time duration passing through the resonance layer, I, is given by the minimum value as, $I = \min(\sqrt{2\pi/n\ddot{\omega}}, 2\pi(n\ddot{\omega}/2)^{-1/3} Ai(0))$, which corresponds to two cases; the simply passing of the resonance layer and the passing near the turning point of a trapped motion (banana tip).

The spatial profile of RF wave electric field is necessary for the accurate calculation of the ICRF heating. The profile of RF wave field is an important factor on the ICRF heating and these profiles affect the particle orbit. We evaluate the RF wave field by the TASK/WM code. TASK/WM solves Maxwell's equation for RF wave electric field, $\mathbf{E}_{RF}$, with complex frequency, ω, as a boundary value problem in the 3D magnetic configuration.

$$\nabla \times \nabla \times \mathbf{E}_{RF} = \frac{\omega^2}{c^2} \bar{\bar{\varepsilon}} \cdot \mathbf{E}_{RF} + i\omega\mu_0 \mathbf{j}_{ext}, \qquad (3)$$

Here, the external current, $\mathbf{j}_{ext}$, denotes the antenna current in ICRF heating. The response of the plasma is described by a dielectric tensor including kinetic effects in a local normalized orthogonal coordinates.

We apply the global simulation code to a LHD configuration (R_{ax} = 3.6m; the in-ward shifted configuration). This LHD configuration conforms the σ-optimized configuration and shows relatively good trapped particle orbit[4]. A significant performance of ICRF heating have also been demonstrated[5-8] and up to 500keV of energetic tail ions have been observed by fast neutral particle analysis (NPA)[9,10] in LHD.

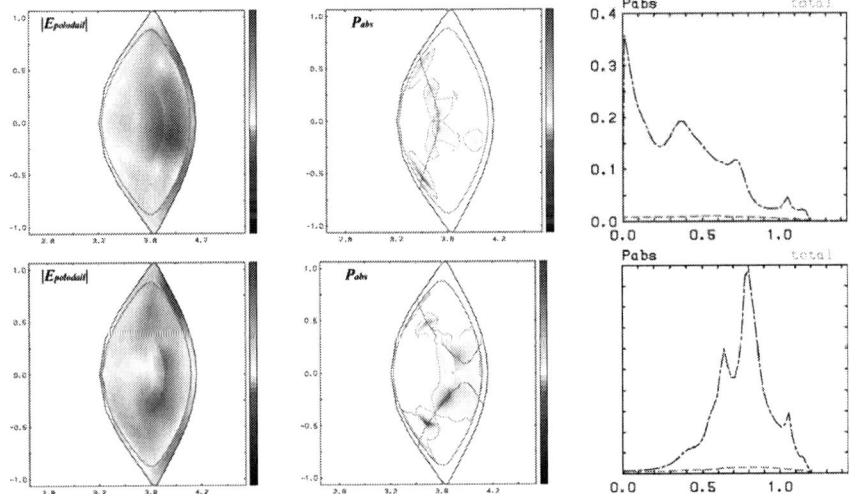

FIGURE 1. Contour plots of the poloidal electric field amplitude (left), RF power absorption (center) in the vertically elongated cross section, and the radial profile of power absorption (right) by TASK/WM code; the on-axis heating case (upper side) and the off-axis heating case (lower side).

The RF resonance position relative to magnetic flux surface has been tested mainly for two cases in the LHD experiments. One is the off-axis heating case in which the resonance surface almost crosses a saddle point of magnetic field at the longitudinally elongated cross section. In the off-axis case the resonance region only exists for $r/a>0.5$. The other is the on-axis heating in which the resonance surface crosses a magnetic axis. The experimentally obtained results have shown the difference in the heating efficiencies and about one order decrement of the energetic particle neutral count detected by natural diamond detector (NDD-NPA) [11].

We, first, applied the TASK/WM to the LHD plasmas in order to evaluate the RF wave electric fields (E_+ and E_-). Figure 1 shows the amplitude of the poloidal electric field (left) and power absorption (center) in the vertically elongated cross section, and the radial profile of the power absorption (right). The obtained results show the stronger RF wave field in the larger side of the major radius since the antenna is set in the outside of the torus.

From this result, a simple RF wave electric fields profile; $E_+ = E_{+0}\tanh((1-r/a)/l)\cos\theta$ with $l=0.2$ is assumed as a first step in the GNET simulation. The other wave field parameters are set as $k_{perp}=62.8 m^{-1}$ and $k_{//}=0$. The amplitude of the wave field, E_{+0}, is changed in the range 0.5kV/m through 1.5kV/m to obtain the dependency on the heating power. The plasma parameters are set to the similar values as the experimental ones. The plasma temperature and density are assumed as $T_s=T_{s0}(1-(r/a)^2)$ with $T_{e0}=T_{i0}=1.6$keV and $n_e=n_{e0}(1-(r/a)^8)$ with $n_{e0}=1.0\times10^{19}m^{-3}$. We solve the drift kinetic equation for the proton minority ions distribution in the helium majority ions. The density ratio of the minority ion is assumed to be 5%.

Figure 2 shows the iso-surface plot of the steady state distribution of the minority ions during ICRF heating obtained by GNET. We plot the flux surface averaged tail ion distribution in the three dimensional space ($r/a, E, \theta_p$), where $a/r, E$ and θ_p are the normalized averaged minor radius, the total energy and the pitch angle, respectively.

The RF wave accelerates minority ions perpendicularly in the velocity space and we can see perpendicularly elongated minority ion distributions. We find a peaked energetic tail ion distribution near $r/a\sim 0.5$ in the off-axis heating case (Fig.2, left). On the other hand we can see no strong peak in the distribution function in the on-axis heating case (Fig.2, left). The energetic particle distribution is broader than that of the off-axis case and the less energetic tail ion is obtained. A small peak can be seen very near the axis.

Figure 3 shows the radial profiles of the RF field power absorption and the energetic ion pressure of the minority ion. The absorption of the RF field shows the strong peak near $r/a=0.5$ in the off-axis case and near the axis in the on-axis case. The peaked pressure profile can be seen in the off axis case and the broader one is in the on-axis case. The higher peak observed in the distribution function of the off-axis case would be due to the stable orbit of the strongly absorbing trapped particles. We cannot see the strong peak due to the unstable orbit of those trapped particles in the on-axis case.

The heat deposition shows the maximum near $r/a=0.5$ in the off-axis case and flat one in the on axis case. The loss of the energetic tail ions actually alters the heating efficiency. The estimated heating efficiency (= the deposit power to the thermal plasma / the absorbed power of RF wave) changes up to about 2MW assuming 5% of minority ion density. The heating efficiencies with 1MW are about 70% for both cases and we can not see the clear difference between the two cases.

To compare with the experimental results we have simulated the neutral count number detected by NDD-NPA using the simulation results. Relatively good agreement is obtained between the experimental and simulation results (Fig.4). Both the computed and the experimental counts have similar dependency on the energy spectrum.

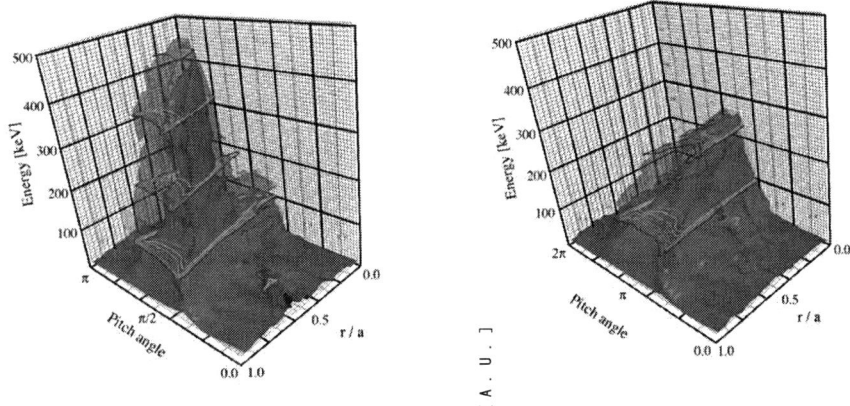

FIGURE 2. Steady state distribution of energetic tail ions in the (r/a, E, pitch angle) space in the off-axis point heating case (left) and on-axis heating case (right).

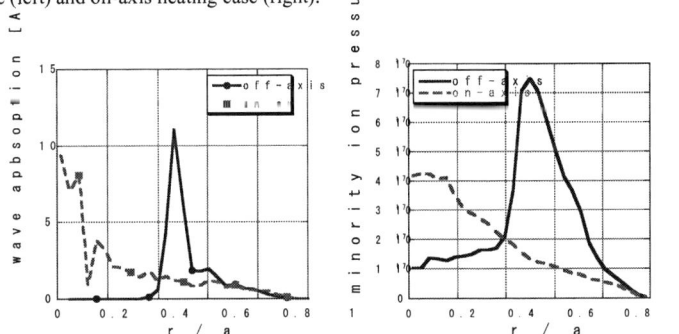

FIGURE 3. Radial profiles of ICRF wave power absorption (left) and energetic ion pressure (right) of the minority ion for the off-axis heating (solid) and the on-axis heating (dotted).

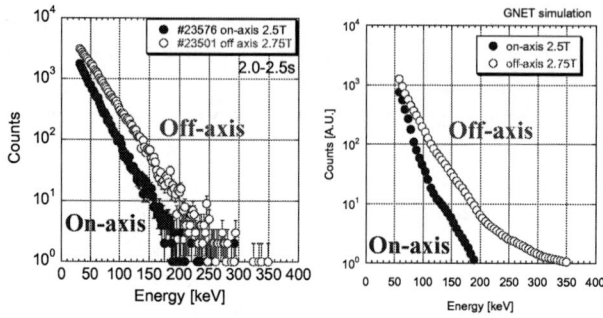

FIGURE 4. Comparisons of the energy spectrum by the NDD-NPA results (left) and the simulation results (right) for the off-axis heating (open) and the on-axis heating (closed).

SELF-CONSISTENT FULL WAVE ANALYSIS

Deviation of velocity distribution from Maxwellian may strongly affect the power absorption of ICRF waves in the presence of energetic ions, current drive efficiency of LHCD, NTM controllability of ECCD. In these cases a self-consistent analysis of wave propagation and absorption including the time evolution of the velocity distribution function and the local dielectric tensor is needed. We have implemented necessary interface code into the integrated modeling code system TASK [12] in order to study the case of ICRF minority heating.

The TASK code system has a modular structure and three modules are mainly related to the self-consistent full wave analysis. The full wave module (TASK/WM) solves Maxwell's equation with a local dielectric tensor calculated in the wave dispersion module (TASK/DP). The obtained wave electric field is used to calculate the bounce-averaged quasi-linear diffusion coefficients in the Fokker-Planck module (TASK/FP), which describes the time evolution of the velocity distribution function $f(v_\parallel, v_\perp, t)$. Finally $f(v_\parallel, v_\perp, t)$ is numerically integrated in velocity space to give the electric susceptibility for minority ions in TASK/DP. This loop has to be repeated until a quasi-steady state is obtained.

The module TASK/DP uses various kinds of plasma models to calculate the electric susceptibility for each particle species; cold plasma, kinetic plasma and gyrokinetic plasma models; Maxwellian and arbitrary velocity distribution functions; nonrelativistic and relativistic versions. For an arbitrary velocity distribution function $f(v_\parallel, v_\perp, t)$, the calculation of susceptibility with the kinetic model requires numerical integration in velocity space which consumes a lot of CPU time. If we calculate both the Hermite and anti-Hermite parts of the susceptibility, the computation time for 50 times 50 velocity-space meshes is about one thousand times longer than the time of usual calculation with the plasma dispersion function. Since the damping rate of the wave is essential, it may be worthwhile to compute only the anti-Hermite part of the susceptibility by numerical integration. In this case, the computation time is about one hundred times longer than that of usual calculation; one order of magnitude reduction compared with the computation including the Hermite part. Parallel processing will efficiently reduce the computation time. We also found that the accuracy of the Hermite part is much lower than that of the anti-Hermite part.

The module TASK/FP solves the bounce-averaged Fokker-Planck equation for the outer-mid-plane velocity distribution function $f(v_{||0}, v_{\perp 0}, t)$. In addition to the quasi-linear term, a nonlinear collision term, a parallel electric field term and a spatial diffusion term are included. The local distribution function $f(v_{||}, v_{\perp}, \theta, t)$ for any poloidal angle θ can be calculated from $f(v_{||0}, v_{\perp 0}, t)$.

The modules exchange the calculated data through the interface module TASK/PL; wave electric field (from WM to FP), local distribution function (from FP to DP), and dielectric tensor (from DP to WM). The universal data interface reduces the number of interface routines.

We have formulated and implemented the interface routines for self-consistent full wave analysis of the minority ion heating. Computation results will be reported very near future.

CONCLUSIONS

We have developed a global simulation code combining two code; GNET and TASK/WM. The GNET code solves a linearized drift kinetic equation for energetic ions including complicated behavior of trapped particles in 5-D phase space and the TASK/WM code solves Maxwell's equation for RF wave electric field with complex frequency as a boundary value problem in the 3D magnetic configuration. The developed code has been applied to the analysis of energetic tail ion transport during ICRF heating in the LHD plasma. A steady state distribution of energetic tail ion has been obtained and the characteristics of distribution in the phase space are clarified. The resonance position dependency on the distribution has been shown and larger tail formation has been obtained in the off-axis heating case. This tendency agrees well with the experimental results. We have compared the GNET simulation results with the experimental results evaluating NDD count number and also obtained similar tendencies.

Self-consistent analysis including the effect of energetic ion distribution on the fast wave propagation by TASK code is under way. Preliminary results of wave propagation and absorption with arbitrary velocity distribution function were discussed.

ACKNOWLEDGMENTS

This work was supported by Grant-in-Aid for Scientific Research from the Japanese Ministry of Education, Culture, Sports, Science and Technology.

REFERENCES

1. S. Murakami, et al, Nuclear Fusion **40** (2000) 693.
2. S. Murakami, et al., Fusion Sci. Technol. **46** (2004) 241.
3. A. Fukuyama, E. Yokota, T. Akutsu, Proc. 18th IAEA Conf. on Fusion Energy (Sorrento, Italy, 2000) **THP2-26**.
4. S. Murakami, et al., Nucl. Fusion **42** (2002) L19.
5. T. Mutoh, et al, Phys. Rev. Lett. **85** (2000) 4530.
6. R. Kumazawa, et al., Phys. Plasmas **8** (2001) 2139.
7. T. Watari, et al. Nucl. Fusion **41** (2001) 325.
8. T. Mutoh, et al., Fusion Sci. Technol. **46** (2004) 175.
9. A.V. Krashilnikov, et al., Nucl. Fusion 42 (2002) 759.
10. T. Saida, et al., Nuclear Fusion **44** (2004) 488.
11. M. Isobe, et al., Rev. Sci. Instrum. **72** (2001) 611.
12. A. Fukuyama, et al., Proc. of 20th IAEA Fusion Energy Conf. (Vilamoura, Portugal, 2004) IAEA-CSP-25/CD/TH/P2-3.

Self-Consistent Full-Wave / Fokker-Planck Calculations for Ion Cyclotron Heating in Non-Maxwellian Plasmas

E. F. Jaeger[a], L. A. Berry[a], R. W. Harvey[b], J. R. Myra[c], R. J. Dumont[d], C. K. Phillips[e], D. N. Smithe[f], D. B. Batchelor[a], P. T. Bonoli[g], M. D. Carter[a], E. D'Azevedo[a], D. A. D'Ippolito[c], J. C. Wright[g]

[a]*Oak Ridge National Laboratory, P.O. Box 2008, Oak Ridge, TN 37831-6169, USA*
[b]*CompX, P.O. Box 2672, Del Mar, CA 92014-5672, USA*
[c]*Lodestar Research Corporation, 2400 Central Avenue P-5, Boulder, CO 80301, USA*
[d]*Association EURATOM-CEA sur la Fusion, CEA/DSM/DRFC, CEA-Cadarache, FRANCE*
[e]*Princeton Plasma Physics Laboratory, P.O. Box 451, Princeton, NJ 08543, USA*
[f]*ATK-Mission Research, 8560 Cinderbed Rd., Suite 700, Newington, VA 22122-85601, USA*
[g]*Plasma Fusion Center, Massachusetts Institute of Technology, Cambridge, Massachusetts 02139, USA*

Abstract. Self-consistent solutions for the wave electric field and particle distribution function are calculated for ion cyclotron heating in non-Maxwellian plasmas. The all-orders wave solver AORSA is generalized to treat non-thermal velocity distributions arising from fusion reactions, neutral beam injection, and wave driven diffusion in velocity space. Quasi-linear diffusion coefficients are derived directly from the wave electric fields and used to calculate velocity distribution functions with the CQL3D Fokker-Planck code. Self-consistent results are obtained by iterating the full-wave and Fokker-Planck solutions.

Keywords: Full-wave, ICRF heating, Fokker-Planck, plasma simulation,
PACS: 52.50.Qt, 52.55.Fa, 52.65 Ff

INTRODUCTION

High-performance burning plasma devices such as ITER will contain significant concentrations of non-thermal plasma species arising from fusion reactions, neutral beam injection, and wave-driven diffusion in velocity space. Initial studies in 1-D [1-3] and experimental results [4] suggest that these non-thermal components can significantly alter wave propagation and absorption in the ion cyclotron range of frequencies (ICRF). In addition, radio frequency (RF) heating of these energetic components can occur at high harmonics of the ion cyclotron frequency where conventional 2-D full-wave models are not valid. In this work, the 2-D all-orders wave solver AORSA [5] is generalized to treat non-thermal velocity distributions. Quasi-linear diffusion coefficients are derived directly from the full-wave RF electric fields and used to calculate non-Maxwellian velocity distributions with the CQL3D Fokker-Planck code [6]. Self-consistent results for the wave electric field and particle distribution function are obtained by iterating the full-wave and Fokker-Planck solutions. Alternately, the quasi-linear diffusion coefficients can be calculated numerically by integrating the Lorentz force equations along particle orbits and obtaining the change in velocity after one toroidal transit of the machine [7].

Time-dependent processes in fusion plasmas are governed by the Maxwell-Boltzmann system of equations. The plasma state is described by a distribution function $f_s(\mathbf{r}, \mathbf{v}, t)$ representing the density of species s in a six dimensional phase space of position and velocity. This function evolves in time according to the Boltzmann equation by convection in the 6-D phase space while under forces exerted by the electric and magnetic fields, $\mathbf{E}$ and $\mathbf{B}$, respectively. For ICRF applications, the wave time scale is by far the fastest time scale in the system. Thus, the fields and distribution function can be separated into a time-average, or slowly varying part, ($\mathbf{E}_0$, $\mathbf{B}_0$, f_s^0), and a time harmonic, or rapidly oscillating part, [$\mathbf{E}(\mathbf{r})e^{-i\omega t}$, $\mathbf{B}(\mathbf{r})e^{-i\omega t}$, $f_s^1(\mathbf{r}, \mathbf{v})e^{-i\omega t}$] where ω is the frequency of the wave. The time-harmonic wave fields are small compared to the equilibrium fields, and we may linearize with respect to these amplitudes. Solving the linearized Boltzmann equation gives the rapidly varying part of the distribution function $f_s^1(\mathbf{r}, \mathbf{v})$ in terms of the equilibrium part f_s^0.

For the rapidly oscillating, time harmonic wave fields, Maxwell's equations reduce to a generalization of the Helmholtz equation,

$$-\nabla \times \nabla \times \mathbf{E} + \frac{\omega^2}{c^2}\left(\mathbf{E} + \frac{i}{\omega \varepsilon_0}\mathbf{J}_p\right) = -i\omega \mu_0 \mathbf{J}_{ant}, \qquad (1)$$

where $\mathbf{J}_{ant}$ is an externally driven antenna current, localized near the plasma edge, that acts as a source for the waves. The fluctuating plasma current $\mathbf{J}_p$ can be derived directly from the rapidly varying part of the distribution function $f_s^1(\mathbf{r},\mathbf{v})$. In general, $\mathbf{J}_p$ is a non-local, integral operator on the wave electric field,

$$\mathbf{J}_p(\mathbf{r},t) = \sum_s \int d\mathbf{r}' \int_{-\infty}^{t} dt' \sigma\left(f_s^0(E), \mathbf{r}, \mathbf{r}', t, t'\right) \cdot \mathbf{E}(\mathbf{r}', t'), \qquad (2)$$

where $\sigma(f_s^0, \mathbf{r}, \mathbf{r}', t, t')$ is the "plasma conductivity kernel."

The numerical solution of Eqs.(1-2) is a very intensive task because of the non-local nature of the plasma current, the geometric complexity of the plasma boundary, and the enormous range of spatial scales that must be treated. New wave solvers called "all-orders spectral algorithms" (AORSA) [5] have been developed that take advantage of computational techniques for today's parallel computers. These solvers include the general integral form of Eqs.(1-2), with no restriction on wavelength relative to orbit size and no limit on the number of cyclotron harmonics. In this paper, AORSA is generalized to treat non-thermal plasma components.

The long time response of the plasma distribution function $f_s^0(\mathbf{r},\mathbf{v},t)$ is obtained from a time-averaged form of the Boltzmann equation known as the bounce-averaged Fokker-Planck equation,

$$\frac{\partial}{\partial t}(\lambda f_0) = \nabla_{\mathbf{u}_0} \cdot \Gamma_{\mathbf{u}_0} + \langle\langle R \rangle\rangle + \langle\langle S \rangle\rangle \qquad (3)$$

where f_0 is the bounce averaged distribution function evaluated at the outer equatorial plane, expressed as a function of one spatial variable that labels a flux surface (ρ) and two velocity space variables that are constants of the motion: energy (u) and midplane pitch angle (θ). In Eq. (3), $\langle\langle R \rangle\rangle$ is a bounce-averaged radial diffusion operator that is set to zero for the calculations in this paper, and $\langle\langle S \rangle\rangle$ is a bounce-averaged particle source/sink operator. The coefficient λ is defined by $\lambda = |u_{\parallel,0}|\tau_b$, where τ_b is the bounce time. The divergence term in Eq. (3) includes two parts: $\nabla_\mathbf{u} \cdot \mathbf{\Gamma}_\mathbf{u} = C(f_0) + Q(f_0, \mathbf{E})$ where $C(f_0)$ is the collision operator, and $Q(f_0, \mathbf{E})$ is the quasi-linear operator [8] describing diffusion of f_0 in velocity space,

$$Q(f_0, \mathbf{E}) = \frac{1}{u_0^2} \frac{\partial}{\partial u_0}\left(B_0 \frac{\partial f_0}{\partial u_0} + C_0 \frac{\partial f_0}{\partial \vartheta_0}\right) + \frac{1}{u_0^2 \sin \vartheta_0} \frac{\partial}{\partial \vartheta_0}\left(E_0 \frac{\partial f_0}{\partial u_0} + F_0 \frac{\partial f_0}{\partial \vartheta_0}\right), \quad (4)$$

where B_0, C_0, E_0, and F_0 are the bounce-averaged quasi-linear diffusion coefficients. Equations (3-4) are solved by the CQL3D code [6], a three dimensional, bounce averaged Fokker-Planck solver in which particle orbits are tied to a flux surface.

CONDUCTIVITY FOR NON-MAXWELLIAN PLASMAS

To obtain self-consistency between the wave fields and the particle distribution function, a procedure of iteration of Eqs. (1-4) is necessary. In this procedure, four different physical models are iteratively coupled: (1) the plasma conductivity for non-Maxwellian distribution functions, (2) a wave solver incorporating this non-Maxwellian conductivity, (3) the quasi-linear operator that drives the non-thermal distribution, and (4) a Fokker-Planck solver to advance the distribution function. Following Stix [9], the normalized plasma conductivity (i.e. susceptibility) for an arbitrary non-relativistic species s is,

$$\chi_s = 2\pi \frac{\omega_p^2}{\omega^2} \left[\sum_{l=-\infty}^{\infty} \int \frac{du_\parallel}{1 - \frac{n_\parallel u_\parallel}{\sqrt{\mu}} - \frac{l\Omega}{\omega}} \int du_\perp U\, \mathbf{S}_l + \hat{\mathbf{e}}_\parallel \hat{\mathbf{e}}_\parallel \int du_\parallel \int du_\perp u_\parallel \left(u_\perp \frac{\partial f}{\partial u_\parallel} - u_\parallel \frac{\partial f}{\partial u_\perp}\right) \right]. \quad (5)$$

The plasma conductivity tensor σ_s is related to the susceptibility χ_s by $\sigma_s = -i\omega\varepsilon_0\chi_s$. The integral over $u_\parallel$ in Eq. (5) is singular at cyclotron resonance, and can be evaluated using the Plemelj relation. The factor U contains velocity space derivatives of the distribution function,

$$U = \frac{\partial f}{\partial u_\perp} - \frac{n_\parallel}{\sqrt{\mu}}\left(u_\parallel \frac{\partial f}{\partial u_\perp} - u_\perp \frac{\partial f}{\partial u_\parallel}\right). \quad (6)$$

In Eqs. (5-6), l is the harmonic number, Ω is the cyclotron frequency, ω_p is the plasma frequency, and $n_\parallel = k_\parallel c/\omega$, where $k_\parallel$ is the wave number parallel to the equilibrium magnetic field. The tensor $\mathbf{S}_l$ is defined as,

$$\mathbf{S}_l = \begin{bmatrix} \frac{1}{2}u_\perp^2 J_{l+1}^2(\xi) & \frac{1}{2}u_\perp^2 J_{l+1}(\xi)J_{l-1}(\xi) & \frac{1}{\sqrt{2}}u_\perp u_\parallel J_{l+1}(\xi)J_l(\xi) \\ \frac{1}{2}u_\perp^2 J_{l+1}(\xi)J_{l-1}(\xi) & \frac{1}{2}u_\perp^2 J_{l-1}^2(\xi) & \frac{1}{\sqrt{2}}u_\perp u_\parallel J_{l-1}(\xi)J_l(\xi) \\ \frac{1}{\sqrt{2}}u_\perp u_\parallel J_{l+1}(\xi)J_l(\xi) & \frac{1}{\sqrt{2}}u_\perp u_\parallel J_{l-1}(\xi)J_l(\xi) & u_\parallel^2 J_l^2(\xi) \end{bmatrix} \quad (7)$$

where the argument of the Bessel functions is $\xi = k_\perp v_\perp / \Omega = (k_\perp u_\perp / \Omega) c / \sqrt{\mu}$. The velocity u is normalized to $v_c = c / \sqrt{\mu}$, c is the speed of light, and $\mu = mc^2/2eE_{norm}$

where E_{norm} is the maximum energy in eV at which the numerical distribution function is evaluated. The velocity components perpendicular and parallel to the magnetic field are $u_\perp$ and $u_\parallel$, respectively, and the distribution function f is normalized to n/v_c^3, where n is the density.

POWER ABSORPTION FOR NON-MAXWELLIAN PLASMAS

The power absorbed by the plasma can be expressed locally as [10-11]

$$P_{RF} = \frac{1}{2} \text{Re} \left\{ \frac{\varepsilon_0 \omega}{i} \sum_{\mathbf{k}_1, \mathbf{k}_2} e^{i(\mathbf{k}_1 - \mathbf{k}_2) \cdot \mathbf{r}} \mathbf{E}^*_{\mathbf{k}_2} \cdot \mathbf{W}_l \cdot \mathbf{E}_{\mathbf{k}_1} \right\} \quad (8)$$

where $\mathbf{W}_l$ is the energy absorption kernel,

$$\mathbf{W}_l = 2\pi \frac{\omega_p^2}{\omega^2} \sum_{l=-\infty}^{\infty} e^{il(\beta_1 - \beta_2)} \mathbf{C}^{-1}(\beta_2) \cdot \left[\int_{-\infty}^{\infty} \frac{du_\parallel}{1 - \frac{n_\parallel u_\parallel}{\sqrt{\mu}} - \frac{l\Omega}{\omega}} \int_0^\infty du_\perp U \, \mathbf{S}' \right] \cdot \mathbf{C}(\beta_1) \, . \quad (9)$$

and $\mathbf{C}(\beta)$ is a rotation matrix that transforms the electric field from local magnetic coordinates to the (E_+, E_-, $E_\parallel$) frame. The summation over Fourier wave numbers $\mathbf{k}_1$ and $\mathbf{k}_2$ in Eq. (8) can be extremely time consuming to evaluate. For example in 2-D, four nested do loops are required, and in 3-D, six nested loops are required. Even for Maxwellians, calculating these summations takes an order of magnitude more time than the wave solution itself. For non-Maxwellians, the time is totally prohibitive. A more efficient way to calculate the power absorption comes about when the velocity space integrals in Eq. (9) are brought outside of the sum over $\mathbf{k}_1$ and $\mathbf{k}_2$. In this case, P_{RF} can be expressed as a product of sums rather than as nested sums,

$$P_{RF} = -\frac{\pi}{2} \frac{\varepsilon_0 \omega_p^2}{\omega} \text{Re} \left\{ \int_0^\infty du_\perp \frac{\sqrt{\mu}}{|n_\parallel|} \pi U \sum_{l=-\infty}^{\infty} \left(\sum_{\mathbf{k}_2} \boldsymbol{\varepsilon}^{*T}_{\mathbf{k}_2} \cdot \mathbf{a}_l^{(2)T} \right) \cdot \left(\sum_{\mathbf{k}_1} \mathbf{a}_l^{(1)} \cdot \boldsymbol{\varepsilon}_{\mathbf{k}_1} \right) \Big|_{u_\parallel, res} \right\} . \quad (10)$$

where ε is the rotated electric field, and $\mathbf{a}_l = \left(u_\perp J_{l+1}(\xi), \, u_\perp J_{l-1}(\xi), \, \sqrt{2} u_\parallel J_l(\xi) \right)$. Because the sums over $\mathbf{k}_1$ and $\mathbf{k}_2$ are separated, there is an enormous savings in computation time, and Eq. (10) can be evaluated for non-Maxwellians in approximately the same time as required to calculate the plasma current.

QUASILINEAR DIFFUSION COEFFICIENTS

To solve the Fokker-Planck equation self-consistently with the full-wave RF solution, the bounce-averaged quasi-linear diffusion coefficients B_0, C_0, E_0, and F_0 must be derived directly from the full-wave RF electric fields. These coefficients can be deduced from P_{RF} in Eq. (10). For example, the local value for B is given by,

$$B = \frac{\varepsilon_0 \omega_p^2}{8 \omega e E_{norm} \Delta u_\parallel} \text{Re} \left[\sum_{l=-\infty}^{\infty} \frac{\pi \sqrt{\mu}}{|n_\parallel|} \left(\sum_{\mathbf{k}_2} \boldsymbol{\varepsilon}^{*T}_{\mathbf{k}_2} \cdot \mathbf{a}_l^{(2)T} \right) \cdot \left(\sum_{\mathbf{k}_1} \mathbf{a}_l^{(1)} \cdot \boldsymbol{\varepsilon}_{\mathbf{k}_1} \right) \Big|_{u_\parallel, res} \right] \quad (11)$$

with similar expressions for C, E, and F. The bounce averaged coefficients can be conveniently expressed in terms of a flux surface average of Eq. (11),

$$B_0 = \lambda \langle\langle B \rangle\rangle_{bounce} = \left\langle \left|\frac{u_{\parallel,0}\psi}{u_\parallel}\right| B \right\rangle_{flux\ surface} \oint \frac{dl_B}{\psi} \quad (12)$$

and similarly for the other coefficients. Unfortunately, these expressions do not always give positive definite results for B_0 and F_0 as required by CQL3D. By including an effective de-correlation rate that limits the phase memory, [i.e. the number of cross terms contributing to the product of sums in Eq. (11)], B_0 and F_0 can be made explicitly positive definite.

The flux surface average heating rate can be written in terms the bounce-averaged distribution function f_0 and quasi-linear diffusion coefficients B_0 and C_0,

$$\langle P_{RF} \rangle_{flux\ surface} = \frac{-4\pi\ eE_{norm}}{\oint dl_B/\psi} \int_0^\infty u_{\perp,0} du_{\perp,0} \int_{-\infty}^\infty du_{\parallel,0} \left(\frac{B_0}{u_0} \frac{\partial f_0}{\partial u_0} + \frac{C_0}{u_0} \frac{\partial f_0}{\partial \vartheta_0} \right). \quad (13)$$

Power absorption profiles obtained with this expression can be compared to those obtained from the flux surface average of Eq. (10) to provide a cross-check on the consistency of the quasi-linear diffusion coefficients calculated by AORSA. Figure 1 shows power absorption profiles calculated for fast deuterium ions (D), minority hydrogen ions (H), and electrons (e) in the National Spherical Tokamak Experiment (NSTX) shot 108251 [4]. In this case, 2.4 MW of high harmonic fast wave power is applied simultaneously with neutral beam injection. In Fig. 1(a), the power density profiles are calculated directly from the full-wave electric fields using the flux surface average of Eq. (10), and in Fig. 1(b), they are calculated from Eq. (13) using the bounce-averaged diffusion coefficients, B_0 and C_0. The observed agreement demonstrates the consistency between the wave electric fields and the quasi-linear diffusion coefficients. In Fig. 1(c), the quasi-linear coefficients are passed to the CQL3D Fokker Planck code which independently calculates the power absorption profile for the fast deuterium ions. The resulting profile agrees closely with that calculated by AORSA.

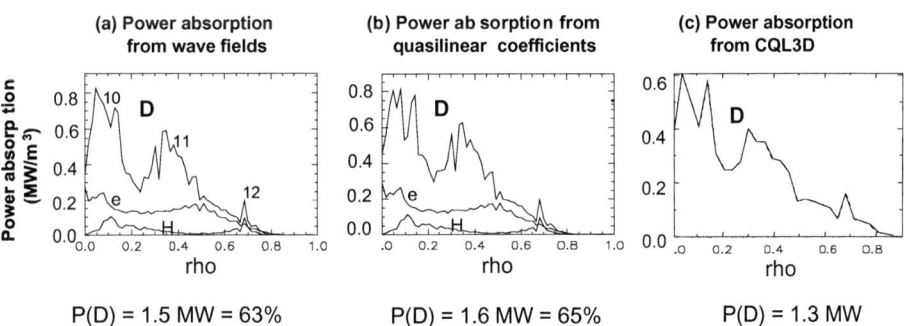

FIGURE 1. Power density calculated from (a) the flux surface average of Eq. (10), (b) Eq. (13), and (c) the CQL3D Fokker-Planck code using quasi-linear diffusion coefficients from AORSA.

SELF-CONSISTENT ITERATIVE SOLUTIONS

In this section, self-consistent wave fields and particle distribution functions are calculated by iterating the full-wave and Fokker-Planck solutions given by AORSA and CQL3D, respectively. Self-consistent results are obtained for neutral beam ions heated by high harmonic fast waves in NSTX [4], and for energetic tritium ions (T) heated at the second harmonic cyclotron resonance in the proposed ITER burning plasma experiment [12].

Power Absorbed by Fast Deuterium Ions in NSTX

Figure 2 shows four iterative steps in the self-consistent solution for the wave fields and distribution function for fast deuterium ions in NSTX shot 108251 with 0.24 MW of high harmonic fast wave power. In the 0th iteration, CQL3D calculates the distribution function for the neutral beam only, with no RF power. Using this distribution function, AORSA calculates the wave electric field, power absorption, and quasi-linear diffusion coefficients for the fast deuterium ions, assuming that the electrons and minority hydrogen are Maxwellian. The quasi-linear diffusion coefficients are then passed to CQL3D and used in the next iteration to recalculate the distribution function. The RF power induces a high energy tail on the distribution function that is evident in this and the following iterations. The iteration procedure is continued until the results stop changing. In Fig. 2, little change is observed in either the distribution function or the absorbed power between the 2nd and 3rd iterations, indicating that the calculation is approximately converged. For higher RF power, the energetic tail continues to grow, and no steady state is reached. This is most likely because the radial diffusion term $\langle\langle R \rangle\rangle$ in Eq. (3) is not included in this calculation.

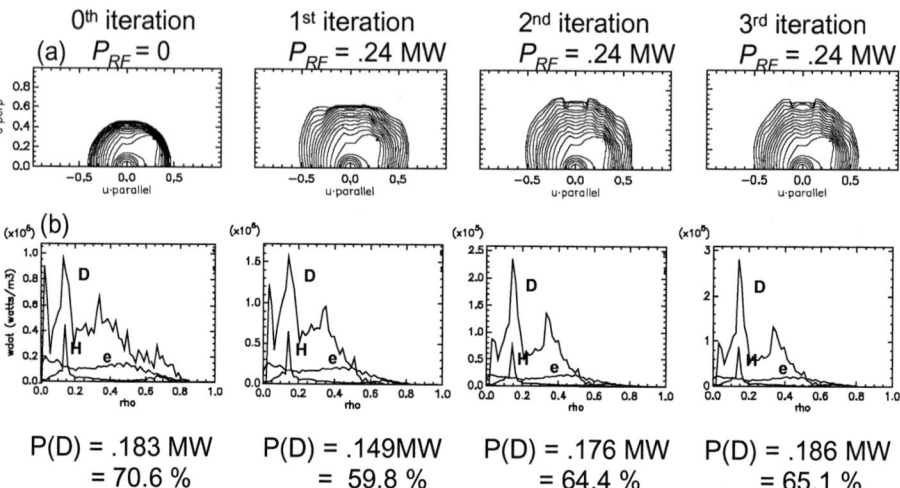

FIGURE 2. Four steps in the iterative solution for (a) the neutral beam distribution function and (b) the RF power absorption in NSTX with 0.24 MW of high harmonic fast wave heating.

Parasitic Absorption by Tritium Ions in ITER

An important problem for ITER is the effect of energetic particle populations on wave propagation, absorption, and current drive in ICRF regimes. Parasitic absorption of power by energetic components such as minority ion species, fusion-generated alpha particles, and fast ions associated with neutral beam injection can reduce direct electron heating and the associated fast wave current drive. Figure 3 shows a calculation of parasitic absorption by tritium ions for a 50-50 D-T mixture in ITER with 20 MW of RF power at 56 MHz. This frequency places the second harmonic tritium resonance left of the magnetic axis near the $\rho = 0.4$ flux surface.

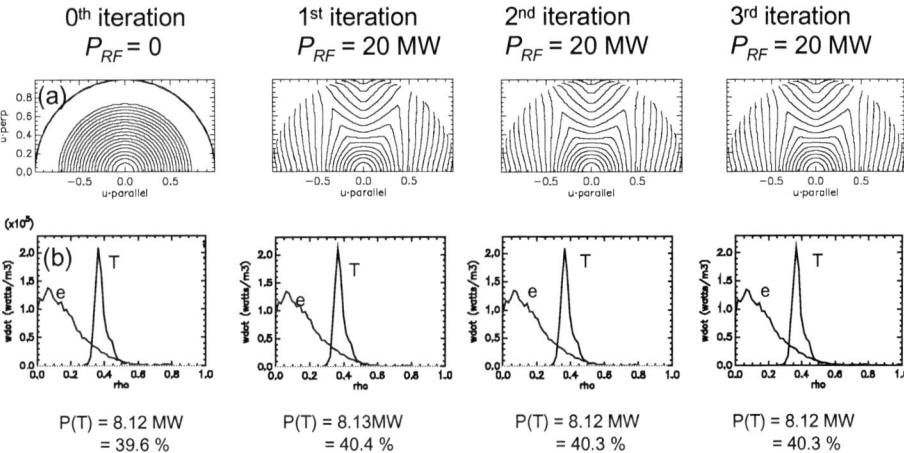

FIGURE 3. Four steps in the iterative solution for (a) the tritium distribution function and (b) the RF power absorption for a 50-50 D-T mixture in ITER with 20 MW of fast wave power at 56 MHz.

In the 0th iteration, the RF power is assumed to be zero, and the tritium distribution function is therefore Maxwellian. AORSA uses this distribution function to calculate the wave propagation, power absorption, and quasi-linear diffusion coefficients, assuming the electrons to be Maxwellian. In the next iteration, an energetic tail develops on the tritium distribution function, particularly near the trapped-passing boundary. This tail remains essentially unchanged in the following iterations as does the total tritium power absorption, $P(T)$. We conclude that for the 50-50 D-T mixture, the energetic tritium tail in ITER is weak and has little effect on the parasitic tritium absorption.

If the tritium density is decreased from 50% to 10% of the total electron density, there is more RF power available per particle, and a more pronounced tail develops on the distribution function. This is shown in Fig. 4, where the energetic tail accounts for a net increase of about 12% in the total tritium absorption between the first and last iterations.

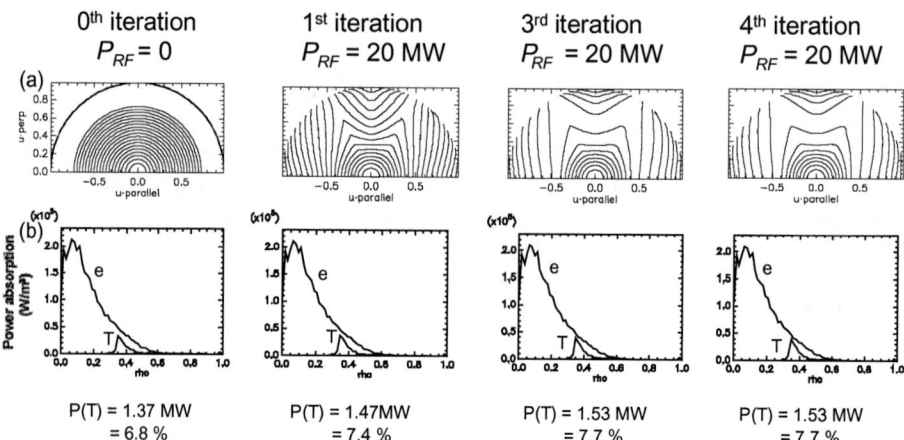

FIGURE 4. Four steps in the iterative solution for (a) the tritium distribution function, and (b) the RF power absorption for a 90-10 D-T mixture in ITER with 20 MW of fast wave power at 56 MHz.

ACKNOWLEDGMENTS

This work was sponsored by SciDAC and the Oak Ridge National Laboratory, managed by UT-Battelle, LLC, for the U.S. Department of Energy under Contract No. DE-AC-5-00OR22725. Numerical calculations carried out with resources of the Center for Computational Sciences at the Oak Ridge National Laboratory and the National Energy Research Scientific Computing Center at the Lawrence Berkeley National Laboratory.

REFERENCES

1. Dumont, R. J., Phillips, C.K., and Smithe, D.N., "ICRF wave propagation and absorption in plasmas with non-thermal populations", in *Controlled Fusion and Plasma Physics*, 26B, paper P-5.051 (29th EPS Conference, Montreux, Jun. 17-21, 2002).
2. Dumont, R.J., Phillips, C.K., and Smithe, D.N., "Effects of non-Maxwellian Plasma Species on ICRF Propagation and Absorption in Toroidal Magnetic Confinement Devices," in *Radio Frequency Power in Plasmas: 15th Topical Conference*, edited by C. B. Forest, AIP Conference Proceedings 694, New York: American Institute of Physics, 2003, pp. 439-446.
3. Dumont, R. J., Phillips, C. K., and Smithe, D. N., *Phys. Plasmas* **12**, 042508 (2005).
4. Rosenberg, A. L., Menard, J. E., Wilson, J. R., et al., *Phys. Plasmas* **11**, 2441(2004).
5. Jaeger, E. F., Berry, L. A., Myra, J. R., et al., *Phys. Rev. Lett.* **90**, 195001-1 (2003).
6. Harvey, R. W., and McCoy, M. G., in *Proceedings of the IAEA Technical Committee Meeting on Advances in Simulation and Modeling of Thermonuclear Plasmas* (IAEA, Montreal, 1992), available through USDOC, NTIS No. DE9300962.
7. Harvey, R. W., et al., contributed paper, this meeting (2005).
8. Kennel, C. F. and Engelmann, F., *Phys. Fluids* **9**, 2377 (1966).
9. Stix, T. H., *Waves in Plasmas*, American Institute of Physics, New York, 1992.
10. Smithe, D. N., *Plasma Phys. Controlled Fusion* **31**, 1105 (1989).
11. Brambilla, M. and Krucken, T., *Plasma Phys. Controlled Fusion* **30**, 1083 (1988).
12. Aymar, R., et al., Nucl. Fusion **41**, 1301 (2001).

Monte-Carlo Orbit/Full Wave Simulation of Fast Alfvén Wave (FW) Damping on Resonant Ions in Tokamaks

M. Choi*, V.S. Chan*, V. Tang[†], P. Bonoli[†], R.I. Pinsker*, and J. Wright[†]

General Atomics, P.O. Box 85608, San Diego, California, 92186-5608 USA
[†]*Massachusetts Institute of Technology, Cambridge, Massachusetts, USA*

Abstract. To simulate the resonant interaction of fast Alfvén wave (FW) heating and Coulomb collisions on energetic ions, including finite orbit effects, a Monte-Carlo code ORBIT-RF has been coupled with a 2D full wave code TORIC4. ORBIT-RF solves Hamiltonian guiding center drift equations to follow trajectories of test ions in 2D axisymmetric numerical magnetic equilibrium under Coulomb collisions and ion cyclotron radio frequency quasi-linear heating. Monte-Carlo operators for pitch-angle scattering and drag calculate the changes of test ions in velocity and pitch angle due to Coulomb collisions. A rf-induced random walk model describing fast ion stochastic interaction with FW reproduces quasi-linear diffusion in velocity space. FW fields and its wave numbers from TORIC are passed on to ORBIT-RF to calculate perpendicular rf kicks of resonant ions valid for arbitrary cyclotron harmonics. ORBIT-RF coupled with TORIC using a single dominant toroidal and poloidal wave number has demonstrated consistency of simulations with recent DIII-D FW experimental results for interaction between injected neutral-beam ions and FW, including measured neutron enhancement and enhanced high energy tail. Comparison with C-Mod fundamental heating discharges also yielded reasonable agreement.

INTRODUCTION

Conventionally, full wave [1] or ray-tracing codes [2] combined with Fokker-Planck code assuming zero banana width have been used to study the interaction between fast Alfven wave (FW) and the plasma. However, these approaches do not take into account both finite orbit drift width of non-Maxwellian ions produced from rf interaction and radial scattering of energetic particles across flux surfaces due to Coulomb collisions and perpendicular heating. Figure 1 shows the 2D full wave code TORIC4 simulation result on Alcator C-Mod hydrogen minority (5%) fundamental heating experiment at 78 MHz (C-Mod 1040415006). It is seen that predicted radial power deposition profile of FW on minority hydrogen ions is very sensitive to initially assumed Maxwellian temperature. For a thermal Maxwellian temperature of $T_H(0) = 3$ keV (red curve), the FW could not penetrate a cut-off region inside the plasma (~2 cm on the low field side from magnetic axis) due to a weak Doppler shift effect. The radial profile is thus off-axis. But, with the assumption of an initial tail temperature of $T_H(0) = 20$ keV (blue curve), a much broader Doppler shifted resonance produces more on-axis heating and stronger power absorption on minority ions.

A Monte-Carlo code, ORBIT-RF [3] provides a capability to investigate the interaction between FW and plasma resonant ions, self-consistently including rf heating and Coulomb collisions. The code solves the Hamiltonian guiding center drift equations [4] in magnetic coordinates $(\psi_p, \theta, \rho_{//}, \zeta)$ in a 2D axisymmetric numerical magnetic equilibrium where parallel and radial drift motions of each ion are directly solved in every time step. Monte-Carlo collision operators for pitch-angle scattering

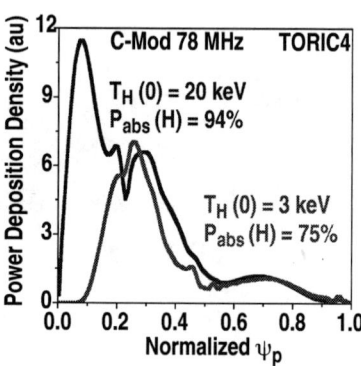

FIGURE 1. Sensitivity of radial power deposition profiles on initial temperatures of resonant ions.

and drag by electron and background ions calculate changes of test ion parallel velocity and pitch angle due to Coulomb collisions [3]. A rf-induced random walk model describing fast ion stochastic interaction with FW is implemented to reproduce quasi-linear diffusion in velocity space, assuming that resonant ions lose their phase information with FW through successive collisions and wave stochasticity before they re-enter the resonance region. A generalized arbitrary harmonic rf diffusion operator calculates perpendicular rf-kicks at resonances including Doppler shifts. Two-dimensional full wave code TORIC4 [1] is coupled to ORBIT-RF to determine the amplitudes of FW fields and their wave numbers in the plasma as a function of (R,Z). Steady-state slowing-down distribution of beam ion species is modeled using a re-injection method of thermalized beam ions [5]. In this work, only a single cyclotron resonance layer is modeled from each simulation. Self-consistent modeling of the interaction between FW heating and slowing-down of energetic ion species from ORBIT-RF offers a capability of identifying qualitatively the experimentally demonstrated strong interaction between FW and resonant ions in both C-Mod and DIII-D tokamaks.

This paper is organized as follows. First, quasilinear rf diffusion operator with arbitrary harmonic absorption is described in detail. In sequence, numerical results from ORBIT-RF are validated against experiments in both C-Mod minority fundamental heating and DIII-D fast wave current drive (FWCD) discharges. Lastly, summary and future plan are given.

QUASI-LINEAR ARBITRARY HARMONIC RF DIFFUSION OPERATOR

When ions pass through an ion cyclotron resonance layer satisfying the condition $\omega_{rf} - k_{//}v_{//} = n\Omega_i$ (n: harmonic number), they may either absorb energy from or lose energy to the wave depending on their phase with the wave polarization. This introduces a random walk motion in addition to a mean change in energy. We introduce a rf-induced random walk model to reproduce quasi-linear diffusion in velocity space through stochastic interaction of ions with the wave. The interaction of ions with FW at Doppler shifted resonances changes ions's velocities in both parallel and perpendicular directions. In this work, the change in $v_{//}$ due to parallel electric field component is ignored. Only perpendicular kicks of magnetic moment (μ) are considered. The increment of magnetic moment ($\Delta\mu_{rf}$) [Eq. (1)] is obtained by calculating a mean value [Eq. (2)] and random variance [Eq. (3)] from the quasi-linear equation governing rf-induced particle diffusion in velocity space.

$$\Delta\mu_{rf} = \overline{\Delta\mu_{rf}} + 2\sqrt{3}\left(R_s - \frac{1}{2}\right)\sqrt{\left\langle\overline{\Delta\mu_{rf}}^2\right\rangle} \quad , \tag{1}$$

where

$$\overline{\Delta\mu_{rf}} = \frac{\pi q^2 l^2 \Omega^2}{m\omega^2 B}|E_+|^2 \times$$
$$\left[\left|J_{l-1} + e^{2i\vartheta_k}\frac{E_-}{E_+}J_{l+1}\right|^2 + \mu\left\{2\left(J_{l-1} + e^{2i\vartheta_k}\frac{E_-}{E_+}J_{l+1}\right)\left(\frac{\partial J_{l-1}}{\partial\mu} + e^{2i\vartheta_k}\frac{E_-}{E_+}\frac{\partial J_{l+1}}{\partial\mu}\right)\right\}\right]\frac{K}{|\dot{w}_l|} \tag{2}$$

$$\left\langle\overline{\Delta\mu_{rf}}^2\right\rangle = 2\mu\left[\frac{\left|J_{l-1} + e^{2i\vartheta_k}\frac{E_-}{E_+}J_{l+1}\right|^2}{2\left(J_{l-1} + e^{2i\vartheta_k}\frac{E_-}{E_+}J_{l+1}\right)\left(\frac{\partial J_{l-1}}{\partial\mu} + e^{2i\vartheta_k}\frac{E_-}{E_+}\frac{\partial J_{l+1}}{\partial\mu}\right)}\right]\overline{\Delta\mu_{rf}} \tag{3}$$

with $w_l = \omega_{rf} - n\Omega_i - k_{//}\rho_{//}\Omega_i$. R_s is a random number between 0 and 1 where the factor $2\sqrt{3}$ is such that

$$\int_0^1 \left[2\sqrt{3}\left(R_s - \frac{1}{2}\right)\right]^2 dR_s = 1 \quad ,$$

assuming uniform probability. J_n is the nth order Bessel function of the first kind, and m is a test ion mass. An operator K is expressed by [6]

$$K = \begin{cases} 1 & \sqrt{2}\tau_{uc} \leq \tau_c \\ \frac{2\pi Ai^2(\zeta)|\dot{Z}_n|}{|\ddot{Z}_n/2|^{2/3}} & \sqrt{2}\tau_{uc} > \tau_c \end{cases} \quad , \tag{4}$$

$$\zeta = -\frac{|\dot{Z}_n|^2}{|2\ddot{Z}_n^2|^{2/3}}, \quad \tau_{uc} = \sqrt{\frac{2\pi}{|\dot{Z}_n|}}, \quad \tau_c = \frac{2\pi Ai(\zeta)}{|\ddot{Z}_n/2|^{1/3}} \quad , \tag{5}$$

where Ai is the Airy function in order to include the case when the successive interactions between a particle orbit and the cyclotron resonances are close to each other. In Eq. (2), E_+ and E_- are left-hand and right-hand polarized components of wave electric field. The $e^{2i\vartheta_k}$ is the phase difference between E_+ and E_-. Radial profiles of $|E_+|^2$, $e^{2i\vartheta_k}$, $k_\perp$ and $k_{//}$ required in Eqs. (2) and (3) are computed from TORIC4. Since TORIC4 calculates unit current wave fields, we rescale the wave fields using actual experimental input power to pass on to ORBIT-RF [7]. Presently, the wave fields from TORIC4 are approximated using a single dominant toroidal and poloidal Fourier components.

VALIDATION OF ORBIT-RF AGAINST EXPERIMENTS

In this section, numerical results from ORBIT-RF are validated against experimental results from both C-Mod minority fundamental heating at 78 MHz and DIII-D FWCD discharges at 60 MHz and 116 MHz. We follow trajectories of ion species by solving the Harmiltonian guiding center drift equation at each time step (Δt is usually ~10^{-5} s) using a Runge-Kutta fourth order integration scheme subject to the FW heating and slowing-down collisions/pitch angle scattering with background ions and electrons. Simulation time is typically several slowing-down times of test ion species, which usually requires several ten thousand transit times of test ions. The grid size used is $(\psi, \theta) = (100, 50)$. In ORBIT-RF, electrons are only considered as neutralizing background. Their role in altering the radial electric field is not considered. For simulations presented in this paper, we used mostly 5000 particles to present Monte-Carlo test ions.

C-Mod Minority Fundamental Heating Discharge at 78 MHz

Recent ICRH minority distribution measurements from the 2004 Alcator C-Mod compact neutral particle analyzer (CNPA) [7] can be directly compared with ORBIT-RF results. In the following discharge, Shot 1040415006, $T_e(0) = T_D(0) = $ ~3 keV, $n_e(0) < 1.0 \times 10^{14}$/cm^3, $B_0 = 5.5$ T, $I_p = 0.8$ MA, and 1 MW of 78 MHz FW is applied to the ~4% D(H) plasma. The symmetric antenna power spectrum with $(0,\pi,\pi,0)$ phasing has a peak at toroidal mode number $N_\varphi = \pm 7$. Figure 2 shows increases of raw voltage from the CPNA [7] and electron temperatures during ICRH. Simulation results for this discharge from TORIC4 (blue curve) and ORBIT-RF (red one) are compared in Fig. 3. The ORBIT-RF result shows much broader radial deposition profile than that from TORIC4 with initial minority ion temperature of 20 keV. The broader profile from ORBIT-RF may be explained by radial diffusion of resonant ions due to successive pitch angle scattering and perpendicular rf heating, as shown in Fig. 4. In Fig. 5, the CNPA experimentally measured on-axis hydrogen-minority perpendicular

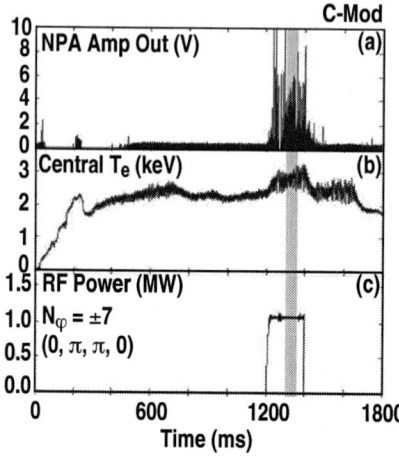

FIGURE 2. (a) Raw voltage from CNPA, (b) central electron temperature, and (c) rf power. The highlighted time band indicates a 50 ms DNB pulse for active neutral particle analysis.

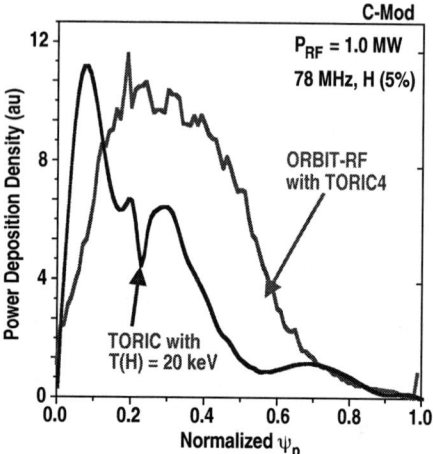

FIGURE 3. Power deposition density from TORIC4 and ORBIT-RF for Alcator C-Mod 78 MHz.

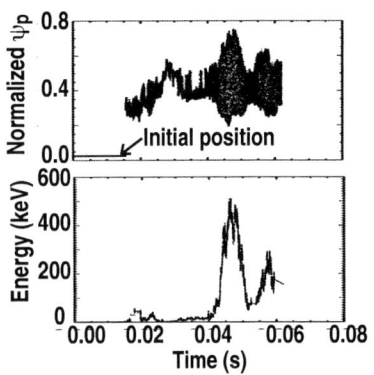

FIGURE 4. Time trajectories of a single resonant ion showing radial scattering due to Coulomb collision and perpendicular rf heating.

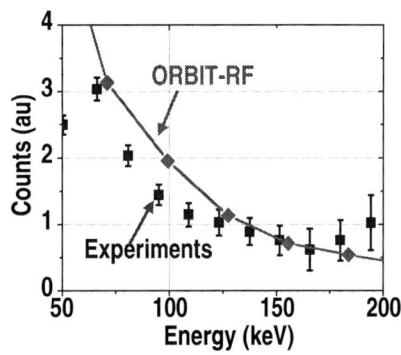

FIGURE 5. Comparison of ORBIT-RF and experiments on C-Mod 78 MHz.

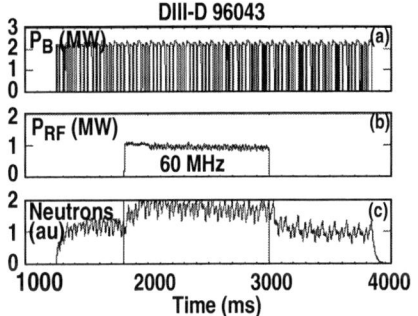

FIGURE 6. (a) Injected beam power (b) rf power $[n_e(0) = 5.0 \times 10^{13}$ cm^{-3}, $T_e(0) = 2.7$ keV, $T_i(0) = 3.5$ keV], (c) neutron flux for DIII-D 60 MHz.

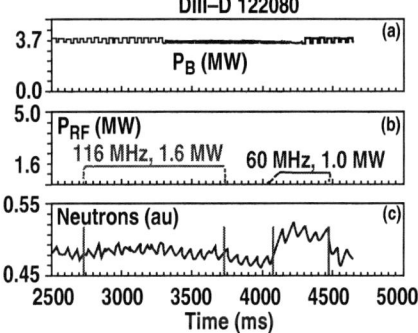

FIGURE 7. (a) Injected beam and (b) rf powers $[n_e(0) = 7.0 \times 10^{13}$ cm^{-3}, $T_e(0) = 1.5$ keV, $T_i(0) = 2.0$ keV], (c) neutron flux and ORBIT-RF for DIII-D 116 MHz.

energy spectrum (blue curve) and ORBIT-RF simulated one (red curve) are compared, showing reasonable agreement.

DIII-D FWCD at 60 MHz and 116 MHz

Figures 6 and 7 show previous (96043.01920) and recent (122080.2700) DIII-D FWCD experimental results. Plasma parameters for these discharges are summarized in Table 1. The results at 60 MHz indicate beam ions acceleration at 4th harmonic resonance [8] while very little interaction of beam ion with 116 MHz is shown at 8th harmonic resonance. This is clearly evident experimentally from neutron reaction rates and electron temperatures that show an increase during rf heating at 60 MHz, while no increase is observed at 116 MHz.

In the simulations, initial distribution of beam ions in velocity space is assumed to be mono-energetic with a full energy component of 80 keV. In real space, they are distributed with a probability $n_f(\psi) \propto (1-\psi^2)^2$. To model a steady-state regime, thermalized beam ions are re-injected at their birth energy (80 keV) into the plasma periodically during a given simulation time. To model actual NB+ICRF experimental situation realistically, NB only heating is first turned on for several slowing-down times

TABLE 1. Plasma parameters at DIII-D FWCD discharges

	96043.01920	122080.2700
$N_e(0)$ (cm^{-3})	5.0×10^{13}	7.0×10^{13}
$T_e(0)$ (keV)	2.7	1.5
$T_i(0)$ (keV)	3.5	2.0

of 80 keV deuterium beam ions. Almost all beam ions have gone through at least one thermalization within this duration to reach a steady-state distribution. Then the FW heating is turned on for a few additional slowing-down times to explore the interaction of beam ions with FW. In Fig. 8, spatial distributions of deuterium beam ions are displayed at t=0, 60000, 90000 transit times. The 60 MHz FW is turned on from t=60000 transit time to 90000 transit time. It is seen that the energetic tails extend up to a few hundreds keV above beam injection energy (80 keV) after the ICRF heating is turned on. These tails are built up in relatively broad layers ($0 < \psi < 0.4$) encompassing Doppler shifted 4th harmonic resonances, as shown in Fig. 8(c).

In Fig 9, radial profiles of beam ion pressure from ORBIT-RF are compared between NB only and NB+ICRF at 60 MHz and 116 MHz cases. Increased peakedness in the beam ion pressure profile observed in the central region of plasma at 60 MHz is reproduced qualitatively from ORBIT-RF. Very little increase of neutron rate at 116 MHz in experimental measurements (Fig. 7) is qualitatively explained in ORBIT-RF results showing no significant difference in the beam ion pressure profile after the rf is turned on.

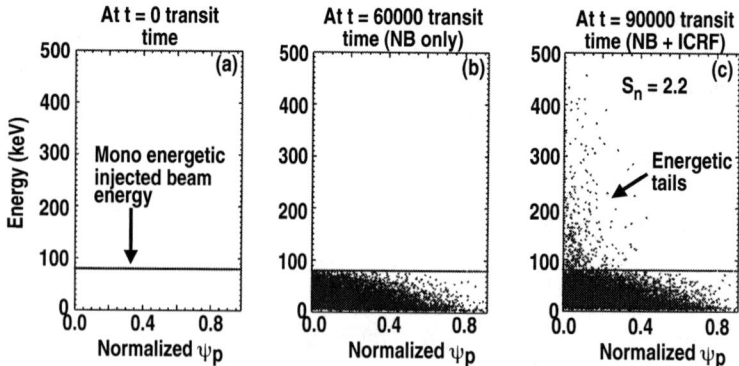

FIGURE 8. Spatial distributions of deuterium beam ions at (a) t=0, (b) t=60000, (c) t=90000 transit time where the 60 MHz ICRF wave is turned on from t=60000 to 90000 transit time.

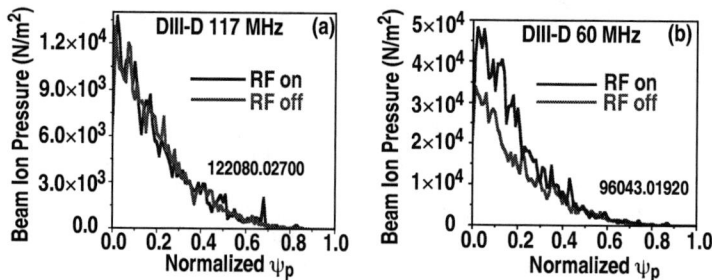

FIGURE 9. Radial profiles of pressure of beam ions as a function of normalized poloidal flux from ORBIT-RF at (a) DIII-D 116 MHz and (b) DIII-D 60 MHz.

SUMMARY AND FUTURE PLANS

ORBIT-RF provides a comprehensive physics package to self-consistently simulate interaction between ICRF and non-Maxwellian fast ions with finite orbit effects.ORBIT-RF with TORIC4 qualitatively reproduces various C-Mod and DIII-D experimental results in fast ion spectrum and neutron enhancement. Specifically, ORBIT-RF reproduces qualitatively the strong wave-beam particle interaction at 60 MHz and lack of wave-particle interaction at 116 MHz in DIII-D. The energetic particle energy spectrum with fundamental heating on C-Mod is also in agreement.

ACKNOWLEDGMENT

Work supported by the U.S. Department of Energy under Grant DE-FG03-95ER54309. Authors would like thank Drs. M. Porklab at MIT and W.W. Heidbrink at UC-Irvine for their valuable discussions for validation work of ORBIT-RF results against experimental measurements in C-Mod and DIII discharges. A special appreciation is given to Dr. M. Brambilla for providing us the latest version of TORIC4.

REFERENCES

1. M. Brambilla, Plasma Phys. Control. Fusion **41**, 1 (1999).
2. R.W. Harvey and M.G. McCoy, Proc. of IAEA TCM on Advances in Simulation and Modeling of Thermonuclear Plasmas, Montreal, 1992; USDOC NTIS document DE93002962].
3. V.S. Chan, et al., Phys. Plasmas **9**, 501 (2002).
4. R.B. White et al., Phys. Fluids **27**, 2455 (1984).
5. M. Choi, et al., "Monte-Carlo orbit/full wave simulation of ion cyclotron resonance frequency (ICRF) wave damping on resonant ions in tokamaks," submitted to Phys. Plasma (2005).
6. S.C. Chiu, et al., Phys. Plasmas **7**, 4609 (2000).
7. V. Tang, et al., Bull. Am. Phys. Soc. **49**, 75 (2004).
8. W.W. Heidbrink, et al., Nucl. Fusion **39**, 1369 (1999).

Numerical Simulation of Ion Cyclotron Heating Experiments with Coupled Maxwell and Quasilinear-Fokker-Planck Solvers

M. Brambilla, R. Bilato, C. Maggi, H.-U. Fahrbach, W. Suttrop, and the ASDEX Upgrade Team

Max-Planck-Institut für Plasmaphysik - Euratom Association Boltzmannstrasse 2, D-85748 Garching, Germany

Keywords: Ion Cyclotron Heating, Quasilinear Distribution Functions.
PACS: 52.50, 52.65

Introduction. The code TORIC [1] solving Maxwell equations in toroidal axisymmetric plasmas in the Ion Cyclotron (IC) frequency range has been integrated in a package which includes: (1) an interface to the experimental data (Grad-Shafranov MHD configuration, density and temperature profiles), (2) the interface QLDCE [2] to a quasilinear Fokker-Planck solver for the electrons, (3) the quasilinear Fokker-Planck solver SSFPQL [3] for the ions, and (4) a subroutine which reevaluates the coefficients of the wave equations taking into account the suprathermal anisotropic tails of minority ions predicted by SSFPQL, so that their effects on wave propagation and absorption can be estimated by iterating TORIC. This package allows somewhat simplified but essentially selfconsistent simulations of heating and current drive (CD) in this frequency domain. Applications to Fast Wave CD have been made in [2]. Here we present the analysis of two IC heating experiments in ASDEX Upgrade (AUG) and in JET.

The code SSFPQL. The code SSFPQL [3] solves the quasilinear equations for ions heated at the fundamental and the first cyclotron harmonic, using the output of TORIC to build the quasilinear diffusion coefficient (QLDC) on each magnetic surface. The main simplifications made by SSFPQL are:

i) The uniform-plasma Kennel-Engelmann quasilinear operator [4] is surface-averaged, neglecting several effects of toroidicity on IC heating (toroidal trapping, finite banana width, losses,...).

ii) The collisional operator is linearized, assuming that the distribution of fast ions reaches steady state by losing energy on the background ions and electrons.

Exploiting assumption (ii), SSFPQL solves directly for the steady-state, and is, therefore, very fast: the distribution functions of two ion species (minority heated at the fundamental, majority at the first harmonic) can be evaluated in less than 20 sec on 100 magnetic surfaces on a laptop. Because of (i), SSFPQL cannot be regarded as a full substitute for a more sophisticated Fokker-Planck solver or for Montecarlo simulations [5], particularly for the most energetic ions. For the bulk of the hot ion populations it nevertheless predicts distributions in good agreement with those measured on AUG.

Iterating TORIC. Evaluating the coefficients of the wave equations for generic non Maxwellian plasmas requires a huge numerical effort, and would increase the execution time of TORIC by orders of magnitude. It is, therefore, a fortunate circumstance that the minority distribution function evaluated by SSFPQL can be approximated with reasonable accuracy by the superposition of two anisotropic Maxwellians (generalizing the well-known analytical approximations obtained by Stix [6])

$$F_m(v_\|, v_\perp) = b_1 \frac{e^{-[(v_\|^2/\alpha_{\|1}^2)+(v_\perp^2/\alpha_{\perp 1}^2)]}}{\pi^{3/2}\alpha_{\perp 1}^2 \alpha_{\|1}} + b_2 \frac{e^{-[(v_\|^2/\alpha_{\|2}^2)+(v_\perp^2/\alpha_{\perp 2}^2)]}}{\pi^{3/2}\alpha_{\perp 2}^2 \alpha_{\|2}} \qquad (b_1+b_2=1) \tag{1}$$

The parameters of this representation are determined by matching the logarithmic slopes and the parallel and perpendicular energy content of the distributions evaluated by SSF-PQL. Accurate matching of the slopes is particularly important, since IC absorption is proportional to velocity derivatives of F_m. With Eq. (1), the coefficients of the wave equations can be expressed in terms of the Plasma Dispersion Function Z. The contribution of the minority species to the coefficients of the wave equations [1] are

$$\delta \hat{L} = -\frac{\omega_{pm}^2}{\omega^2} \sum_k b_k \left[-\frac{x_0}{\alpha_{\|k}} Z\left(\frac{x_1}{\alpha_{\|k}}\right) \right] \tag{2}$$

$$\delta \hat{\lambda}_i^{(2)} = \frac{1}{2} \frac{\omega_{pm}^2}{\Omega_{cm}^2} \frac{v_{thm}^2}{c^2} \sum_k b_k \alpha_{\perp k}^2 \left[-\frac{x_0}{\alpha_{\|k}} Z\left(\frac{x_2}{\alpha_{\|k}}\right) \right] \tag{3}$$

where $x_n = (\omega - n\Omega_{c\alpha})/k_\| v_{th\alpha}$ and the other notations are standard. Iterating TORIC including these contributions involves a greater number of evaluations of the function Z (each poloidal Fourier component of the electric field has its own $k_\|$); the algorithm for this purpose, however, is quite efficient.

Applications. As an application, we present the simulation of two IC heating experiments in ASDEX Upgrade (AUG) and in JET [7]. Although the scenarios were similar (Hydrogen minority in Deuterium, at low concentration well within the minority regime; the main parameters are summarized in Table 1), and the coupled power densities almost identical, strong electron heating was observed in JET, but not in ASDEX Upgrade. The simulations reproduce well the observations, and allow to ascribe the different outcome to the lower collisionality and somewhat lower Hydrogen concentration of the JET plasma.

Figures 1 and 2 show the power deposition profiles evaluated by TORIC, and the power collisionally transferred to the electrons and the parameters of the minority distribution function evaluated by SSFPQL, in AUG and JET respectively. The absorbed power density in the central region is comparable in the two devices: the much larger plasma volume in JET is compensated by the larger total power available, and by a better focussing of the waves in the central region. In AUG the predicted peak effective temperature (logarithmic derivative of $F_m(E)$) is $\simeq 35$ keV in the perpendicular, and $\simeq 15$ keV in the parallel direction. The former value agrees well with charge exchange measurement available up to about 100 keV. In JET the predicted peak values are much higher, and more anisotropic: $T_{\perp \text{eff}} \simeq 320$ keV, $T_{\|\text{eff}} \simeq 60$ keV. Except for minor

TABLE 1. Main parameters used in the simulations.

	AUG # 19314	JET # 52095
Major radius	1.67 m	2.95 m
Plasma radius	0.47 m	0.85 m
Plasma composition	6% H in D	4% H in D
Central magnetic field	1.97 T	2.77 T
Ohmic current	836 kA	2645 kA
Central density*	$6.53 \; 10^{19}$ m^{-3}	$3.8 \; 10^{19}$ m^{-3}
Central electron temperature	4.35 keV	8 keV
Central ion temperature	4.33 keV	8.0 keV
Applied frequency	30.5 Mhz	42.0 Mhz
Representative toroidal wavenumber	$n_\varphi = 12$	$n_\varphi = 24$
Position of minority resonance (r/a)	0.197 (h.f.s.)	0.070 (h.f.s.)
Estimated total power coupled	4 MW	9 MW

* The experimental profiles for n and T have been used for ASDEX Upgrade; for JET the profiles of ref. [7] have been approximated analytically.

differences in the power deposition profiles, these results are consistent with the analysis made in [7]. As a consequence of the high effective tail temperatures, in the core of JET $\simeq 80\%$ of the power absorbed by the minority is thermalized on the electron ($\simeq 65\%$ integrated over the entire plasma). The corresponding figures for AUG are $\simeq 45\%$ in the core and $\simeq 35\%$ integrated. The difference is due to the larger power per minority ion available in JET (both the total density and the minority concentrations being lower), and to the lower collisionality (by about a factor 4) of the JET plasma. Simulations scanning the total power and the minority concentration in the two devices confirm this interpretation.

Comments and conclusions. The main source of inaccuracy in these simulations is the fact that SSFPQL, in common with all surface averaged Fokker-Planck solvers, neglects the finite radial width of the orbits of the most energetic ions. Thus, while iterating TORIC with the minority distributions evaluated by SSFPQL shows some Doppler broadening of the power deposition profiles, the actual broadening is likely to be larger due to finite orbits effects. Nevertheless, simulations with the combined TORIC and SSFPQL codes reproduce well the experimental results, and help in their interpretation, with a modest numerical effort.

REFERENCES

1. M. Brambilla, *Plasma Phys. Contr. Fusion* **41**, 1, (1999).
2. R. Bilato, M. Brambilla, I. Pavlenko, and F. Meo, *Nucl. Fusion* **42**, 1085, (2002).
3. M. Brambilla, *Nucl. Fusion* **34**, 1121, (1994).
4. C.F. Kennel, and F. Engelmann, *Phys. Fluids*, **9**, 2377, (1966).
5. T. Hellsten, T. Johnson, J.C. Carlson, L.-G. Erikson, J. Hedin, M. Laxaback, and M. Mantsinen, *Nucl. Fusion* **44**, 181, (2004).
6. Stix T.H., *Nucl. Fusion* **35**, 737, (1975).
7. W. Suttrop, R. Budny, J.C. Cordey, G. Gowers, M. Mantsinen, et Al., *28th EPS Conf. on Contr. Fusion and Plasma Physics*, ECA Vol. 25A, 989, (2001).

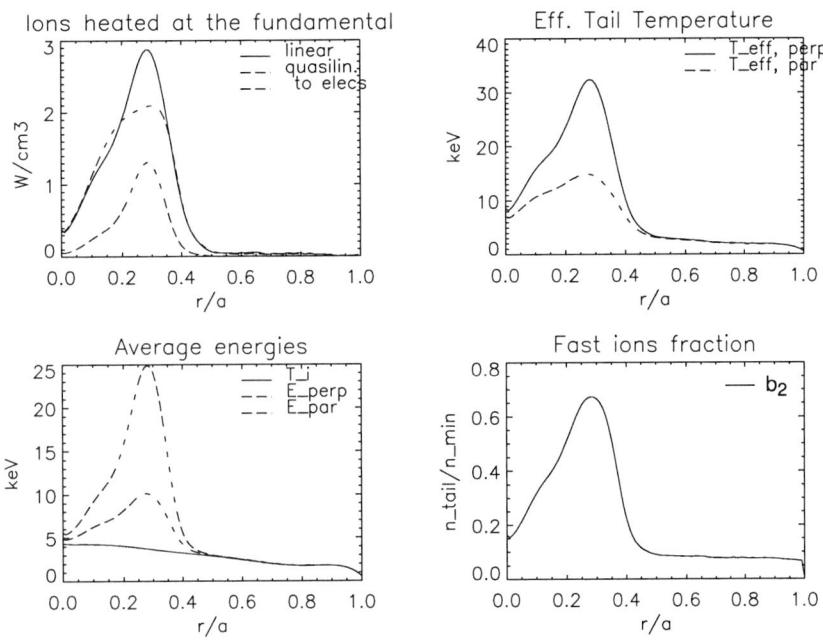

FIGURE 1. Results from TORIC and SSFPQL for ASDEX Upgrade.

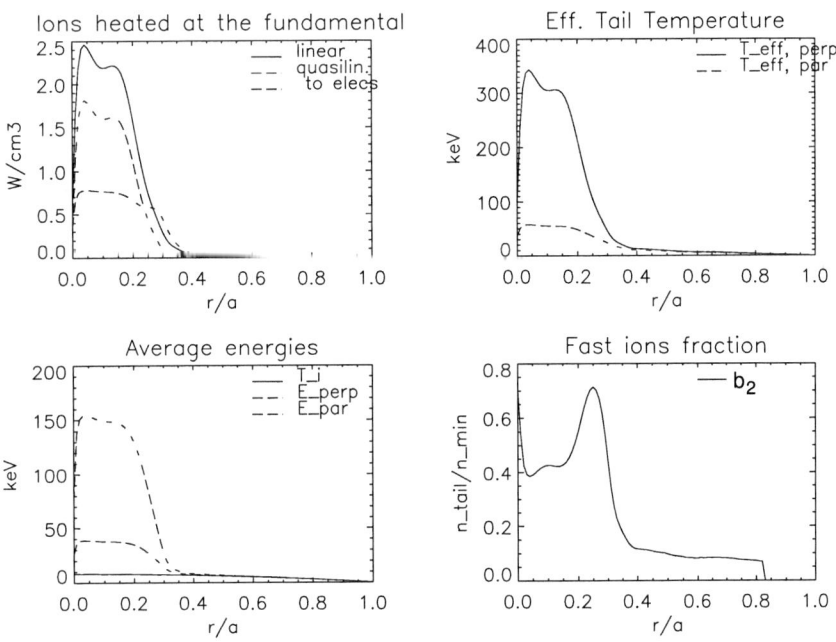

FIGURE 2. Results from TORIC and SSFPQL for JET.

Positive Quasi Linear Operator Formulation

L. A. Berry and E. F. Jaeger and the
SciDAC RF-Plasma Interactions Team

Oak Ridge National Laboratory, Oak Ridge TN USA

Abstract. Expressions for the RF quasi-linear operator are biquadratic sums over the Fourier modes (or FLR equivalent) that describe the RF electric field with a kernel that is a function of the two wave vectors, $\bar{k}_L$ and $\bar{k}_R$, in the sum. As a result of either an implicit or explicit average over field lines or flux surfaces, this kernel only depends on one parallel wave vector, conventionally $k_R^{\square}$. When $k^{\square}$ is an independent component of the representation for E, the sums are demonstrably positive. However, except for closed field line systems, $k^{\square}$ is dependent on the local direction of the equilibrium magnetic field, and, empirically, the absorbed energy and quasi-linear diffusion coefficients are observed to have negative features. We have formally introduced an independent $k^{\square}$ sum by Fourier transforming the RF electric field, (assuming straight field lines) using a field-line-length coordinate. The resulting expression is positive. We have modeled this approach by calculating the quasi linear operator for "modes" with fixed $k^{\square}$. We form these modes by discretizing $k^{\square}$ and then assigning all of the Fourier components with $k^{\square}$ that fall within a given $k^{\square}$ bin to that $k^{\square}$ mode. Results will be shown as a function of the number of bins. Future work will involve implementing the expressions derived from the Fourier transform and evaluating the dependence on field line length.

Keywords: quasi linear, Fokker Planck, RF plasma interactions.
PACS: 52.25.Dg, 52.25.Os, 52.35.Hrv, 52.35.Mw52.65.

INTRODUCTION

The slow evolution of the time-averaged distribution function is typically calculated from a Fokker-Planck equation that uses the gyro-averaged rf quasi linear (QL) operator as presented by Kennel and Engelmann [1] as a source term [2]. Attempts to apply this formalism to the RF fields calculated from the All Orders Spectral Algorithm (AORSA) have forced the reexamination of the assumptions of the Kennel-Engelmann formulation. In this paper, these assumptions will be examined and a more general formulation developed.

ANALYSIS

We begin at an intermediate point in the analysis using the methodology of Reference 5, and derive the quasi linear equation:

$$\frac{\partial f}{\partial t} + v_\| \frac{\partial f_1}{\partial l} = \frac{\pi q^2}{2} \sum_{n,k',k} \nabla_v \cdot \begin{pmatrix} a^2 & ab \\ ab & b^2 \end{pmatrix} \frac{1}{|k_\||} \delta\left(\frac{\omega}{k_\|} - v_\| - \frac{n\omega_c}{k_\|}\right) \text{Re}\left(\Psi^*_{n,k'} \Psi_{n,k}\right) \cdot \nabla_v f \quad (1)$$

where n is the cyclotron harmonic number; k and k' are Fourier sum indices (multidimensional, as appropriate); $a = \left(1 - \frac{k_\| v_\|}{\omega}\right)$; $b = \frac{k_\| v_\perp}{\omega}$; and

$\Psi_{n,k} = \left[v_\perp \vec{E}_k^+ J_{n-1}(z) + v_\perp \vec{E}_k^- J_{n+1}(z) + \sqrt{2} v_\| \vec{E}_k^\| J_n(z)\right] e^{in\beta}$. In the preceding expression, $z = \frac{k_\perp v_\perp}{\omega_c}$ and $E_k^\pm = \frac{1}{\sqrt{2}}\left(E_k^x \pm i E_k^y\right)$ (x, y, and parallel refer to local, Stix coordinates). The angle β is the angle between k_x and k_y. Equation 1 (in CGS units) uses the same notation as in [3], but is cast in the form given by Stix [4] for the purpose of comparison. It differs in two respects from the analysis presented by Stix. First, based on a bounce-average ordering [2], the parallel gradient term on the LHS has been retained. The subscript "1" on the distribution function for this term indicates that is first order in the bounce time compared to f (a function only of constants of the motion along a field line) which is zero order. Second, based on a gyro-phase average of the dissipative component of the RHS of the second order equation for f in [5], a biquadratic sum over modes has been retained. This form thus does not make use of the random phase approximation (RPA) for perpendicular modes. The use of the RPS is not appropriate for computer simulations that employ Fourier basis functions such as AORSA or poloidal mode expansion such as TORIC. For these basis functions, the RF fields for a "mode", e.g. a fast wave, are comprised of a sum over basis functions that are highly correlated for which the RPA is not correct.

The parallel gradient term is annihilated by performing a bounce average and yields the usual QL form on the LHS of Eq. 1. The effect on the RHS depends on the symmetry of the system. For a tokamak without poloidal field, the toroidal angle, φ, is both a symmetry coordinate and a parallel coordinate along a field line. For this geometry, we then obtain the form

$$\bar{D}_{QL} = \frac{\pi q^2}{2} \sum_{n,k',k} \begin{pmatrix} a^2 & ab \\ ab & b^2 \end{pmatrix} \frac{1}{|k_\||} \delta\left(\frac{\omega}{k_\|} - v_\| - \frac{n\omega_c}{k_\|}\right) \text{Re}\left(\Psi^*_{n,k'} \Psi_{n,k}\right)$$
$$= \frac{\pi q^2}{2} \sum_{n,k_\|} \begin{pmatrix} a^2 & ab \\ ab & b^2 \end{pmatrix} \frac{1}{|k_\||} \delta\left(\frac{\omega}{k_\|} - v_\| - \frac{n\omega_c}{k_\|}\right) \sum_{k_\perp} \Psi^*_{n,k} \sum_{k_\perp} \Psi_{n,k} \quad (2)$$

and $\bar{D}$ is positive. The key to this result is that there is an independent sum over parallel modes (i.e., a parallel coordinate) that can be placed outside of an independent sum over perpendicular modes.

For a tokamak with poloidal field, the procedure is much less clear and not as satisfying. For the previous case, the bounce average both simplified the LHS of Eq. 1, and resolved the Fourier sum into coherent interactions between toroidal modes. For the present case, the mode coupling must first be specified before performing the bounce average, resulting in two parallel integrals: first over the wave structure to define a $k_\parallel$ spectrum, and then over the resulting resonances to annihilate the LHS f_1 term. For both AORSA and TORIC, the parallel mode structure is not based on a parallel coordinate, but is defined locally by taking the component of the wave vector parallel to $\vec{B}$ for each poloidal (TORIC) or "R" and "Z" mode (AORSA fields will be used for the remainder of this analysis). Thus there is no independent parallel mode number, and $k_\parallel = k_\parallel^{N,m,n}$ for toroidal mode number N, "R" mode number m, and "Z" mode number n.

We now integrate along the tangent to $\vec{B}$ a (yet) unspecified distance $\pm L/2$ finding

$$\tilde{D}_{QL} = \frac{\pi q^2}{2} \sum_{n,k',k} \begin{pmatrix} a^2 & ab \\ ab & b^2 \end{pmatrix} \frac{\sin \tfrac{1}{2}\left(k_\parallel^{N,m,n} - k_\parallel^{N',m',n'}\right)}{\tfrac{1}{2}\left(k_\parallel^{N,m,n} - k_\parallel^{N',m',n'}\right)} \frac{1}{|k_\parallel^{N,m,n}|} \delta\left(\frac{\omega}{k_\parallel^{N,m,n}} - v_\parallel - \frac{n\omega_c}{k_\parallel^{N,m,n}}\right) \operatorname{Re}\left(\Psi_{n,k'}^* \Psi_{n,k}\right). (3)$$

For very large L each mode is independent, $k' = k$ for all indices, and we recover the Kennel-Engelmann result

$$\tilde{D}_{QL} = \frac{\pi q^2}{2} \sum_{n,k} \begin{pmatrix} a^2 & ab \\ ab & b^2 \end{pmatrix} \frac{1}{|k_\parallel^{N,m,n}|} \delta\left(\frac{\omega}{k_\parallel^{N,m,n}} - v_\parallel - \frac{n\omega_c}{k_\parallel^{N,m,n}}\right) \operatorname{Re}\left(\Psi_{n,k}^* \Psi_{n,k}\right).$$

For small L, we regain the form comparable previous AORSA analyses:

$$\tilde{D}_{QL} = \frac{\pi q^2}{2} \sum_{n,k',k} \begin{pmatrix} a^2 & ab \\ ab & b^2 \end{pmatrix} \frac{1}{|k_\parallel^{N,m,n}|} \delta\left(\frac{\omega}{k_\parallel^{N,m,n}} - v_\parallel - \frac{n\omega_c}{k_\parallel^{N,m,n}}\right) \operatorname{Re}\left(\Psi_{n,k'}^* \Psi_{n,k}\right). \quad (3)$$

Because of the additional cross terms compared the case for a tokamak with no poloidal field, this result is not, by its structure alone, positive and the Fokker-Planck solutions for distribution functions based on this operator can have instability. Whether this instability is due to numerical issues or an underlying breakdown in the assumptions for quasi linear analysis is an open issue.

One approach to overcoming the numerical issue, is to (somewhat arbitrarily) construct an independent $k_\parallel$ spectrum. We do this by observing that the effect of the $\sin(x)/x$ factor in the above is to emphasize those (m,n) modes with $k_\parallel$ differences less than $2/L$. We thus replace the $\sin(x)/x$ function with an impulse function of unity amplitude and $2/L$ wide, and assign a midpoint $k_\parallel$ value. to the (m,n) modes that fall within that bin. The (m,n) sum over modes is then restricted to the modes within the bin. Figure 1 displays the energy dissipation for an NSTX case using this formulation. For ~20 bins with a width of N/R, about 10 cm, the energy deposition was essentially the same as previous calculations, but lacked the small negative regions that indicated instability for the Fokker Planck calculations. For a large number of bins, the Kennel-Engelmann result was regained, but the total power was almost a factor of two low compared to $\vec{J}\cdot\vec{E}$. For 200 bins, the total power was about right, but details of the deuterium fast ion absorption were missed.

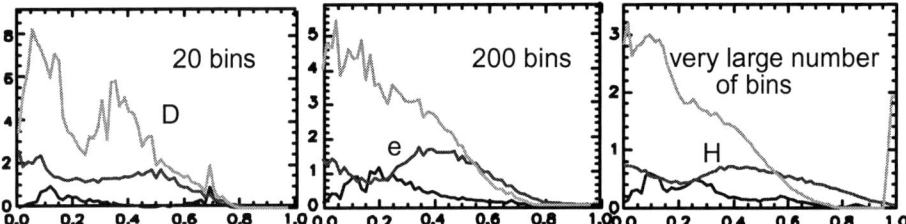

FIGURE 1. Energy deposition (10^5 w/m^3) for a NSTX case [6] with D beams into an H plasma as a function of normalized minor radius, rho for different numbers of $k_\parallel$ bins. Deuterium beam ions are shown in turquoise, electrons in blue, and hydrogen background in blue.

One can argue that the L ten of ~10 cm for NSTX is something like the distance along a field line before the change in the direction of the field line results in a significant change in $k_\parallel$. However, for other systems, the use of ~20 bins did not produce agreement between the binned, positive expression for $\bar{D}_{QL}$ and more the quantitatively correct, but-not-always positive result of Eq. 5. Thus binning, based on present understanding of how to establish a bin-width, is not a robust procedure. .

We plan to test the expression for $\bar{D}_{QL}$ in Eq. 3 for robustness to changes in L as well as to implement the formal Fourier transform the fields in the straight-line coordinate. The resulting $\bar{D}_{QL}$ will be positive, but again, robustness to variations in L will be an issue. Finally, we will assess the possibility that assumptions for quasi linear theory are not valid for some cases

ACKNOWLEDGMENTS

This work was supported by the SciDAC and the U.S. Department of Energy (DOE) under Contract DE-AC05-00OR22725 with UT-Battelle, LLC. The calculations in this paper have been carried out using the resources of the National Energy Research Scientific Computing Center and the Lawrence Berkeley National Lab and the Center for Computational Sciences at Oak Ridge National Lab. The work has benefited from discussions with Harold Weitzner regarding quasi linear theory.

REFERENCES

1. C. F. Kennel and F Engelmann, *Phys. Fluids* **9,** 2377 (1966).
2. R. W. Harvey and M. G. McCoy, *The CQL3D Fokker-Planck Code,* (Available though USDOC/NTIS doc. DE93002962), and M. Choi, V. S. Chan, *et al., Radio Frequency Power in Plasmas: 15th Topical Conference on Radio Frequency Power in Plasmas,* edited by Cary B. Forest, American Institute of Physics, New York (2003) p86.
3. E. F. Jaeger, this conference
4. T. H. Stix, *Waves in Plasmas* (American Institute of Physics, New York, 1992) p498.
5. J. R. Myra, L. A. Berry, *et al., Phys. of Plasmas* **11,** 1786 (2004).
6. D. N. Smithe, *Plasma Phys. Controlled Fusion* **31,** 1105 (1989) and references therein.
7. A. L. Rosenberg, J. E. Menard, *et al., Phys. Of Plasmas* **11,** 2441 (2004).

Velocity-Space Diffusion Coefficients Due to Full-Wave ICRF Fields in Toroidal Geometry

R.W. Harvey, F. Jaeger[*], L.A. Berry[*], N.M. Ershov[†], A.P. Smirnov[†],
P. Bonoli[‡], J.C. Wright[‡], D.B. Batchelor[*], E. D'Azevedo[*], M.D. Carter[*],
D.N. Smithe[§]

CompX, P.O. Box 2672, Del Mar, CA 92014-5672

Abstract

Jaeger et al.[1] have calculated bounce-averaged QL diffusion coefficients from AORSA[2] full-wave fields, based on non-Maxwellian distributions from CQL3D Fokker-Planck code[3]. A zero banana-width approximation is employed. Complementing this calculation, a fully numerical calculation of ion velocity diffusion coefficients using the full-wave fields in numerical tokamak equilibria has been implemented to determine the finite orbit width effects. The un-approximated Lorentz equation of motion is integrated to obtain the change in velocity after one complete poloidal transit of the tokamak. Averaging velocity changes over initial starting gyro-phase and toroidal angle gives bounce-averaged diffusion coefficients. The coefficients from the full-wave and Lorentz orbit methods are compared for an ITER DT second harmonic tritium ICRF heating case: the diffusion coefficients are similar in magnitude but reveal substantial finite orbit effects.

Jaeger *et al.*[1] have obtained a new calculation of Wdot in the AORSA full-wave code with non-Maxwellian dielectric, which has been rearranged to obtain velocity-space spatially-local rf quasilinear diffusion coefficients [2]. The coefficients are bounce-averaged(BA), assuming zero-banana-width charged-particle trajectories. The resulting BA coefficients are used to calculate the rf radial power deposition profile, and the results compare well with previous methods for calculation of Wdot power deposition[2].

The QL diffusion coefficients from AORSA have been imported into the CQL3D Fokker-Planck code[3] and provide a means to calculate 2D-in-velocity non-Maxwellian distributions directly including the effects of the AORSA the full-wave calculation. By iteration back and forth between the AORSA and CQL3D codes, full-wave fields are obtained that are self-consistent with the QL modification of the distribution functions.

In this work, we report initial results of a direct comparison between the zero-banana-width, full-wave QL diffusion coefficients from AORSA and numerically computed diffusion coefficients obtained by integrating the Lorentz force equation for the gyro-orbits of ions in computational equilibria magnetic field data and the AORSA electromagnetic fields. The object is to provide a means for examination of finite orbit width effects and

---

[*]ORNL, Oak Ridge, Tenn.
[†]Moscow State Univ., Russia
[‡]PSFC, MIT, Boston, Mass.
[§]ATK-Mission Research

nonlinear effects. We have previously calculated such "Lorentz-orbit diffusion coefficients" in analytic equilibria using EM fields derived from the TORIC full wave code[4]. The present work enables direct comparison of the Lorentz-orbit coefficients with the AORSA derived coefficients for the same equilibria.

The general velocity space diffusion equation, averaged over times which are long compared to the particle gyro-, bounce- and toroidal symmetrization- times, is [5, 6]

$$\frac{\partial f}{\partial t}|_{RF} = \frac{\partial}{\partial J_i}\left(\overline{D_{ij}}\frac{\partial f}{\partial J_j}\right), \quad \overline{D_{ij}} \equiv \frac{\langle \Delta J_i \Delta J_j \rangle}{2\Delta t}, \quad i,j = 1:3,$$

where $J_1 = (m_e/Ze)(m_i u_\perp^2/2B)$, $J_2 = \oint v_\parallel dl$, and $J_3 = p_\phi$ are action variables with associated cyclic angle variables, $\theta_1 =$ gyrophase, $\theta_2 =$ bounce phase, $\theta_3 =$ toroidal angle ϕ. The average $\langle \Delta J_i \Delta J_j \rangle$ in the above equation is taken over the cyclic variables. The time Δt is the average time $\Delta t = \tau_b(2\pi/\Delta\phi)$ to return to a particular toroidal angle increment $\Delta\phi$ of the toroidal phase. The bounce time is τ_b. We transform to the coordinates used in CQL3D, that is, pitch angle θ_0 and momentum-per-mass u_0 of the electrons evaluated at the minimum magnetic field (B) point on each flux surface. This gives the equation for f at the minimum B point to be compared with the CQL3D FP form and verifies that the CQL3D momentum diffusion coefficient $D_{u_0 u_0}$ in CQL3D is simply $\langle \Delta u_0^2 \rangle / 2\Delta t$, where the $\langle \rangle$-average is taken as above for the action variables.

The averages were taken over changes $\langle \Delta u_0^2 \rangle$ from tritium ions starting at 8 toroidal locations near the minimum B point on the flux surface at radius $\rho = 0.356a$, uniformly spaced over one toroidal period of the ICRF wave, mode number $n_\phi = 19$. For each toroidal starting position, the phase average is over 8 uniformly phased ions with gyrocenters at the toroidal starting location. The diffusion coefficient was calculated for starting velocities at 32 parallel velocities and 16 perpendicular velocities. Total number of trajectories for a given radial flux surface is 8*8*16*32=32,768.

A fourth order Runge-Kutta scheme was used to integrate the ion equations of motion in toroidal geometry with fields depicted in Fig. 1, for the 50-50 DT 2nd harmonic case. Step size equal to 1/80 gyro-period was used to obtain adequate convergence of the results. 20 CPU hours is required on a 2.4 GHz Zeon-based PC for a single flux surface. [Substantial speed up is possible, for example, by following guiding center orbits except in the vicinity of cyclotron resonance.]

Figure 2 shows four typical orbits for this ITER case. Perpendicular energy is E_0=1MeV, and parallel velocity is varied in 4 equal steps from minus to plus $(E_0/2m_{Tritium})^{1/2}$. The width of the lines is due to the gyro-motion. The orbit calculation has been benchmarked against orbits calculated independently with the MCGO Monte Carlo Guiding Center Orbit code[7, 8], which has a Lorentz-orbit option. The two co-current launched ions have orbits inwards of their initial poloidal flux value, and the counter-launched ions are more outwards. Of these orbits, only the co-transiting ion intersects the $\omega = 2\omega_{cT}$-layer. That is, the finite banana width effects reduce the region of co-launch velocity space that reaches cyclotron resonance.

Figure 3 shows the diffusion coefficient from the orbit calculations as outlined above, and fig. 4 gives the zero-banana-width AORSA diffusion coefficients for the same starting flux surface. The AORSA diffusion coefficients have been linearly interpolated on to the Lorentz-orbit velocity grid. The most prominent finite orbit width effect evident in these figures is the absence of positive-$v_\parallel$ resonant ions at higher energies. From Fig.

2, this is evidently due to the inwards radial drift from co-current launched ions which causes these ions not to reach the cyclotron resonant surface. There appear to be further gross differences between the two diffusion coefficient calculations, for example, the larger peaks in the Lorentz-orbit calculation.

The categorization of the new effects in the finite orbit calculation of the diffusion will have to await all-radii calculations and detailed analysis of the finite orbit resonant velocity kicks compared to the analytic zero-banana width calculations. In the present calculation, coherence is maintained between up to four wave-particle resonances per bounce (for the trapped particles). This coherence can contribute to larger variations as a function of initial velocity, as seen in Fig. 3.

Future work includes improved centering of the ion orbits on each flux surface: that is, ions are assigned to the flux surface bin encompassing their time-averaged radial location. All-radii diffusion coefficients with be used in the CQL3D Fokker-Planck code. Although CQL3D has a zero-banana width approximation, the collisional effects are correct to first order in (banana width)/(plasma scale length). Attention will be paid to measuring toroidal momentum imparted by the RF to the ions.

The present, single-bounce Lorentz-orbit calculation of wave-particle diffusion coefficients also complements the Monte Carlo determination of ICRF[10] in that smooth distribution functions can be obtained from CQL3D although the finite orbit effect on collisions is not fully accounted for as in [10]. The smooth distribution functions are particularly helpful in the calculation of full-wave code dispersion and absorption.

References

[1] E. F. Jaeger *et al.*, Invited talk I-03, this meeting (2005).

[2] E.F. Jaeger, L.A. Berry *et al.*, Phys. of Plasmas **8**, 1573 (2001).

[3] R.W. Harvey and M.G. McCoy, "The CQL3D Fokker-Planck Code", Proc. of IAEA TCM on Advances in Simulation and Modeling of Thermonuclear Plasmas, Montreal, 1992, p. 489-526, IAEA, Vienna (1993); NTIS document DE93002962.

[4] P.T. Bonoli, M. Brambilla, *et al.*, Phys. of Plasmas **7**, 1886 (2000).

[5] R.W. Harvey *et al.*, 15thTopical Conference on Radio Frequency Power in Plasmas, Moran, Wy., edited by C. B. Forest, AIP Conf. Proc. 694, p. 471, 2003.

[6] A.N.Kaufman, Physics of Fluids **15**, 1063(1972).

[7] L.-G. Eriksson, P. Helander, Physics of Plasmas **1**, 308 (1994).

[8] R.W.Harvey, D.K.Bhadra, and S.C.Chiu, "Calculations of Neutral Beam Power Deposition for Doublet III". Poster 3B5, Sherwood Theory Meeting, Tucson (1980).

[9] H.E. St.John, R.W.Harvey, *et al.*, "Finite Banana Width Effects on Charge Exchange Spectra", Bull. Amer. Phys. Soc. 8, Part II, poster 6T9, p. 1059 (1982).

[10] M. Choi *et al.*, Invited talk I-04, this meeting (2005).

Acknowledgments

This work is supported by the USDOE RF-SciDAC project, grant DE-FC02-01ER54649.

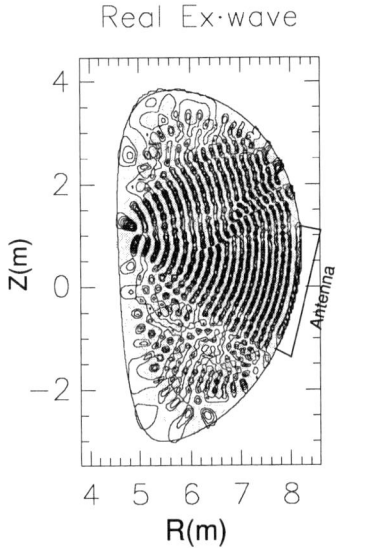

Figure 1: ICRF electric field from AORSA, for an ITER 50-50 DT second harmonic heating tritium case.

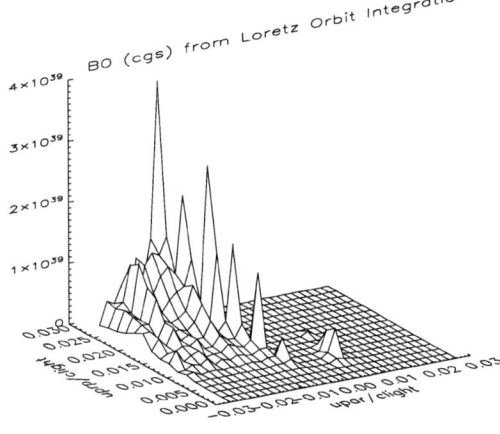

Figure 3: B0 diffusion coefficient $(v_\| \tau_B u_0^2 D_{u_0 u_0})$ obtained as bounce, gyro-phase, and toroidal angle average by direct integration of the Lorentz force equation of motion. Orbits start at $\rho = 0.361a$.

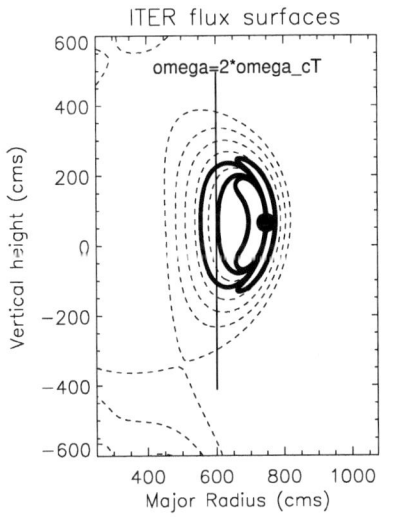

Figure 2: Tritium gyro-orbits starting from plasma radius $\rho = 0.361a$ with perpendicular energy 1 MeV, parallel velocity varying in equal steps from minus to plus $v_0 = (2E_0/m_{Trit})^{1/2}, E_0 = 1\text{MeV}$. The orbits start at the black dot.

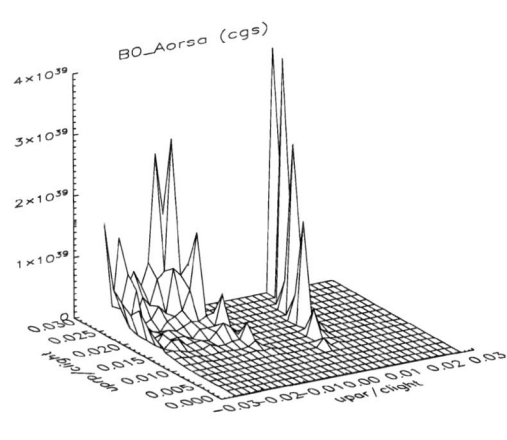

Figure 4: Bounce-averaged B0 diffusion coefficient $(v_\| \tau_B u_0^2 D_{u_0 u_0})$ calculated in AORSA code[1] in zero-banana width approximation. This is for the $\rho = 0.361a$ flux surface.

Effects of Finite Orbit Width and RF-Induced Spatial Diffusion on Ion Cyclotron Emission

T. Hellsten, T. Bergkvist, T. Johnson and M. Laxåback

Alfvén Laboratory, Association VR-Euratom, Sweden

Abstract. The theory of ion cyclotron emission, ICE, in tokamak plasmas has been revised by including the effects of finite orbit width and RF-induced spatial transport in the wave-particle interactions. Two mechanisms for excitation of edge localised magnetosonic modes are discussed. An inverted distribution function of suprathermal ions near the plasma edge is driving the modes. Counter current propagating waves can be excited by interacting with barely co passing ions. Co current propagating waves interacting at the inner leg only can drive the modes unstable by throwing the fast ions out of the plasma.

Keywords: Ion Cyclotron Emission, ICRH
PACS: 52.25.Tx, 52.55.Pi, 52.50.Qt, 52.35.Bj, 42.20.Dq

INTRODUCTION

Emission of waves in the ion cyclotron frequency range has been observed in several tokamaks [1-3]. The emission spectra are characterised by narrow peaks corresponding to the fundamental or harmonic cyclotron frequencies of high-energy ions at the low field side edge. In JET, emission related to fusion products has been found to correlate with the fusion reactivity [2]. In plasmas heated with ICRH emission peaks appear when the power exceeds a threshold, and is delayed with a slowing down time after the application of the ICRH [4]. In TFTR supershot experiments, with NBI, the spectrum changed during the discharge. In the early part of the heating phase, high amplitude peaks were seen corresponding to unshifted ion cyclotron resonances located about 30cm outside the plasma, whereas at a later time the peaks were lower and the frequency higher, the corresponding cyclotron resonances were located about 30cm inside the plasma [3]. The former peaks disappear as the density profiles peaks up in the supershots. In L-mode plasmas the peaks remain during the discharge. The high intensity of the emission is consistent with instabilities driven by suprathermal high-energy ions, originating either from thermonuclear reactions, neutral beam heating or from ion cyclotron heating [2]. An explanation of the emission has been given in terms of ion cyclotron instability caused by a local inverted energy distribution function from high-energy ions with trapped drift orbits extending out to the plasma edge on the low field side [5-7]. In those studies the effects of the orbit width and RF-induced spatial diffusion by waves with finite toroidal mode numbers were not fully taken into account. The theory of ion cyclotron emission, ICE, in tokamak plasmas has been revised by including the effects of finite orbit width and RF-induced spatial transport in the wave-particle interactions.

Because of the weak drive only weakly damped modes can be excited, such modes can be found at the plasma edge, if the density profile is not too peaked [12]. If the modes can be localised also at the low field side edge, the damping of the eigenmode can be reduced by avoiding stable interactions at smaller major radius.

WAVE-PARTICLE INTERACTIONS IN TOROIDAL PLASMAS

Edge localised eigenmodes can be formed provided that the density profile is not too peaked [12]. Radial localisation, in minor radius, for waves with large poloidal mode numbers [8] and poloidal localisation occurs for large toroidal mode numbers [9]. The higher the density is, and the flatter its profile is, the stronger the mode can be localised at the edge. The appearance of modes localised at the low field side edge follows from the dispersion relation of the magnetosonic wave

$$k_\perp^2 = \frac{\omega^2}{c^2}\varepsilon_{\perp\perp} - k_\parallel^2 - \frac{\omega^4}{c^4}\varepsilon_{\lambda\perp}^2 \bigg/ \frac{\omega^2}{c^2}\varepsilon_{\perp\perp} - k_\parallel^2, \qquad (1)$$

where $\perp$ denotes the direction of the component of the wave vector that is perpendicular to the magnetic field and λ the direction perpendicular both to the magnetic field and wave propagation. For modes with high $|n_\phi|$ one has $k_\parallel \approx n_\phi/R$, and the wave propagation becomes restricted to $R = R_0(1+\varepsilon\cos\theta) > |n_\phi|c/\omega\sqrt{1/\sigma(\text{Re}\,\varepsilon_{\perp\perp} - |\text{Im}\,\varepsilon_{\perp\lambda}|)}$, where $\sigma = sign(\text{Re}\,\varepsilon_{\perp\perp} - n_\phi^2/R^2)$.

Decorrelated interactions with a single mode in a toroidal geometry describe a one dimensional diffusion process of the guiding centre orbit invariants along the characteristics [10]

$$\Delta P = \frac{n_\phi}{\omega}\Delta W \qquad \Delta\Lambda = (\Lambda_r - \Lambda)\frac{\Delta W}{W} \qquad (2)$$

where Δ denotes the changes due to wave particle interactions, W the energy, $\Lambda = \mu B_0/W$, $\Lambda_r = n\omega_{c0}/\omega$, P_ϕ the canonical momentum, μ the magnetic moment, n the harmonics of the cyclotron frequency and ω_{c0} is the cyclotron frequency at the magnetic axis. Cyclotron instabilities can occur when the distribution function increases with energy along a subset of the characteristics. For instability the sum of the interactions, with all resonant ions and the background damping, have to give rise to a net increase of the mode energy. In a toroidal plasma the resonance condition $\omega = n\omega_c + k_\parallel v_\parallel$ can be satisfied in a large volume in phase space, because of the variation of the cyclotron frequency with the magnetic field. In general, the region along the characteristics, where the resonance condition is satisfied, is limited by two Doppler shifted cyclotron resonances merging into a tangential one.

When the Doppler shifted cyclotron resonance is located at the low field side of the plasma only modes with $k_\parallel > 0$ can interact with ions having $v_{//} < 0$ and vice versa. When an ion loses energy to the wave, its Doppler shifted resonance moves, in general, towards the high field side [11]. Note, even though the energy decreases during the interaction the condition for the Doppler shifted resonance is fulfilled at a smaller major radius, and will in general not restrict the interaction. However, the interaction can be restricted at an upper energy along the characteristic as the Doppler shifted resonances merge into a tangential one at the low field side of the orbit [13].

The turning points of ions on trapped orbits, undergoing cyclotron interactions, are displaced across the flux surfaces and their poloidal angle changes, also the width of the orbits changes. The displacement of an orbit expressed in the change in the magnetic flux at constant R for an interaction localised at $R = R_c$ becomes

$$\Delta \psi = \frac{R}{eZ\omega v_\parallel}(\omega - n\omega_c - k_\parallel v_\parallel) \qquad (3)$$

After an interaction the orbit will still pass through the points where the wave particle interaction took place, although for a finite change in energy the Doppler shifted resonance will be displaced.

We restrict our analysis to interactions with magnetosonic modes localised at the low field side edge, for which $k_\parallel \approx n_\phi/R$. By studying how the drift orbits change as their invariants move along the characteristics for cyclotron interactions, one can draw conclusions whether the distribution function is expected to increase with energy or not. The first mechanism describes interactions with modes having $n_\phi > 0$, counter current propagating modes. When the interaction reduces the particle energy the main effect is a displacement of the inner leg towards the centre, whereas the outer leg remains more or less constant, since the interactions take place at the outer leg of the orbit. The distribution function is therefore expected to be decreasing with energy along this part of the characteristic. Following the characteristic further in the direction of increasing energy the trapped orbit will be transformed into a co passing one. The number of such barely co passing ions will be decreasing with decreasing energy along this part of the characteristic, and hence the interactions will be unstable and give rise to emission. For net excitation of the mode it is important that the interactions with trapped ions, appearing on the same characteristics in the invariant space (W, Λ, P_ϕ), are reduced, which is possible, if the unshifted cyclotron resonances, or harmonics of them, are located sufficiently far out on the low field side.

The second mechanism describes interactions with co-current propagating waves, $n_\phi < 0$, take place at the inner leg of the trapped orbit. If the Doppler shifted resonance only intersects the inner leg, the outer leg will be displaced outwards as the ion loses energy to the wave. A trapped ion orbit localised close to the edge losing sufficient energy can then be lost by intersecting a limiter or the wall. The loss of ions with lower energy results in a local inverted energy distribution, which is necessary for instability. However, if the wave interacts both at the inner and outer leg with the mode, the outmost part of the orbit on the low field side will be displaced inwards, and the particle will not be lost. Note, the turning points move towards the low field side as the particle gain energy by the interaction. In accordance with the discussions above for interactions with waves having $n_\phi < 0$, we expect the number of ions to decrease at the upper energy range along the characteristic for waves with $n_\phi < 0$ resulting in a distribution function, which decreases in energy. Thus, the distribution function will have a maximum along its characteristics. Interactions taking place along the characteristics where the distribution function decreases with energy will stabilise the mode, whereas interactions taking place where the distribution function increases with energy will destabilise the mode. If the unshifted cyclotron resonance is displaced sufficiently far out on the low field side, the interactions with ions that stabilise the mode are avoided [10]. Because of the loss of ions by wave particle interactions an inverted distribution function can be maintained. In the absence of losses, the trapped orbit will eventually be converted into a counter passing one near the inner leg. Since the number of fast ions is expected to increase towards the centre, the distribution function is expected to be stable against such interactions in the absence of losses.

CONCLUSIONS AND DISCUSSIONS

Two mechanisms for cyclotron emission are discussed. One is based on interacting with ions on barely co current passing orbits with edge localised magnetosonic waves propagating counter to the plasma current. The other mechanism is based on interactions between co-current propagating waves and fast ions on trapped orbits, when only the inner leg of the orbit intersects the Doppler shifted resonance, which result in losses of ions as they give energy to the wave. The losses of the ions, when giving energy to the wave, result in an inverted distribution function driving the emission.

The strong interaction at tangential resonances and the possibility of preferential interaction between modes with finite toroidal mode numbers and resonant high-energy ions in phase space, where the distribution function is increasing with energy along the characteristics, provide an explanation of the narrow emission peaks with frequencies corresponding to the cyclotron frequencies or harmonics at the low field side of the plasma. The strong interaction at tangential resonances provides a mechanism for net excitation eventhough the mode may resonate at other harmonics at smaller major radii.

The change in the emission spectra during the TFTR supershots is qualitatively consistent with a change in the localisation of the magnetosonic modes due to the change in the density profile. The more peaked the density profile becomes the further the modes extend into the plasma and the stronger they are damped. If the drive by the beam ions is sufficiently strong it may overcome the stronger damping of a mode localised further into the plasma, and result in emission.

REFERENCES

1. Clark, W.H.M., in Heating in Toroidal Plasmas Proc. 4th Int. Symp. Rome, 1984), Vol.1 (1984) 385.
2. Cotrell *et al.*, Nucl Fus **33** (1993) 1365.
3. Cauffman, S., *et al.*, Nucl. Fusion **35** (1995) 1597.
4. Cottrell, Phys Rev. Lett **84** (2000) 2397.
5 Belikov, V. S., Kolesnichenko, Ya. I., Sov. Phys. Tech. Phys. **20** (1976) 1146.
6 Coppi, B., Physics Letters **A172** (1993) 439
7 Dendy, R. O., *et al.*, Nucl Fusion **35** (1995) 1733
8 Coppi, B. *et al.*, Phys. Fluids **29** (1986) 4060.
9 Hellsten, T and Laxåback, M., Phys. of Plasmas **10** (2003) 4371.
10 Eriksson, L.-G. and Helander, P., Physics of Plasmas **6** (1999) 513.
11 Hellsten, T. *et al.*, Nucl. Fusion **44** (2004) 892.
12 Mahjan, S.M. and Ross D.W., Phys. Fluids **26** (1983) 2561.
13 Johnson, T., *et al.*, this conference.

Analysis of a quasilinear model for ion-cyclotron interactions in tokamaks

T. Johnson*, T. Hellsten* and L.-G. Eriksson[†]

*KTH Association Euratom/VR, SE-10044 Stockholm, Sweden
[†]Association EURATOM-CEA, CEA/DSM/DRFC, CEA-Cadarache, F-13108 St. Paul lez Durance, France

Abstract. The quasilinear diffusion coefficient deviates significantly from the lowest order Larmor radius scaling $D \propto v_\perp^{2n}$. This is not only caused by the finite Larmor radius effects, but also by the inhomogeneous electric field polarisation and the changes of the guiding centre orbits. The regions with strong interaction and the boundaries for resonant interaction are identified. At these boundaries the quasilinear diffusion coefficient becomes discontinuous. A new Monte Carlo scheme has been developed to treat problems with discontinuous diffusion coefficients.

Keywords: Quasilinear diffusion, resonance condition, Monte Carlo methods
PACS: 52.50.Gj

The orbit averaged Fokker-Planck equation for ions interacting with ion cyclotron resonance frequency waves can be written as [1]

$$\frac{\partial f_0}{\partial t} = <C(f_0) + Q(f_0)> \tag{1}$$

where f_0 is the orbit averaged distribution function, C is the collision operator, Q is the quasilinear operator for particle-wave interactions, and $<...>$ denotes the average over orbits. Here we express f_0 in terms of the adiabatic invariants of the guiding centre motion $\{E, \Lambda, P_\phi; \sigma\}$, where E is the energy, $\Lambda \equiv B_0 \mu/E$, μ is the magnetic moment, B_0 is the magnetic field at the magnetic axis, P_ϕ is canonical toroidal angular momentum, and σ is a label separating orbits with the same invariants, $\{E, \Lambda, P_\phi\}$. The quasilinear operator then takes the form

$$<Q(f_0)> = \Sigma_{n,n_\phi,\omega} \mathscr{L} D \mathscr{L} f_0 \tag{2}$$

$$\mathscr{L} = \frac{\partial}{\partial E} + \frac{\Lambda_{res} - \Lambda}{E}\frac{\partial}{\partial \Lambda} + \frac{n_\phi}{\omega}\frac{\partial}{\partial P_\phi} \tag{3}$$

$$D = \frac{1}{4\tau_b} \Sigma_{\dot{v}=0} \left|Zev_\perp \left(\mathscr{E}_+ J_{N-1} - \mathscr{E}_- J_{N+1}\right) \Pi \right|^2 \tag{4}$$

The sum in $<Q>$ is to be performed over the harmonic cyclotron resonances, n, the toroidal mode numbers, n_ϕ, and the angular wave frequencies ω. The quasilinear diffusion coefficient, D, is a sum over all points along the orbit where the the phase difference, v, between the gyro motion and the wave oscillation is stationary ($\dot{v} = 0$). The variables used above are $\Lambda_{res} = n\Omega_0/\omega$, Ω_0 is the cyclotron frequency at the magnetic axis, τ_b is the bounce, or transit time, the argument of the Bessel function $J_{n\pm 1}$ is $k_\perp v_\perp/\omega$, and $\mathscr{E}_+$ and $\mathscr{E}_-$ are the left and right hand polarised electric field components, respectively. The phase integral, $\Pi(v)$, can be calculated by expanding the $v(t)$ around the local resonance $\{t|\dot{v}=0\}$. Expanding $v(t)$ to second order yields $|\Pi(v)| \approx \sqrt{2\pi/\ddot{v}}$. When $\ddot{v}$ becomes small a more detailed evaluation is used [2, 3, 4].

The diffusion coefficient D includes a sum over all local resonances $\dot{v}=0$. To find the position of the local resonances we use a magnetic field of the form $B = B_0 R_0/R$ and assume $k_\parallel \approx n_\phi/R$. The local resonance condition can then be written as a third order polynomial equation in pitch angle ξ

$$0 = \dot{v}(t) \equiv \omega - n\Omega - k_\parallel v_\parallel = \omega\left(1 - \frac{\Lambda_{res}}{\Lambda}(1-\xi^2)\left(1+\frac{\xi}{\xi_0}\right)\right) \quad (5)$$

where $\xi_0 \equiv n\Omega_0 R_0/(n_\phi v)$. This equation has at most two solutions for $\xi \in [-1, 1]$. It has a double root for $\xi = \xi^+$, where $\xi^+ = -\xi_0/3(1 - \sqrt{1+3/\xi_0^2})$. This double root appears at the maximum value of Λ for which the particle can be resonant. Thus, a particle can only be resonant when $\Lambda \leq \Lambda^+$, where $\Lambda^+ \equiv \Lambda_{res}(1-\xi^{+2})(1+\xi^+/\xi_0)$.

EVALUATION OF THE DIFFUSION COEFFICIENT

A test particle interacting with *a single wave mode*, i.e. a wave defined by a single toroidal mode number and a single frequency, will be accelerated such that the invariants of motion are constrained to characteristic lines [3, 5, 6, 7]. The characteristic line that pass through the point $I_0 = \{E_0, \Lambda_0, P_{\phi 0}\}$, can be parameterised by the energy E

$$\Lambda(E; I_0) = \Lambda_0 + (\Lambda_{res} - \Lambda_0)\frac{E - E_0}{E} \quad (6)$$

$$P_\phi(E; I_0) = P_{\phi 0} + \frac{n_\phi}{\omega}(E - E_0) \quad (7)$$

For illustration we evaluate the diffusion coefficient along these characteristics for the scenario of minority ^{3}He heating in a ^{4}He plasma with JET-like parameters: 3.45 T, $n_e = 2.3 \times 10^{19}$ m^{-3}, $T_i = 3$ keV, $R_0 = 3$ m, and minor radius 1 m. The wave frequency 37 MHz is chosen to have the cyclotron layer crossing the magnetic axis, i.e. $\Lambda_{res} = 1$. To reduce the complexity the wave field amplitude is assumed to be homogeneous while the polarisation $\mathscr{E}_+/\mathscr{E}_-$, as well as the perpendicular wave number, $k_\perp$, are given by a warm quasi-homogeneous plasma model. The presence of a ^{3}He population with density $n_{He3} = 0.01 n_e$ makes $k_\perp$ range from $22-35$ m^{-1} and $|\mathscr{E}_+/\mathscr{E}_-|$ range from $0.1-0.85$ between the unshifted cyclotron resonance and the location of the cold plasma ion-ion Alfvén resonance.

In figure 1 various quantities characterising the drift orbits and the cyclotron interactions are followed as the invariant $\{E, \Lambda(E; I_0), P_\phi(E; I_0)\}$ ($E_0 = 10$ keV and $\Lambda_0 = 0$) are traced along the two characteristics with toroidal mode numbers $n_\phi = -15$ and $n_\phi = +15$. The orbits on the two characteristics, as pictured in 1a) and 1e), are very different. For $n_\phi = +15$ the effects of the inverted RF-pinch and the orbit broadening result in an outward drift of the orbit, which eventually hits the wall at a major radius of 4 m. For $n_\phi = -15$ the RF-pinch confines the ion, which is resonant up to an energy of about 3 MeV, where the resonance becomes tangential to the orbit in the midplane.

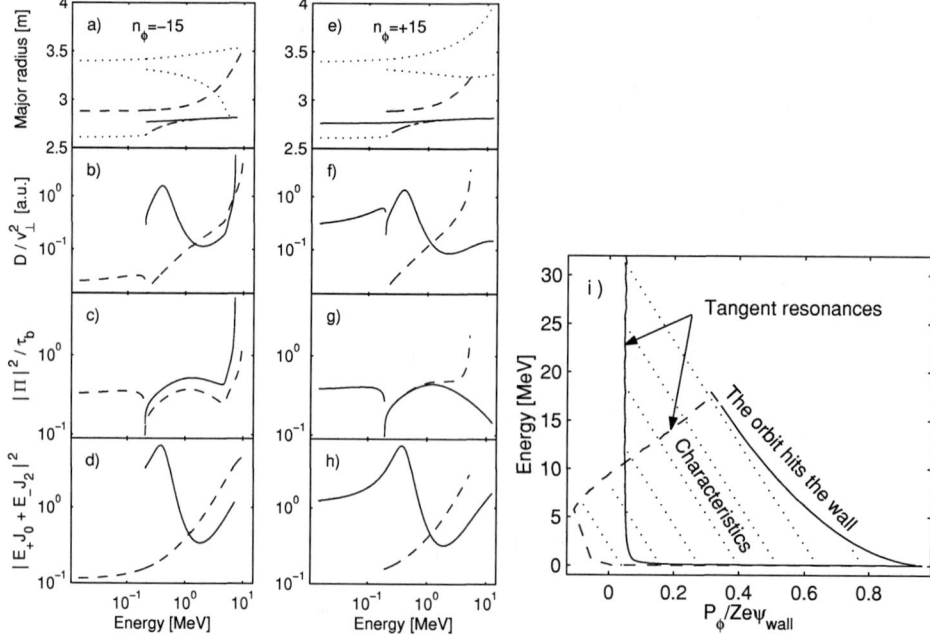

FIGURE 1. In figure *a*) to *h*) various quantities characterising the drift orbits and the cyclotron interactions are followed as the invariant $\{E, \Lambda(E; I_0), P_\phi(E; I_0)\}$ ($E_0 = 10$ keV and $\Lambda_0 = 0$) are traced along the two characteristics with toroidal mode numbers $n_\phi = -15$ and $n_\phi = +15$. The solid lines represent the inner resonances and the dashed the outer resonances. In *a*) and *e*) the major radii where the orbit crosses the midplane is shown as dotted lines and of the turning point as dotted-dashed lines. The figure *i*) shows the plane given by $\Lambda = E/(E - E_0)$ ($E_0 = 10$ keV) in the $\{E, \Lambda, P_\phi\}$-space. Orbits with invariants within the region bounded by the dashed and solid lines exhibit local resonances, where Eq. (5) has solutions.

The phase integral $|\Pi|^2/\tau_b$ is shown in 1*c*) and *g*). This factor vanishes at the trapped passing boundary at $E \approx 90$ keV, where the bounce time becomes infinite. On the other hand $|\Pi|^2/\tau_b$, and consequently also the diffusion coefficient, almost diverges at the tangential resonances. Close to, but below, the tangent resonances the distribution function should be relatively flat, with an energy dependence very far from a Maxwellian.

The polarisation of the electric field is strongly inhomogeneous, with the $\mathcal{E}_+$ component differing by up to a factor 8 between the high and low field side of the unshifted resonance. For the low energy passing ions in figure 1 the factor $|\mathcal{E}_+ J_{n-1} + \mathcal{E}_- J_{n+1}|^2$ therefore differ by an order of magnitude between the inner and the outer resonance obtained with $n_\phi = +15$ and $n_\phi = -15$, respectively. The difference have a maximum at about 130 keV coinciding with the maximum of $|\mathcal{E}_+/\mathcal{E}_-| \approx 0.85$ on the inner resonance.

The resonance condition (5) is satisfied on most, but not all orbits. The regions of resonant orbits are illustrated in figure 1*i*). At the higher energy boundaries of these regions resonances are located in the midplane (unless the region is bounded by orbits intersecting the wall). These resonances are sometimes called tangential since the orbit is tangential to the surface $R = R_0 \Lambda/(1-\xi^2)$, where ξ satisfies the resonance condition (5).

A MONTE CARLO METHOD FOR DISCONTINUOUS DIFFUSION COEFFICIENTS

Numerical solutions of Eq. (1) can be obtained using Monte Carlo test particle methods. However, at tangent resonances the diffusion coefficients (4) becomes discontinuous, causing the standard Euler schemes to fail (the discontinuity can be resolved by evaluating the orbit averages over the complete orbit and not just at local resonances). To treat problems with discontinuous diffusion coefficients a new scheme is proposed. The Euler schemes for the energy $\mathbf{E}_i(t+\Delta t)$ of test particle i during interactions with a single wave mode can be written as

$$\mathbf{E}_i(t+\Delta t) = \mathbf{E}_i(t) + \mathscr{L}D\Delta t + \chi_i\sqrt{2D\Delta t} \qquad (8)$$

where χ_i is a random variable with unit variance and zero mean value. However, if the step takes the test particle across a discontinuity, then we propose that the test particle is "partially reflected". This means that if the particle is taken, by Eq. (8), from a region A with diffusion coefficient D^A into a region B with lower diffusion coefficient $D^B < D^A$, then the particle may either be reflected back into A or transmitted into B. The probability for reflection should be $\sqrt{D^B/D^A}$. If on the other hand $D^A < D^B$, then the particle should be transmitted with probability 1. If E_b is the energy at the discontinuity, then a transmitted, or reflected, particle should step according to

Transmitted: $\mathbf{E}_i(t+\Delta t) = E_b + \left(\mathbf{E}_i(t) + \mathscr{L}D\Delta t + \chi_i\sqrt{2D\Delta t} - E_b\right)\sqrt{D^B/D^A}$

Reflected: $\mathbf{E}_i(t+\Delta t) = 2E_b - \left(\mathbf{E}_i(t) + \mathscr{L}D\Delta t + \chi_i\sqrt{2D\Delta t}\right)$

CONCLUSIONS

The quasilinear diffusion coefficient has been shown to depend more strongly on the energy than the $D \propto v_\perp^2 |J_{N-1} - J_{N+1}\mathscr{E}_-/\mathscr{E}_+|$ scaling suggests; in fact the diffusion coefficient deviates from this scaling by more than an orders of magnitude. The reason is that the polarisation of the electric field makes the ion experience different magnitudes of left hand polarised waves component at different Doppler shifts. Furthermore, the phase integral $|\Pi|^2/\tau_b$ depends on the topology of the guiding centre orbit and the position of the local resonances. At tangent resonances the diffusion coefficient becomes discontinuous. To treat problems with discontinuous diffusion coefficients a new Monte Carlo scheme has been developed.

REFERENCES

1. L.-G. Eriksson and P. Helander, *Physics of Plasmas*, **1** 308–314 (1994)
2. G. D. Kerbel and M. G. McCoy, *Physics of Fluids*, **28** 3629–3653 (1985)
3. A. Bécoulet, D. J. Gambier, and A. Samain, *Physics of Fluids B*, **3** 137–150 (1991)
4. P. U. Lamalle, *Plasma Physics and Controlled Fusion*, **39** 1409–1460 (1997)
5. L.-G. Eriksson, et al., *Physics of Plasmas*, **6** 513–518 (1999)
6. T. Hellsten, et al., *Nuclear Fusion*, **44** 892 (2004)
7. J. Hedin, T. Hellsten, and L.-G. Eriksson, *Nuclear Fusion*, **40** 1819 (2000)

A simple method to account for drift orbit effects when modeling radio frequency heating in tokamaks

D. Van Eester

Laboratory for Plasma Physics, Association "EURATOM - Belgian State", ERM/KMS, Trilateral Euregio Cluster, Brussels, Belgium

Abstract. In the last years tremendous progress was made in modeling radio frequency heating in tokamaks. Not only the adopted models have gradually become more realistic, also the present generation of computers has allowed to study wave-particle interaction effects with previously unattainable detail. In the present paper a semi-analytical method is adopted to evaluate the dielectric response of a plasma to electromagnetic waves in the ion cyclotron domain of frequencies accounting for drift orbit effects in an axisymmetric tokamak [1]. The method relies on subdividing the orbit into elementary segments in which the integrations can be performed analytically or by tabulation, and it hinges on the local bookkeeping of the relation between the variables defining an orbit and those describing the magnetic geometry. Although the method allows computation of elementary building blocks for either the wave or the Fokker-Planck equation, the focus here is on the latter. Using the coefficients evaluated using the proposed semi-analytical method, a 3-D Fokker-Planck code was developed which accounts for the radial width of the guiding center orbits and thus not only describes RF induced velocity space diffusion, but equally accounts for the RF induced radial drift. Preliminary results of this new 3-D Fokker-Planck code are presented. The adopted numerical resolution relies on a subdivision of the integration domain in tetrahedres. This specific shape of the elementary volumes allows imposing the boundary conditions (in particular the nonlocal conditions across the curved trapped/passing boundary connecting one trapped to two passing orbits) elegantly. The particular chosen shape also readily permits zooming in on regions where more detail is required. Casting the equation in its weak Galerkin form, it is solved relying on the finite element technique. Unless special attention is devoted to the optimization of the inversion of the system of linear equations resulting from projecting the 3-D Fokker-Planck equation onto proper base functions, the computer memory and time required is excessive. Off-the-shelf algorithms permitting speedy and accurate inversion of systems of linear equations with sparse matrices proved to be extremely useful to cope with this difficulty.

Keywords: radio frequency heating, tokamaks, drift orbit effects
PACS: 51.10.+y, 52.20.Dq, 52.50.Qt, 52.65.Ff

INTRODUCTORY REMARKS ON RF HEATING IN TOKAMAKS

Three constants of motion $\vec{\Lambda}$ define an unperturbed orbit in a tokamak. The Fokker-Planck equation governing the combined effects of RF heating of a population of charged particles and their slowing down as well as scattering under the influence of Coulomb collisions can be written in the general form $\frac{\partial F_o}{\partial t} = \frac{\partial F_o}{\partial t}|_{RF} + \frac{\partial F_o}{\partial t}|_{Coll} + S - L(F_o) = \nabla \cdot \vec{H} + S - L(F_o)$ in which $\frac{\partial F_o}{\partial t}|_{RF}$ is the RF diffusion operator, $\frac{\partial F_o}{\partial t}|_{Coll}$ is the Coulomb collision operator, $\vec{H}$ takes the form $\mathbf{D}.\nabla F_o + \vec{F} F_o$, S is a source and L a loss term. Multiplying the equation by a test function G, integrating over a volume of interest V and integrating by parts to remove the highest order derivatives from the flux term yields the weak Galerkin form of the equation, $\frac{\partial}{\partial t}\int_V d\vec{\Lambda}JGF_o = \int_S d\vec{S}.G\vec{H} - \int_V d\vec{\Lambda}J\nabla G.\vec{H} + \int_V d\vec{\Lambda}JGS - \int_V d\vec{\Lambda}JGL(F_o)$, in which the surface term only involves the component of the outward flux $\vec{H}$ normal to the surface S surrounding V.

Making a number of simplifications (for details and notation, see [1]), the general expression for the dielectric response can be written as a bounce integral. Rather than the bounce angle, one can use the usual poloidal angle θ as the independent variable. Each orbit, characterized by a set of constants of the motion $\vec{\Lambda}$, is broken into small segments in which all relevant functions are varying only moderately but in which the resonance crossing is captured accurately. This yields an expression of the form $\int d\vec{x}d\vec{v}G\frac{\partial F_o}{\partial t}\bigg|_{RF} = -\sum_{n,N}\int d\vec{\Lambda}H(G)H(F_o)\tilde{Q}_{RF}(\vec{\Lambda})$ in which, for $\vec{\Lambda} = (p_\varphi, x, v) = (\Psi - \frac{2\pi mR_oB_{oo}\varphi v_{//}}{qB_o}, (\frac{v_\perp}{v})^2\frac{B_{o,ref}}{B_o}, v)$, one finds $H(F_o) = \frac{1}{v\bar{m}}\left(\frac{\partial F_o}{\partial v} - \frac{2}{v}[x + \frac{B_{ref}m_gq}{m\omega}]\frac{\partial F_o}{\partial x}\right) - \frac{2\pi m_d}{q\omega}\frac{\partial F_o}{\partial \bar{p}_\varphi}$,
$\tilde{Q}_{RF}(\vec{\Lambda}) = \int_{orbit} d\theta J_{\tilde{p}_\varphi vx\theta\varphi\phi}\sum_{m,m'}\frac{1}{2}Re[q^2(2\pi)^2\frac{L_{m,n,N}L_{m',n,N}e^{i(m-m')\theta}}{i(k_{//\tilde{m},n}v_{//} + N\Omega - \omega)}]$, $L_{...}$ is the Kennel and Engelmann operator and $\tilde{m} = (m+m')/2$. Following the same reasoning, a similar term can be found for the wave equation.

EVALUATING THE ELEMENTARY INTEGRALS

Assuming the guiding center orbits to be straight lines along which the "background" temperature, density and magnetic field required to evaluate the equation's RF and collisional diffusion and drag coefficients for the Fokker-Planck equation are constant, it was demonstrated in [2] that by subdividing the domain in velocity space into sufficiently small portions, the dielectric tensor needed in the wave equation can be evaluated semi-analytically for an arbitrary distribution function, hence allowing a self-consistent wave + Fokker-Planck description of the RF heating. In the present paper, the same simple computational philosophy is adopted for inhomogeneous plasmas but rather than the parallel velocity the poloidal angle is used as independent variable in the integration.

The integrals over the small poloidal angle intervals are of the form $I = \int_{\theta_i}^{\theta_{i+1}}\frac{G}{\alpha}e^{iM\theta} \approx \frac{G_i + G'_i[\tilde{\theta} - \theta_i]}{\alpha'_i}I_1 + \frac{G'_i}{\alpha'_i}I_2$ in which $\tilde{\theta} = \theta_i - \alpha_i/\alpha'_i$, with $\alpha = k_{//}v_{//} + N\Omega - \omega$ and $' = d/d\theta$. The integrals I_1 and I_2 are readily found. For $M \neq 0$ one gets e.g. $I_1 = e^{iM\tilde{\theta}}\big\{[Ci(M[\theta_{i+1} - \tilde{\theta}]) - Ci(M[\theta_i - \tilde{\theta}])] + i[Si(M[\theta_{i+1} - \tilde{\theta}]) - Si(M[\theta_i - \tilde{\theta}])]\big\}$. Substituting the usual resonant denominator by a more sophisticated expression D, effects of decorrelation can be taken into account. Also a proper treatment of the absorption at the turning points can be incorporated by merely adding a supplementary "switch-off" function S in the integrand. For a sufficiently refined grid, the upgraded expression for $\tilde{Q}_{RF}$ is a sum of terms of the form $J_{\tilde{p}_\varphi vx\theta\varphi\phi}\sum_{m,m'}\frac{1}{2}Re[q^2(2\pi)^2L_{m,n,N}L_{m',n,N}e^{i(m-m')\theta_i}\int_{\theta_i}^{\theta_{i+1}}d\theta DS]$. Through $d\alpha/d\theta$, a distinction is made between turning points in the R- direction where the magnetic field reaches an extremum along the orbit but the poloidal velocity stays finite when crossing the equatorial plane, and (trapped particle) turning points at the banana tips where the poloidal velocity goes through zero. Stronger absorption is expected at the latter but even when the velocity goes through zero, the amount of power the particle receives when crossing the resonance does not diverge.

3-D FOKKER-PLANCK MODELING USING TETRAHEDRES

Near the magnetic axis, nearly all particles are passing. Further away from the axis, gradually more trapped orbits appear. Aside from this (dominant) dependence on the toroidal angular momentum, the trapped/passing boundary depends on the energy of the particle when radial drifts are accounted for. Consequently, the trapped/passing surface is a curved surface. Across this surface the particle flux needs to be conserved to guarantee the overall particle conservation in cases when all orbits are confined. As the incoming fluxes from families of 2 (one co- and one counter-) passing orbits merge into single trapped orbits, this boundary condition is also nonlocal. Rather than cutting up the configuration space along coordinate surfaces (and having 3-D "cubes" as elementary volumes), it is then more elegant to use a grid that is nonuniform, and to adopt tetrahedres as elementary volumes.

In each of the tetrahedres with corner points A, B, C and D a local coordinate system is set up. To span the 3-dimensional volume of the tetrahedre, 3 independent coordinates ζ_1, ζ_2 and ζ_3 can be used to parametrize each of the directions $\vec{DA}$, $\vec{DB}$ and $\vec{DC}$, respectively. It is useful to define a fourth (dependent) coordinate, ζ_4, such that the value of a function H inside or on the edge of the tetrahedre is given through simple interpolation by $H(\vec{\zeta}) = \sum_{i=1}^{4} H_i \zeta_i$ in which the H_i are the values at the four corner points, each having 1 coordinate that is 1 while the other 3 are zero. To accomplish this, ζ_4 must be related to the former 3 by $\sum_{i=1}^{4} \zeta_i = 1$ and $0 \leq \zeta_i \leq 1$ holds. In each tetrahedre, $\vec{\Lambda} - \vec{\Lambda}_D = \mathbf{B}.\vec{\zeta}$ in which the matrix $\mathbf{B}$ has the vector components of $\vec{DA}$, $\vec{DB}$ and $\vec{DC}$ as columns so that the volume of the parallelepiped formed by the 3 vectors, $\vec{DA} \times \vec{DB}.\vec{DC}$, is the determinant of $\mathbf{B}$ and the Jacobian of the local transformation $\vec{\Lambda} \to \vec{\zeta}$. The volume integral of a function H inside the tetrahedre is then $\int_V d\vec{\Lambda} H = \vec{DA} \times \vec{DB}.\vec{DC} \int_0^1 d\zeta_1 \int_0^{1-\zeta_1} d\zeta_2 \int_0^{1-\zeta_1-\zeta_2} d\zeta_3 H$. Similar expressions for the surfaces integrals can readily be found.

Adopting "hat" functions, composed of the ζ_i, that are 1 at a grid point and 0 at all its neighbors as test functions G yields a set of linear equations relating the values of the distribution function at the grid points. The resulting sparse linear system is solved using PSPARSLIB [3]. An example is provided in the figure. The adopted geometry is that of JET but an up-down symmetric equilibrium has been imposed, and parameters typical for central minority hydrogen heating are chosen. The obtained distribution function shows various intuitive ingredients: a Maxwellian subpopulation at small energy, tail formation at large energy (more pronounced on the energy-density-like $v^2 F_o$ plot) and ("rabbit ear") tangent resonance effects.

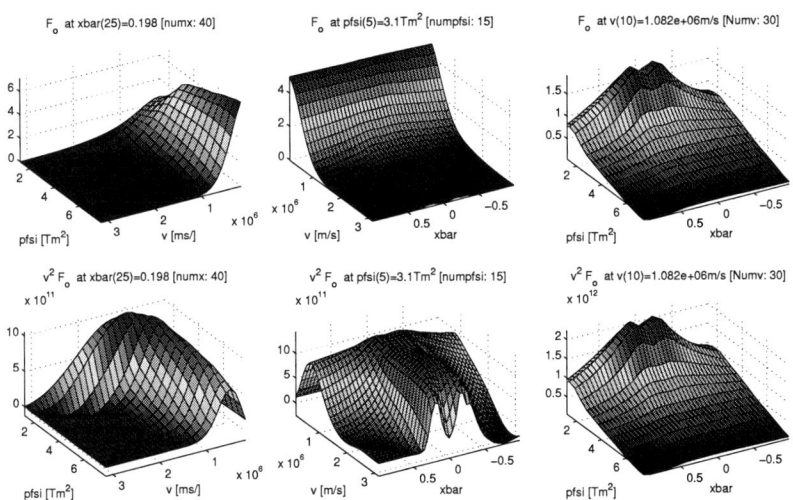

FIGURE 1. Distribution function F_o (top row) and $v^2 F_o$ of RF heated hydrogen minority ions heated at their fundamental cyclotron frequency, at fixed x (left), $\bar{p}_\varphi$ (middle) and v (right).

SUMMARY AND FINAL REMARKS

A newly written 3-D bounce averaged Fokker-Planck code was presented. The coefficients of the equation account for drift orbit effects and are computed semi-analytically making use of the relation between the magnetic surface labeling variable and the toroidal angular momentum, thus avoiding having to integrate the equation of (unperturbed) motion. The finite radial width of the orbit yields RF as well as Coulomb collisionally induced radial diffusion. Tangent resonance as well as decorrelation effects are incorporated through simple multiplicative factors. The governing equation is solved in its weak Galerkin variational form. The subdivision in tetrahedres (and its equally tested 2-D equivalent adopting triangles) is a powerful method for tackling other equations. To minimize memory and computing time, the PSPARSLIB library was used to solve the linear system resulting from the projection of the equation onto the base functions. This library allows inverting the large size matrices accompanying the 3-D Fokker-Planck problem. PSPARSLIB permits distributing the task over several processors without having to do more than some elementary bookkeeping of the equivalent sparse matrix in the so-called Harwell-Boeing form, which only stores nonzero elements and uses pointers to relate the stored values and the places where these values belong in the actual matrix.

REFERENCES

1. D. Van Eester, *Pl. Phys. Contr. Fusion*, **47**, 459-481 (2005), and references therein.
2. D. Van Eester, *Plasma Phys. Control. Fusion*, **36**, 1327-1354 (1994).
3. Y. Saad, G.-C. Lo, S. Kuznetsov, *PSPARSLIB: A Portable Library for parallel Iterative Solvers*, Proc. Parallel Computing Technologies (**PaCT-95**), St. Petersburg, Russia, 1995.

A Full-Wave Solution of the Maxwell's Equations in 3D Plasmas

P. Popovich, N. Mellet, L. Villard, W. A. Cooper

Centre de Recherches en Physique des Plasmas
Association EURATOM - Confédération Suisse
EPFL, 1015 Lausanne, Switzerland

Abstract. We present a new global solver for the Maxwell's equations in stellarator plasmas (LEMan). The 3D geometrical effects are fully taken into account, no assumption on the wavelength is made. The full cold plasma dielectric tensor including finite electron mass is implemented, extension to the kinetic model is under development. The wave equation is discretised with finite elements (linear or cubic) radially and Fourier decomposition in the poloidal and toroidal angles. Special care is taken to correctly treat the problem near the magnetic axis. Unicity of the solution, gauge condition and the energy conservation are ensured. Some results and benchmarks for the Alfvén and ion cyclotron frequency ranges are presented.

Keywords: full-wave, Alfvén, ICRF, three-dimensional geometry.
PACS: 52.35.Bj, 52.35.Hr

Optimization of the low-frequency plasma heating systems or use of active diagnostic methods such as the MHD spectroscopy requires the knowledge about the electromagnetic (E/M) wave propagation in plasmas. Another motivation for the wave propagation studies is the possible interaction of the wave with the fast particles that can potentially have a negative effect on the confinement properties of the future fusion device. For stellarator applications, fully 3-dimensional tools are indispensable due to the strong effect of the geometry on the properties of the low-frequency wave propagation. Geometrical effects and the inhomogeneity of the plasma on the spatial scales comparable to the perturbation wavelength necessitate the development of full-wave algorithms for the wave equation solution. At the Alfvén and the ion-cyclotron frequencies, the WKB approximation breaks down and the ray tracing techniques are generally no longer applicable. A global solution of the wave equation is then required, which takes into account the finite spatial extent of the plasma, the effects of the absorption of the incident wave, the reflection from the walls, etc. The methods based on a global solution of the Maxwell's equations in 3D inhomogeneous plasmas have been successfully developed and applied for many years [1-3].

The LEMan code represents a global solver of the linearised set of the Maxwell's equations formulated in terms of the E/M potentials:

$$\begin{cases} \nabla^2 \vec{A} + k_0^2 \vec{\varepsilon} \cdot \vec{A} + ik_0 \vec{\varepsilon} \cdot \nabla \phi = -4\pi/c\ \vec{j}_{ext}, \\ \nabla \cdot (\vec{\varepsilon} \cdot \nabla \phi) - ik_0 \nabla \cdot (\vec{\varepsilon} \cdot \vec{A}) = -4\pi \rho_{ext} \end{cases}$$

The potential formulation of the wave equation is used instead of the traditional field formulation to avoid the previously described [4] numerical pollution effect, potentially dangerous for the standard finite element discretization.

The initial underlying equilibrium for the propagation problem is obtained with the VMEC code [5]. The TERPSICHORE code [6] is then used to project this equilibrium onto the Boozer magnetic coordinate frame and to calculate the metrics of the geometry. The perturbations of the equilibrium are excited with an antenna described through the external current density distribution $\vec{j}_{ext}(s,\theta,\phi)$. In the present version, the antenna can be chosen as a helical stripe with one or several prescribed Fourier harmonics (in θ and ϕ) or a poloidally (or toroidally) localized antenna. The second type of the antenna is used primarily for the ICRF studies to investigate the effect of the wave launching direction on the wave propagation and absorption.

The full cold plasma model is presently implemented in LEMan [7]. It retains the finite electron inertia, and so can account for the effect of the conversion to the short wavelength oscillations at the Alfvén resonant surface. The classical cold plasma dielectric tensor $\hat{\varepsilon}$ does not describe energy dissipation. An introduction of a "friction force" to the equations of the particle motion results in a small imaginary part in the frequency ω in $\hat{\varepsilon}$. This imaginary part plays the role of the absorption in our model.

Recent developments of the model include the partial implementation of finite temperature effects incorporating Landau damping. Only the 0th-order terms in the Larmor radius are retained in the Vlasov equation, which is a reasonable approximation for the electrons, but has limited applicability for the ions. A Maxwellian is presumed for the distribution function. This model is fully implemented and tested in 1D geometry. As opposed to the cold plasma, an introduction of the finite temperature effects allows for a new solution, the Kinetic Alfvén wave. This converted wave can propagate in the direction opposite to the quasielectrostatic wave present in the cold plasma model. The LEMan version with the kinetic tensor has been successfully benchmarked against the ISMENE code [8] calculations in a cylindrical geometry. Both the propagation direction and the wavelength of the converted KAW are in good agreement between the two codes, and also agree with the analytical dispersion relation. The implementation of this model in 2D and 3D geometry is complicated by the coupling between the Fourier harmonics in the dielectric tensor. Extension to hot plasma in 3D is still under development.

The wave equation (1) is cast into the Galerkin form to reduce it to the first order PDE. This system is then discretized with a combination of the finite elements in the radial direction and Fourier harmonics in the poloidal and toroidal variables (Boozer angles). This representation is particularly efficient for the Alfvén frequency range. The polynoms used for the finite elements are well suited to describe the rapid variation of the wavefields across the magnetic surfaces. At the same time, a relatively small gradient along the surfaces requires only a small number of Fourier components. Another advantage is the simplicity of the $k_{//}$ calculation (direct access to the angular derivatives), which is important for the future extension of the plasma model.

The error measure for the numerical solution is directly provided by the energy conservation law. It is easy to see that the Coulomb gauge $\nabla \cdot \vec{A} = 0$ everywhere in the plasma volume directly follows from the wave equation (1) for any divergence-free antenna if it is imposed on the boundary. Therefore, the true solution of the continuous problem is exactly divergence-free, while the numerical solution can only converge to the $\nabla \cdot \vec{A} = 0$ condition. On the other hand, the Galerkin form can be shown to correspond exactly to the energy conservation law, apart from the additional terms containing $\nabla \cdot \vec{A}$. The residual of the physical power balance can thus be used as a measure of the convergence of the numerical solution.

The LEMan code has been extensively tested in various geometries. Numerous verifications have been performed in 1D cylindrical geometry: vacuum waveguide modes, Alfvén wave, fast magnetosonic wave; comparisons with analytical expressions have been done when possible. All the modes have been recovered with a one-to-one correspondence to the analytical solutions, no spurious solutions have been observed.

Alfvén spectrum calculations have been done in various 2D geometries with selected symmetries (tokamak, mirror, helix). Positions of the Alfvén resonances, gaps and corresponding eigenmodes have been found according to the dominating coupling terms in the equilibrium. Benchmarks in the tokamak configuration against the 2D LION code [9] show a good agreement between the two codes for the Alfvén resonant surface positions; TAE frequency agrees within 3%. The frequency of the TAE obtained with the LEMan code has also been compared to the measurements of the low-frequency activity during the JET discharge #52206 [10]. The calculated values correspond within 8% to the experimental frequency of the n=1 TAE modes. Part of the difference between the results is due to the slight differences in the underlying equilibriums reconstructed using different codes, 2D CHEASE [11] for the LION code and JET configuration, and 3D VMEC for LEMan.

As an example of the application in a fully 3D geometry, we present here the calculations performed for the LHD stellarator (Fig. 1). Due to the 3D coupling, the Alfvén resonance branches form gaps near the crossings of the cylindrical branches. These positions (dashed lines) are obtained with a separate code which is designed to calculate the Alfvén continuum branches in a 3D geometry by directly searching the zeros of the dispersion function for the Alfvén wave [12]. The resonance radial positions found by this direct dispersion relation solution and by the global LEMan code are in a good agreement, which provides an additional verification of the LEMan results, now in 3D geometry. The discretization is done with 100 radial elements and 135 Fourier harmonics, mode amplitudes on the edge of the Fourier box do not exceed 0.5% of the maximum value, global/local power balance is satisfied within 0.05% / 5%. Runtime on one NEC-SX5 processor is 1600 s for one frequency.

While the discretization implemented in LEMan is very efficient for the Alfvén frequency range, ICRF is a more challenging scenario for the Fourier representation. We have applied the code in the JET geometry for two ion species plasma (70% deuterium and 30% hydrogen) at the frequencies close to the ion-ion hybrid resonance. When launched from the high-field side (HFS), the wave is almost completely absorbed at the hybrid resonance. In the case of the low-field side (LFS) antenna, the wave is mostly reflected from the cutoff located on the LFS of the resonance and, superposing with the incident wave, forms eigenmodes. The frequency scan of the plasma response, almost monotonic for the HFS and peaked for the LFS propagation, qualitatively agrees with the LION code simulations performed for a similar configuration. For these calculations, the grid is composed of 150 radial elements and 45 poloidal harmonics. These parameters are sufficient to obtain well converged results with the energy conservation satisfied with an accuracy of ~0.01%.

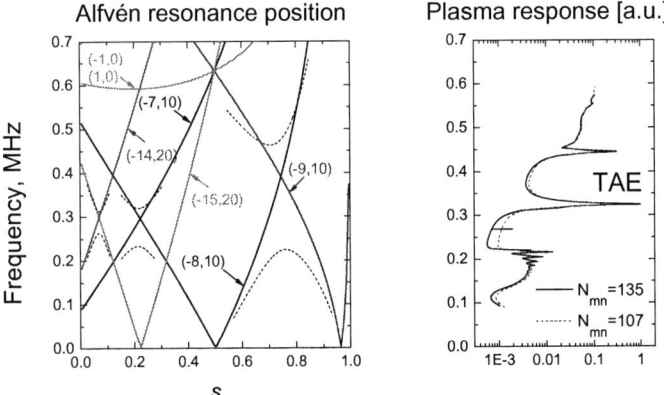

FIGURE 1. Left: Cylindrical Alfvén continuum branches (solid lines) and gaps near the crossings in the LHD geometry (dashed lines). Right: Normalised plasma response (total absorbed power) to a (-7,10) helical antenna perturbation calculated for two sizes of the Fourier table.

ACKNOWLEDGMENTS

We thank Dr. S.P. Hirshman for the use of the VMEC code. This work was partly supported by the Swiss National Science Foundation. Some of the calculations were performed on the NEC-SX5 platform at the Centro Svizzero di Calcolo Scientifico, Switzerland.

REFERENCES

1. E. F. Jaeger, L. A. Berry, E. D'Azevedo, D. B. Batchelor, M. D. Carter, K. F. White, H. Weitzner, *Phys. Plasmas*, **9**(5):1873 (2002).
2. A. Fukuyama, and T. Akutsu, *43rd Annual Meeting of the APS Division of Plasma Physics, Long Beach, California* (2001).
3. V. Vdovin, T. Watari, A. Fukuyama, *Laboratory report NIFS-469, Japan* (1996).
4. A. Jaun, K. Appert, J. Vaclavik, L. Villard, *Comput. Phys. Comm.*, **92**:153 (1995).
5. S. P. Hirshman, *Phys. Fluids*, **26**(12): 3553 (1983).
6. D. V. Anderson, W. A. Cooper, R. Gruber, S. Merazzi, U. Schwenn, *Int. J. Supercomput. Appl.*, **4**(3):34 (1990).
7. T.H. Stix, *Waves in Plasmas*, American Institute of Physics, New York, 1992.
8. K. Appert, T. Hellsten, J. Vaclavik, L. Villard, *Comput. Phys. Comm.*, **40**:73 (1986).
9. L. Villard, K. Appert, R. Gruber, J. Vaclavik, *Comput. Phys. Report*, **4**:95 (1986).
10. D. Testa, G.Y. Fu, A. Jaun, A. Fasoli, O. Sauter and JET-EFDA contributors, *Nucl. Fusion* **43**:479, (2003).
11. H. Lütjens, A. Bondeson, A. Roy, K., *Comput. Phys. Comm.*, **69**:287 (1992).
12. N. Mellet, *Diploma project report*, CRPP, EPFL (2004).

Parametric Decay During HHFW on NSTX

J. R. Wilson[1], S. Bernabei[1], T. Biewer[1,3], S. Diem[1], J. Hosea[1], B.LeBlanc[1], C. K. Phillips[1], P. Ryan[2], D. W. Swain[2]

[1]*Princeton Plasma Physics Laboratory, Princeton, NJ USA*
[2]*Oak Ridge National Laboratory, Oak Ridge, TN*
[3]*Present address MIT, Cambridge, MA*

Abstract. High Harmonic Fast Wave (HHFW) heating experiments on NSTX have been observed to be accompanied by significant edge ion heating ($T_i \gg T_e$). This heating is found to be anisotropic with $T_{perp} > T_{par}$. Simultaneously, coherent oscillations have been detected with an edge Langmuir probe. The oscillations are consistent with parametric decay of the incident fast wave ($\omega > 13\omega_i$) into ion Bernstein waves and an unobserved ion-cyclotron quasi-mode. The observation of anisotropic heating is consistent with Bernstein wave damping and the Bernstein waves should completely damp in the plasma periphery as they propagate toward a cyclotron harmonic resonance. The number of daughter waves is found to increase with rf power and to increase as the incident wave's toroidal wavelength increases. The frequencies of the daughter wave are separated by the edge ion cyclotron frequency. Theoretical calculations of the threshold for this decay in uniform plasma indicate an extremely small value of incident power should be required to drive the instability. While such decays are commonly observed at lower harmonics in conventional ICRF heating scenarios they usually do not involve the loss of significant wave power from the pump wave. On NSTX an estimate of the power loss can be found by calculating the minimum power required to support the edge ion heating (presumed to come from the decay Bernstein wave). This calculation indicates at least 20-30% of the incident rf power ends up as decay waves.

Keywords: RF Heating, Parametric Decay, Spherical Torus
PACS: 52.50.Qt, 52.55.Hc

INTRODUCTION

High Harmonic Fast Wave (HHFW) heating and current drive has been explored on NSTX since its inception. The interest in HHFW is motivated by the need for electron heating and current drive to aid in steady-state non-inductive operation, a crucial component in demonstrating the attractiveness of the spherical tokamak (ST) reactor concept. HHFW was predicted to be an attractive option to meet this need[1], particularly for the high beta plasmas that naturally occur in ST's. As previously reported [2], efficient electron heating and current drive via HHFW have been found on NSTX. Some anomalies were observed in the details of the results, particularly in the dependence on antenna phasing. Faster wave phasings, smaller toroidal mode number, were found to be less effective than expected [3]. With the installation of a new diagnostic to measure the edge ion temperature and plasma rotation [4], a new piece of puzzling evidence was found. The edge ion temperature (over ~ 10 – 15 cm of the minor radius inside the periphery of the plasma), as measured by Doppler broadened line emission from both impurity and majority ion species, was seen to increase during the rf pulse. This temperature increase indicated a Maxwellian heating in the perpendicular component of the ion distribution function. Direct ion heating from the fast wave is not expected at such a high harmonic and low edge temperature. In searching for an explanation of this heating, which was reminiscent of that observed in direct launch Ion Bernstein Wave heating (IBW) [5], an edge probe was installed on NSTX and connected to a spectrum analyzer. Frequency peaks that were consistent with IBW originating via parametric decay were observed. In the following we will review parametric decay theory as applied to NSTX HHFW heating and then describe in detail the experimental observations. The decay mechanism postulated here has been observed in laboratory plasmas [6,7]

PARAMETRIC DECAY THEORY

The theory of parametric decay in the ion cyclotron range of frequencies has been elaborated in a paper by Porkolab [8]. If one examines the Vlasov equation in the presence of a linear rf wave

$$\frac{\partial f}{\partial t} + \vec{v} \cdot \nabla_x f + \frac{q}{m}\left(\frac{\vec{v}\times\vec{B}}{c} + \vec{E} + \vec{E}_0 \cos\omega_0 t\right) \cdot \nabla_v f = 0 \qquad (1)$$

the solutions of the zero order orbit equations are:

$$x_d = -\frac{q}{m}\frac{E_{0\perp}}{\omega_0^2-\Omega^2}\cos\omega_0 t, \quad y_d = \frac{\Omega}{\omega_0}\frac{q}{m}\frac{E_{0\perp}\sin\omega_0 t}{\omega_0^2-\Omega^2}, \quad z_d = -\frac{q}{m\omega_0^2}E_{0\|}\cos\omega_0 t \qquad (2)$$

As can be seen, the particle drift orbits are mass dependent, therefore the ions and electrons will drift apart creating a zeroeth order charge separation oscillating at the driving frequency ω_0. This charge separation represents a source of free energy that can drive instabilities in the plasma. Looking at the first order perturbation to Eq. 1 the ansatz

$$f_k^{(1)}(\vec{r},\vec{v},t) = F(\vec{r},\vec{u},t)\exp(-i\vec{k}\cdot\vec{x}_d) \qquad (3)$$

is introduced. This assumption is essentially a transformation to the oscillating frame of reference in which there is no induced drift. Also, the driving field is assumed to have infinite wavelength, $k_0=0$. The first order equation is then

$$\frac{\partial F}{\partial t} + \vec{u}\cdot\nabla_x F + (\vec{u}\times\vec{\Omega})\cdot\nabla_u F = -\frac{q}{m}E_k^{(1)}\cdot(\nabla_u f_0)e^{i\vec{k}\cdot\vec{x}_d} \qquad (4)$$

$\vec{k} \cdot \vec{x}_d$ can be written as $\mu \sin(\omega_0 t - \beta)$ where

$$\mu = \frac{q}{m}\left[\frac{E_{0\parallel}k_z}{\omega_0^2} + \frac{E_{0y}k_y}{\omega_0^2 - \Omega^2}\right)^2 + \frac{\left(E_{0x}k_y\right)^2 \Omega^2}{\left(\omega_0^2 - \Omega^2\right)^2 \omega_0^2}\right]^{1/2} \quad (5)$$

and β is a phase angle which is unimportant. The solution to the first order equation (4) can be found by the method of characteristics, i.e.

$$F(r,u,t) = \frac{q}{m}\int_{-\infty}^{t} dt' \nabla_{x'}\phi(r',t') \cdot \nabla_{u'} f_0 e^{i\mu \sin(\omega_0 t - \beta)} \quad (6)$$

By expanding the exponential in a Bessel series

$$\exp[i\mu\sin(\omega_0 t - \beta)] = \sum_{n=-\infty}^{\infty} J_n(\mu)\exp[in(\omega_0 t - \beta)] \quad (7)$$

the integration of (6) is simplified. The charge density is given by

$$\rho = q\int du F(r,u,t)$$

Because of the oscillating exponential in (6) the charge density also exhibits oscillation with beat frequencies $n\omega_0$. Further, since the electrons and ion charge densities oscillate by different amounts the earlier mentioned charge separation appears. By Fourier transforming, the following expression for the charge density is obtained:

$$\rho_\sigma(k,\omega) = q\int d^3 u F(\vec{k},\vec{u},\omega) = -\frac{k^2}{4\pi}\sum_n J_n(\mu)e^{-in\beta}\chi_\sigma(\omega,k)\phi(\omega + n\omega_0, k) \quad (8)$$

where χ is the linear susceptibility. Substituting into Poisson's equation yields

$$\phi(k,\omega) = \frac{4\pi}{k^2}\sum_\sigma \sum_n J_n(-\mu)e^{-in\beta}\rho_\sigma(\omega + n\omega_0, k) \quad (9)$$

combining (8) and (9) yields the dispersion relation. Clearly, it is an infinite set of equations of order n. Making the simplest non-trivial assumption, for $\mu<1$, we can keep only three modes ω, $(\omega \pm \omega_0)$. Further, assuming that $(\omega, \mathbf{k})$ is a low frequency ion mode and $|\omega \pm \omega_0| \gg \omega$ the three coupled equations can be solved to yield [9]

$$\varepsilon(\omega^-, k^-)\varepsilon(\omega, k) = -\frac{1}{8}|\mu_i - \mu_e|^2 \left[(\chi_i - \chi_i^-)(\chi_e - \chi_e^-)\right] \quad (10)$$

where $\varepsilon(\omega,k) = 1+\Sigma\chi_j(\omega,k)$ is the dielectric constant, χ_j is the susceptibility of species j, and μ is the parametric coupling constant defined in equation (5). In the following the side band mode $\omega^- = \omega - \omega_0$, will be assumed to be an ion Bernstein wave and the low frequency mode may be either another Bernstein wave or a quasi-mode (a dissipative mode where $\omega - \Omega \sim k_\| v_{ti}$ or $\omega = k_\| v_{te}$ which exists only when driven by the pump). Examining μ for the parameters of NSTX, $\omega \sim 13\Omega_i$, and it is clear that the dominant term is the second one, due to the polarization drift

$$\mu_\sigma \sim \frac{eZ_\sigma}{m_\sigma} \frac{E_y^2 k_y^2}{\left(\omega_0^2 - \Omega_\sigma^2\right)^2}$$

and if the upper sideband is a resonant IBW wave

$$\varepsilon(\omega^-) = -i(\gamma_0 + \gamma_2)\left|\frac{\partial \varepsilon}{\partial \omega_R^-}\right|$$

where $\gamma_2 = \varepsilon_{\text{Im}}(\omega^-)/\left|\partial\varepsilon/\partial\omega_R^-\right|$ is the damping rate at the side band then the dispersion relation (10) yields

$$(\gamma_0 + \gamma_2)\left|\frac{\partial \varepsilon}{\partial \omega_R^-}\right| = \text{Im}\frac{1}{4}\left[\frac{eZ_i}{m_i} \frac{(E_y k_y)^2}{(\omega_0^2 - \Omega_i^2)^2} \left(\frac{(\chi_e(\omega) - \chi_e(\omega^-))(\chi_i(\omega) - \chi_i(\omega^-))}{\varepsilon(\omega,k)}\right)\right] \quad (11)$$

The electron coupling term has been dropped since $\omega/\Omega_e \ll 1$. The instability will grow when the right hand side is positive. The first possibility is $\varepsilon(\omega,k) = 0$, corresponding to a resonant lower frequency wave. It is very difficult to match both frequency, $\omega + \omega^- = \omega_0$, and wavenumber, $k + k^- = 0$ for two resonant IBW waves. The other way to get the right hand side large is for $\chi_i(\omega)$ to be large. This can happen when $\omega = n\Omega_I$, where n is an integer. This is the ion-quasi mode, quasi since $\varepsilon(\omega) \neq 0$. The quasi mode has no k dependence so satisfying k matching is trivial. With χ_i large only the electron susceptibilities remain and for $\omega = \Omega_I$ they can be written as

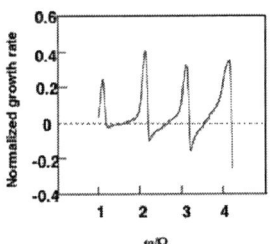

Figure 1. Parametric decay normalized growth rate for NSTX edge plasma

$$\chi_e(\omega) - \chi_e(\omega^-) = i\sqrt{\pi}\omega\frac{\left(1-k_\perp^2\rho_e^2\right)}{k_\| v_{te}}$$

which will be positive when $1-(k_\perp\rho_e)^2$ is positive, which is easily satisfied. In fig. 1 we show a numerical evaluation of eq. 11 for NSTX edge parameters, $T_e = T_i = 50$ eV, $n_e = 2 \times 10^{18}$ m^{-3}, B = 2.3 kG, $k_\| = \Omega_i/0.7v_{te}$, $f_0 = 30$ MHz and an assumed $E_y = 1$ statvolt/cm. This value of $k_\|$ was found to maximize the growth rate at $\omega = \Omega_i$. Different values of $k_\|$ maximize the growth rate at the harmonics. The first four harmonics are seen to be unstable for NSTX parameters.

EXPERIMENTAL RESULTS FROM NSTX

The NSTX HHFW system has been described previously [10]. A Langmuir probe was inserted between two of the antenna modules and is placed in the scrape-off plasma in the shadow of the rf protection tiles. The probe was connected via a high frequency fiber optic link to a spectrum analyzer. A filter is placed in-line to attenuate the 30 MHz rf by ~30 dB in order to protect the analyzer. The spectrum analyzer was swept in frequency during the rf pulse. A typical spectrum obtained during HHFW heating on NSTX is shown in fig. 2.

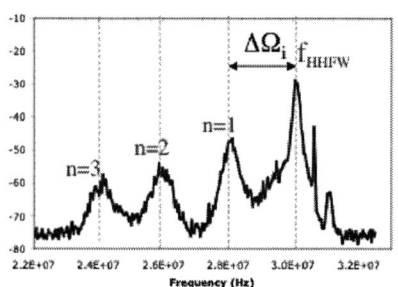

Figure 2. Frequency spectrum from Langmuir probe

Three lower sidebands are clearly seen. The frequency separation between peaks corresponds to the edge ion cyclotron frequency. When the toroidal magnetic field was lowered from 4.5 kG to 3.0 kG the frequency separation decreased as expected if the side bands for IBW decay waves. The number of sidebands is found to increase with applied rf power; for example in He discharges with $k_T = 14$ m^{-1}, the second peak appears for powers greater than 0.65 MW and the third for powers greater than 1.3 MW. At fixed rf power the number of side bands increases with decreasing toroidal wave number applied to the HHFW antenna, suggesting that the fast wave electric field in the surface of the plasma increases with the launched wavelength. A side band above the pump frequency is occasionally seen at the highest applied rf power levels.

The edge ion temperature measurement is obtained from toroidal and poloidal viewing arrays of the Edge Rotation Diagnostic (ERD), that measures the Doppler broadened emission from both HeI and CIII ions [4]. By projecting these views onto the perpendicular and parallel directions a bi-Maxwellian temperature is found as shown in fig. 3 for helium ions. The perpendicular temperature increases strongly with rf power while the parallel temperature remains close to the local electron temperature. This perpendicular ion temperature increase is observed in the outer, ~10-15 cm, region of the plasma [11].

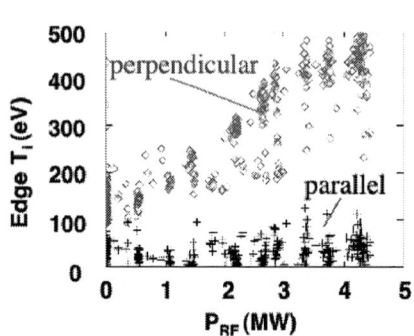

Figure 3 Perpendicular and Parallel Edge ion temperature as a function of rf power

DISCUSSION

The probe spectra indicate the presence of IBW waves in the NSTX edge plasmas. Detection of decay peaks is not un-common in ICRF experiments [12-14]. Only in the case of direct Bernstein wave excitation [13] has it been inferred that significant amounts of power are involved. By solving the IBW dispersion relation for the observed frequencies it is possible to predict where the IBW power will be deposited in the plasma. The IBW wave is predicted to propagate inward until it reaches a cyclotron harmonic where all of the power will be damped. For NSTX this means that the heating should take place in the outer ~10 cm of the plasma in agreement with the ERD observations. The power in the low frequency quasi-mode is locally damped since the mode is non-propagating but, since the partitioning of power between the two side bands is proportional to their frequency, less than ten percent of the decay power is in the quasi-mode. The IBW wave deposits its energy into the low energy ($<v_{ti}$) part of the ion distribution function, hence the distribution function is expected to remain essentially Maxwellian. The IBW wave heats the perpendicular part of the ion distribution function, so a bi-Maxwellian response can be expected. This was seen in earlier direct IBW experiments [9] and, again, is in agreement with the ERD measurement. It is nearly impossible to get an estimate of the power lost into decay waves from the probe measurements; an absolute calibration of the probe response would be required the probe would need to be scanned everywhere in the edge plasma. Therefore, in order to get an estimate of the amount of rf power that is being channeled into the decay waves, the following ansatz is made: all the decay power flows to the ions and the ions then lose all their energy to the cooler electrons via collisions and the electrons lose all of their energy out of the plasma in a time short compared to the ion relaxation time (in these plasmas this time is of order 1 ms). By

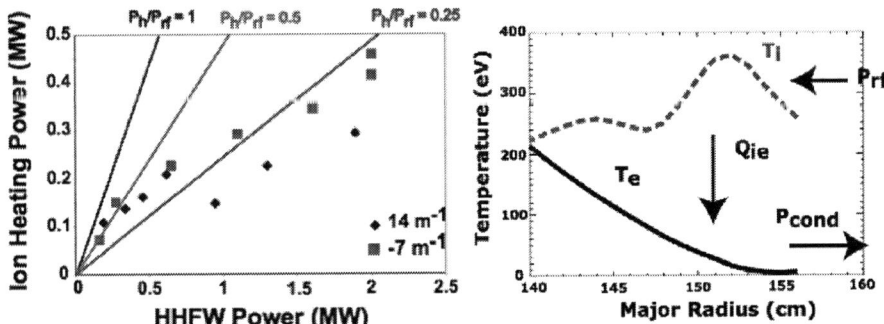

Figure 4. Rf power going into edge ion heating (a) Ion heating power for two antenna phasings as a function of rf power. (b) Power is calculated by integrating Q_{ie} versus radius and assuming power flows as in figure.

this ansatz an estimate of any one of these processes yields a value for all three. Since the collisional power transfer from the ions to electrons:

$$Q_{ie}(r) = \frac{3m_e}{m_i}\frac{n(r)k_B}{\tau_e}(T_i(r)-T_e(r)) \quad (12)$$

where $\tau_e = \frac{3\sqrt{m_e}(k_B T_e)^{3/2}}{4\sqrt{2\pi}n_e \Lambda e^4}$ is the collisional relaxation time, can be calculated knowing only the radial dependence of the temperature difference and density, it is the easiest of the three to obtain. In fig. 4 we show sample temperature profiles and values of the integral of $Q_{ie}(r)$ over the outer part of the plasma where T_i exceeds T_e for helium plasmas.

It can be seen that at high power levels as much as 25% of the applied rf is showing up in edge heating power. The deviation from a straight line fit in fig4a may be indicative of either an overestimate of the edge heating power at low applied power levels due to using the full density in the calculation or an underestimate at high values of the applied heating power if some of the edge ion energy leaves the plasma directly without transferring to the electrons. To some extent the later effect must be occurring since the ERD measurements also indicate a change in the edge rotation consistent with an edge electric field that would be produced from ions being lost directly from the edge.

SUMMARY

Heating of the perpendicular component of the ion distribution function for the outer part of the NSTX plasma has been observed during HHFW heating and current drive experiments. An interpretation in terms of parametric decay of the incident fast wave into ion Bernstein waves and an ion-quasi mode is presented. Frequency components consistent with IBW are observed on a spectrum analyzer connected to a langmuir probe inserted in the scrap-off plasma. The theory is presented and a low threshold in power is calculated for NSTX parameters. The ion Bernstein wave is expected to damp in the outer 10-15 cm of the plasma radius and heat the ions perpendicular component in agreement with the experimental observations.

ACKNOWLEDGMENTS

This work supported by DOE Contract No. DE-AC02-76CH03073.

REFERENCES

1. Ono, M., "High harmonic fast waves in high beta plasmas", *Physics of Plasmas* **2**, 4075 (1995).
2. Wilson, J. R. *Physics of Plasmas* **10**, 1733 (2003).
3. Hosea, J.C. this conference
4.. Biewer, T. M. Bell, R., Feder, R. et al. *Rev. Sci. Instrum.* **75**, 650 (2004).
5. Ono, M. et al. *Phys. Rev. Lett.* **60**, 294 (1988).

6. Ono, M. Porkolab, M. and Chang, R.P.H., *Phys. Rev. Lett.* **38**, 962 (1977).
7. Skiff, F. Ono, M., and Wong, K.L., *Physics of Fluids*, **27**, 1051 (1984).
8. Porkolab, M., *"Symposium on Plasma Heating and Injection"* Varenna Italy 1972.
9. Porkolab, M., *Fusion Eng. Design* **12**, 93 (1990).
10. Ryan, P. M., et al., *Fusion Engineering and Design* **56-57**, 569-573 (2001).
11. Biewer, T. M. et al. *Physics of Plasmas* **12**, 056108 (2005)
12. Nieuwenhove, R van, et al., *Nuclear Fusion* **28**, 1603 (1988).
13. Pinsker, R. I., et al. *Radio Frequency Power in Plasmas*, 314 (1989).
14. Rost, J. C. et al., *Physics of Plasmas,* **9**, 1262 (2003).

Absorption of Fast Waves at Moderate to High Ion Cyclotron Harmonics on DIII-D

R.I. Pinsker*, M. Porkolab[a], W.W. Heidbrink[b], Y. Luo[b], C.C. Petty*,
R. Prater*, M. Choi*, F.W. Baity[c], E. Fredd[d], J.C. Hosea[d],
R.W. Harvey[e], A.P. Smirnov[e], M. Murakami[c], and M.A. Van Zeeland[f]

*General Atomics, P.O. Box 85608, San Diego, California, 92186-5608 USA
[a] Massachusetts Institute of Technology, Cambridge, Massachusetts, USA
[b] University of California, Irvine, California, USA
[c] Oak Ridge National Laboratory, Oak Ridge, Tennessee, USA
[d] Princeton Plasma Physics Laboratory, Princeton, New Jersey, USA
[e] CompX, Del Mar, California, USA
[f] Oak Ridge Institute for Science Education, Oak Ridge, Tennessee, USA

Abstract. The absorption of fast Alfvén waves (FW) by ion cyclotron harmonic damping in the range of harmonics from fourth to eighth is studied theoretically and with experiments in the DIII-D tokamak. A formula for linear ion cyclotron absorption on Maxwellian ion species is used to estimate the single-pass damping for various cases of experimental interest. It is found that damping on fast ions from neutral beam injection can be significant even at the eighth harmonic if the fast ion beta and the background plasma density are both high enough. The predictions are tested in several L-mode experiments in DIII-D with FW power at 60 MHz and at 116 MHz. It is found that 4th and 5th harmonic absorption of the 60 MHz power on the beam ions can be quite strong, but 8th harmonic absorption of the 116 MHz power appears to be weaker than expected. Possible explanations of the discrepancy are discussed.

INTRODUCTION

A prerequisite for efficient fast wave current drive (FWCD) is that direct electron damping of the fast wave in the core of the plasma must dominate all other absorption mechanisms. Other damping mechanisms that may be important include ion cyclotron damping on either thermal or nonthermal populations such as result from neutral beam injection or alpha particles in a burning plasma, edge losses from far-field sheaths (reviewed by Myra [1]), or parametric decay instabilities in the outer part of the plasma (discussed in this context by Wilson, et al. [2]). In the present work, we consider ion cyclotron harmonic damping on a fast ion population from neutral beam injection, which has previously been shown to constitute an important competing power absorption mechanism in FWCD experiments on DIII-D [3]. The aim of this investigation is to compare observed ion cyclotron harmonic damping in DIII-D with theoretical predictions, in order to be able to quantitatively predict the power partitioning between direct electron damping and ion cyclotron damping for future applications of FWCD.

For the frequencies used for FWCD experiments in DIII-D (60–120 MHz) and the range of toroidal fields used, the cyclotron harmonic number is in the range of f/f_{ci} = 4-12 for deuterium (half this for hydrogen). As is well known [4,5], the fraction of the wave power absorbed by ion cyclotron damping as the fast wave propagates through a particular cyclotron harmonic layer in a slightly inhomogeneous magnetic field depends the density of the absorbing ion species and the ratio of the wavelength of the fast wave to the gyroradius ρ_s of the absorbing ion species s. This ratio may be expressed as $k_\perp \rho_s$, and typical numerical values range from $k_\perp \rho_s \sim 0.1$ for thermal hydrogen at the second harmonic at $B_T = 2$ T to $k_\perp \rho_s \sim 4$ for absorption on an injected deuterium beam at the 8th harmonic at $B_T = 1$ T.

The outline of the paper is as follows. The first part addresses the linear theory of ion cyclotron absorption in a uniform magnetized plasma, with isotropic Maxwellian velocity distributions. While ignoring the self-consistent effect of the FW power on the ion distribution, such a simplified estimate of the damping strength may be appropriate for situations in which the neutral beam power significantly exceeds the FW power. The most important effect of relaxing the uniform plasma assumption can be addressed with ray tracing, so a brief discussion of the characteristics of fast wave ray trajectories follows. The latter part of the paper constitutes a short survey of recent experiments in DIII-D in which ion cyclotron harmonic absorption on injected beams was studied at two FW frequencies and at various harmonic numbers.

THEORY

Porkolab has given a formula for the optical depth due to ion cyclotron absorption of the fast wave on a hot ion species (designated with the subscript s) [5]:

$$2\eta = \frac{\pi}{2}\left(\frac{n_{hot}}{n_i}\right)\left(\frac{\omega_{pi} R_M}{c}\right)(\ell-1)^2 e^{-\lambda_s}\left\{I_{\ell-1}(\lambda_s) + \frac{\lambda_s}{\ell}\left[I_\ell(\lambda_s) - I_{\ell-1}(\lambda_s)\right]\right\} \quad (1)$$

in which the argument of the modified Bessel functions is $\lambda_s \equiv (1/2)(k_\perp \rho_s)^2$, the ion cyclotron harmonic is $\ell = f/f_{cs}$, ω_{pi} is the ion plasma frequency for the bulk ions, n_{hot} and n_i are the densities of the hot and bulk (thermal plus hot) ion species, R_M is the major radius of the ion cyclotron resonance layer (assuming the magnetic field strength satisfies $R_M B_0$ = constant). The optical depth is defined by the power transmission coefficient for one traversal of the cyclotron resonance layer of $T \sim \exp(-2\eta)$ and hence a fraction $A = 1 - T$ of the incident power absorbed there. This formula assumes that only the left-circularly polarized component of the wave electric field accelerates ions, which is nearly true for absorption at the fundamental ion cyclotron resonance. The simplest approximations for the fast wave dispersion $\omega^2 = (k_\perp^2 V_A^2)$ and polarization $[|E_+|^2/|E_y|^2 = (\ell-1)^2]$ have been used, and the effect of non-zero $k_\parallel$ has been neglected. The hot species is taken to have an isotropic Maxwellian distribution; several authors have examined the effect of relaxing this assumption, as in [6]. Arbitrarily large values of $k_\perp \rho_s$ are correctly treated.

We have relaxed a number of the simplifying assumptions to obtain the following, more complete form for this optical depth. The resulting expression is:

$$2\eta = R_M \frac{\ell \omega_{ps}^2}{4 n_\perp f c} e^{-\lambda_s}\left[\left(\frac{R - n_\parallel^2}{S - n_\parallel^2}\right)^2 \Delta_-^\ell(\lambda_s) + \left(\frac{L - n_\parallel^2}{S - n_\parallel^2}\right)^2 \Delta_+^\ell(\lambda_s) + \frac{2\lambda_s}{\ell}\left\{\Delta_+^\ell(\lambda_s) - \Delta_-^\ell(\lambda_s)\right\}\right] \quad (2)$$

in which the cold plasma Stix symbols for a multi-ion species plasma R, L, and S have their usual definitions[4], and the combinations of modified Bessel functions have the definitions $\Delta_+^\ell(x) \equiv I_\ell(x) - I_{\ell+1}(x)$ and $\Delta_-^\ell(x) \equiv I_{\ell-1}(x) - I_\ell(x)$. The perpendicular index of refraction for the FW, $n_\perp = k_\perp c/\omega$ either can be taken from a solution of the full local dispersion relation, or the cold plasma approximation in which the slow and FWs have been decoupled can be used, which (introducing the Stix symbol D) yields $n_\perp^2 \approx [D^2 - (n_\parallel^2 - S)^2]/(n_\parallel^2 - S)$.

In the derivation of Eq. (2), the assumption of isotropic Maxwellian ion distribution functions has been retained. Non-zero parallel wavenumber is accounted for, and the contributions to the damping from the left-hand circularly polarized (the first term in the square bracket), right-hand circularly polarized (the second term in the bracket) and the cross-terms are all included. Although in the limit of small $\lambda_s \ll 1$ the E_+ term is dominant in the square bracket, the E_- term increases more quickly with λ_s so that at some value of λ_s it becomes larger in magnitude than the E_+ term. This is illustrated in Fig. 1, in which the three terms in the square bracket are compared as a function of λ_s for second and sixth harmonics. It is evident that treating only the E_+ contribution [as in Eq. (1)] considerably underestimates the ion cyclotron damping for moderate to large values of λ_s. The contribution due the third term in the square bracket, which results from the residual linear polarization, can be safely neglected for moderate to high harmonic numbers.

In the comparison shown in Fig. 1, the approximations for the polarization factors $|E_+|^2/|E_y|^2 \approx (\ell - 1)^2$ and $|E_-|^2/|E_y|^2 \approx (\ell + 1)^2$ have been used to isolate the λ_s dependence of the square bracket, i.e., the effect of non-zero $n_\parallel$ and of plasma parameters other than λ_s and $\ell = f/f_{cs}$ have been neglected. The dependences of the polarization factors on these parameters can be quite significant. For example, in cases typical of DIII-D parameters, the polarization factors $[(R - n_\parallel^2)/(S - n_\parallel^2)]^2$ and $[(L - n_\parallel^2)/(S - n_\parallel^2)]^2$ decrease by about a factor of two as $n_\parallel$ increases from 0 to 10.

Equation (2) has been used to estimate the single pass damping for three different cases relevant to DIII-D FW experiments. In Fig. 2, a comparison is presented of the optical depth 2η for three different cases at a fixed magnetic field of 1.97 T. The first condition, from the FWCD study reported in [7], consisted of a 1.1 MA low density

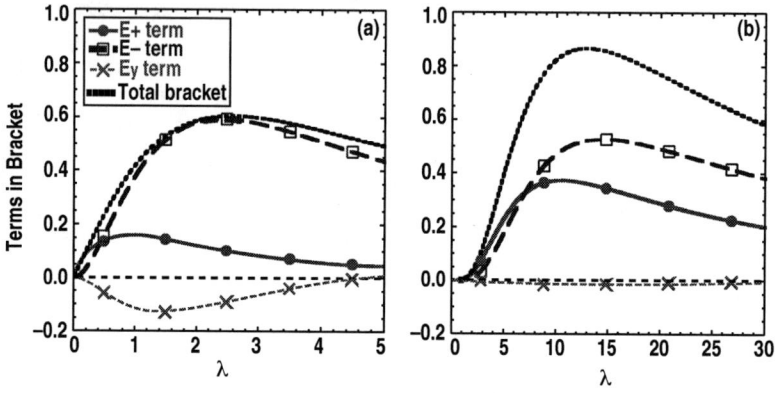

FIGURE 1. Comparison of terms in the square bracket of Eq. (2) for the (a) 2nd harmonic and (b) 6th harmonic. Note that the E_+ term dominates only at small values of λ.

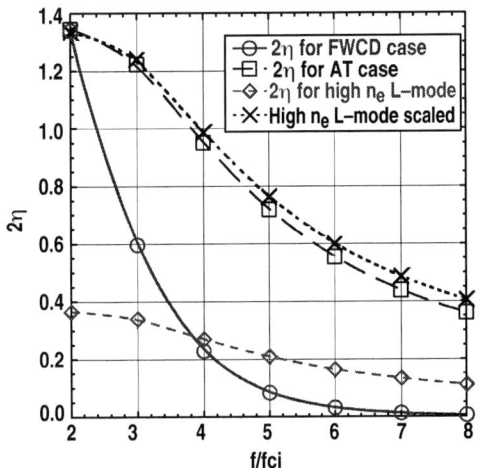

FIGURE 2. Predicted optical depth for absorption on fast ions from second to eighth harmonic for three different sets of plasma parameters given in the text.

[$n_e(0) = 2.6 \times 10^{13}$ cm^{-3}] L-mode discharge with 2.7 MW of 80 keV deuterium neutral beam (NB) injection. The central fast ion density was $n_f = 1.5 \times 10^{13}$ cm^{-3}, and the effective temperature of the fast ion population was $T_{eff} = 24.4$ keV, where the effective temperature is defined by $W_f = (3/2) n_f T_{eff}$, in which W_f is the stored energy density in the fast ions. We consider the harmonic number range $\ell = 2, 3, \cdots 7, 8$. In agreement with the conclusions of Ref. [7], the predicted 8th harmonic absorption is negligible, while the ~18% single pass damping predicted at 4th harmonic is comparable to the expected direct electron absorption.

The second condition studied is representative of H-mode Advanced Tokamak (AT) discharges with 9.8 MW of NB power in which central FWCD is desired for control of the central safety factor, q_0. The parameters are taken from an experimental case that was run without FW power; the central density in this discharge was $n_e(0) = 6.3 \times 10^{13}$ cm^{-3} and the central fast ion density was $n_f = 0.95 \times 10^{13}$ cm^{-3}. The effective temperature of the beam, 22.4 keV, is approximately the same as in the low density L-mode case. The higher central density yields values of λ_s that are about 2.5 times higher than in the low density case for a given harmonic, and this results in substantially higher absorption for $\ell \geq 3$ despite the lower fast ion density. Furthermore, the decrease in absorption with increasing harmonic number is slower, so that ~30% single-pass absorption at the 8th harmonic is predicted. Very substantial degradation in the FWCD efficiency due to cyclotron absorption on the beam is therefore expected at all harmonics in the range considered.

The third case is from a recent experiment, discussed in more detail below, in which FW absorption at 4th and 8th harmonics was studied in a high density L-mode with 5.0 MW of NB heating. The central electron and fast ion densities were $n_e(0) = 6.8 \times 10^{13}$ cm^{-3} and $n_f = 0.26 \times 10^{13}$ cm^{-3}; the effective beam temperature was 21.4 keV. Since the central density and beam energy were about the same as in the high density H-mode case, the values of λ_s and the dependence of the absorption on ℓ were similar in the two cases. The single-pass absorption is lower owing to the lower fast ion density in the L-mode case. This can be seen by scaling up the values of 2η

by the ratio of the beam densities, which yields values of 2η that are nearly the same for the H-mode case and the scaled-up high density L-mode case.

These estimates have been made with a zero-dimensional model. To determine the corrections due to the effects of propagation in a toroidal equilibrium, the GENRAY ray-tracing code has been used [8]. One very important correction results from the evolution of $n_\parallel$ along the ray trajectory, which in turn stems from the non-constancy of the wave's poloidal mode number along the ray trajectory and the magnetic shear. As the ray propagates, the polarization factors in Eq. (2) and hence the strength of the damping upon traversing a resonance layer are strongly affected by the local $n_\parallel$. This phenomenon largely explains the differences in damping seen in the ray-tracing results from different crossings of a given resonance, even at comparable normalized radii. A number of such resonance traversals have been examined in detail [8], and the linear ion cyclotron resonance damping from GENRAY [which should be equivalent to Eq. (2)] is indeed found to be in good agreement with the results of Eq. (2) in which the local values of $n_\parallel$ and $n_\perp$ are used.

EXPERIMENTS

DIII-D experiments performed in 1998-1999 demonstrated strong damping of 60 MHz FW power on beam ions at the 4th harmonic [9]. The acceleration of beam ions by the rf power was observed directly with active neutral particle charge exchange analyzers and indirectly from enhanced fusion neutron emission and from equilibrium reconstruction. Fifth harmonic absorption of FW power at 83 MHz was found to be relatively weak. Introduction of a thermal hydrogen minority decreased the power absorbed by the deuterium beam ions, as the strong 2nd harmonic absorption on the hydrogen reduces the power available to the deuterium.

Recent DIII-D experiments extended these results in several directions. In the same 60 MHz, $\ell = f/f_{cD} = 4$ regime of the earlier work, a new diagnostic technique was used to observe the rf-driven acceleration of the deuterium beam ions. The technique, described in [10], is based on observation of the Doppler-shifted D_α spectrum from the charge exchange between the fast ions and the injected neutral beam. The spectra observed using this diagnostic indicate acceleration of the beam above the injection energy, and the fast ion tail correlates with enhanced neutron emission. Detailed observations with this diagnostic of the effect of the 4th harmonic acceleration on the beam will be presented in a forthcoming publication.

DIII-D deuterium discharges in general have a background hydrogen fraction of H/D < 2%. To demonstrate that residual (thermal) hydrogen absorption at half the harmonic number does not play a dominant role in the 4th harmonic case, a condition was studied in which the $\ell = 5$ deuterium harmonic for $f = 60$ MHz passed through the magnetic axis and no hydrogen harmonic was present in the core of the discharge. This condition was otherwise similar to the 4th harmonic condition studied in [9], with the equivalent of 1.2 MW of 80 keV neutral beams and 0.8 MW of coupled 60 MHz FW power. Time histories from this discharge are shown in Fig. 3. The neutron emission rate is compared with the predicted rate based a zero-dimensional model [9] that includes classical beam slowing-down physics and uses measured plasma parameters but does not include acceleration of the beam. The model successfully predicts the neutron rate before and after the FW power is injected, but the neutron emission rate is enhanced by a factor of 2.0 (as defined in Ref. [9]) during combined FW and NB injection. This result is confirmed by more sophisticated calculations using the

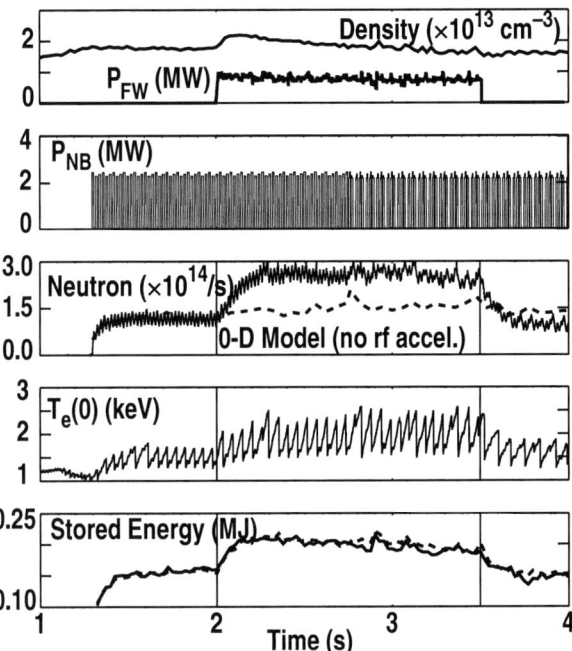

FIGURE 3. Time history of low density L-mode discharge with fifth harmonic heating (60 MHz, B_T = 1.55 T). The two small disturbances in the stored energy, central electron temperature and density around 3 seconds are due to "dithers" into H-mode.

measured profiles in the ONETWO and TRANSP codes. This value of the neutron enhancement is comparable to that obtained in the best cases with 60 MHz at 4th harmonic (at 1.25 times higher magnetic field) in otherwise very similar plasma conditions. The single-pass absorption predicted by Eq. (2) at 5th harmonic in this case is about double what is predicted for 4th harmonic at the same FW frequency; the higher value of λ_s at the lower magnetic field more than compensates for the decrease in the Bessel functions with increasing order at fixed values of the argument. This is consistent with Ref. [9], where it was concluded that 5th harmonic absorption was weak compared with 4th, because that conclusion referred to comparison at fixed magnetic field and varying the frequency (Fig. 2).

In contrast to a comparison of absorption at different harmonics at fixed FW frequency, Fig. 2 showed that the single-pass absorption indeed decreases with increasing harmonic number when the magnetic field is fixed. While the low density L-mode experiments reported in Ref. [7] showed experimentally that the 8th harmonic absorption was negligible compared to 4th harmonic at ~2 T, one expects that the much slower decline in absorption with harmonic number associated with higher values of λ_s, obtained at higher plasma densities, should lead to significant 8th harmonic absorption under both high density L-mode and H-mode conditions. An experiment was carried out to test this prediction, in which a high density beam-heated L-mode was used as the target plasma and FW power was coupled both at 60 MHz and 116 MHz. L-mode was chosen in order to maximize the coupled FW power (due to the associated high antenna loading resistance and the corresponding low antenna voltages.) The relatively low electron temperature (compared to H-mode) and the shorter slowing-down time led to a much lower fast ion density in the high density L-mode case, and hence the smaller single-pass absorption described above. Still, the predicted single-pass absorption at 8th harmonic, 116 MHz is about 12% per pass (depending sensitively on the value of $n_\parallel$ at the intersection of the ray with the resonance layer), which is larger than the value predicted for the 4th harmonic low density L-mode case and about the same as the 5th harmonic, 60 MHz case discussed above. Efficient absorption and accompanying neutron enhancements were expected for both the 4th and 8th harmonic absorption in the high density L-mode condition.

These expectations were not born out in the experimental result, shown in Fig. 4, where the time history of a discharge with coupled FW power levels of 1.6 MW at 116 MHz and 1.1 MW at 60 MHz are compared with a reference shot with no FW power. Although a substantial neutron enhancement (~1.4 in comparison with the reference shot) is seen for the 60 MHz power, only a very small enhancement is seen for the 8th harmonic heating at 1.45 times the FW power. The increment in stored energy is comparable for the two FW frequencies, despite the difference in power levels. The increase in stored energy due to the 116 MHz power appears across the electron and ion temperature profiles, while the deposition of the 60 MHz power is more centrally peaked.

Possible explanations for the apparent discrepancy between the linear theory, which predicts fairly strong absorption for the 8th harmonic in this case, and the

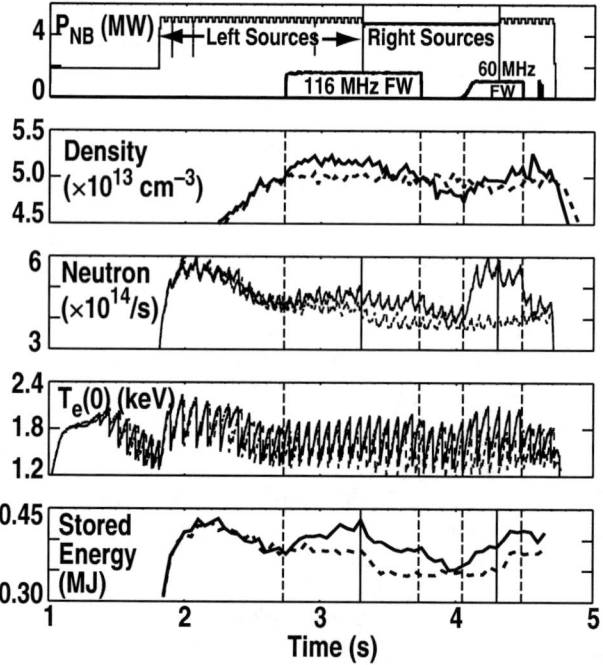

FIGURE 4. Time history of high density L-mode discharge comparing 4th (60 MHz) and 8th harmonic (116 MHz) FW heating at 1.85 T. Dotted vertical lines show rf on and off times; solid vertical lines demarcate the periods where "left" (Rtan=115 cm) and "right" (Rtan=76 cm) ion sources are used. Dashed traces from comparison case with no FW.

experimental result, which gave little evidence for strong central absorption, are (1) stronger edge losses for both frequencies in this high density L-mode condition, so that the edge losses dominate the weaker 8th harmonic absorption and only reduce the central 4th harmonic absorption; (2) losses that are stronger for the higher harmonic, e.g. edge-localized parametric decay such as is observed in a similar regime in NSTX (Ref. [2]); (3) the simple linear theory may be inadequate in this situation. In support of the third possibility, it should be noted that a much more sophisticated nonlinear model, based on the combination of the TORIC and ORBIT-RF codes appears to predict almost no absorption for the 8th harmonic in this experimental condition [11]. Understanding the differences between the results predicted by this model and those of linear theory is the topic of ongoing work.

SUMMARY AND CONCLUSIONS

We have presented a new theoretical form for the linear ion cyclotron harmonic damping which, though it is equivalent to existing formulations (e.g. [4]), has physical meaning more easily assigned to each term. The effects on the cyclotron absorption due to ray propagation in a toroidal equilibrium have been evaluated and found to be quite significant. The theory predicts that absorption at high ion cyclotron harmonics can be important in discharges at high beta with substantial fast ion populations. Strong 4th and 5th harmonic absorption has been observed in DIII-D L-mode experiments with 60 MHz FW power, with direct evidence of the rf-driven beam acceleration obtained with the Doppler-shifted D_α charge exchange recombination diagnostic. Contrary to the theoretical expectations, absorption at 7th and 8th harmonics appears to be weak in high density L-mode discharges with 116 MHz fast waves. Resolving this discrepancy is the aim of ongoing study; the subject is relevant not only to experiments combining neutral beams and fast waves, but also to future burning plasma experiments that use FWCD, where parasitic damping on the alpha particles may be an issue.

ACKNOWLEDGMENT

Work supported by the U.S. Department of Energy under Cooperative Agreement DE-FC02-04ER54698.

REFERENCES

1. J. Myra, these proceedings.
2. J.R. Wilson, et al., these proceedings.
3. C.C. Petty, et al., in *Radio Frequency Power in Plasmas (Proc. 12th Topical Conf., Savannah, GA, 1997)* (AIP, New York, 1997) p. 225.
4. T.H. Stix, *Waves in Plasmas* (AIP, New York, 1992).
5. M. Porkolab, in *Advances in Plasma Physics (Thomas H. Stix Symposium, 1992)* (AIP, New York, 1994) p. 99.
6. R.J. Dumont, C.K. Phillips, and D.N. Smithe, in *Radio Frequency Power in Plasmas (Proc. 15th Top. Conf., Moran, Wyoming, 2003)*, (AIP, New York, 2003) p. 439.
7. C.C. Petty, et al., Plasma Phys. Control. Fusion **43**, 1747 (2001).
8. R. Prater, et al., these proceedings.
9. W.W. Heidbrink, et al., Nucl. Fusion **39**, 1369 (1999).
10. W.W. Heidbrink, et al., Plasma Phys. Control. Fusion **46**, 1855 (2004).
11. M. Choi, et al., these proceedings.

Electron Energy Confinement For HHFW Heating and Current Drive Phasing on NSTX

J. C. Hosea[1], S. Bernabei[1], T. Biewer[1], B. LeBlanc[1], C. K. Phillips[1],
J. R. Wilson[1], D. Stutman[2], P. Ryan[3], D. W. Swain[3]

[1]*Princeton Plasma Physics Laboratory, Princeton, NJ USA*
[2]*Johns Hopkins University, Baltimore, MD*
[3]*Oak Ridge National Laboratory, Oak Ridge, TN*

Abstract. Thomson scattering laser pulses are synchronized relative to modulated HHFW power to permit evaluation of the electron energy confinement time during and following HHFW pulses for both heating and current drive antenna phasing. Profile changes resulting from instabilities require that the total electron stored energy, evaluated by integrating the midplane electron pressure Pe(R) over the magnetic surfaces prescribed by EFIT analysis, be used to derive the electron energy confinement time. Core confinement is reduced during a sawtooth instability but, although the electron energy is distributed outward by the sawtooth, the bulk electron energy confinement time is essentially unaffected. The radial deposition of energy into the electrons is noticeably more peaked for current drive phasing (longer wavelength excitation) relative to that for heating phasing (shorter wavelength excitation) as is expected theoretically. However, the power delivered to the core plasma is reduced considerably for the current drive phasing, indicating that surface/peripheral damping processes play a more important role for this case.

Keywords: RF Heating, Electron Energy and Confinement Time, Spherical Torus
PACS: 52.50.Qt, 52.55.Hc

INTRODUCTION

HHFW heating should occur via electron heating for ion temperatures less than ~ 2 keV [1] for the experiments considered here so that the incremental electron energy divided by the electron energy confinement time can serve as a good indicator of the RF power that is deposited in the bulk of the plasma in this case ($\Delta P_{RFB} = \Delta W_e/\tau_{We}$). With the introduction of a second Thomson scattering laser on NSTX [2] it is now possible to measure τ_{We} and W_e by synchronizing the two laser pulses relative to the modulated HHFW RF power. This then permits the comparison of bulk RF power deposition for different antenna phasing to determine if phasing (i.e., launched spectrum) affects the efficiency of the power reaching the core plasma. The equation governing this technique is

$$W_e = W_0 - [W_0 - W_F] \times [1 - \exp(-t/\tau_{We})] \qquad (1)$$

where W_0 is the energy at the starting point of the rise or fall of energy (t = 0), and W_F is the final energy that would be reached after several confinement times. By placing one of the laser pulses near the end of the RF pulse on and off periods, and placing the other laser pulse in between, three measurements are obtained from which W_0, W_F and τ_{We} can be derived in principal for each on and each off period.

ELECTRON PRESSURE RESPONSE FOR HEATING AND CO-CURRENT DRIVE PHASING — $k_\phi = 14$ m^{-1} AND – 7 m^{-1}

Discharge conditions were selected for providing a near constant density condition over the time of interest of HHFW modulation and the $P_e(r = 0)$ time traces were observed with the lasers synched with the RF pulses as indicated in Fig.1. Initial

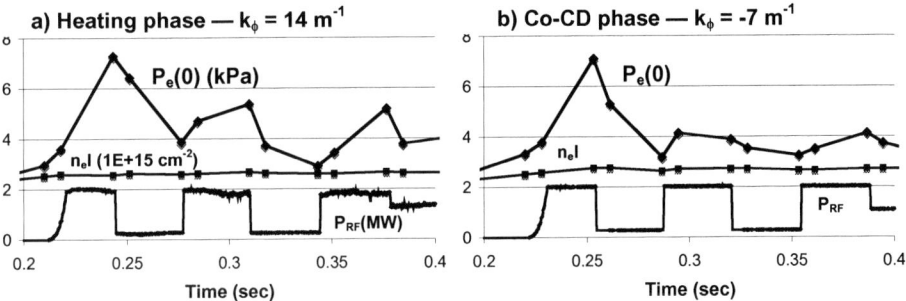

Figure 1. $P_e(r = 0)$ versus time for a) $k_\phi = 14$ m^{-1} (Shot 112699) and b) $k_\phi = -7$ m^{-1} (Shot 112705). Gas = He, I_P = 0.6 MA, B_T = 0.45 T for both cases.

attempts to apply Eq. 1 to the three measured $P_e(0)$ values for each rise or decay period for the RF pulses were not generally successful (e.g. note that $\tau_{Pe0} \approx \infty$ after the first RF pulse in Fig. 1a and is undefined during the second RF pulse of Fig. 1b) due to fluctuations of $P_e(0)$ caused by MHD instabilities and/or radial displacements.

The P_e radial profiles at the ends of the 1st and second 2nd pulses are given in Fig 2

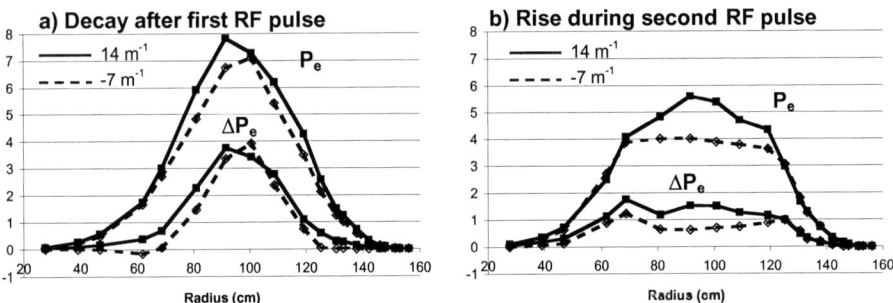

Figure 2. P_e versus radius at end of RF pulse and ΔP_e during a) the decay after the first RF pulse and b) the rise during second RF pulse. Discharge conditions as in Fig. 1.

along with the ΔP_e values over the associated decay and rise periods. For the first decay period the profiles are peaked and the ΔP_e profiles indicate that the RF deposition profile is somewhat narrower for the smaller k_ϕ as expected theoretically [3]. Minor profile perturbations can change $P_e(0)$ (the point near 100 cm) in Fig 2a and a sawtooth instability hollows out the pressure profile in Fig. 2b for the $k_\phi = -7$ m^{-1} case (the laser time is only 0.2 ms after the crash). To compensate for profile changes, the total electron energy contained in the plasma can be calculated versus time in order to determine values of τ_{We}.

ELECTRON STORED ENERGY AND ESTIMATED RF POWER DEPOSITION FOR $k_\phi = 14$ m^{-1} AND -7 m^{-1}

In order to determine $W_e(t)$, the Thomson scattering electron pressure $P_e(r, t)$ measurements taken on the midplane of the plasma are integrated over the EFIT magnetic surface defined volumes [4]. Generally, the $P_e(r)$ profile on the midplane does not exactly match the $P_{Total}(r)$ profile obtained with EFIT (the EFIT profile is usually somewhat broader and shifted somewhat outward in major radius) so the inner and outer (from R_o) values of $P_e(r)$ are each integrated over the EFIT defined volumes and the resulting W_e values are then averaged. $W_e(t)$ values thus obtained are given in Fig. 3 and compared with the corresponding $W_{EF}(t)$ values. A flattening of W_e during

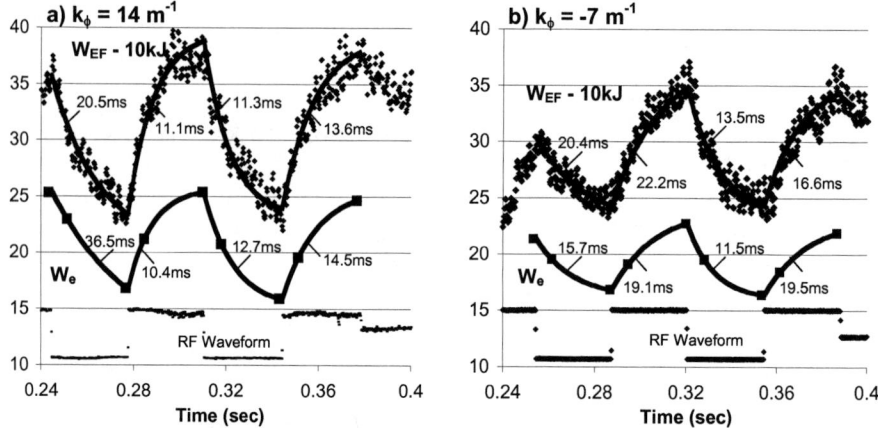

Figure 3. Electron stored energy W_e and total EFIT stored energy W_{EF} versus time for a) $k_\phi = 14$ m^{-1} and b) $k_\phi = -7$ m^{-1}.

the second RF pulse for the -7 m^{-1} case is not observed as it was for $P_e(0)$ (see Fig. 1) and τ_{We} can be calculated straightforwardly. The resulting τ_{We} values for the W_e are qualitatively similar to the corresponding τ_{WEF} values for W_{EF}.

An estimate of the core power deposition, ΔP_{RFD}, to the electrons to produce the observed W_e values during the RF pulses can be obtained from $\Delta W_{eF}/\tau_{We}$ where ΔW_{eF} is the difference in final W_{eF} values (Eq. 1) with and without the RF pulse, and similarly from $\Delta W_{EFF}/\tau_{WEF}$ for the EFIT total stored energy. Table 1 summarizes the power estimates for the second and third RF pulses. These estimates indicate that the

TABLE 1. Estimate of power delivered to the bulk plasma from W_e and W_{EF}.

	ΔW (kJ) $\Delta W_{eF}/\Delta W_{EFF}$	τ (ms) τ_{We}/τ_{WEF}	ΔP_{RFD} (MW) $\Delta P_{RFDe}/\Delta P_{RFDEF}$	$\eta = \Delta P_{RFD}/\Delta P_{RF}$ η_e/η_{EF}
14 m^{-1} 2nd pulse	15.1/19.4	10.4/11.1	1.45/1.75	0.84/1.01
14 m^{-1} 3rd pulse	10.6/16.0	12.7/13.6	0.834/1.18	0.48/0.68
-7 m^{-1} 2nd pulse	7.9/15.1	19.1/22.2	0.413/0.680	0.24/0.39
-7 m^{-1} 3rd pulse	7.2/12.6	11.5/16.6	0.626/0.759	0.36/0.44

RF power reaching the electrons is on average about 3/4th that going to the bulk ($\eta_e \div \eta_{EF}$), and that the delivered power to the electrons is substantially less than that launched from the antenna, especially in the -7 m^{-1} co-current drive case.

CONCLUSIONS

The electron energy confinement time obtained from an integration of $P_e(r)$ over the EFIT magnetic surface defined volumes tracks reasonably well the total energy confinement time obtained from EFIT analysis for both the heating 14 m^{-1} and co-current drive -7 m^{-1} cases. However, considerable RF power does not reach the core of the plasma, especially in the longer wavelength -7 m^{-1} case. Many processes are possibly contributing to this "surface" power loss – surface wave excitation, RF sheath dissipation, and parametric decay wave excitation to name a few. The presence of decay waves was detected in these cases and edge power loss, attributable to helium ion heating via the Bernstein wave, was determined from analysis of ERD (edge rotation diagnostic) measurements to be several hundred kilowatts and to increase with wavelength (16%, 23% of P_{RF} loss for 14 m^{-1}, -7 m^{-1}, respectively) [5,6]. The dramatic difference in apparent surface power loss between the two phasing cases considered here, suggests that accurate modeling of these cases should help to resolve the dominant loss mechanism(s) at play.

ACKNOWLEDGMENTS

This work is supported by DOE Contract No. DE-AC02-76CH03073.

REFERENCES

1. Ono, M., "High harmonic fast waves in high beta plasmas", *Physics of Plasmas* **2**, 4075 (1995)
2. LeBlanc, B.P., et al., *Rev. of Scientific Instum.* **74**, 1659 (2003)
3. Phillips, C.K., et al., *Bulletin of APS* **49**(8), Savanna, Ga, Nov. 2004, p.
4. 223Sabbagh, S.A., et al., *Nuclear Fusion* **41**, 1601 (2001
5. Biewer, T. M., et al., *Bulletin of APS* **49**(8), Savanna, Ga, Nov. 2004, p324; to be published in Physics of Plasmas 2005.
6. Wilson, J.R., et al., This conference.

Investigation of HHFW and NBI Combined Heating in NSTX*

B.P. LeBlanc[a], R.E. Bell[a], S. Bernabei[a], T.M. Biewer[b] J.C. Hosea[a], J.R. Wilson[a]

[a]*Princeton Plasma Physics Laboratory, NJ;* [b]*Plasma Science and Fusion Center, MIT, MA*

Abstract. A series of experiments was conducted to investigate the combined utilization of HHFW and NBI auxiliary heating in NSTX plasmas. A modest increase of the total stored energy coincident with a near doubling of the neutron production rate is observed when NBI heating is added to HHFW in L-mode plasmas. An increase in the core electron temperature is also observed. On the other hand, essentially no stored energy augmentation nor neutron production rate enhancement is observed when applying HHFW during the "H" phase of NBI driven H-mode plasmas. Spectroscopic measurements of the edge carbon line radiation indicate an unpredicted ion temperature increase, suggesting that edge effects are reducing the amount of HHFW power reaching the plasma core.

Keywords: NSTX HHFW NBI
PACS: 52.50 Qt; 52.50 Gj

INTRODUCTION

High-Harmonic Fast-Wave (HHFW) constitutes an integral component of the NSTX auxiliary heating program, where it complements neutral beam injection (NBI). With a principal goal of current drive[1], HHFW has already demonstrated electron heating when applied to ohmic plasmas[2] and can also induce H-mode operation[3] Details of the 30-MHz rf system can be found elsewhere[4]. In the following, we describe experimental attempts at combining HHFW and NBI heating, done with the goal of expanding NSTX's operational envelope. The deuterium plasmas discussed here have a toroidal field of 0.44 T at the geometric center and a plasma current of 0.8 MA with a flattop starting at 0.2 s. The outer gap is 3-4 cm and the beam energy has been reduced to 70 keV (from typical 80-90 keV) to prevent strong plasma-antenna interaction. The antenna phasing creates launch spectra with $k_{//} = 14$ m^{-1} or $k_{//} = 7$ m^{-1}.

HHFW IN NBI L-MODE PLASMAS

Figure 1 shows temporal overlays of plasma parameters for two discharges: one using combined HHFW and NBI heating is displayed with solid lines; the other using NBI only is shown with dashed lines. The neutron production rate, S_n, and the heating waveforms are shown with arbitrary scaling. The H_α traces are also shown for reference. A power ramp initiates the HHFW pulse, reaching 2.9-MW flattop approximately 0.1 s before the NBI onset, which has a flattop power of 1.4 MW; the

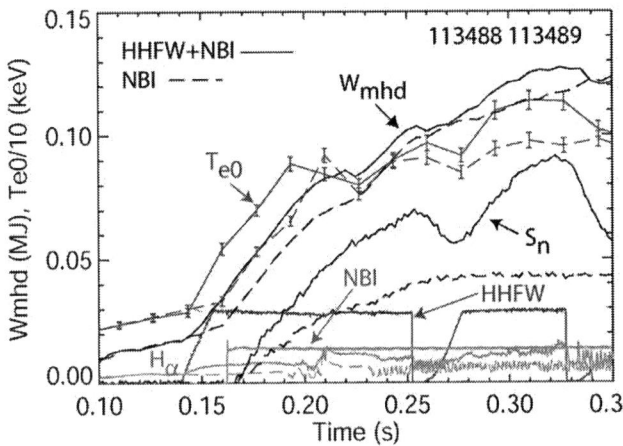

Figure 1. Plasma parameter T_{e0}, W_{mhd}, S_n, H_α and heating powers HHFW, NBI time evolutions: *solid lines*, discharge with combined HHFW and NBI; *dashed lines*, reference NBI-only discharge. $k_\parallel = 14$ m^{-1}.

reference NBI-only discharge has the same NBI. Both the central electron temperature, T_{e0}, and the stored energy W_{mhd} respond to HHFW heating prior to NBI. T_e is obtained from Thomson scattering[5] and W_{mhd} from EFIT[6] equilibrium calculations. The NBI-only plasma has a similar rate of rise for T_{e0} and W_{mhd} except for being slightly delayed. There is a higher neutron production when HHFW is combined with NBI: S_n increases by a factor ≈ 2 compared to the NBI-only case. Keeping in mind that the neutron production is dominated by "beam target" nuclear reactions for these "low" temperature plasmas, the near doubling of S_n indicates an interaction between the HHFW and the fast ions of NBI origin[7]. Observation of the H_α traces reveals that both discharges dither into H mode at $t ≈ 0.2$ s, before entering longer lasting H phases at $t ≈ 0.24$ s.

Figure 2 shows a comparison between the T_e profiles for these two discharges at $t = 0.193$ s, slightly before the T_{e0} increase saturation. While the profiles overlay well in the edge regions, the core region shows a marked increase for the HHFW heated plasma, with T_{e0} being 0.2 keV above the reference profile. It generates a T_e profile with internal-transport-barrier like ∇T_e as seen at $R ≈ 55$ cm and $R ≈ 135$ cm.

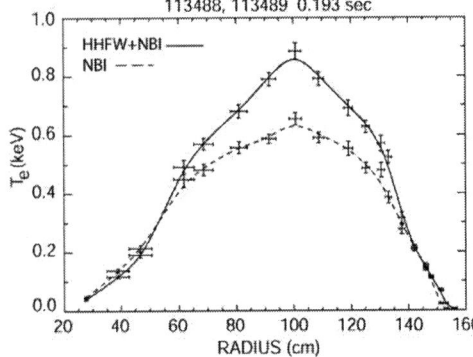

FIGURE 2. T_e profile overlay at $t = 0.31$ s for combined HHFW and NBI, *solid line*; NBI heating, *dashed line*. $k_\parallel = 14$ m^{-1}.

HHFW IN NBI H-MODE PLASMAS

Figure 3 shows temporal overlays similar to Fig. 1, but for cases where HHFW power is applied after an H-mode transition produced by NBI. $k_\parallel = 7$ m^{-1}. The early NBI heating makes use of two sources to ensure H-mode access. HHFW starts at 0.24 s after the H-mode transition time of ≈ 0.2 s. The maximum NBI and HHFW

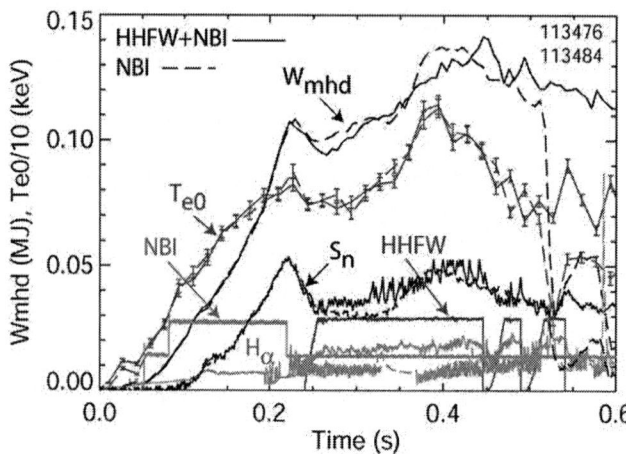

FIGURE 3. Plasma parameter T_{e0}, W_{mhd}, S_n, H_α and heating powers HHFW, NBI time evolutions: *solid lines*, discharge with HHFW applied to a NBI driven H-mode; *dashed lines*, reference NBI-driven H-mode discharge. $k_\parallel = 7m^{-1}$.

powers are ≈ 3.0 MW. One can see that, in contrast to the previous case, parameters of this discharge are essentially not modified by the application of HHFW. The T_e, W_{mhd} and S_n overlay well, suggesting that the HHFW power does not reach the main plasma column. In particular, S_n, which appeared to be a sensitive indicator of HHWF core penetration in the previous case, does not show significant increase, except for times 0.26-0.35 s, when S_n is higher for the HHFW heated discharge.

We can see in Fig. 4 an overlay of T_e profiles taken at $t = 0.310$ s roughly in the middle of the enhanced neutron production interval. No significant electron heating is observed. While a small amount of HHFW power appears to briefly reach the plasma core causing a modest neutron signal enhancement, no bulk electron heating occurs. Comparisons with other discharges suggest that edge effects, *e.g.* fueling, might be responsible for this small neutron production enhancement.

But edge measurements show HHFW does heat ions. One can see in Fig. 5 a time evolution of T_e and T_i measured at major radius $R \approx 145$ cm. T_i is obtained by edge spectroscopy on the carbon impurity[8]. Similarly solid lines correspond to HHFW combined with NBI, while dashed lines correspond to NBI-only reference plasma. The ion temperature shown here corresponds to a "poloidal" sightline, and is higher than that from a "toroidal" sightline (not shown) during HHFW operation. Parametric decay of the pump wave into IBW has been suggested as a means by which power is delivered to the edge ions. More details can be found elsewhere[9].

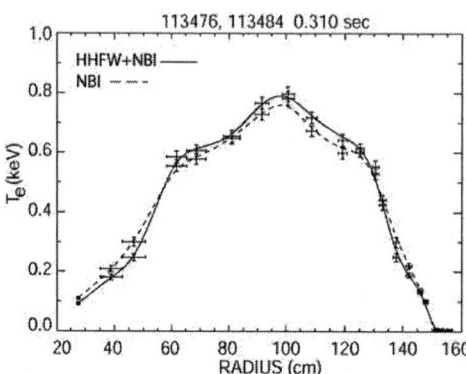

FIGURE 4. T_e profile overlay at $t = 0.31$ s: *solid line*, combined HHFW and NBI, heating; *dashed line*, NBI heating. $k_\parallel = 7m^{-1}$.

CONCLUSIONS

The combination of HHFW and NBI presents challenges. Depending on

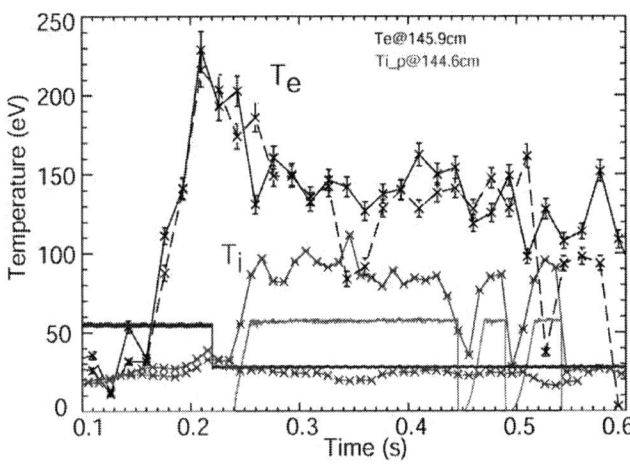

FIGURE 5. T_e and T_i time evolutions at $R \approx 145$ cm: *solid lines*, combined HHFW and NBI heating; *dashed lines*, NBI heating. $k_{//} = 7\text{m}^{-1}$.

condition, a modest or no increase of W_{mhd} is observed although the HHFW power is greater or equal to that of NBI. This result can be partly explained by edge parasitic rf power absorption through parametric decay and the generation of IBW, which deposit power into the edge ions. Based on helium discharges, a few tens of percent of HHFW incident power can be diverted to the low-confinement peripheral region. The presence of NBI generated fast ions creates further hurdles: Fast ions are accelerated by HHFW which, while increasing the neutron production, reduces the power available for direct HHFW electron heating and potential current drive. Furthermore the high NBI energy in relation to the magnetic field strength necessitates special care in order to reduce plasma-antenna interaction. More work is needed before a complete understanding of the physics involved can be obtained, and a strategy is developed to improve on the performance of these plasmas.

ACKNOWLEDGMENTS

This work is supported by U.S. DOE contract DE-AC02-76CH03073.

REFERENCES

[1] P.M. Ryan et al., 15th Topical Conference on Radio Frequency Power in Plasmas, AIP Proc. 694
[2] B.P. LeBlanc et al., Nuclear Fusion **44** (2004) 513
[3] B.P. LeBlanc el al., 15th RF Conference, AIP Conference Proceedings 694, p.201
[4] J.R. Wilson et al., Phys. Plasmas, Vol. 10, No. 5, (2003) 1733-1738
[5] B.P. LeBlanc et al, Rev. Sci. Instrum. **74** (2003) 1659
[6] S. A. Sabbagh, et al., Nuclear Fusion 41 (2001) 1601.
[7] A. L. Rosenberg, et al., Phys. Plasmas **11**, 2441 (2004)
[8] T.M. Biewer et al., Rev. Sci. Instrum. **75** (2004) 650-654
[9] T.M. Biewer el al., PPPL-4027, Dec. 2004 and J.R. Wilson at this conference

Development of RF Tools and Scenarios for ITER on JET

J.-M. Noterdaeme[1,2], M. Mantsinen[3a], V. Bobkov[1], A. Ekedahl[4], L.-G. Eriksson[4], P.U.Lamalle[5], A. Lyssoivan[5], J. Mailloux[6], M.-L. Mayoral[6], F. Meo[7], I. Monakhov[6], K. Rantamaki[3b], A. Salmi[3a], M. Santala[3a], S. Sharapov[6], D. Van Eester[5], JET-EFDA Task Force H and JET-EFDA Contributors[*]

[1]*Max Planck Institute for Plasmaphysics, EURATOM Association, D-85748 Garching, Germany,* [2]*Gent University, EESA Department, B-9000 Gent, Belgium,* [3]*Euratom Association-Tekes,*[a] *HUT,*[b]*VTT, Helskini, Finland,* [4]*Association Euratom-CEA, CEA, Cadarache, France,* [5]*Association Euratom-Belgian State, LPP, Trilateral Euregio Cluster, EMR-KM,S Brussels, Belgium,* [6]*Euratom-UKAEA Fusion Association, Culham, UK,* [7]*Association Euratom-Denmark, Riso, Denmark* *See the Appendix of J.Pamela et al., Fusion Energy 2004 (Proc. 20th Int. Conf. Vilamoura, 2004) IAEA, Vienna (2004)

Abstract. The improvement of LH coupling with local puffing of D_2 gas, which made operation at ITER relevant distances (10 cm) and with ELMs a reality, has been extended to ITER- like plasma shapes with higher triangularity. With ICRF, we developed tools such as (1) localized direct electron heating using the ^{3}He mode conversion scenario for electron heat transport studies, (2) the production of ^{4}He ions with energies in the MeV range by 3 ω_c acceleration of beam injected ions at 120 keV to investigate Alfven instabilities and test α diagnostics, (3) the stabilisation and destabilisation of sawteeth and (4) ICRF as as a wall conditioning. Several ITER relevant scenarios were tested. The (^{3}He)H minority heating scenario, considered for the non-activated start-up phase of ITER, produces at very low concentration energetic ^{3}He which heat the electrons indirectly. For $n_{3He}/n_e > 2\%$, the scenario transforms to a mode conversion scenario where the electrons are heated directly. The (D)H minority heating is not accessible as the concentration of C^{6+} dominates the wave propagation and always leads to mode conversion. The minority heating of T in D is very effective heating for ions and producing neutrons. New results were obtained in several areas of ICRF physics. Experimental evidence confirmed the theoretical prediction that, as the larmor radius increases beyond 0.5 times the perpendicular wavelength of the wave, the second harmonic acceleration of the ions decreases to very small levels. An exotic fusion reaction (pT) must be taken into account when evaluating neutron rates. The contribution of fast particles accelerated by ICRF to the plasma rotation was clearly identified, but it is only part of an underlying, and not yet understood, co-current plasma rotation. Progress was made in the physics of ELMs while their effect on the ICRF coupling could be minimized with the conjugate-T matching scheme. The addition of 3 dB couplers is a step in increasing the power capability of the ICRF heating on JET in ELMy plasmas. The installation of an ITER-like, ELM resilient antenna in 2006 will further improve this and be a test of the ICRF scheme for ITER.

Keywords: tokamak, LH, ICRF, ICRH, heating, coupling, plasma production, rotation
PACS: 52.55.Fa, 52.50.Qt, 52.35.Hr

INTRODUCTION

The focal points of the Task Force Heating (TF H) at JET are plasma heating, current drive and plasma rotation as well as the optimization of the systems. The very intense 2003-2004 campaigns included reversed I_p, B_T operation, the use of tritium in trace amounts, and of H and He as majority gas. We emphasized the development of tools and scenarios for progress towards ITER, and collected some interesting physics results only obtainable in JET.

LH COUPLING

The coupling of the LH can be improved by using CD_4 as additional gas. Up to 3 MW was coupled at distances of up to 10 cm. However, the possible use of CD_4 in ITER raised concerns about T co-deposition. D_2 as an additional gas had not been successful. Prior to the 2003 campaign, the pipe which extends poloidally along the launcher and through which the gas is delivered, was modified. The top holes, whose poloidal location was near the top of the launcher, but magnetically above the field lines connected to the launcher, were plugged. Since, D_2 can be used as effectively as CD_4. Fig. 1 compares two pulses, one with CD_4 and one with D_2 gas injection. The CD_4 gas flow rate was 12×10^{21} el/s; the D_2 gas flow rate was smaller by 1/3 (8×10^{21} el/s). The discharge with CD_4 injection had the last closed flux surface (LCFS) at 9 cm and a coupled LH power of 2.5 MW; the one with D_2 injection, the LCFS at 10 to 10.5 cm and a power of 3 MW. The ELMs were smaller in the discharge with D_2 (due to the gas puffing), but the plasma performance of both discharges was similar [1] [2, 3]. LH was now also coupled to high triangularity ($\delta \sim 0.4$-05) ITB plasmas. D_2 injection (4×10^{21} el/s) to improve the LH coupling was used together with strong D_2 and Ne puffing for ELM control. The plasma-launcher distance was 7 cm and the coupled LH power 2 MW (in combination with 18 MW of NBI and 2.5 MW of ICRF)[4, 5].

Bright spots have been observed on the divertor apron during LH operation. They are due to the impact of fast particles, created in front of the LH grill mouth and travelling along magnetic field lines. The heat load increases with LH power, edge density and temperature but decreases when the LH grill is operated at a larger distance[6, 7]. This is beneficial for ITER, where the distance between LCFS and LH grill will be large.

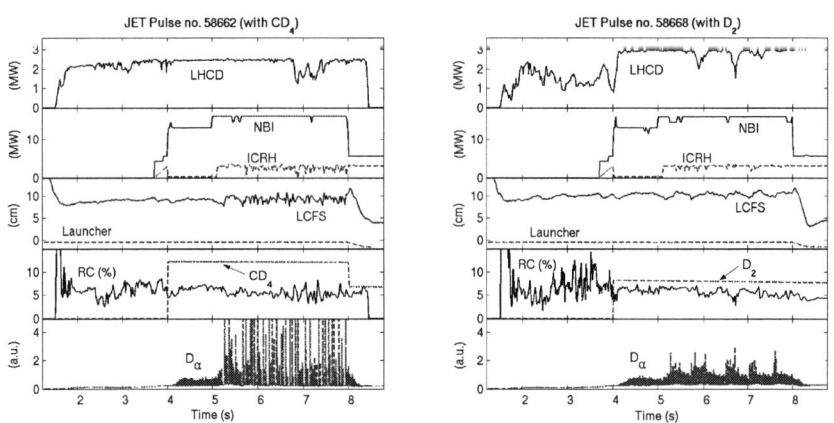

FIGURE 1. Comparison of CD_4 and D_2 injection to improve the LH coupling

ICRF TOOLS

Four major tools were developed.

One is ^{3}He-D mode conversion, which provides in JET a very localized e-heating method[8]. It was used extensively for the investigation of electron heat transport [9, 10] and also pushed the ITB discharges on JET, in combination with minority heating to some of the highest performance (T_i = 27 keV, T_e = 13 keV)[9]. With the second tool MeV energy ^{4}He ions can be produced by acceleration at 3 ω_c of beam injected ^{4}He (at 120 kV)[11]. In a recent extension of the third tool, where Ion Cyclotron Current Drive (ICCD) is used to influence sawteeth[12, 13], we showed for the first time that it is possible to destabilize long sawteeth initially stabilized by fast ions. The stabilization was obtained by central H minority heating, the destabilization by the application of off-axis ICCD. This has potential application for the avoidance of neoclassical tearing modes. The fourth tool, whose development was only started, is the application of ICRF to condition a machine in the presence of a permanent magnetic field[14]. We discuss here only in some more detail the second and fourth topics.

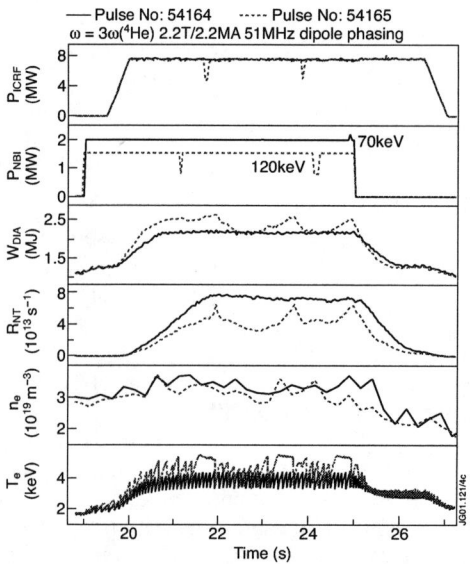

FIGURE 2. Acceleration of beam injected ^{4}He ions (at 70 keV and 120 keV) with 3 ω_c.

Production of ^{4}He at MeV Energy without Activation

Figure 2 compares two discharges with injection of ^{4}He ions at 70 keV and 120 keV (at 2 MW and 1.5 MW respectively). An ICRF power of 8 MW at 3 ω_c of ^{4}He was applied. The discharge with the 120 keV shows stabilization of the sawteeth, an indication of a high fast particle pressure in the center. Gamma ray diagnostics [15] [16]confirm that a steady-state distribution of ^{4}He ions with

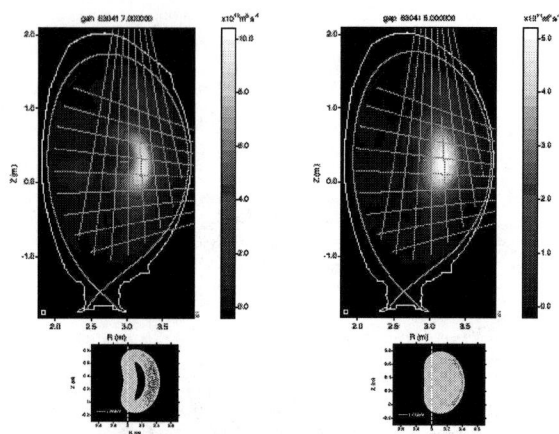

FIGURE 3. Gamma ray emissivity showing the location of the fast ^{4}He particles for two shapes of the q profile (monotonic-left, and reversed q-right) and comparison with the calculated orbits with those q- profiles

energies above 1 MeV have been achieved. The calculated fraction of fast particle (n_H/n_e) is 10^{-3}, close to the record DT discharge with 4×10^{-3}, but with a neutron rate lower by 4 orders of magnitude. This method thus lends itself perfectly, both on JET and on ITER to test α-diagnostics[17].

The technique was used to investigate Alven eigenmodes produced by the fast particles and can provide, together with γ-ray images, qualitative information on the confinement of fast particles for different types of q-profiles. Figure 3 shows the difference in location of the γ-ray emission for fast ^{4}He in monotonic and reversed q-profiles. The emission are in agreement with the calculated orbits of ^{4}He ions (E=1.9 MeV), taking into account the q-profile[17].

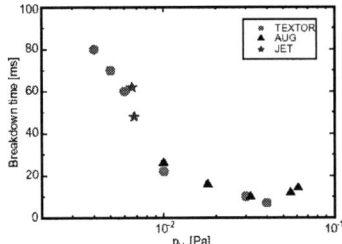

FIGURE 4. Gas breakdown time as a function of He pressure for several machines

ICRF for Vessel Conditioning in the Presence of a Magnetic Field

Conditioning of the vessel walls using glow discharge, as is usually done, will be prevented in superconducting machines by the presence of the magnetic field. After similar experiments in TEXTOR, Tore Supra and ASDEX Upgrade, ICRF was used on JET to produce an RF discharge[14]. The ^{4}He gas pressure was in the 7×10^{-3} Pa range. The gas breakdown time (time between application of the power and appearance of H_- light) was 50 to 60 ms. Since similar values were obtained in smaller machines at similar antenna voltages (8-12 keV), frequencies (30-34 MHz) and at the same pressures (see Fig. 4) the breakdown time is independent of machines size. The radial distribution of the plasma density (measured through multi-channel far infrared interferometry) could be varied by adding H to the He. This provides a way to scan the position of highest density for example across a divertor. High energy fluxes of H and D atoms with energy $E_H < 60$ keV and $E_D < 25$ keV were observed. The partial pressure of the masses 2, 3, 4, 6, 18, 20 all increase for about 100 s by an order of magnitude after an RF pulse of 245 kW for 4 s.

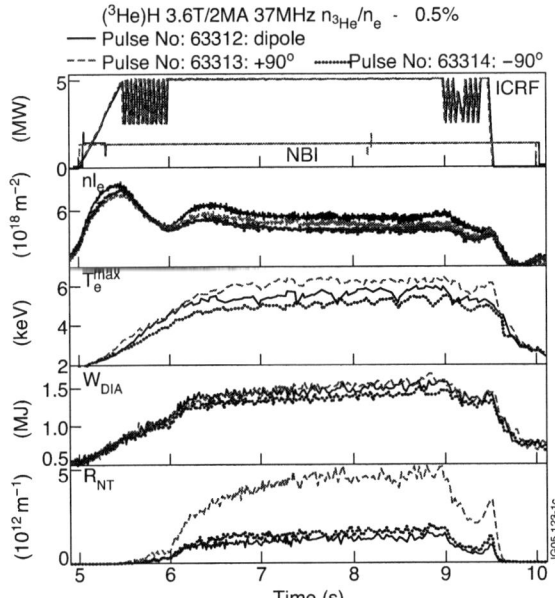

FIGURE 5 (^{3}He)H minority scenario with different phasings

ICRF SCENARIOS

The flexibility of the JET ICRF system, the possibility to use tritium, and the good confinement of energetic particles are key capabilities of JET. Campaigns where H was used as majority gas, or T in trace amounts provided the opportunity to investigate ITER relevant scenarios.

^{3}He minority in H, D minority in H, and T minority in D are three scenarios of a somewhat unusual type, called "heavy minority scenario", since the minority has a larger M/Z ratio than the majority. In this case, the fast wave, launched from the low field side, first encounters the mode conversion, before encountering the cyclotron resonance of the minority.

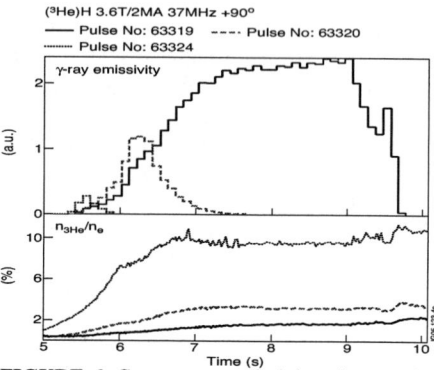

FIGURE 6. Gamma ray emissivity, disappearing as the ^{3}He concentration raises above 2 %

^{3}He minority regime in H

The frequency for ^{3}He minority in H is the same as for the main ITER scenario (2 ω_c of T in D) and can be used in ITER both to test the ICRF system at its intended frequency and to heat the plasma in the non-active phase. In order to explore the ^{3}He minority regime and its transition to the mode conversion regime, a precise control of the ^{3}He concentration is useful. The feedback control of the ^{3}He concentration had been developed at high concentration values to investigate the mode conversion regime used for e- heating and heat transport studies (mentioned above). An extension of the feedback control allowed the required precise control of very low ^{3}He concentration. Figure 5 shows three discharges with similar density ($n_{e0}=3 \times 10^{19}$ m^{-3}), ^{3}He concentration (< 1 %), and power (NBI 1 MW, ICRF 5 MW), but with different phasing. The higher electron temperature as well as the higher neutron rates with the +90° phasing (wave propagates co-current) are clear indication of the pinch effect, by which the fast particles produced by the ICRF are transported towards the center. Higher tail energies, with tails of up to 0.3 MeV, are obtained. As the ^{3}He concentration is increased, the tail temperature decreases, and at 2 %, a very abrupt transition to the mode conversion regime occurs. This is shown in Fig. 6 where the γ-ray emissivity (produced by fast ^{3}He reactions on ^{9}Be and ^{12}C) disappears when the ^{3}He concentration is above 2%[18, 19].

D minority in H

D minority in H has also been investigated but was shown not to be accessible because the presence of C in the machine. The C^{6+} ion has the same Z/M as D. Its presence is, for the wave propagation, as an additional sixfold concentration of D pushing the scenario directly into the mode conversion regime[19].

Heating of Low Concentrations of T

Both minority heating of T and second harmonic heating of trace T was investigated. Minority heating of T is a quite challenging scenario in JET as it requires operation at the highest magnetic fields (B_T = 3.9 – 4T) and at the lowest RF frequency (23 MHz), where the available power is low (1.5 MW) and the coupling is poor. With T concentrations of up to 3 %, energetic tails of 80 to 120 keV were obtained, a very efficient energy range of neutron production and ion heating[18].

Second harmonic of tritium, which is the main ICRF scenario on ITER was attempted but further investigation will require campaigns where the T concentration is higher.

ICRF PHYSICS

Second harmonic heating

Second harmonic heating is a finite larmor radius effect (FLR). If the magnitude of wave field is constant over the motion of a particle along its cyclotron motion, then, because the field will accelerate and decelerate the ion along its orbit, no net energy can be transferred between wave and ion. If the diameter of the cyclotron motion (2ρ) of the ion is approximately equal to half the wavelength of the wave, the wave amplitude will have changed sign along the second half of the ion orbit and the energy transfer is maximized. The energy transfer goes again to zero, if the wavelength is approximately equal to the diameter of the cyclotron motion. The net transfer of energy thus stops for a particular value of $k_{perp}\rho$. This affects strongly the distribution function of the fast particles and was investigated by varying kperp through its dependence on density. Figure 7 shows the measured distribution functions at high and low density and at high and low power. Whereas the discharge with high power has more energetic ions than the one at low power, the shape of the distribution function is the same. On the other hand, when the ratio $k_{perp}\rho$ was modified by changing the density, the distribution function, and in particular the energy at which the distribution function drops rapidly (no more acceleration of the ions) changed markedly. The shapes were shown to be in agreement with theory, when finite larmor radius effects [20] are taken into account.

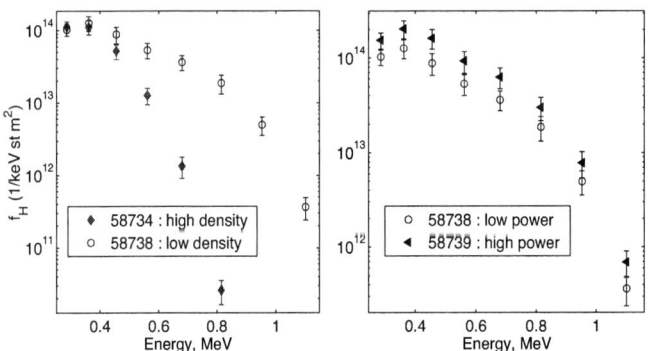

FIGURE 7. Perpendicular proton energy distribution for two densities (left) and two power levels (right)

ELMs

ELMs cause a strong variation of the plasma density in front of the antenna, and can be a source of significant problems both for matching and voltage stand-off[21]. By using a 3 dB coupler, the first problem can be alleviated[22]. Two of the four antennas on JET will be connected in this way in 2005. An alternative method is to use the conjugate matching scheme: in a proof of principle experiment, the connection of

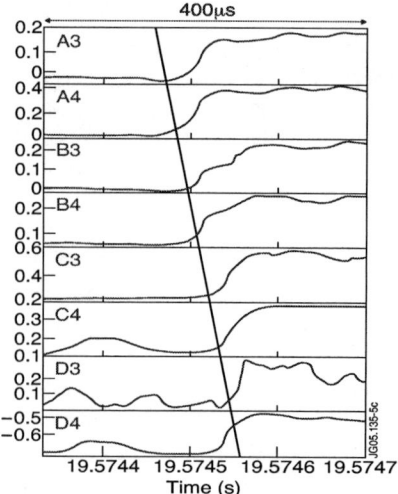

FIGURE 8. Start of change in antenna coupling due to an ELM for the different straps of the antennas. Antennas are labeled A, B, C ,D. Antenna A and C are 180° apart. The straps within an antenna numbered 1, 2, 3, 4. Only 3 and 4 are shown.

two straps of one antenna were reconfigured. With those straps the power to the plasma was little affected by the ELMs[23]. In 2006, the scheme will be extended to the other two antennas.

The variation of the coupling with the ELM arrival at the antenna provides some interesting information on the evolution of an ELM. The time delay of this variation between the toroidally spaced antennas gives information about the toroidal development of an ELM. Figure 8 shows how an ELM appears successively in front of the different antennas, takes about 100 μs for a full toroidal rotation, and rotates in the counter-current direction[24].

pT fusion

During the trace T experiment, the endothermic fusion reaction T+ p -> n + 3H + Q with Q= -764 keV was investigated. It can act a source of neutrons which must be taken into account in the analysis of DD and DT neutrons[25].

Plasma Rotation

Tokamak plasmas show a significant toroidal rotation even with only a small momentum input. A systematic analysis of plasma rotation with ICRF was done on JET[26, 27]. Recent results have identified the component predicted by theories that rely on fast particle effects for the rotation drive. Plasmas with waves injected in the co- and counter-current direction show a different rotation profile. While the difference in the profile is in agreement with theory, its magnitude is moderate and comes on top of a still unexplained dominant co-current rotation[28].

ACKNOWLEDGMENTS

This work was performed under the European Fusion Development Agreement. It is a pleasure to acknowledge the extraordinary dedication of all that make it possible to get interesting work done on this machine.

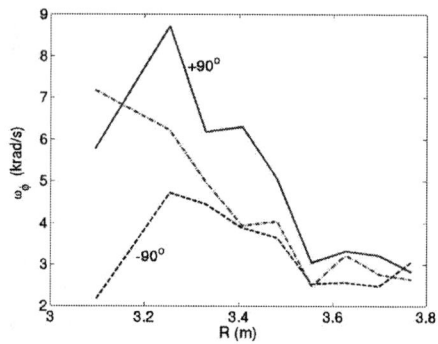

FIGURE 9. Toroidal rotation for three discharges: ICRF with +90° phasing, LH and ICRF with −90° phasing. The difference in rotation between co- and counter-current phasing is in agreement with theory, the discharge with LH is for control.

REFERENCES

1. A. Ekedahl et al., "Long distance coupling of lower hybrid waves in ITER relevant edge conditions in JET reversed shear plasmas", in *RF Power in Plasmas,* (Moran, Wyoming, USA, 2003), Vol. 694, (C.B. Forest ed.), AIP (May 19-21, 2003) 227-234.
2. A. Ekedahl, "Demonstration of ITER Relevant LHCD Operation: large distance coupling in JET and long pulse operation in Tore Supra", in *31th EPS Conf. on Controlled Fusion and Plasma Phys.,* (London, 2004), Vol. ECA Vol. 28G, EPS O1-01.
3. A. Ekedahl, "Long Distance Coupling of Lower Hybrid Waves in JET Plasmas with Edge and Core Transport Barriers", *Nucl.Fusion* (2005), accepted for publication.
4. J. Mailloux et al., "ITER Relevant Coupling of Lower Hybrid Waves in JET", in *20th IAEA Fusion Energy Conference,* (Vilamoura, 2004), Vol. IAEA-CSP-25/CD, IAEA EX/P4-28.
5. J. Mailloux, "Lower Hybrid waves in ITER relevant operational scenarios in JET", in *16th RF Power to Plasmas,* (Park City, 2005), AIP, this conference.
6. K.M. Rantamaki et al., "Hot Spots generated by LH Waves on JET", in *30 th Conf. on Controlled Fusion and Plasma Physics,* (St. Petersburg, 2003), Vol. ECA 27A, (R. Koch et al. ed.), EPS P-1.190.
7. K.M. Rantamaki et al., "Bright spots generated by lower hybrid waves on JET", *Plasma Phys.and Controlled Fusion* (2005), accepted for publication.
8. M. Mantsinen et al., "Localized bulk electron heating with ICRF mode conversion in the JET tokamak", *Nucl. Fusion* **44** (2004) 33-36.
9. D. Van Eester et al., "Recent 3He radio frequency heating experiments on JET", in *RF Power in Plasmas,* (Moran, Wyoming, USA, 2003), Vol. 694, AIP (2003) 66-73.
10. P. Mantica, "Power Modulation experiments in JET ITB Plasmas", in *31 EPS Conference on Controlled Fusion and Plasma Physics,* (London, 2004), Vol. ECA Vol. 28G, EPS P-1.154.
11. M. Mantsinen et al., "Alpha Tail Production with Ion Cyclotron Resonance Heating of He4 Beam Ions in JET Plasmas", *Phys. Rev. Letters* **88** (2002) 105002-1-105002-4.
12. M.-L. Mayoral et al., "Neo-classical Tearing mode control through sawtooth destabilisation in JET", in *Controlled Fusion and Plasma Physics,* (Montreux (CH), 2002), Vol. ECA 26B, EPS (2002) P-1.026.
13. M. Mantsinen et al., "Analysis of ion cyclotron heating and current drive at $\omega=2\ \omega_{cH}$ for sawtooth control in JET plasmas", *Plasma Phys. Control. Fusion* **44** (2002) 1521-1542.
14. A. Lyssoivan et al., "Studies of ICRF Discharge Conditioning (ICRF-DC) on ASDEX Upgrade, JET and TEXTOR", in *16th Conference on RF Power to Plasmas,* (Park City, 2005), Vol. AIP Conf. Proceedings, AIP, this conference.
15. V. Kiptily et al., "γ–ray diagnostics of energetic ions in JET", *Nucl.Fusion* **42** (2002) 999.
16. V. Kiptily et al., "Gamma-ray Imaging of D and 4He ions accelerated by ion-cyclotron-resonance heating in JET Plasmas", *Nucl.Fus.* (2005) accepted for publication.
17. S. Sharapov, "Experimental studies of instabilities and confinement of Energetic Particles on JET and on MAST", in *20th IAEA Fusion Energy Conf.,* (Vilamoura, 2004), Vol. IAEA-CSP-25/CD, IAEA EX/5-2Ra.
18. P.U. Lamalle et al., "Expanding the operating space of ICRF on JET with a view to ITER", in *20th IAEA Fusion Energy Conference,* (Vilamoura, 2004), Vol. IAEA-CSP-25/CD, IAEA EX/P4-26 and to appear in Nuclear Fusion.
19. M.L. Mayoral, "ICRF Heating for the non-activated phase of ITER: from inverted minority to mode conversion regime", in *16th Conf. on RF Power to Plasmas,* (Park City, 2005), AIP Conference Proceedings AIP, this conference.
20. A. Salmi, "JET Experiments to assess finite Larmor radius effects on resonant ion energy distribution during ICRF heating", in *31 EPS Conference on Plasma Physics,* (London, 2004), Vol. ECA 28G, EPS P-5.167.
21. J.-M. Noterdaeme et al., "Status and development of the ICRF antennas on ASDEX Upgrade", in *15th Radio Frequency Power in Plasmas,* (Moran, 2003), Aip Conference Proceedings Vol. 694, 154-157.
22. J.-M. Noterdaeme, "Matching to ELMy Plasmas", in *22nd Symposium on Fusion Technology,* (Venice, 2004), to appear in Fusion Eng. Des.
23. I. Monakhov et al., "Test of load-tolerant external conjugate-T matching system for A2 ICRF antenna at JET", in *22nd Symposium on Fusion Technology,* (Venice, 2004), to appear in Fusion Eng. Des.
24. V. Bobkov et al., "Studies of ELM toroidal asymmetry using ICRF antennas at JET and ASDEX Upgrade", in *31st EPS Conf. on Plasma Phys.,* (London, 2004), Vol. ECA Vo. 28G, EPS P-1.141.
25. M. Santala, "pT fusion by RF-heated protons in JET trace tritium discharges", in *31 EPS Conference on Controlled Fusion and Plasma Physics,* (London, 2004), Vol. ECA 28G, EPS P-5.163.
26. J.-M. Noterdaeme et al., "Spatially resolved toroidal plasma rotation with ICRF on JET", *Nuclear Fusion* **43** (2003) 274-289.
27. L.-G. Eriksson et al., "Bulk plasma rotation in the presence of waves in the ion cyclotron rance of frequencies", in *15 Top Conf on RF Power in Plasmas,* (Moran, Wyoming, USA, 2003), Vol. 694, (C.B. Forest ed.), AIP (2003) 41-49.
28. L.-G. Eriksson et al., "Plasma rotation induced by directed waves in the ion cyclotron range of frequencies", *Phys. Rev. Letters* **92** (2004) 235001-1 to 4.

Long Pulse Plasma Heating Experiment by Ion Cyclotron Heating in LHD

T. Seki, T. Mutoh, R, Kumazawa, K. Saito, T. Watari, Y. Nakamura,
M. Sakamoto[1], T. Watanabe, S. Kubo, T. Shimozuma, Y. Yoshimura,
H. Igami, K. Ohkubo, Y. Takeiri, Y. Oka, K. Tsumori, M. Osakabe,
K. Ikeda, K. Nagaoka, O. Kaneko, J. Miyazawa, S. Morita, K. Narihara,
M. Shoji, S. Masuzaki, M. Goto, T. Morisaki, B.J. Peterson, K. Sato,
T. Tokuzawa, N. Ashikawa, K. Nishimura, H. Funaba, H. Chikaraishi,
N. Takeuchi[2], T. Notake[2], H. Ogawa[3], Y. Torii[4], F. Shimpo, G, Nomura,
M. Yokota, C. Takahashi, A. Kato, Y. Takase[5], H. Kasahara[5],
M. Ichimura[6], H. Higaki[6], Y.P. Zhao[7], J.G. Kwak[8], H. Yamada,
K. Kawahata, N. Ohyabu, K. Ida, Y. Nagayama, N. Noda, A. Komori,
S. Sudo, O. Motojima and LHD Experimental Group

National Institute for Fusion Science, Toki 509-5292, Japan
[1]*Kyushu University, Kasuga 816-8580, Japan*
[2]*Nagoya University, Faculty of Engineering, Nagoya 464-8601, Japan*
[3]*Graduate University for Advanced Studies, Hayama 240-0162, Japan*
[4]*Kyoto University, Institute of Advanced Energy, Uji 611-0011, Japan*
[5]*University of Tokyo, Tokyo, Japan*
[6]*University of Tsukuba, Tsukuba, Japan*
[7]*Institute of Plasma Physics, Academia Sinica, Hefei 230031, P.R. China*
[8]*Korea Atomic Energy Research Institute, Daejeon 305-600, Korea Rep.*

Abstract. It is very important to demonstrate the ability to sustain the plasma in a steady state on the Large Helical Device (LHD), which has external helical magnetic coils and is a super-conducting device. The long pulse discharge experiment was carried out using the ion cyclotron range of frequencies (ICRF) heating mainly. The plasma discharge of 31 minutes and 45 seconds was achieved by a total injected heating energy of 1.3GJ. Swing of the magnetic axis to scatter the local heat load on the divertor plate was one of the key methods for the steady state operation. The repetitive hydrogen pellet injection was tried successfully to fuel the minority hydrogen ions for long pulse operation.

Keywords: ICRF heating, LHD, steady state operation

INTRODUCTION

Feature of the helical device is that the no plasma current is necessary to confine the plasma. The Large Helical Device (LHD), which has super-conducting helical windings, has a potential for a steady state operation [1]. One of the main missions of the LHD experiment is to demonstrate the ability to sustain the plasma for a long time.

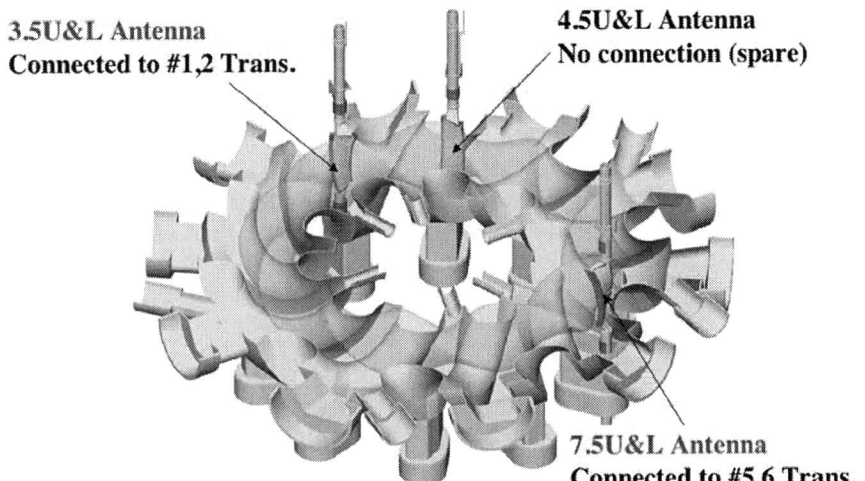

FIGURE 1. Layout of the ICRF antennas in LHD.

Ion Cyclotron Range of Frequencies (ICRF) heating has been carried out successfully in LHD [2-6]. It started using short-pulse of a few seconds and the pulse length was extended with improvement of LHD and the ICRF system for the steady state operation [7-9]. The discharge length of 150 seconds was achieved in 2002. The duration time was limited by the uncontrollable density increase caused by influx of the hydrogen. Then, some of the divertor graphite plates, which is thought one of the source of outgas, is replaced by the superior one. The steady state experiment was selected as one of the important study for the 2004 experimental campaign.

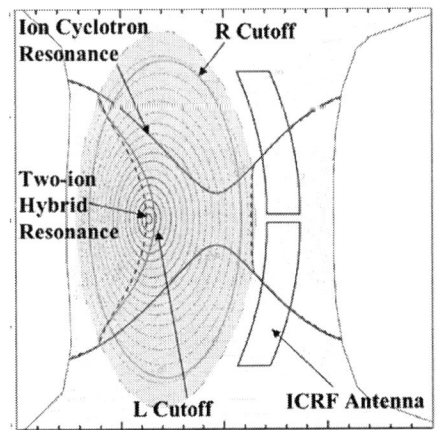

FIGURE 2. Positions of resonance and cutoff layers in a poloidal cross-section on LHD.

EXPERIMENTAL SETUP

The LHD is the largest helical device and the nominal major and minor radius are 3.9 m and 0.6 m, respectively and the magnetic field strength is 3 T. It has an electron cyclotron heating (ECH) system and a neutral beam injection (NBI) heating system adding to ICRF heating system. The ECH system can inject an RF power of around 100 kW for steady state at a frequency of 84 GHz. The NBI system can inject a power of several hundred kW for a pulse length of several tens to hundred seconds at a beam-energy of around 110 keV.

For the ICRF system, the number of the steady state RF transmitter was increased and four transmitters can be operated simultaneously. Their operation tests have been done using the dummy load at a power and pulse length of 0.5 MW / 30 minutes or 1 hour and 0.25 MW / 1 hour. Each transmitter is connected to the four loop antennas of the six installed antennas in LHD as shown in Fig.1. The antennas are installed from the upper and lower port and located at the outboard side of the torus. They are movable in the radial direction by 15 cm. The experiments were carried out in the conditions of the minority heating, which was the best in the short pulse ICRF heating experiments. The position of the ion cyclotron layers is shown in Fig.2. The helium plasma with the minority hydrogen ions is used. The wave frequency is 38.47 MHz, the magnetic axis position is 3.6m and the magnetic field at the axis is 2.75T.

EXPERIMENTAL RESULTS

Key Factors for Steady State Discharge

For the steady state operation, many things are improved in this experimental campaign. Before the experiment, some of the divertor plate was replaced improved one, which was expected that outgas from the divertor graphite tile was extremely reduced. The temperature measurement of the ICRF antenna is newly adapted. The

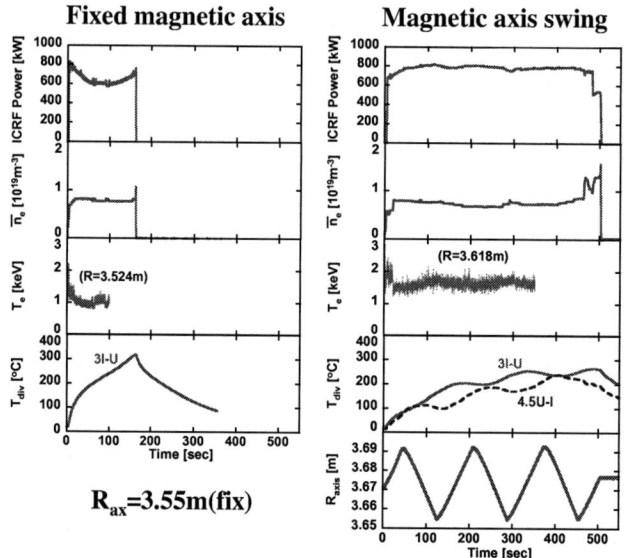

FIGURE 3. Comparison between the fixed magnetic axis (left figures) and magnetic axis swing (right figures). ICRF power, line-averaged electron density, electron temperature, and temperature at divertor plate are shown. The position of magnetic axis is also shown in the axis swing case.

temperature increase is more than 100 °C in an ICRF power of 250 kW and a pulse length of 5 seconds at an antenna-plasma gap of 7 cm, which is the antenna position used for the former experiments. The temperature rise is 10 °C at a gap of 12 cm. The plasma loading resistance is 3 ohms at the antenna position. Then, the antenna position is set to an antenna-plasma gap of 12 cm. The automatic impedance matching is adapted to the liquid stub tuner system [10-11]. The reflected power increased gradually during the ICRF long pulse operation. The net injected ICRF power decreased and a reflection interlock terminated the RF pulse when the ratio of the reflected power to the forward power exceeded 20 %. The increased reflected power was suppressed by the feedback control of the liquid surface level during the long pulse operation.

In the plasma experiment, swing of the magnetic axis position is important method. Figure 3 compares the case of the fixed axis and the axis swing. When the axis is fixed, the temperature at the divertor plate continues to increase as shown in the bottom of the left figures. 3I-U is the name of the position of divertor plate, upper part of the 3I port. When the magnetic axis is swung, the temperature rise at the 3I-U plate is reduced at the outward axis position as shown in right figures of Fig.3. The temperature rise at the other divertor plate, 4.5U-I (inner part of 4.5U port), shows the inversed trend. Then, the heat load at the divertor plate is scattered by the magnetic axis swing [12-13].

Longest Plasma Discharge

The plasma discharge more than 30 minutes was achieved by integration of the key factors. The time histories of the plasma parameters of the longest discharge are shown in Fig.4. The plasma duration time is 31 minutes and 45 seconds. The plasma

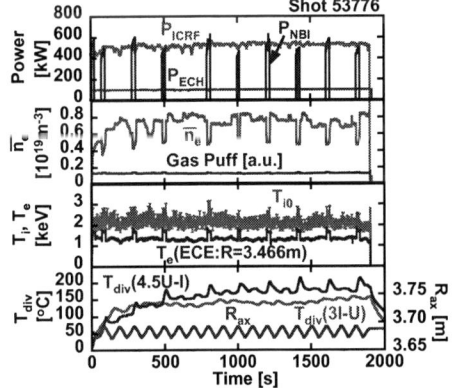

FIGURE 4. Time history of the plasma parameters of the longest plasma discharge. Heating power, line-averaged electron density, ion and electron temperatures, temperatures at the divertor plates, and position of the magnetic axis are shown.

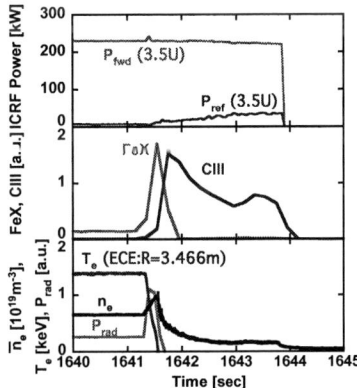

FIGURE 5. Time evolution of the plasma parameters when the discharge was terminated uncontrollably. Forward and reflected ICRF power, intensity of FeX and CIII signals, line-averaged electron density, electron temperature, and radiation loss are shown.

is sustained by an ICRF power of about 0.52 MW mainly. An ECH power of around 100 kW and an averaged NBI power of about 0.06 MW are also used for the support of the discharge. The total injected heating energy is 1.29 GJ. The line averaged electron density was about $0.8 \times 10^{19} \mathrm{m}^{-3}$ and the central ion temperature was around 2 keV. The position of the magnetic axis was swung 18.5 times between 3.67 m and 3.7 m. The electron density was controlled to keep constant with a helium gas puffing feedback. The measured divertor temperature was less than 250 °C. The temperatures at the feedthrough and the transmission line at the maximum current reached 50 and 30 °C at the end of the discharge. They were tolerable level in our system.

Behavior of End of Discharge

Figure 5 shows the example of the time history, which the discharge was terminated uncontrollably. At first, the intensity of FeX signal increases rapidly and rise of the electron density and the radiation loss and decrease of the electron temperature are follows. The intensity of CIII signal also increases by the lower electron temperature. After that, the ICRF power was turned-off manually judging from the plasma state on the real time monitor. This indicates that the iron impurities are flowed in the plasma. The abnormal termination is often in connection with the spark, which is seen on the real time plasma monitor. The spark may be caused by the high-energy ions generated by the ICRF heating.

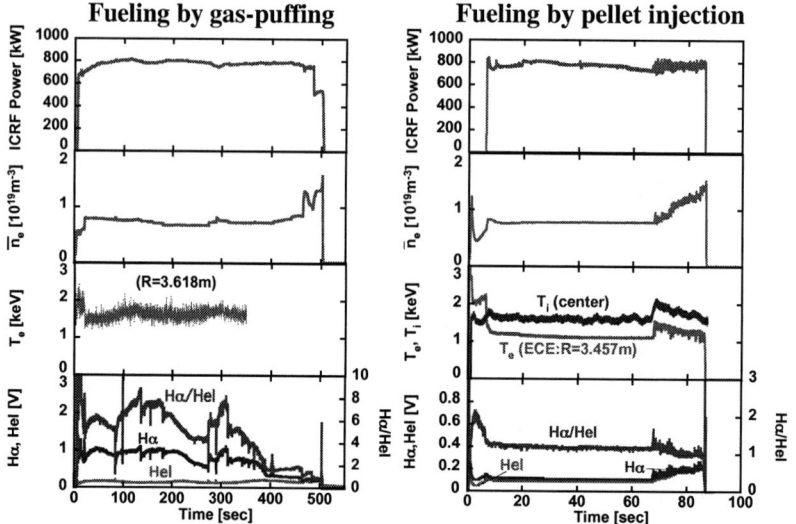

FIGURE 6. Comparison of the time evolution of the plasma parameters between the gas-puffing (left figures) and pellet injection (right figures). ICRF power, line-averaged electron density, ion and electron temperatures, intensity of Hα and HeI signals, and ratio of Hα to HeI are shown. Gas-puff injects a helium gas continuously to keep the density constant. Repetitive injection of a hydrogen ice pellet starts at 67.46 second.

DISCUSSION

Particle Fueling during Long Pulse Discharge

During the long pulse operation, the density feedback control was utilized by the helium gas puffing. The hydrogen minority ions flowed in the plasma from the vacuum vessel wall. To keep lower level the minority ratio, no active fueling of hydrogen ions was used for the ICRF heating experiment. Then, the ratio of the hydrogen minority ions to the helium majority ions was gradually decreased as shown in the left bottom figure of the Fig.6. This will cause the problem to the ICRF heating, when the pulse length is extended furthermore. As a method of the particle fueling for the long pulse operation by the ICRF heating, the repetitive hydrogen pellet injection was tried (Fig.6 right figures). Relatively small size of the hydrogen pellet (2.5 mmϕ x 2.5 mmϕ) was injected at a frequency of 5 Hz. The electron density increased gradually and the minority ratio increased slightly at the early stages of the injection. The ion and electron temperatures increased also. The plasma confinement may be improved and the ICRF heating still looks like working during the pellet injection. The time-axis is extended in Fig.7 (a). The plasma loading resistance responds to the each pellet injection, where the minority ratio increases rapidly. Figure 7(b) shows the electron density profile as a function of the normalized plasma minor radius. The pellet penetrates into the plasma peripheral region because the size of pellet is small. However, the electron density at inner region increases and the profile becomes center-peaked during the pellet injection.

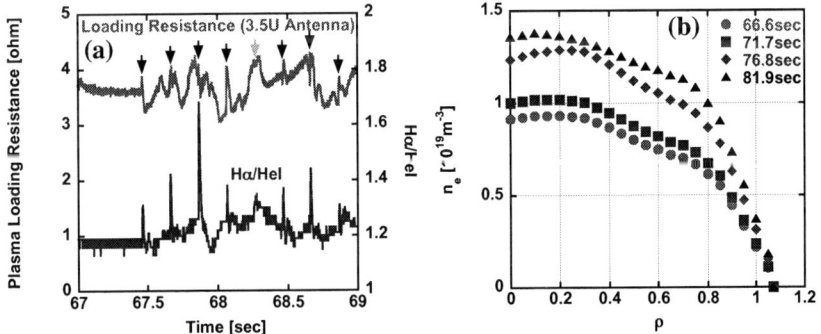

FIGURE 7. (a) Response of the plasma loading resistance and Hα/HeI to the pellet injection. (b) Evolution of profile of the electron density before (66.6 sec.) and during the pellet injection as a function of a normalized plasma minor radius.

The power absorption by the ICRF heating before and during the pellet injection was calculated using the three-dimensional full wave code, TASK/WM [14]. Before the pellet injection, it is assumed that the central electron density is 1.0×10^{19} m^{-3} and the density profile is $1-\rho^4$, where ρ is the normalized minor radius and the minority ratio is 10 %. During the pellet injection, it is assumed that the central electron density is 1.3×10^{19} m^{-3} and the density profile is $1-\rho^3$ and the minority ratio is 15 %. The result is shown in Fig.8. The wave power is absorbed by the hydrogen ions near the cyclotron resonance layer at the plasma peripheral region. The profile of the wave absorption is almost same between before and during the pellet injection. The absorbed power increases during the pellet injection at all the toroidal angles. The total absorbed power increases double by the pellet injection in this calculation. The higher density and minority ratio result in the more hydrogen absorption. In the former experiments, the plasma discharge by the ICRF heating was terminated by the pellet injection using the larger size of the pellet. The repetitive pellet injection with the optimized pellet size is one of the promising tools as a particle fueling for the long pulse ICRF heating.

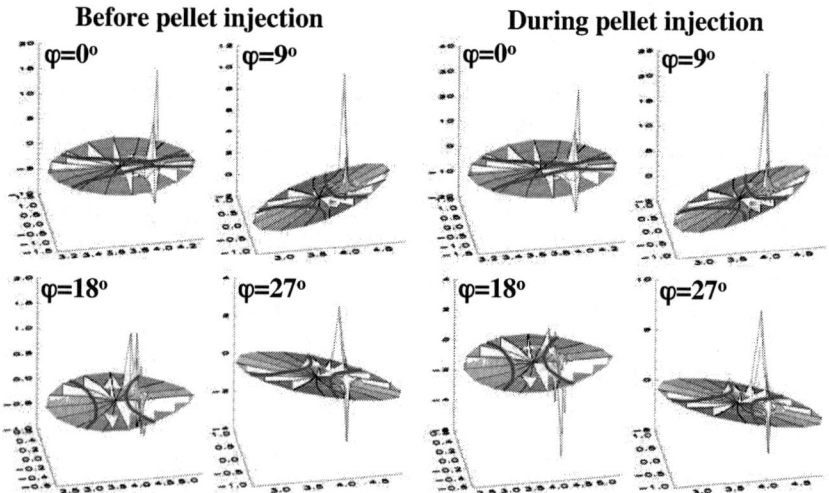

FIGURE 8. Calculated absorption power before and during the pellet injection using the three-dimensional full wave code. Profile of absorbed power on the poloidal cross section is shown at the four different toroidal angles of 0, 9, 18, 27 degree.

CONCLUSIONS

Long pulse plasma discharge more than 30 minutes was achieved in LHD using the ICRF heating mainly. The line-averaged electron density was 0.8×10^{19} m^{-3} and the central ion temperature was 2 keV. The total input heating energy reached 1.3 GJ. Some improvement was carried out for the ICRF system. The antenna position was determined by the strengthened temperature measurement of the antenna. The reflected ICRF power was reduced during the power injection by changing the liquid surface level of the liquid stub tuner. The temperature rise of the divertor plate was suppressed by the swing of the magnetic axis during the plasma operation. The discharge was terminated uncontrollably by influx of an iron impurity. It is important for the further long pulse operation to investigate the source of the metal impurities and prevent influx of them. The repetitive hydrogen pellet injection is a promising method as a particle fueling for the steady state ICRF heating.

ACKNOWLEDGMENTS

The authors would like to thank all the scientist and technical staff at the National Institute for Fusion Science for their support during this work.

REFERENCES

1. Motojima, O., et al., Nucl. Fusion **43**, 1674 (2003).
2. Mutoh, T., et al., Physical Review Letter **85**, 4530 (2000).
3. Kumazawa, R., et al., Physics of Plasmas **8**, 2139 (2001).
4. Watari, T., et al., Nucl. Fusion **41**, 325 (2001).
5. Saito, K., et al., Nucl. Fusion **41**, 1021 (2001).
6. Seki, T., et al., J. Plasma and Fusion Research, SERIES **5**, 478 (2002).
7. Kumazawa, R., et al., J. Plasma and Fusion Research, SERIES **3**, 352 (2000).
8. Saito, K., et al., 15th Top. Conf. (2003), AIP Conf. Proc. **694**, 58.
9. Mutoh, T., et al., Nucl. Fusion **43**, 738 (2003).
10. Kumazawa, R., et al., Review of Scientific Instruments **70**, 555 (1999).
11. Saito, K., et al., Review of Scientific Instruments **72**, 2105 (2001).
12. Masuzaki, S., et al., Nucl. Fusion **42**, 750 (2002).
13. Morisaki, T., et al., Contrib. Plasma Phys. **42**, 321 (2002).
14. Fukuyama, A., et al., IAEA Conference, Sorrento, IAEA-CN77/THP2/26 (2000).

Modeling of Discharges with Fast Wave Power in DIII–D

R. Prater,[1] M. Choi,[1] R.W. Harvey,[2] J.C. Hosea,[3] C.C. Petty,[1] R.I. Pinsker,[1] M. Porkolab,[4] and A.P. Smirnov[5]

[1]*General Atomics, San Diego, California USA*
[2]*CompX, Del Mar, California, USA*
[3]*Princeton Plasma Physics Laboratory, Princeton, New Jersey, USA*
[4]*Massachusetts Institute of Technology, Cambridge, Massachusetts, USA*
[5]*Moscow State University, Russia*

Abstract. In order to facilitate the planning and analysis of experiments in which fast wave power is applied to discharges in DIII–D, a code has been developed to run conveniently a ray tracing code (GENRAY) and/or a full wave code (TORIC) for experimental or planned conditions. The code gathers information about the kinetic profiles (T_e, T_i, n_e, Z_{eff}) and the neutral beam injection and calls the ONETWO transport code to obtain the density and equivalent Maxwellian temperature profiles of the energetic beam ions. Using all these data and the wave frequency and toroidal and poloidal spectra characteristic of any DIII–D antenna, GENRAY or TORIC can be run. For ray tracing some self-consistency checks on the absorption and current drive physics may be made using the output files, and comparisons with analytic expressions have been made. A goal of this program is the development and validation against experiment of computational models for FWCD which will be suitable for use on ITER.

Heating and current drive by fast magnetosonic waves (FWs) are in use on many of the present generations of tokamaks and are projected for use on the International Thermonuclear Experimental Reactor (ITER). A careful validation of the theory of FWs is needed in order to provide guidance for applications of FW power in present applications and for projection from present experiments to future scenarios, and this is therefore an objective of the FW program worldwide.

The DIII–D experiments use a combination of three FW systems. One system is set to 60 MHz, and the four-strap antenna with $\pi/2$ phasing has an equivalent toroidal wavelength of 90 cm. The other two systems operate near 116 MHz and have identical antennas with equivalent toroidal wavelength of 74 cm. Previous experiments have shown that the peak of the $n_\parallel$ spectrum is 4.7 and 3.1 for these respective systems when the effect of coupling through the plasma edge is included.

Many computer codes have been developed for modeling the fast wave interactions with plasma. Key physics issues are the coupling of the antenna to the plasma, what electric fields are generated in the plasma, and how the constituent components of the plasma react to the fields. Full wave codes use Fourier expansions to express the electric fields in the plasma cross-section, while ray tracing codes treat the waves as non-interacting rays propagating through the plasma. Ray tracing misses important physics effects like interference and diffraction and, additionally, a rigorous method of determining the starting locations and angles of the rays in the plasma is an unsolved problem. However, the ray tracing process is very useful for understanding some aspects of the physics, and in many cases ray tracing is significantly faster than full wave analysis.

For experiments on DIII–D it was found desirable to have a convenient way to run ray tracing and full wave computer codes for planning and analysis of experiments. To this end, a code was written in the IDL language which first obtains the kinetic profiles — n_e, T_e, T_i, and Z_{eff} — from the MDSplus data storage system for any equilibrium along with the neutral beam power, beam energy, and beam geometry. With these data, the ONETWO transport code is called to calculate the profiles of beam ion density n_{hot} and beam energy density W_{hot}. An equivalent Maxwellian temperature profile for the beam ions is estimated as $\langle T_{hot}\rangle = 2W_{hot}/3n_{hot}$, typically 20 to 25 keV for the 80 keV beams used in DIII–D. ONETWO also calculates the impurity density given the Z_{eff} profile, and a minority ion concentration, usually a trace 1–2% of H in D plasmas with D beams. Other information input by the user is the choice of which transmitter/antenna system should be evaluated, and on this basis the frequency and antenna geometry data are input. For purposes of planning experiments, the IDL code allows the user to scale any of these data to any new values to evaluate the effects of alternative scenarios.

Two fast wave codes may be called: the ray tracing code GENRAY [1] and the full wave code TORIC [2]. These codes use a large number of input parameters and switches which are input through a namelist file. Our approach here is to start with a file containing default inputs and to modify a few values to use the kinetic profiles and rf system information described above and to modify a few other values input by the user if desired. For the GENRAY code, the $k_\parallel$ spectrum is automatically set to peak at the $n_\parallel$ values given above. Lacking a clear model for the poloidal spectrum, the poloidal component of the peak of the spectrum is set to zero. In this paper, only the GENRAY calculations will be discussed. GENRAY has numerous options for the dispersion relation and absorption model. In this work, the cold plasma model is used for determining the indices of refraction and the wave polarization, and the Chiu model [3] is used for absorption.

Previous studies of FWCD in DIII–D [4] were performed in low density L–mode discharges in order to maximize the current drive. Use of power at higher frequency (typically the eighth harmonic at 116 MHz) was found more effective than at lower frequency (60 MHz and fourth harmonic) due to reduced parasitic damping by the hot D beam ions. Application of FWCD to AT H–mode discharges, which are much higher in both electron and hot ion densities, introduces new physics. The calculation of absorption by ions involves modified Bessel functions $I_p(\lambda)$, where p is the order, $\lambda = 1/2(k_\perp \rho_i)^2$, ρ_i is the ion Larmor radius, and $k_\perp \approx \omega/V_A$ where V_A is the Alfvén speed. High density makes V_A small and high harmonic number makes ω large, so the argument of the Bessel function can be as large as 10 to 15 in these experiments at 116 MHz, as opposed to the usual limit of small λ. For such large values of λ, another conventional limit is violated, namely that the effect on absorption by ions of the terms due to E_- cannot be neglected in comparison to the E_+ terms, and in fact they may dominate [5].

The projection onto a poloidal plane of the trajectory from GENRAY of the central ray propagating in a high performance equilibrium is shown in Fig. 1(a) for the 60 MHz case and in Fig. 2(a) for the 116 MHz case. The width of the ray is proportional to the normalized absorption rate $(1/P)dP/ds$, where P is the power remaining in the ray and s is the arc length. In the 60 MHz case the ray experiences

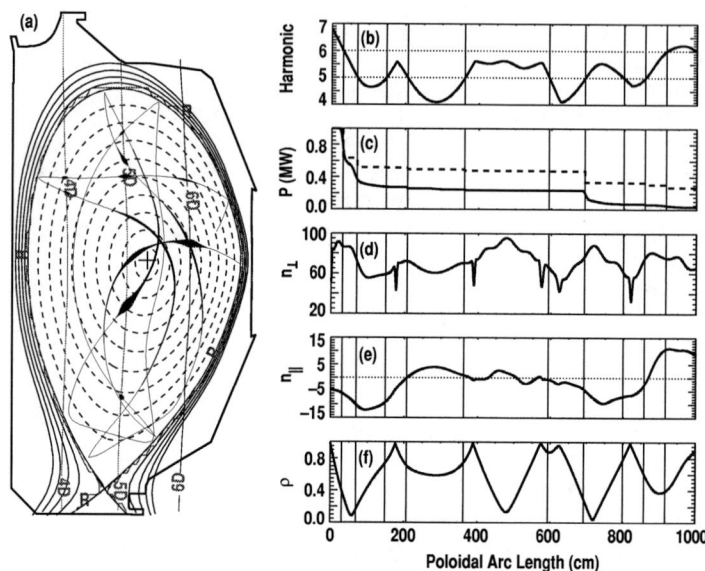

FIGURE 1. (a) Projection of a GENRAY ray onto a poloidal cross-section of a typical high performance H–mode plasma. The width of the ray is proportional to $(1/P)dP/ds$, and the cyclotron harmonics for deuterium are shown. The frequency is 60 MHz, $n_e(0) = 5.2\times10^{13}\,\mathrm{cm}^{-3}$, $n_{hot}(0) = 0.39\times10^{13}\,\mathrm{cm}^{-3}$, $T_e = 2.9$ keV, $T_{hot} = 25.1$ keV, $T_i = 4.22$ keV, and $Z_{eff} = 1.75$. The D plasma has 1.5% minority H component and the C density is adjusted to give the Z_{eff}. (b) Harmonic ℓ of the D ions, (c) power remaining in the ray which starts as 1 MW and dashed line representing the Porkolab-Pinsker formula for power lost crossing a harmonic due to the hot ions, the indices of refraction (d) $n_\perp$ and (e) $n_\parallel$, and (f) the normalized minor radius, all plotted as a function of poloidal arc length. Harmonic crossings are vertical lines.

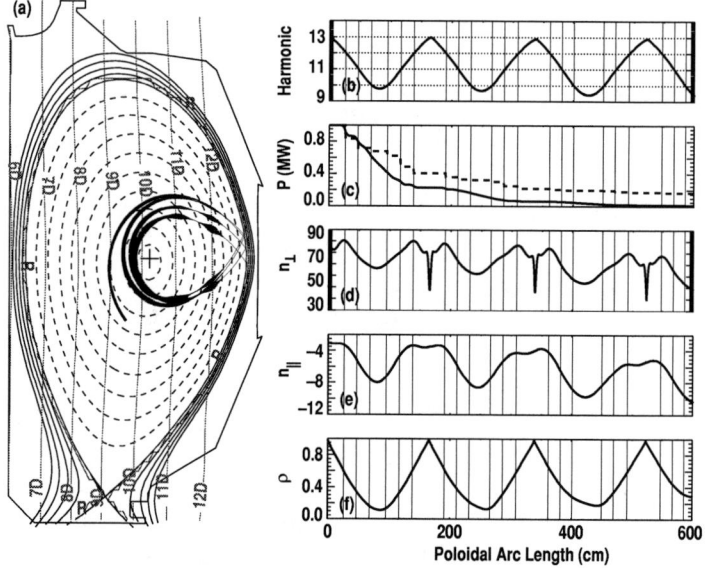

FIGURE 2. Same as Fig. 1, but for 116 MHz.

most of the cross-section, while in the 116 MHz case the ray travels in a small region of the available space attached to the outboard midplane.

In Figs. 1(b–f) and 2(b–f) some ray data are plotted as a function of the poloidal arc length. Figures 1(c) and 2(c) show the power remaining in the wave as calculated by GENRAY, while the dashed line is an analytic evaluation [5] using the power deposited on the fast ions. Here the absorption per crossing is $e^{-2\eta}$, where

$$2\eta = \frac{R_B \omega_{pi-hot}^2 e^{-\lambda_{hot}}}{4 n_\perp f c} \left\{ \ell \Delta_-^\ell (\lambda_{hot}) \left(\frac{R - n_\parallel^2}{S - n_\parallel^2} \right)^2 \right.$$
$$\left. + \ell \Delta_+^\ell (\lambda_{hot}) \left(\frac{L - n_\parallel^2}{S - n_\parallel^2} \right)^2 + 2\lambda_{hot} \left[\Delta_+^\ell (\lambda_{hot}) - \Delta_-^\ell (\lambda_{hot}) \right] \right\} ,$$

and R_B is the magnetic field scale length, ω_{pi-hot} is the ion plasma frequency for the hot ions, ℓ is the harmonic number, R, L, and S are the Stix quantities or their hot plasma equivalents, and $\Delta_-^\ell(\lambda) = I_{\ell-1}(\lambda) - I_\ell(\lambda)$ and $\Delta_+^\ell(\lambda) = I_\ell(\lambda) - I_{\ell-1}(\lambda)$. The figures show that this is an underestimate of the total damping, because there is also damping on electrons in both cases, and significantly on the hydrogen minority in the 60 MHz case. Nevertheless, the damping on the hot ion minority is the major damping mechanism in both cases and the analytic theory gives a reasonable estimate.

Equally interesting is the behavior of $n_\parallel$, which changes sign in rather chaotic behavior in the low frequency case but which stays in a comparatively narrow range in the high frequency case. This suggests that the fraction of power damped on electrons in the high frequency case will be much more effective at driving current. The GENRAY calculations viewed in this manner provide substantial insight into the wave damping and current drive physics, and more effective experiments can be designed.

ACKNOWLEDGMENT

Work supported by US Department of Energy under DE-FC02-04ER54698, DE-AC02-76CH03073, and DE-FG02-90ER54084.

REFERENCES

1. A. P. Smirnov and R. W. Harvey, "The GENRAY Ray Tracing Code," CompX Report CompX-2000-01 (2001).
2. M. Brambilla, Plasma Phys. Contr. Fusion **44**, 2423 (2002).
3. S. C. Chiu et al., Nucl. Fusion **29**, 2175 (1989).
4. C. C. Petty et al., Plasma Phys. and Contr. Fusion **43**, 1747 (2001).
5. R. Pinsker et al., "Absorption of Fast Waves on Fast Ions at Moderate to High Ion Cyclotron Harmonics on DIII–D," Proc. 16th Top. Conf. on Radio Frequency Power in Plasmas, Park City, UT 2005.

Toroidal Rotation and ICRF Heating in NBI-driven Discharges in JET

J.S. deGrassie*, L-G Eriksson[†], and J-M Noterdaeme[¶]
and JET EFDA Contributors[‡]

General Atomics, P.O. Box 85608, San Diego, California, 92186-5608 USA
[†]*Association EURATOM-CEA, CEA/DSM/DRFC, CEA-Cadarache,
F-13108 St. Paul lez Durance, France*
[¶]*Max-Planck IPP-EURATOM Association, Boltzmann-Str. 2, D-85748, Garching, Germany
and University Gent, EESA Department, Gent, Belgium*

Abstract. The addition of rf heating to an NBI-driven target discharge is observed to reduce the toroidal rotation frequency. Experiments on this effect were performed on JET using (H)-D and (^{3}He)-D minority ICRH to vary the bulk electron to ion heating ratio. However, to lowest order, there is no clear difference in the two heating scenarios. We apply a recent model of Nishijima et al. [4] based upon the degradation of confinement with auxiliary power, and find that these JET data are in reasonable agreement with it.

In general, the application of rf heating to a tokamak discharge with an established toroidal rotation driven by neutral beam injection (NBI) results in a reduction of the magnitude of this rotation [1–4]. One explanation is that the additional heating power increases the turbulent transport of toroidal momentum [2–4]. Ion temperature gradient turbulence is predicted to be enhanced with greater T_e/T_i, the electron to ion temperature ratio, and this would appear to qualitatively fit with DIII–D experiments in which direct-electron rf heating is applied [1–3].

A series of experiments has been performed on JET designed to test the effect of T_e/T_i upon this reduction in toroidal speed by utilizing two different minority ICRH scenarios in order to vary the ratio of bulk electron to bulk ion heating [5]. In bulk ion D discharges, the standard JET minority H ICRH results in strong electron heating. The other scenario selected is minority ^{3}He with the object of reducing the electron heating in favor of bulk ion heating. Post-experiment modeling with the PION code [6] indicates that this was only partially successful. The change in T_e/T_i was not large, and the basic response of the plasma rotation to added ICRH appears insensitive to the heating scenario used, to lowest order. There may be subtle profile differences, but these cannot be definitively extracted from the data.

Since enhanced auxiliary heating with added ICRH is the common factor in either JET scenario, we will test the recent model of Nishijima et al. [4] used to explain a similar slowing observed in ASDEX-U. Briefly, this model postulates a decrease in energy and toroidal momentum confinement times (assumed to be equal) as $1/\sqrt{P_{aux}}$, where P_{aux} is the total auxiliary heating power. Incrementally, added P_{rf} increases P_{aux} but does not supply any significant toroidal torque, so there is an incremental decrease in the momentum confinement time and, hence, momentum itself. But P_{rf} does supply heating power, so there is a net gain in total energy.

[‡]See the Appendix of J. Pamela et al. Fusion Energy 2004 (Proc. 20th Int. Conf., Vilamoura, 2004) IAEA, Vienna (2004).

The set of data from these JET sessions includes both L– and H–mode discharges, and target discharges with co- and counter-NBI, relative to the direction of toroidal plasma current. All have $B_T = 3.4$ T, and $I_p = 1.8$ MA. For the (H) heating scenario $f = 51$ MHz, launched on the four-strap antennas with $0\pi\pi 0$ phasing, while for (^{3}He) $f = 33$ MHz with $0\pi 0\pi$ phasing. The phasing difference is due to technical reasons. In each case, the fundamental resonance passes near the magnetic axis, R ~ 3.0 m.

A typical toroidal rotation response in the JET co-NBI, L–mode discharges is shown in Fig. 1. Two shots are displayed, one for each ICRH scenario. In Fig. 1(a) we show T_i near the core ($\rho \cong 0.17$), the toroidal rotation frequency, ω_ϕ, at the same location, and the rf power profile, P_{rf}, while in Fig. 1(b) are T_e ($\rho \cong 0.17$), n_{el}, the line-averaged electron density, and the NBI power, P_{NBI}. The clear signature of a reduction in ω_ϕ is seen with application of P_{rf}, recovering after the rf pulse. In contrast, the thermal energies, indicated by the temperatures, rise steadily throughout the rf pulse. There is an indication of greater T_i at this location for the (^{3}He) discharge. It appears that there is a small increase in the thermal confinement time throughout the shot since the temperatures return to a higher value after the rf pulse than before, at the same P_{NBI}. This is also indicated by the return of ω_ϕ to a slightly larger value, on average.

In Fig. 1 we have divided ω_ϕ by the factor of 8.3 determined by a computation of the NBI torque to power ratio, as described in Ref. [3]. If this scaled value of ω_ϕ were equal to T_i, then the $\tau_{\phi i}$ parameter, defined in Ref. [3], would be 1, as generally seen in the core in DIII–D. Here in JET, this parameter is about twice this value for these discharges.

We will evaluate the dataset in terms of the ASDEX model [4]. The global (volume integrated) toroidal angular momentum, L, and thermal energy, W, are described by

$$\begin{aligned} L &= N\tau = sP_{NBI}\tau \\ W &= P\tau = (P_{NBI} + P_{rf})\tau \end{aligned} \quad (1)$$

where τ is a common confinement time for both. This model neglects the ohmic heating power as small. The beam injected torque is N and the ratio of N to P_{NBI} is s, nominally equal to $2R_{tan}/V_b$, where R_{tan} is the beam trajectory tangency major radius and V_b is the beam particle speed. The confinement time is modeled to decay with auxiliary power as $\tau = C/\sqrt{P}$, where C is a constant for fixed target discharge conditions. So W increases with P_{rf} as $W = C(P_{rf} + P_{NBI})^{1/2}$, while L decreases with

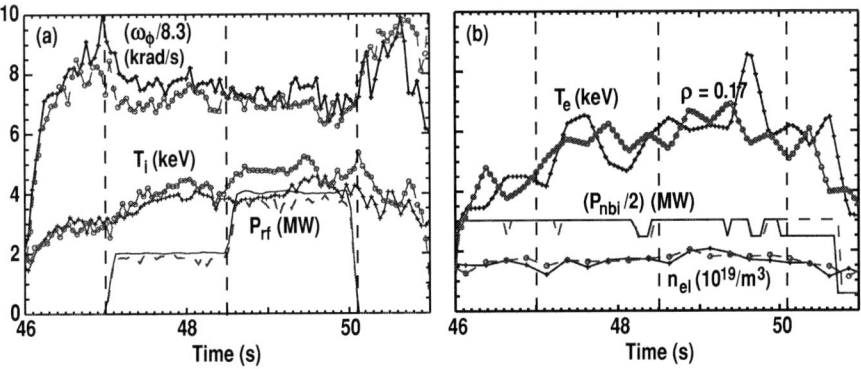

FIGURE 1. (a) ω_ϕ, T_i, and P_{rf} versus time. (H)-D pulse 55664, n_H/n_D ~ 2% (+, - - -) and (^{3}He)-D pulse 55666, n_{He}/n_D ~ 7% (○,—) (b) T_e, n_{el}, and P_{NBI}.

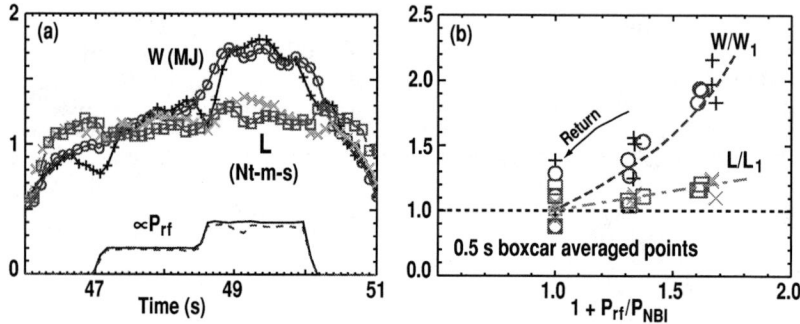

FIGURE 2. (a) Volume integrated thermal energy, W, and toroidal mechanical angular momentum, L, versus time. (H)-D (+,x), (^{3}He)-D (O, □). (b) W and L scaled to values with P_{NBI} only, prior to P_{rf}, versus $1+P_{rf}/P_{NBI}$.

P_{rf} as $L = CsP_{NBI}/(P_{rf}+P_{NBI})^{1/2}$. L and W are computed from the data by doing the volume integrals of the mechanical momentum density, $n_i M_i R^2 \omega_\phi$, and energy density, $(3/2)(n_i T_i + n_e T_e)$, respectively. Here we assume that the ion density n_i is equal to n_e, and that ω_ϕ, T_e, T_i, and n_e are flux functions, and we replace R^2 in the L integral by R_0^2, where R_0 is the major radius of the magnetic axis.

In Fig. 2(a) we plot the time histories of W and L for the same two discharges as in Fig. 1. Both show an increase in these global quantities with rf power, in spite of the core reduction in ω_ϕ seen in Fig. 1(a). Note that the value of s for the JET NBI mix in these two discharges is $s = 1\mu s = 1$ Nt-m-s/MJ, and we see then from Fig. 2(a) that the global confinement times of W and L are indeed very similar. Figure 2(b) shows 0.5 s averaged values of W and L scaled to their initial (averaged) value prior to rf turn-on, plotted versus $1+P_{rf}/P_{NBI}$. Although L/L_1 does increase with P_{rf}, this increase is much less than that seen in W/W_1 with P_{rf}. We conclude that qualitatively the ASDEX model predicts the difference in response of W and L with P_{rf}, but here there is an overall bias toward an increase in each, which is probably due to an increase in τ throughout the discharge, that is, $C = C(t)$. After the rf pulse, W clearly returns to a value above the starting value, at the same NBI power, and this is seen to a lesser extent in L, as shown in Fig. 2(b).

In order to apply the ASDEX model to this entire dataset of JET discharges we define a parameter, A, by taking the ratio of L to W, as defined above. That is,

$$A = \left[(P_{NBI}+P_{rf})/sP_{NBI}\right](L/W) = \left[(1+P_{rf}/P_{NBI})/s\right](L/W) , \quad (2)$$

motivated by the discussion following Eq. (1). This serves to remove C from each discharge and leaves only the power dependence of τ. In computing A for a discharge, we compute the actual, possibly time dependent, value of s given the NB injectors used for the specific discharge. The data is time-averaged for 0.25 s to generate a data point. The ASDEX model predicts $A = 1$ for all values of P_{rf}/P_{NBI}. Actually, in Ref. [4] this model is applied only to changes due to rf within a discharge and does not require the conclusion that $L/W = s$ in a steady NBI-only portion of a discharge.

The resultant values of A for 22 discharges, taken in three separate sessions spanning nearly two years, are shown in Fig. 3, where we plot A versus $1+P_{rf}/P_{NBI}$. Each session falls clearly into its own band of points, with A relatively independent of P_{rf}/P_{NBI}, again supporting the Nishijima et al. explanation of the reduction in L with P_{rf} [4].

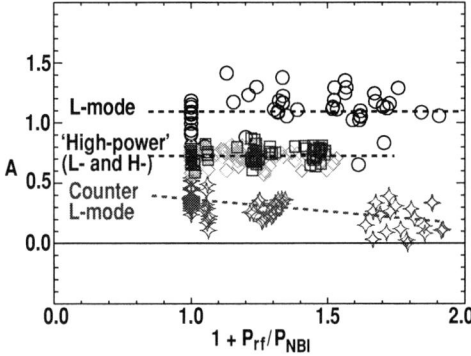

FIGURE 3. Parameter 'A' versus $1+P_{rf}/P_{NBI}$. 'L–mode' consists of 4 co-NBI pulses. 'High-power' consists of 12 co-NBI pulses with some L– and H–mode cases. 'Counter' consists of 6 counter-NBI pulses in L–mode. The data points are 0.25 s boxcar averages in a pulse.

The cause of the separate bands of points is revealed by the set of data taken with counter-NBI, the lowest values of A. As is well-known, a tokamak has nonzero L even with $P_{NBI} = 0$, that is, there is an 'intrinsic' rotation, L_0, which is not negligible [7–10]. L_0 is typically in the direction of I_p, but it can be opposite. For the counter discharges, it is observed that there is an L_0, which would be negative if shown in Fig. 3 because it is in the direction of I_p, opposite to the direction of toroidal NBI in this case. (L is positive in the direction of toroidal NBI in Fig. 3.) In one discharge, the early NBI rotation data indicates that $L_0 \sim -0.25$ Nt-m-s near the start of the rf pulse. Thus, L and W should be replaced in Eq. (1) by $L-L_0$ and $W-W_0$, where W_0 would logically be the Ohmic heating energy. This would raise the A values for the counter-NBI discharges by ΔA, $0.15 < \Delta A < 0.35$. There are also L_0 values, now positive, for the other data in Fig. 3, which would lower A for these sets. Care must be taken to purposely measure L_0 with short NBI pulses. Including the fact that $L_0 = L_0(W)$ [7,10] also complicates the details of applying the ASDEX model, although its basic plausibility is consistent with these JET results.

ACKNOWLEDGMENT

Work supported in part by the U.S. Department of Energy under Contract DE-FC02-04ER54698.

REFERENCES

1. J.S. deGrassie et al., Proc. 13th Top. Conf. on RF Power in Plasmas, Annapolis (1999) p. 140.
2. J.S. deGrassie et al., Proc. 26th EPS Conf. on Contr. Fusion and Plasma Phys., Maastricht, Vol. **23J**, 1189 (1999).
3. J.S. deGrassie et al., Nucl. Fusion **43**, 142 (2003).
4. D. Nishijima et al., Plasma Phys. Control. Fusion, **47**, 89 (2005).
5. V.P. Bhatnagar et al., Nucl. Fusion **33**, 83 (1993).
6. L.-G. Eriksson and T. Hellsten, Phys. Scr., **55**, 70 (1995).
7. J.E. Rice et al., Phys. Plasmas **11**, 2427 (2004), and references therein.
8. L.-G. Eriksson, E. Righi, and K.-D. Zastrow, Plasma Phys. Control. Fusion, **39**, 27 (1997).
9. J.-M. Noterdaeme et al., Nucl. Fusion **43**, 274 (2003).
10. J.S. deGrassie et al., Phys. Plasmas **11**, 4323 (2004).

ICRH Current Ramp Discharges and Alfven Cascades in Alcator C-Mod

M. Porkolab,[1] E. Edlund,[1] J. Snipes,[1] S. Wukitch,[1] N. Basse,[1] P. Bonoli,[1] C. Boswell,[1] C. Fiore,[1] N. Gorelenkov,[2] A. Hubbard,[1] G. J. Kramer,[2] L. Lin,[1] Y. Lin,[1] E. Marmar,[1] and G. Schilling[2]

[1]*MIT Plasma Science and Fusion Center, Cambridge, MA 02139, USA*
[2]*Princeton Plasma Physics Laboratory, Princeton, NJ 08540, USA*

Abstract. Current ramp experiments with intense ICRF power injected early in the ramp phase in Alcator C-Mod have been carried out. The goal of these experiments is to produce suitable reversed shear (RS) target plasmas for future Advanced Tokamak (AT) plasma research. Future plans call for off-axis injection of Lower Hybrid current drive (LHCD) to maintain the RS plasmas while increasing beta with additional ICRH[1]. In the present experiments evidence of RS q-profiles has been demonstrated in the ramp stage of the discharge with the observation of Alfven-Cascades (or Reversed Shear Alfven-Eigenmodes or RSAE) driven by the energetic ICRF ion tail. The frequencies are in agreement with MHD code predictions. Evidence of sawtooth delay has also been observed by increasing the injected ICRF power.

Introduction

In past experiments[2,3] we have demonstrated that by injecting 2-3 MW of ICRF power in the D(H)-minority heating mode, the plasma can be heated to temperatures of T_e = 4 -5 keV, T_i =3 keV, n_e=1x10^{20}m^{-3} before sawteeth set in at t = 0.22-0.24 sec. Typical plasma currents were 800 kA at B=5.3T, corresponding to q_{95} = 4.5, and the discharges were typically inner wall limited or inner diverter nose limited. In some cases diverted discharges were also used, however, the goal was to keep the density low and obtain reversed shear discharges at high electron temperatures (T_e = 5 keV) which would then serve as suitable target plasma for off-axis lower hybrid current drive (LHCD) for advanced tokamak (AT) studies.[1] In past experiments, both central and off axis ICRH have been tested[2,3], with the aim of achieving high T_e target plasmas and delaying sawteeth as long as possible. To summarize previous experimental campaigns, off-axis ICRF did not help significantly in delaying sawteeth phenomena, possibly due to the limitations imposed by the evolution of Z_{eff}. Furthermore, previous attempts to measure the q-profile using the early version of the MSE diagnostic were unsuccessful.[3] In the present experiments, we obtained evidence of of sawtooth delay as the RF power was increased. Furthermore, in the early (pre-sawtooth) phase of the discharges strong Alfven wave cascade (AWC, or Reversed Shear Alfven Wave Eigenmodes) activity has been observed with phase contrast imaging (PCI) diagnostic as well as with magnetic pick-up coils.[4] The AWC is believed to be generated by gradients of the energetic proton minority tail in reversed shear discharges with energies approaching $V_p/V_A \approx 1$. A theoretical interpretation of the evolution of the frequency spectrum allows one to interpret the q_{min} evolution of these discharges.[5] We present data indicating that mildly reversed shear current profiles dominate these discharges in the early phase of their evolution before sawtooth phenomena occurs. The results substantiate previous modeling of the time evolution of RS q-profiles as predicted by E-fit q-profile analysis.[2] LHCD injection in this early RS phase of the discharges with $q_{min} \geq 2$ offers effective means to initiate high beta RS plasmas for AT studies.

Experimental Results

A typical discharge evolution during current ramp with intense central ICRF heating at 80 MHz in the D(H) minority regime at 5.3 T is shown in Fig. 1. We see that as the current is ramped up to 800kA, and ICRF power at the 3 MW level is injected, central electron temperatures of 5 keV and edge electron temperatures of 0.5 keV are achieved even in inner wall limited L-mode discharges. At the same time, from the neutron emission rate the central ion temperature is estimated to be in excess of 3 keV. As indicated in Fig. 2, sawtooth activity is delayed until ~ 0.3 sec by a combination of ohmic and ICRF heating. The delay

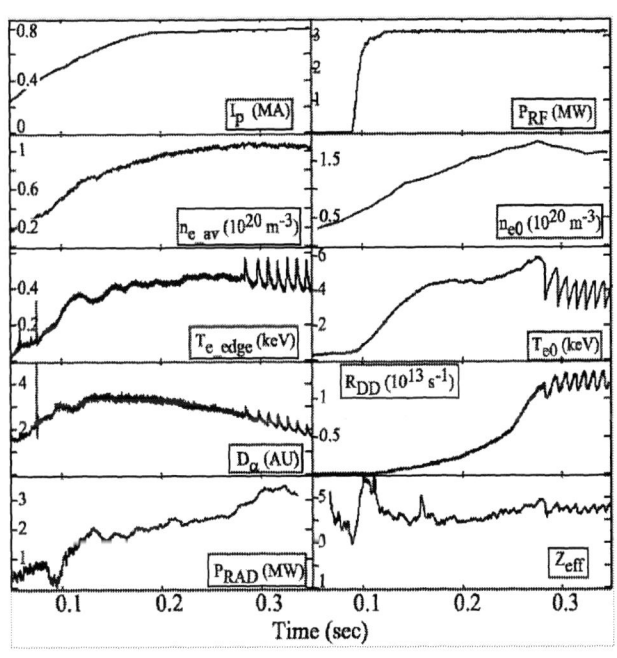

Figure 1: Time sequence of discharge parameters during current ramp with intense ICRH.

time is increasing as the RF power is increased, possibly indicating a favorable temperature dependence (since the sawtooth delay is inversely proportional to the resistive diffusion time of the ohmic electric field, $r^2 T_e^{3/2}/Z_{eff}$ where r is the plasma radius). However, as indicated by the above formula, the potential increase of impurities as the ICRF power is raised would negatively impact the positive temperature dependence of sawtooth delay. To further complicate things, we see that at the highest power (3.2MW in this case) the sawtooth period is also increasing, possibly due to sawtooth stabilization by the energetic H minority tail. Hence the delay in sawtooth activity may partially be explained by stabilization effects. Nevertheless, the recent observation of Alfven Cascades by both magnetic loops and PCI (see Figs. 3, 4) indicate that up to 0.18 second reversed shear should be present, gradually turning toward flat q profiles while typical target electron temperatures of the order 5 keV are achieved. Given that the L/R time varies as a^2, where a is the minor radius, we see that 0.2-0.3 sec in C-Mod is reasonable when we compare our results with the typical 2-3 sec current penetration times during ramp-up in machines such as DIII-D, with a minor radius of 3 times that of C-Mod.

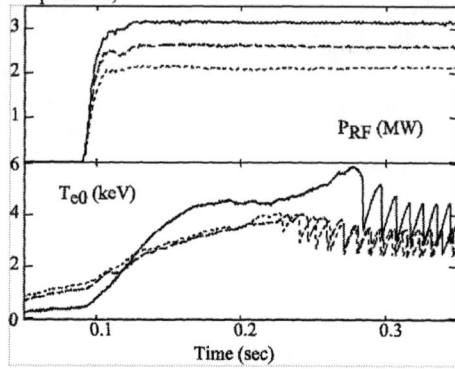

Figure 2: Time sequence of T_e showing sawtooth delay as the RF power is increased. Also note the tendency to stabilize sawteeth at the highest power.

The presence of Alfven cascades (AC) has been interpreted as evidence of reversed shear. A small amount of reversed shear is sufficient to deform the shear Alfvén continuum in such a way as to form a region where a mode can exist essentially free of interaction with the continuum. The drive for these cascade modes is the pressure gradient of the ICRH fast ions. The eigenmodes are localized radially near the minimum value of the safety factor q(r), and as the q_{min} value evolves in time, modes with different (m, n) values sweep (chirp) upward in frequency as:[5]

$$\omega(t) = \left| \frac{m}{q_{min}(t)} - n \right| \frac{V_A}{R_0} + \Delta\omega \quad (1)$$

Modeling of mode frequencies as observed by either magnetic pickup loops or the PCI diagnostic shows good agreement with the above formula. In Fig. 3(a) we show experimental measurements of AWC using the PCI diagnostic, and in Fig. 3(b) we show predictions by the MISHKA MHD code[6] indicating excellent agreement of the time evolution of the frequency of these modes. In Fig. 4 we plot predicted values of q_{min} corresponding to this shot. Finally, it is noted that the minimum frequency $\Delta\omega$, may be interpreted as the geodesic deformation of Alfven continuum at low frequency[7]. The Princeton NOVA-K code[8] has been used with good success to model the evolution of some of the chirping modes on Alcator C-Mod, including the low frequency deformation

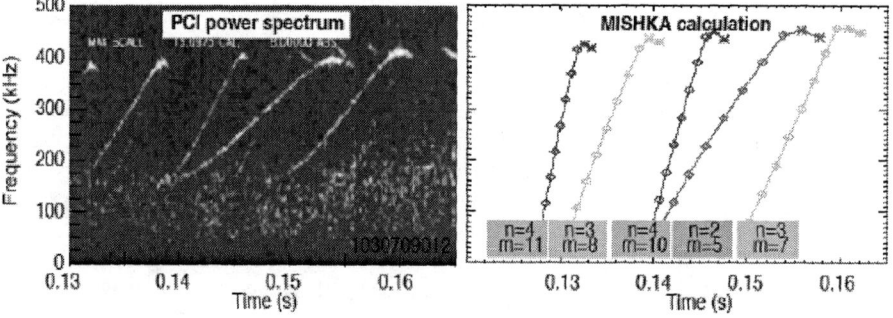

Figure 3: (a) Alfven wave cascade as observed by the PCI diagnostic, (b) modeling of the modes by the

of the continuum (see Fig. 5, which is a different shot from Figs. 3, 4). Other measurements (not shown) indicate that as q_{min} decreases to below 2, its temporal rate of decrease slows down and the q profile also tends to flatten out (the cascades disappear). Ultimately, q_{min} (near the center) decreases to below unity and sawteeth appear.

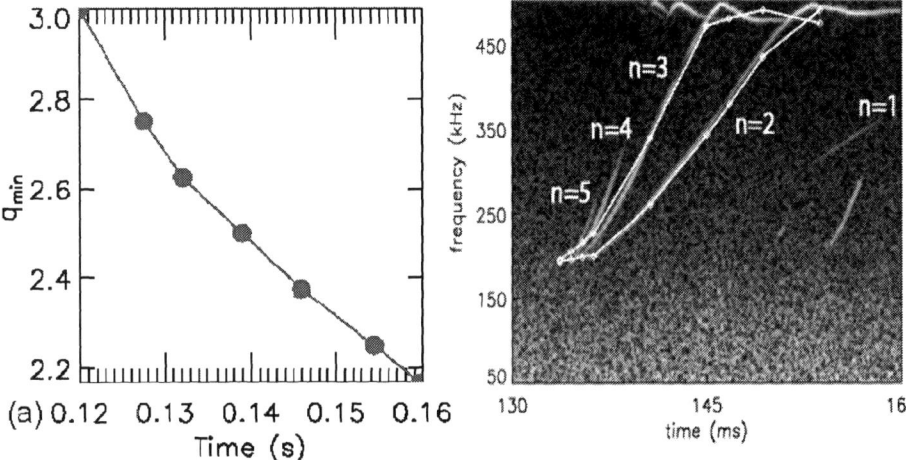

Figure 4: Time evolution of q_{min} as determined form the Alfven cascade spectrum and the MISHKA code during current ramp and ICRH.

Figure 5: The Nova-K code results (white lines) show good agreement with the observed Alfven cascade spectrum measured by the PCI when the finite pressure is included (cut-off at 175 kHz).

Summary

We see that in the early phases of the discharge the Alfven Cascades indicate the presence of reversed shear, corroborating EFIT modeling of similar ramping shots[2]. The plasma parameters obtained in these experiments offer excellent target plasmas for driving off-axis current drive by lower hybrid waves with the goal of obtaining quasi-steady state RS plasmas to carry out advanced tokamak plasma studies.

References

1. P.T. Bonoli, M. Porkolab, J. Ramos, et al, Plasma Phys. Contr. Fusion, **39**, 223 (1997).
2. M. Porkolab,et al, Proc. 24th EPS Conf. On Contr. Fusion and Plasma Physics, Berchtesgaden, 1997, Vol. 21A, Part II, pp. 569. J.A. Snipes, et al, Plasma Phys. Contr. Fusion **42**, 381 (2000).
3. M. Porkolab, et al, "15th Topical Conference on RF Power in Plasmas", Moran, Wy., 2003, Ed. C.B. Forest, AIP publication 0-7354-0158-6/03.
4. E. Edlund, et al, Bull. Am. Phys. Soc. **49**,73(2004); also, J. Snipes, et al, Physics of Plasmas **12**, 056102 (2005).
5. S.E. Sharapov et al, Phys. Letts. **A289**, 127 (2001).
6. A. B. Mikhailovskii, G.T.A. Huysmans, W.O.K. Kerner and S.E Sharapov, Plasma Phys. Rep.,**23**, 844 (1997).
7. M.S. Chu, J. M. Greene, L.L. Lao et al., Phys. Fluids B **4**, 3713 (1992).
8. C.Z.Cheng and M.S. Chance, Journal of Comp. Phys.**71**, 124 (1987).

Work supported by US DoE, Office of Fusion Energy Sciences. The contribution of the Alcator C-Mod team to this work is gratefully acknowledged.

Edge Minority Heating Experiment in Alcator C-Mod

S.J. Zweben[a], J.L. Terry[b], P. Bonoli[b], R. Budny[a], C.S. Chang[c], C. Fiore[b],
J. Hughes[b], Y. Lin[b], R. Perkins[a], M. Porkolab[b], G. Schilling[a]. S. Wukitch[b]
and the Alcator C-Mod Team

[a] Princeton Plasma Physics Laboratory, Princeton NJ 08540 USA
[b] MIT, Cambridge Mass USA
[c] New York University, New York

Abstract. An attempt was made to control global plasma confinement in the Alcator C-Mod tokamak by applying ICRH heating power to the plasma edge in order to deliberately create a minority ion tail loss. In theory, an edge fast ion loss could modify the edge electric field and so stabilize the edge turbulence, which might then reduce the H-mode power threshold or improve the H-mode barrier. However, the experimental result was that edge minority heating resulted in no improvement in the edge plasma parameters or global stored energy, at least at power levels of $P_{RF} \leq 5.5$ MW. Some analysis of these results is presented and some ideas for improvement are discussed.

Keywords: ICRH, H-mode, Alcator C-Mod, minority heating
PACS: 52.55.Fa, 52.50.Qt, 52.25.Fi

The present experiment was motivated by suggestions of Chang [1] and Perkins [2] that the plasma edge E_r could be changed if a relatively small population of ICRH minority tail ions was created at the plasma edge and deliberately lost to the wall. For example, the radial ion loss current corresponding to 1 MW of tail ion loss at $T_{tail} \approx 10$ keV would be ≈ 100 A, i.e. comparable to the radial current needed to create H-modes in the biased limiter experiments [3]. If the edge electric field could be controlled in this way, it might be possible to reduce the H-mode power threshold or improve the H-mode barrier in future magnetic fusion devices.

This paper describes an attempt to use minority edge heating to control the edge electric field and H-mode transition in the Alcator C-Mod tokamak at MIT. Only the basic experimental results and conclusions are presented here; further details concerning the results, modeling, and references are elsewhere [4].

All the plasmas used in for these experiments had a major radius of R=0.67 m, a minor radius a=0.23 m, a plasma current of I=0.6 MA (except three at 0.4 MA), and central electron temperatures and densities typically $T_e(0) \approx 1$ keV and $n_e(0) \approx 1 \times 10^{14}$

cm^{-3}. The initial ICRH edge heating experiments were done at 78 MHz with $P_{RF} \leq 2.3$ MW, and in a second set the ICRH power was increased up to $P_{RF} \leq 5.5$ MW by the addition of ICRH power at 80.0 and 80.5 MHz. All the plasmas were made with a deuterium majority (except one shot with a helium majority), and all had a $\approx 5\%$ hydrogen minority concentration as measured by the H_α/D_α line ratio.

The hydrogen minority resonance position was varied from the inner edge to the outer edge on a shot-to-shot basis by varying the toroidal magnetic field. At the lowest field of B=3.68 Tesla the ICRH resonance was at $R_H = 47.4$ cm, i.e. 2 cm inboard of the inner separatrix at the midplane and 3.5 cm from the inner wall. At the highest field of B=6.75 Tesla the ICRH resonance was at $R_H = 85.8$ cm, i.e. 3.5 cm outboard of the outer separatrix and 4.5 cm from the outer ICRH antenna-protection limiter at the midplane. For both cases the resonant surfaces extended into the scrape-off layer above and below the midplane due to the curvature of the flux surfaces, and in the high field case the resonance intersected the RF antenna.

Figures 1(a) and (b) show the dependence of the total stored plasma energy on the resonance location and ICRH power for the magnetic field scan with 78 MHz only. For neither the inner nor the outer edge ICRH heating was there any significant increase in stored energy compared to the Ohmic plasma at these fields. In contrast, for the central heating cases at B=5-6 T the plasma stored energy increased by almost a factor of two with 1 MW of heating, corresponding to an H-mode transition (circled points). Fig. 1(c) shows that there was also no increase in the stored energy for inner edge heating cases when ICRH power level was increased to $P_{RF} \leq 5.5$ MW with combined 78 MHz, 80 MHz, and 80.5.

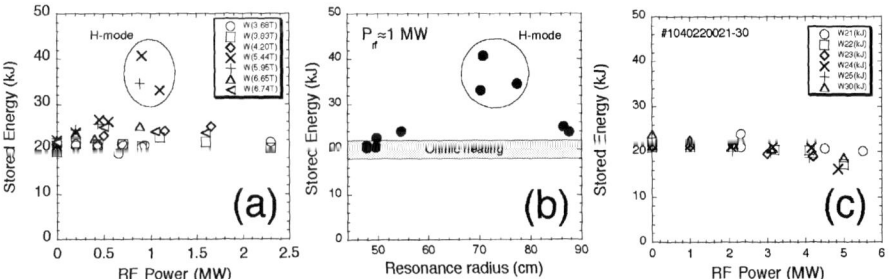

Fig. 1: The effect ICRH minority heating on the total plasma stored energy. The circled points are for central heating which created an H-mode, the others points are for inner or outer edge heating. Part (c) is a power scan for inner edge heating only.

Figure 2 shows the dependence of the edge plasma density and temperature on the ICRH resonance location during the scan of Figs. 1(a,b) based on the edge Thomson scattering data averaged over a 2 cm wide region just inside the separatrix. For neither the inner nor the outer edge heating there was there any significant increase in edge parameters compared to the Ohmic plasma just before the ICRH was applied (although there were slight increases for the outer edge heating case). In contrast, with central heating at B=5-6 T the plasma edge parameters increased as expected for an H-mode in C-Mod (circled points).

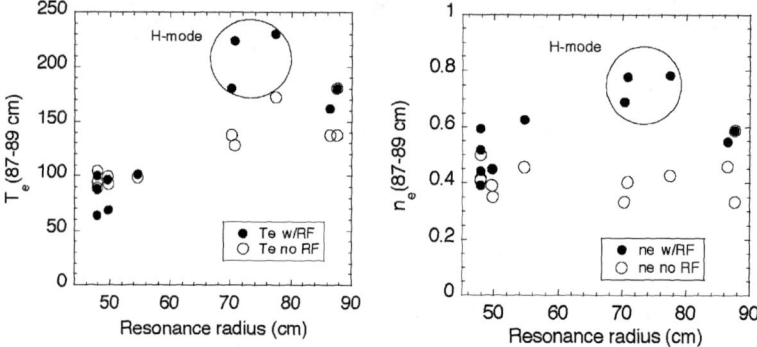

Fig. 2: The effect of ICRH minority heating on the edge electron temperature and density. The circled points are for central heating in which an H-mode was produced. The other points are for inner or outer edge heating (units are eV and 10^{14} cm^{-3}).

Several variations on this scenario were tried to improve these null results. The ICRH antenna phasing was varied from the co-current to the counter-current direction (instead of balanced as for Figs. 1 and 2), the plasma current was reduced from 0.6 MA to 0.4 MA, and a He majority plasma was tried instead of D majority (to vary the edge conditions). However, in all cases the stored energy slowly decreased with increasing ICRH power. Thus the experimental results were quite clear; namely, that edge heating caused no increase in the plasma confinement. The only effect of this edge heating was to slightly reduce the total stored energy, most likely due to recycling and/or impurity influx from the increased plasma-wall interaction.

Some modeling of this experiment was attempted using the TORIC ICRH code in TRANSP, with typical results shown in Fig. 3. For conditions of this experiment the code predicted that there would be a significant minority tail ion loss generated when the minority resonance was located at the inner (or outer) edge. In this 1.4 MW inner edge heating case the tail ion loss power was ≈350 kW at a peak tail ion temperature of ≈ 5 keV, corresponding to a calculated loss ion current of ≈ 60 Amps.

However, several potentially important effects were not modeled in this code, e.g. "parasitic" absorption of ICRF waves in the edge plasma, charge exchange of tail ions, and radial diffusion of tail ions in the edge. In particular, analytical estimates of the single-pass absorption of the ICRH in the edge showed that only a few percent of the ICRH power would be absorbed per pass since the edge ion temperature was so low. Thus any parasitic absorption at this level could significantly reduce the net power for creating a minority tail in the edge.

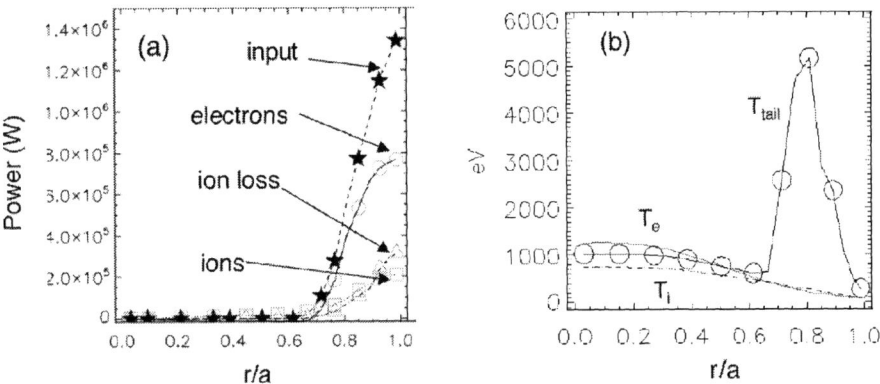

Fig. 3: Simulation of an inner edge minority heating case with the TORIC/TRANSP code. The code predicts an ion loss current of ≈ 60 Amps for this 1.4 MW case.

Unfortunately, it is not possible at this stage to make a quantitative comparison of the (null) experimental results with the theoretical modeling, since the modeling had such large uncertainties, and since there was no direct measurement of the ion loss or radial electric field in this experiment. It is quite possible that the tail ion loss was less than expected, and so the resulting change in E_r was just not large enough to affect the plasma confinement.

The outcome of this experiment might be improved by reducing the edge density further to increase the tail ion energy, by increasing the edge ion temperature to improve the first-pass absorption of ICRH waves, or by reducing the SOL thickness to increase the fast ion orbit loss. The conditions would more easily occur in a device like JET, in which high edge ion temperatures have already been measured even without edge ion heating.

Acknowledgements: We thank J. Hosea, D. Majeski, D. McCune, C.K. Phillips, R. Wilson, D. Rasmussen, and D. Darrow for help with this experiment. This work was supported by DOE contract DE-AC02-76CHO3073 (PPPL) and USDOE Cooperative Agreement DE-FC02-99ER54512 (MIT).

References:
1) C.S. Chang et al, Phys. Plasmas 11, 2649 (2004)
2) F.W. Perkins, private communication (2001)
3) G. Van Oost et al, Plasma Phys. Cont. Fusion **45**, 621 (2003)
4) S.J. Zweben et al, PPPL Report PPPL-4059 (Mar. 2005)

ICRF Heating for the Non-Activated Phase of ITER: From Inverted Minority to Mode Conversion Regime

M.-L. Mayoral[1], P.U. Lamalle[2], D. Van Eester[2], P. Beaumont[1], E. De La Luna[3], P. De Vries[1], C. Gowers[1], R. Felton[1], J. Harling[1], V. Kiptily[1], K. Lawson[1], M. Laxåback[4], E. Lerche[2], P. Lomas[1], M.J. Mantsinen[5], F. Meo[6], J.-M. Noterdaeme[7,8], I. Nunes[9], G. Piazza[10], M. Santala[5] and JET-EFDA Contributors*

[1] *UKAEA/EURATOM Fusion Association, Culham Centre, Abingdon, OX14 3DB, United Kingdom*
[2] *Laboratory for Plasma Physics, Association EURATOM-Etat Belge, Partner in the Trilateral Euregio Cluster, Royal Military Academy, Brussels, Belgium*
[3] *Asociación /EURATOM -CIEMAT, Laboratorio Nacional de Fusion, Spain*
[4] *Association EURATOM-VR, Swedish Research Council, Sweden*
[5] *Association /EURATOM -Tekes, Helsinki University of Technology, Finland*
[6] *Association Euratom-RISO, Riso National Laboratory, Denmark*
[7] *Max-Planck-Institut für Plasmaphysik- /EURATOM Assoziation, Germany*
[8] *Gent University, Department EESA, Belgium*
[9] *Associação EURATOM / IST, Instituto Superior Técnico, Portugal*
[10] *Forschungszentrum Karlsruhe, IKET, Germany*

See the appendix of J. Pamela et al., Fusion Energy 2004 (Proc. 20th Int. Conf. Vilamoura, 2004) IAEA, Vienna (2004)

Abstract. In the initial phase of ITER H plasmas will be used in order to avoid activating the machine. The reference ICRF heating scenarios rely on minority species such as Helium (^{3}He) or deuterium (D). These schemes' distinctive feature comes from the presence of the fast magnetosonic wave ion-ion hybrid resonance/cut-off pair, between the antennas and the minority cyclotron layer. In order to document these unusual heating schemes, ICRF experiments were carried out recently on JET. First, the use of ^{3}He ions in H plasmas was investigated with a sequence of discharges in which 5 MW of ICRF power was coupled to the plasma and the ^{3}He concentration was varied from below 1% up to 10%. The inverted minority heating regime was observed at low concentrations (up to ~2%). Energetic tails in the ^{3}He distribution were observed with effective temperatures up to 300 keV and central electron temperatures up to 6 keV. At around 2%, a sudden transition was reproducibly observed to the mode conversion regime, in which the ICRF fast wave couples to short wavelength modes, leading to efficient direct electron heating and central electron temperature up to 8 keV. All these experiments systematically used power modulation techniques to assess the radial profiles of the wave absorption by the electrons. Secondly, experiments to study the ICRF heating of D minority ions in H were performed. This heating scheme proved much more difficult since modest quantities of C^{6+} impurity, which has the same Z/A ratio than the D minority ions, led us directly into the mode conversion regime. This effect preventing any absorption by D ions at minority cyclotron layer, could make the (D)H scenario not suitable for the non-active phase of ITER.

Keywords: ICRF, inverted minority heating, mode conversion.
PACS: 52.50.Qt; 52.60.Qt

INTRODUCTION

It is envisaged that in the initial phase of ITER [1] operations, H plasmas will be used not to activate the machine during its commissioning stages. The two relevant ion cyclotron resonance frequency (ICRF) scenarios foreseen in H plasmas are based on the heating of ^{3}He ions minority, referred as (^{3}He)H or D ions minority, referred as (D)H. The peculiarity of these so-called inverted scenarios comes from the fact that the ion minority species have a smaller charge to mass ratio than the ion majority species, i.e. $Z_{min}/A_{min} < Z_{maj}/A_{maj}$. In which case, the fast wave (FW) dispersion relation places the ion-ion hybrid layer R_{ii} (corresponding to the FW resonance at $n_{//}^2 = S$), and its associated cut-off $R_{Lcut\text{-}off}$ (corresponding to $n_{//}^2 = L$) between the low field side (LFS) ICRF antenna and the minority ion cyclotron layer. The quantities L and S are defined by Stix [2][3], $n_{//} = k_{//}c/\omega$ is the refractive index parallel to the magnetic field, where $k_{//}$ is the parallel wave number.

Mode conversion (MC) of the FW to hot plasma waves occurs near R_{ii} when finite temperature plasma effects are considered. The nature of these short wavelength waves depends on the plasma [4]. They can be kinetic Alfvén waves, ion Bernstein waves (IBW) and electromagnetic ion cyclotron waves (ICW) depending on the relative importance of temperature and poloidal field effects [5][6][7]. Their main characteristic is to damp strongly on electrons by electron landau damping (ELD) due to a strong up-shift of $k_{//}$ associated with the presence of a finite poloidal magnetic field [8].

Minority heating (MH) occurs when the FW energy is absorbed at a minority ion cyclotron resonance layer R_{IC}. The resonance condition can be written as $\omega - n\omega_{c_i} - k_{//}v_{//} = 0$, where ω is the ICRF wave frequency, ω_{c_i} is the minority ion cyclotron frequency, n is the harmonic number and $v_{//}$ is the minority ion parallel velocity. The FW absorption by the minority ions results in the formation of a high-energy population. The background population heating takes place on a rather long time scale of the fast minority ions collisional slowing-down time. When the energy of the fast minority ions is above (below) the critical energy E_{crit} [2], the heating goes predominantly to electrons (ions).

In standard scenarios with the $Z_{min}/A_{min} > Z_{maj}/A_{maj}$, the FW encounters R_{IC} first, the pair R_{ii}/ $R_{Lcut\text{-}off}$ being located further on the high field side (HFS). MH is the main heating scenario until the minority concentration reaches a critical value above which the fast wave electric field component which rotates in the same direction as the minority ions, is significantly reduced at the minority layer. Consequently as the single-pass minority damping decreases significantly, the FW can be mode-converted at the ion-ion layer. In JET, optimal concentration for MC regime in D plasmas with ^{3}He ions was found in the range of 12-20% [9]. In the inverted scenario, above a certain minority concentration, when the MH regime single-pass is very low, the MC regime is expected to dominate and has been well documented on other tokamaks[10][11]. Nevertheless, below this level, it is not clear if and for which minority concentration the FW will be mode-converted or absorbed by the minority ions. Indeed, even if the $R_{ii}/R_{Lcut-off}$ pair is encountered first by the FW, MH regime can occur if the minority ions have parallel temperature high enough to Doppler broaden the resonance out of $R_{Lcut-off}$. The set of experiments described in the next sections, were recently performed on JET and especially designed to study the inverted MH conditions and the transition to MC regime with ^{3}He or D minority ions in H plasmas [12].

ICRF HEATING OF H PLASMAS WITH ^{3}He MINORITY IONS

Experimental Set-up

The experiments were carried out in H plasmas with a magnetic field in the range of 3.3 to 3.6T and a plasma current of 2 MA. Up to 5 MW of ICRF power was applied at a frequency of 37 MHz using the four JET A2 ICRF antennas [13] positioning the fundamental ^{3}He cyclotron resonance layer in the plasma center at around 2.9m. Dipole phasing (0 π 0 π) of the antennas was used to launch waves with a symmetric toroidal mode number spectrum ($|n_\Phi| \approx 27$ at the maximum of the antenna power spectrum); +90° (and –90°) phasings were used to predominantly launch co-current (and counter-current) wave momentum (with $|n_\Phi| \approx 13$ at the maximum of the antenna power spectrum). In order to deduce experimentally the electron power deposition profiles, the electron temperature response to modulation of the ICRF power was performed using Fast Fourier Transform (FFT) and Break-In-Slope (BIS) analysis [14][15][16]. As discussed in the introduction, the minority ions concentration n_{3He}/n_e, is one the critical parameter influencing the FW damping. Hence, its control and evaluation were crucial for these experiments. As ^{3}He gas is lost through transport, keeping the concentration at a specific level requires puffing extra ^{3}He gas into the machine during the discharge [9]. A very efficient way to do so is to use the JET real time central controller in order to link a measurement of the ^{3}He density to the opening of the gas injection valve. This technique developed in the last few years [16] has been successful used in the here experiments presented to control ^{3}He concentrations as low as 1.8 %. The ^{3}He concentration was estimated from the effective charge of ions, the relative concentration of the majority and minority ions measured via the respective light in the divertor and the relative intensity of the impurity lines. It has to be noted that for experiments requiring $n_{3He}/n_e < 1\%$, RTCC was not used and only a carefully chosen puff of ^{3}He was made just before the ICRF heating phase. Finally, information on the presence of fast ICRF-accelerated ions was obtained from gamma-ray (γ-ray) spectrometry analysis [17] and low and high-energy neutral particle analysis (NPA)[18].

^{3}He Minority Heating Regime

To establish the feasibility of ^{3}He MH regime in inverted scenarios a set of discharges was performed with ^{3}He concentration below 1 %. Fig. 1 shows an overview of three discharges with $n_{3He}/n_e < 1\%$ and different phasings of the ICRF waves: dipole, +90° and −90°. Clear signatures of an efficient ^{3}He MH regime were obtained with mainly electron heating. A maximum electron temperature T_e^{max} of 6.2 keV reached with +90° phasing and $P_{ICRF} = 5$ MW.

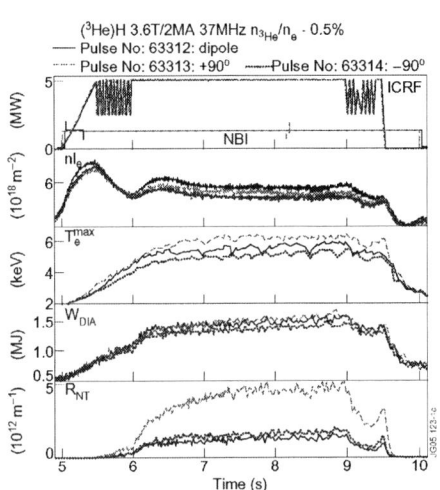

FIGURE 1. Overview of three discharges in the minority heating regime: 63312 (dipole), 63313(+90°) and 63314 (−90°)

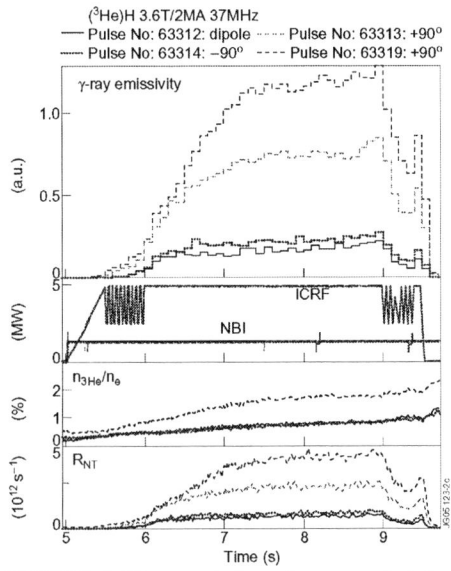

FIGURE 2. Time evolution of γ-ray emissivity, ICRF and NBI power and [^{3}He] for pulse 63312 (dipole), 63313 (+90°), 63314 (−90°) with $n_{3He}/n_e < 1\%$ and 63319 (+90°) with n_{3He}/n_e up to 1.8%

The presence of a fast ^{3}He ion population led to an increase of the neutron rate R_{NT} which in H plasmas, comes from the nuclear reaction between the Beryllium impurities and ^{3}He ions, i.e. ^{9}Be(^{3}He,n)^{11}C. A much higher neutron rate was obtained for the pulse 63313 (+90° phasing) indicating higher energy ^{3}He population, compared to the pulses 63312 (dipole phasing) and pulse 63314 (−90° phasing) which have similar ICRF power, ^{3}He concentration and electron density. The fast ^{3}He ions energy content W_{fast} was estimated up to 60% higher in the +90° case than in the −90° and dipole cases. The higher T_e, W_{fast}, R_{NT} obtained with +90° phasing is consistent with the expected wave induced pinch effect [19][20]. Indeed, the +90° phasing asymmetric spectrum leads to an inward drift of the fast ions turning point, as opposed to the outward drift obtained with −90° phasing. More information on the fast ^{3}He ions population was obtained from the γ-rays produced by the nuclear reaction between ^{3}He ions with energy above 0.9 MeV and ^{9}Be impurity ions and ^{3}He ions with energy above 1.3 MeV and the ^{12}C impurity [17]. The presence of a higher energy ^{3}He tail with +90° phasing was confirmed by the higher γ-ray emissivity (represented on Fig. 2) for the shots 63313 than for the shots 63312 (dipole phasing) or 63314 (−90° phasing). In the shot 63319, also represented on Fig. 2, the ^{3}He concentration was doubled compared to that of the shot 63313. The changes in the γ-ray signals and neutron rate confirm a higher fast ion population but with a lower temperature tail, as the emissivity and neutron production is only increased by a factor of 1.7. In this shot, it was estimated that the tail temperature was $T_{tail} \sim 0.3 \pm 0.1 MeV$.

Transition From Minority Heating To Mode Conversion Regime

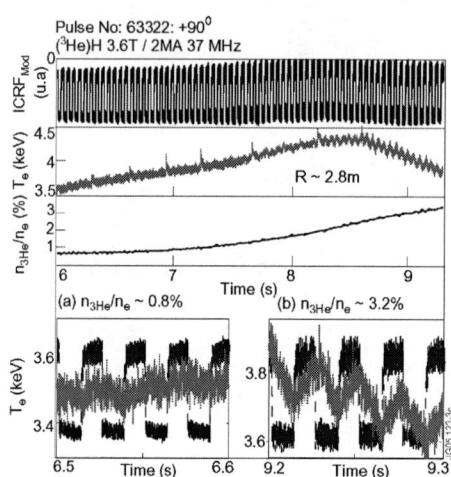 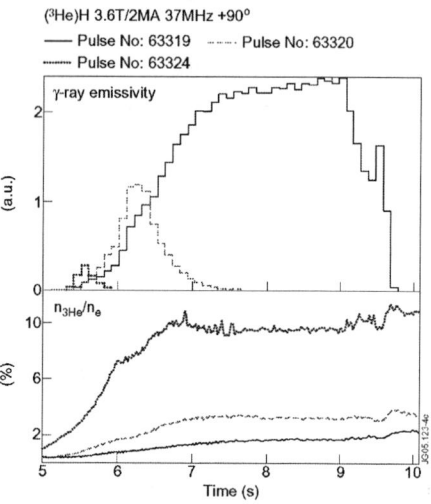

FIGURE 3. Electron temperature response to ICRF power modulation for (a) $n_{3He}/n_e \sim 0.7\%$ and (b) $n_{3He}/n_e \sim 3\%$

FIGURE 4. Time evolution of the γ-ray emissivity for three pulses with different ^{3}He concentration.

When the ^{3}He concentration was increased above 2% a reproducible transition from the MH to the MC regime was observed. This transition to a different heating regime was seen on different signals. First, as n_{3He}/n_e was increased, the T_e response to the ICRF power modulation gradually changed, indicating a change in the power deposition. In Fig. 3, the T_e time evolution at 3.09m and the ^{3}He concentration are displayed for pulse 63322 for which the ICRF modulation was applied throughout the ICRF power flat-top. One can clearly see a dramatic change in T_e when n_{3He}/n_e becomes larger than 2%. In the MH regime, indirect electron heating by fast ions has a delay with respect to the ICRF power modulation due to the finite slowing-down time of fast ions on the electrons. By contrast, in the MC regime the electron heating is a direct process by ELD and TTMP. Indeed, one can see, that with $n_{3He}/n_e \sim 3\%$, the T_e time response was prompter, consistent with stronger direct electron heating. In the $n_{3He}/n_e \sim 0.7\%$ case, the period of the modulation of 0.05 s compared with an estimated fast ^{3}He slowing down time of about 0.2s prevented us, as expected, to see any straightforward T_e response. A sudden change in the fast ^{3}He ion population was also observed. The γ-ray emissions, which required ^{3}He ions with energies above 0.9 MeV, is represented on Fig. 4 for the shots 63319, 63320 and 63324 differing only by their ^{3}He concentration. In discharge 63319, a high signal was collected throughout the pulse as the ^{3}He concentration stayed below 1.8%. In pulse 63320, as n_{3He}/n_e

increased above 2% at around 6.3s the γ-ray signal began to decrease rapidly. In shot 63324, almost no signal was collected.

Mode Conversion Regime

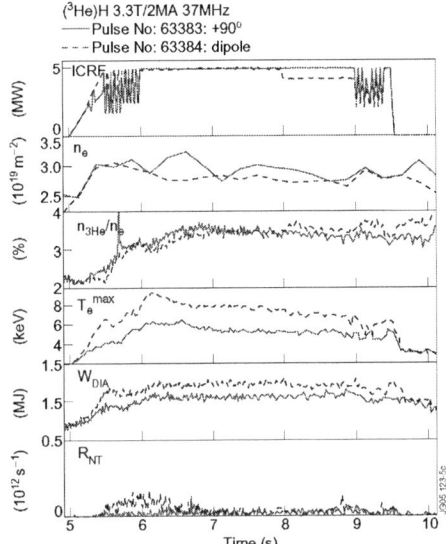

FIGURE 5. Overview of two discharges in the mode conversion regime and two different phasings 63383 (+90°),(63384 dipole).

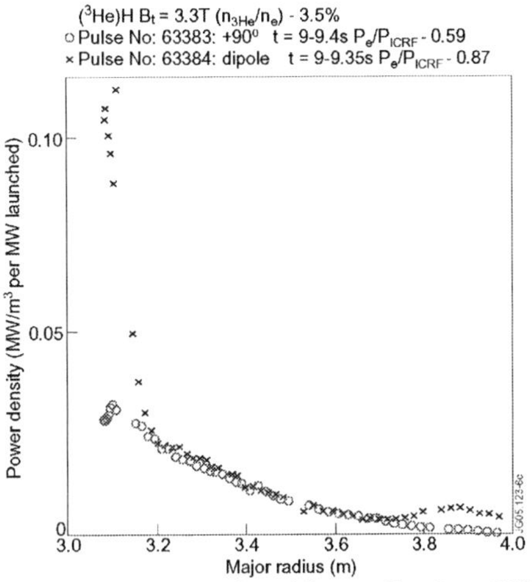

FIGURE 6. Electron deposition profiles from BIS analysis for central MC and two different phasing 63383 (+90°), 63384 (dipole).

In order to maximise the central electron temperature with $n_{3He}/n_e \sim 3.5\%$, B_t was lowered to 3.3T, positioning the ion-ion hybrid layer at ~ 2.8 m instead of ~ 3.2 m in the previous set of experiments. As illustrated on Fig. 5, for the dipole phasing case T_e up to 8 keV was reached. The corresponding electron power deposition profiles given by the BIS analysis are shown on Fig. 6. One has to point out that due to the T_e diagnostic line of sight [21] which is below the plasma center, electron deposition profiles cannot be obtained between 2.8 and 3.05m. Nevertheless, one can see clearly a much higher deposition profile in the dipole case with an integrated ICRF power fraction to the electron up to 87%. This higher electron absorption with a phasing corresponding to a higher $k_{//}$ is far from obvious. Nevertheless, a first explanation can be found by looking at the ratio between the wave phase velocity $v_\Phi = \omega/k_{//}$ and the electron thermal velocity $v_{the} = \sqrt{T_e/m_e}$ as when v_ϕ is of the order of v_{the}, strong localized electron damping of the mode converted wave is expected. By assuming a 3 keV target temperature

TABLE 1: TOMCAT integrated electron power deposition (double wave transit)

T_e (keV)	[^{3}He] (%)	+90°		Dipole	
		P_e (%)	P_{3He} (%)	P_e (%)	P_{3He} (%)
3	2	37.9	7.1	11.6	7.6
5	2	36.9	11.8	10.7	15.6
8	2	34.3	19.9	12.7	24.4
3	3.5	36.4	5.7	76.1	2.3
5	3.5	35.8	8	71.6	10.5
8	3.5	33.8	12.5	42.9	44.9

plasma, one finds that the dipole case (with $k_{//} \sim |n_\phi|/R_{ant}$ and $R_{ant} \sim 4$) gives $v_\phi / v_{the} \sim 1.5$ as compared with the less favorable +90° case ($k_{//} \sim 3.3$ m^{-1}) for which $v_\phi / v_{the} \sim 3$. Finally, simulations made with the 1D code TOMCAT [22], support this higher electron power deposition in the dipole case. As illustrated on Table 1, for $n_{3He}/n_e = 3.5\%$, the integrated power going to electrons is systematically higher for the dipole case than for the +90° case with a maximum absorption for $T_e = 3\text{-}5$ keV.

ICRF HEATING OF H PLASMAS WITH D MINORITY IONS

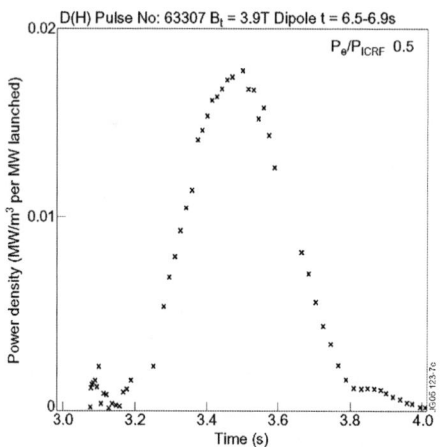

FIGURE 7. Typical electron deposition profiles from break-in slope analysis obtained in (D)H plasma.

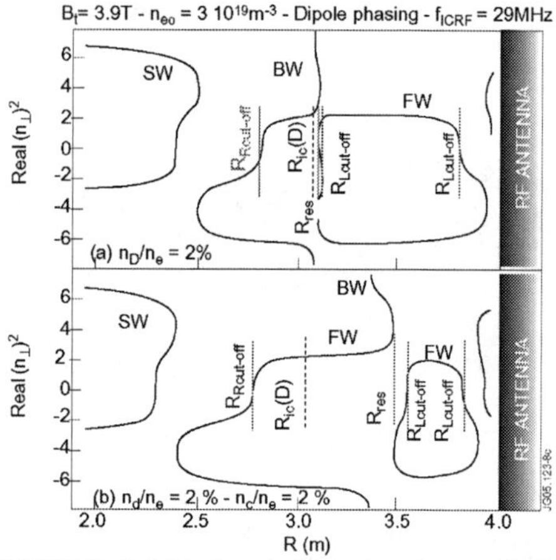

FIGURE 8 Cold plasma dispersion diagram in H plasma with 2% of D and (a) no C, (b) 2% of C

This set of experiments were carried at a magnetic field of 3.9T, a plasma current of 2 MA and a central electron density of 3 10^{19} m^{-3}. In order to position the fundamental D cyclotron resonance in the plasma center (~ 3.1 m) a ICRF wave frequency of 29 MHz was used. This frequency being at the limit of the generator capability, only up to 2.5 MW of ICRF power was applied. The D concentration obtained and estimated around 1 to 2 % arose solely from the "natural" D desorption from the wall. Surprisingly, no D fast ions were detected, neither by the neutral particle analyzer nor the γ-rays emission. No neutrons were produced and the maximum electron temperature obtained was around 3 keV. A BIS analysis performed on several shots revealed, as shown on Fig. 7, a rather peaked electron power deposition with its maximum at 3.5 m and a full width at half maximum of around 30 cm. This off-axis electron deposition was explained by the presence in the plasma of 2-3% of carbon C^{6+}, which has the same cyclotron layer as the minority D ions. As illustrated one Fig. 8, the fast wave propagation is then similar to the one that could be obtained at much higher equivalent D concentration, directly leading into the mode conversion regime. One has to point out also the sensitivity of the mode conversion layer position to small change in the C concentration. Finally, simulations of this heating scenario with the 2D full-wave code CYRANO shown that from C concentration above 0.5%, significant MH of D ions can be ruled out.

CONCLUSIONS

Inverted minority scenarios relevant for the non-activated phase in H have been tested at JET. Due to the Doppler broadening of the ^{3}He cyclotron resonance, ^{3}He minority heating was successfully achieved for n_{3He}/n_e below 2%. When n_{3He}/n_e was increased further, the mode conversion regime dominated. By positioning the conversion layer in the plasma center, electron temperature up to 8 keV was obtained using dipole phasing. D minority heating was also tested. This was not successfull as the presence of carbon impurity led directly to far off-axis fast wave mode conversion. This effect, which is expected to take place from C concentration as low as 0.5%, prevented any absorption by D ions at minority cyclotron layer. As the mode conversion layer position is very sensitive to any change in the C concentration, the (D)H scenario could be found unusable for the non-active phase of ITER.

ACKNOWLEDGMENTS

This work was performed under the European Fusion Development Agreement, and funded partly by the UK Engineering and Physical Sciences Research Council and by EURATOM.

REFERENCES

[1] ITER Physics Basis Editors, ITER Physics Expert Group Chairs and Co-Chairs and ITER Joint Central Team and Physics Integration Unit, *Nucl. Fusion* **39**, 12, (1999).
[2] T.H. Stix, Waves in plasmas (New York: American Institut of Physics).
[3] T.H. Stix, *Plasma Phys.* **14**, 367 (1972).
[4] F.W. Perkins, *Nucl. Fusion* **17(6)**, 1197-1224, (1977).
[5] Y. Lin et al,. *Plasma Phys. and Control. Fusion* **45**, 1013-1026, (2003).
[6] E. Nelson-Melby et al., *Phys. Rev. Lett.* **90(15)**, 155004, (2003).
[7] E. F. Jaeger et al., *Phys. Rev. Lett.* **90(19)**, 195001, (2003).
[8] A.K. Ram and A. Bers, *Phys. Fluids B* **3**, 1059-10069, (1991).
[9] M.J. Mantsinen et al., *Nucl. Fusion* **44**, 33-46, (2004).
[10] P. Bonoli et al., *Phys. Plasmas* **4(5)**, 1174-1182, (1997).
[11] B. Saoutic et al., *Phys. Rev. Lett.* **76(10)**, 1647-1650, (1996).
[12] P. Lamalle et al., Proceedings of 20th IAEA Fusion Energy Conference, Vilamoura, Portugal, 1-6 November 2004, submitted for publication in *Nucl. Fusion*.
[13] A. Kaye et al., *Fusion Eng. Des.* **24**, 1-21, (1994).
[14] D.J. Gambier et al., *Nucl.Fusion* **30**, 23-24, (1990).
[15] D. Van Eester, *Plasma Phys. and Control. Fusion* **46**, 1675-1697, (2004).
[16] D. Van Eester et al., Proceedings of 15th Conference on Radio-Frequency Plasmas, Moran, Wyoming, 19-21 May 2003, edited by Cary B. Forest, American Institut of Physics, Melville, New–York, Vol. 694, (2004).
[17] V. Kiptily et al., *Nucl. Fusion* **42**, 999–1007 (2002).
[18] A.A. Korotkov, A. Gondhalekar and A.J. Stuart, *Nucl. Fusion* **37**, 35 (1997).
[19] L.-G. Eriksson et al., *Phys. Rev. Lett.* **81(6)**, 1231-1234, (1998).
[20] M.J. Mantsinen et al., *Phys. Rev. Lett.* **89(11)**, 115004, (1998).
[21] E. de la Luna et al.,*Rev. of Scientific Instruments* **75(10)**, 3831-3834,(2004)
[22] D. Van Eester and R. Koch, *Plasma Phys. and Control. Fusion* **40**, 1949-1675, (1998).

Current and Flow Driven in Thin Layers

Harold Weitzner

Courant Institute of Mathematical Sciences, New York University
251 Mercer St., New York, NY 10012, USA

Abstract. Current and flow drive at arbitrary frequency is described with the methods of low collisionality or neoclassical transport theories. The basic formulation and a possible scaling law for the phenomena are given. The underlying ideas behind the analysis and the nature of the intricate system of equations are sketched.

INTRODUCTION

This note describes the calculation of a particular form of RF-induced flow and current. The analysis is based on a low collisionality expansion of coupled electron and ion Fokker-Planck equations, with external RF sources. The plasma state sustained by the RF fields is axisymmeteric and time independent. The low collisionality expansion is similar in spirit to a neoclassical expansion, but the low collisionality expansion is a more realistic characterization of fusion relevant plasmas. The flows and currents are driven by a relatively large, time independent, electrostatic potential in a thin layer and an associated electric field. The plasma is near local thermodynamic equlibrium, with the electron and ion distribution functions close to Maxwell-Boltzmann distributions. For convenience, it is assumed that the RF field are given by geometrical optics and that there is no resonance in the RF fields, although the individual particles may be resonant. The analysis is not otherwise sensitive to the frequency range, so that this treatment applies to high harmonic ion flow drive, lower hybrid or ECR current drive. Many other forms of current and flow drive are possible, but without strongly non-Maxwellian plasmas, the options are far fewer. This note offers one self-consistent possibility.

THE MODEL

To give the results with any specificity, we must explain the scaling assumptions, the related form of the kinetic equations, and give their consequences. Electron and ion equations are non-dimensionalized separately with the units of distance and velocity being the appropriate species larmor radius ρ_s and thermal speed v_s^{th}. Thus, the unit of time is the inverse of the species larmor frequency. The macroscopic distance scale for the equilibrum is $L \gg \rho_s$. We introduce $\epsilon_s = \rho_s/L$, and assume $m_e/m_i \sim \epsilon_i^2 \sim \epsilon_e$ and $T_i \sim T_e$, For the general approach see Ref. 1 which lays out the scaling and methods. We assume that in a layer of thickness $\sqrt{\rho_i L}$ there is an electrostatic potential varying on this distance scale and of magnitude $(\sqrt{\rho_i L})^{-1}$. This potential yields electric fields of O(1) in the ion equation and order $\epsilon_e^{1/2}$ in the electron equation. In Ref. 2 we have given the forms of the collision integrals for systems with such flows. We gave the nonlinear ion-ion self-collision operator C_{ii}, the linear ion-electron collision operator C_{ie}, which exchanges energy and momentum. For electrons we gave the electron-electron self-collision operator C_{ee}, and the linear electron-ion collision operator, the latter being the sum of three terms, the usual pitch angle scattering operator, C_{PA}, a momentum exchange operator, C_M, proportional to the mean ion flow velocity, and an energy exchange operator C_E. In the region of relatively large flow $C_{ii} \sim \epsilon_i^2$, $C_{ie} \sim \epsilon_i^3$, $C_{ee} \sim \epsilon_e^{3/2} \sim C_{\text{PA}}$, while $C_M \sim \epsilon_e^2$ and $C_E \sim \epsilon_e^{5/2}$. To carry out the low collisionality – or neoclassical – expansion we must construct a magnetic movement adiabatic invariant. In Ref. 3 we gave the construction of the invariant when the fields varied on the ion larmor radius distance scale. Here, we do not use such a severe condition, we assume only that the steady state electrostatic potential vary on the distance scale $\sqrt{\epsilon_i} L = \rho_i/\sqrt{\epsilon_i}$.

The applied RF fields are large enough to sustain the given flows, but not so large as to move the system far from local thermodynamic equilibrium. In particular we choose $\mathbf{E}_{\text{RF}} \sim \epsilon_i^{3/2}$, which in the electron equation becomes $\mathbf{E}_{\text{RF}} \sim \epsilon_e^{5/4}$. Another paper with joint authors, Ref. 4, carried out a transport calculation of this type in a simpler geometry. The calculation of the quasilinear RF operator is different from the usual cases as there is a mean flow which is neither small, nor $\mathbf{E} \times \mathbf{B}/B^2$. In addition to this standard drift there is also a term proportional to $\mathbf{B}$. The quantities that we use from the quasilinear wave operator are

$$T_1 = \frac{1}{2}\text{Re} \int \frac{(\mathbf{v}-\mathbf{u})^2}{2} d\mathbf{v} \frac{\partial}{\partial \mathbf{v}} \cdot \left[(\bar{\mathbf{E}}_{\text{RF}} + \mathbf{v} \times \bar{\mathbf{B}}_{\text{RF}})f_{\text{RF}}\right] d\mathbf{v}$$

and

$$T_2 = \frac{1}{2}\text{Re}\int \mathbf{v}\frac{\partial}{\partial \mathbf{v}} \cdot \left\{(\bar{\mathbf{E}}_{\text{RF}} + \mathbf{v} \times \bar{\mathbf{B}}_{\text{RF}})f_{\text{RF}}\right\}d\mathbf{v}$$

and a simple calculation shows that

$$T_1 = -\frac{1}{2}\text{Re}\left\{(\bar{\mathbf{E}}_{\text{RF}} + \mathbf{u} \times \bar{\mathbf{B}}_{\text{RF}}) \cdot \mathbf{J}_{\text{RF}}\right\}$$

and

$$T_2 = \left(\frac{\mathbf{k}}{\omega + \mathbf{k} \cdot \mathbf{u}}\right)T_1$$

T_1 and T_2 involve only the particle resonances with the applied fields, while the full RF quasilinear operator also has non-resonant contributions.

SINGLE PARTICLE MOTION

We require a few explicit results on single particle motion in order to clarify the calculation. We represent the magnetic field in terms of the flux coordinate ψ, safety factor Q, poloidal angle θ and toroidal angle ϕ. The electrostatic potential has the form $\epsilon_s^{-1/2}\Phi_{-1}(\psi/\epsilon^s) + \Phi_0(\psi, \theta)$ where $s = 1/2$ for ions and $1/4$ for electrons. If we introduce a new set of canonically conjugate variables (p_j, q_1), $j = 1, 2, 3$ then q_1 is the ignorable toroidal angle, p_1 the constant corresponding momentum, essentially ψ; q_3 p_3 and p_2 are components of velocity and q_2 is essentially θ. In particular we have $\psi = \epsilon p_1 + \epsilon q_3$, p_2 is the velocity parallel to $\mathbf{B}$, and $\epsilon_s r q_3$ the velocity parallel to $\hat{\phi}$. The Hamiltonian in terms of the initial variables $\mathbf{r}, \mathbf{v}$ has the form $H_s = \frac{1}{2}\mathbf{v}^2 - \Phi'_s(\psi/\epsilon^s)\epsilon^{1-2s}(\epsilon r v_\phi) - \frac{1}{2}(\epsilon r v_\phi)^2 \Phi''_s(\psi/\epsilon^s)\epsilon^{2-3s} + \Phi_0(\psi, \theta)$. Such a Hamiltonian leads to a mean flow in the $\hat{\phi}$ direction $\Phi'_s(\epsilon r)\epsilon^{1-2s}[1 + (\epsilon r)^2 \Phi''_s(\epsilon r)\epsilon^{2-3s}]^{-1}$. Further, a distribution function of the form $f_S = \dfrac{N_s(\epsilon^{1-s}p_1)}{[2\pi T_s(\epsilon^{1-s}p_1)]^{3/2}}$ $\exp[-H/T_s(\epsilon^{1-s}p_1)]$ is a local Maxwell-Boltzmann distribution plus corrections of order ϵ^{1-s}.

THE SOLUTION OF THE KINETIC EQUATION

The expansion of the kinetic equation order by order is relatively straightforward. One obtains in each order a drift kinetic equation on a given flux surface. If one refers to the form of the leading order distribution function one sees that there are five unknown functions N_s, T_s, $s = i, e$ and Φ_{-1}. One constraint is charge neutrality. Thus, we seek four other constraints. As argued in Ref. 1 these constraints should be

$$0 = \int (C_s + \mathrm{RF}_{\mathrm{QL}}) d\mathbf{r} d\mathbf{v}$$

$$0 = \int H(C_s + \mathrm{RF}_{\mathrm{QL}}) d\mathbf{r} d\mathbf{v}$$

where $\mathrm{RF}_{\mathrm{QL}}$ is the full quasilinear RF operator and the integral is restricted to the domain $\psi/\epsilon^s - \epsilon^{1-s} q_3 \leq \mathrm{const}$. We can expand these constraints order by order as done in Ref. 2 for a similar problem. For electrons, the constraints are simple and

$$\epsilon_e^{3/4} \frac{\partial}{\partial \psi_e/\epsilon_e^{3/4}} \int Q(\psi) \frac{(\epsilon r)^2}{B_T} d\theta n(\theta, \psi) \hat{\phi} \cdot [\mathbf{w}_e - \mathbf{u}_i (m_e/m_i)^{1/2}] \tau_{ei} = 0$$

$$\frac{m_e}{m_i} \int Q(\psi) \frac{\epsilon r d\theta}{B_T} n \tau_{ei}^{-1} [3(T_i - T_e) + 2\mathbf{u}_i^2] + \int \frac{\epsilon r Q d\theta}{B_T} T_1 = 0 \ .$$

The function $\mathbf{w}_e$ is given by $n\mathbf{w}_e = 3T_e^{3/2}(\pi/2)^{1/2} \int f_e(\mathbf{v}/v^3) d\mathbf{v}$. The ion constraints are more intricate but we find fairly easily

$$\epsilon_i \left(\frac{\partial}{\partial \psi/\epsilon_i^{1/2}} \right)^2 \int \frac{(\epsilon r)^3 Q}{B_T} d\theta \int d\mathbf{v} C_{ii}(f_i)[(v_\phi - u_\phi^i)]^2 + \frac{\partial}{\partial \psi/\epsilon_i^{1/2}} \int \frac{(\epsilon r)^2 Q d\theta T_2}{B_T} = 0$$

and

$$-\epsilon_i^{1/2} \frac{\partial}{\partial \psi/\epsilon_i^{1/2}} \int \frac{(\epsilon r)^2 Q}{B_T} d\theta \int d\mathbf{v} C_{ii}(f_i) H(v_\phi - u_\phi^i) + \int (d\mathbf{v}) C_{ie}(f) H\left(\frac{\epsilon r Q}{B_T}\right) d\theta$$

$$-\frac{1}{2} \epsilon_i^{1/2} \int (Q d\theta (\epsilon r)^3 / B_T)(v_\phi - u_\phi^i)^2 (d\mathbf{v}) \Phi''_{-1} C_{ii}(f_i) + \int T_1 (\epsilon r Q d\theta / B_T) = 0 \ .$$

Provided there is energy exchange between the RF field and either ions or electrons the RF fields act as sources to maintain the electrostatic potential over the thin layer. Details will be given in another publication.

REFERENCES

1. Harold Weitzner, *Phys. Plasmas* **1**, 3942 (1994).

2. Harold Weitzner, *Phys. Plasmas* **7**, 3330 (2000).

3. Harold Weitzner and Choong-Seock Chang, *Phys. Plasmas* **12**, 012106 (2005).

4. Harold Weitzner, Lee A. Berry, E. Fred Jaeger, and Donald A. Batchelor, *Phys. Plasmas* **7**, 564 (2000).

Investigation of Mode-Transformed Ion Cyclotron Waves at the Ion–Ion Hybrid Layer

R. Bilato and M. Brambilla

Max-Planck-Insitut für Plasmaphysik – EURATOM Association – Boltzmannstr. 2, D85748 Garching, Germany

Keywords: mode transformation, mode conversion, ion-ion hybrid resonance, ion cyclotron wave
PACS: 52.55.Fa, 52.50.Qt

INTRODUCTION

The Perkins theoretical prediction [1] of mode-transformation of externally launched fast waves (FW) into shear Alfvén waves (SAW) [2] (also known in this range of frequencies as quasi–electrostatic ion cyclotron waves (ICW)) near the ion–ion hybrid resonance has been invoked to explain the experimental results on TMO1 and TFR, as discussed by Jacquinot and co-workers [3]. Recently, for the first time in Alcator C-Mod it has been possible to confirm experimentally the occurrence of mode transformation, and to determine the wavelength of the mode-converted waves [4]. The agreement with numerical simulations [5] done with the full–wave code TORIC [6] is a further proof of the nature of these waves. In addition, there is evidence through the effects on the sawtheeth period that the ICWs affect the local current density [7]. As a consequence, these waves have captured the attention as a means to control the plasma current density profile and possibly to suppress plasma turbulence through the generation of sheared plasma flows [8, 7].
The quantitative investigation of this mode transformation is possible only by means of 3–dimensional full–wave codes, since the gradients due to the geometry play a central role. To gain physical insight in this process, however, it is useful to analyze a simplified one–dimensional wave equation, as suggested by Perkins. In the next section, by assuming that the most important spatial gradient is the horizontal variation of the toroidal magnetic field, we approximate the 2-dim wave equation valid in Tokamak geometry with a 1-dim problem. This model takes into account the poloidal component of the confining magnetic field $\vec{B}$ and the variation of the module of $\vec{B}$ along magnetic field lines, as required for the existence of mode–transformation to the ICW branch. From the 1-dim wave equation derived in this way one can deduce the transmission and reflection coefficients for the compressional and shear waves [9]. It is also possible to take into account Finite Larmor Radius effects extending the considerations [9] to include the lowest order ion Bernstein wave.
Let us first analyze under which conditions this mode-transformation takes place. We assume for simplicity that the module of the confining magnetic field varies as $B(X) = B_0/(1+X/R_0)$ where $-a \leq X \leq a$, with R_0, a respectively the major and minor radii. In this scenario $n_\parallel^2 \ll \omega_{pi}^2/\Omega_{ci}^2$, with $n_\parallel$ the parallel refractive index, ω_{pi} and Ω_{ci} the ion plasma and cyclotron frequencies. The positions of the ion–ion resonances and cut-offs

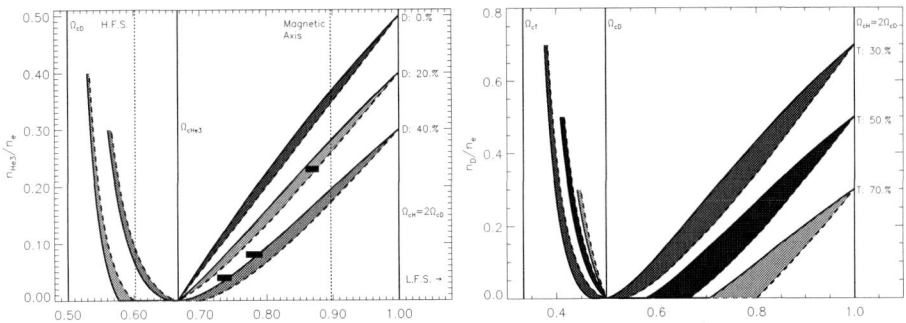

FIGURE 1. Operation diagram for the mode transformation in (a) H–D–^{3}He and (b) D–T–H plasmas, with the H concentration chosen to guarantee the charge neutrality, and $h = \omega/\Omega_{cH}$. The ion densities are normalized to the electron density. The shadowed areas are the evanescence layers for a given concentration of D in (a) and of T in (b). The dashed and solid boundaries of these layers correspond respectively to the cutoff and the resonance. In (a) the dashed vertical lines refer to the plasma extension and the black rectangles to the concentrations of the Alcator C-Mod experiments [4].

can be approximated by the solutions of the equations [2]:

$$\sum_i \frac{A_i\, v_i}{1 - (A_i/Z_i)^2\, h^2} = 0, \quad \sum_i \frac{A_i\, v_i}{1 - (A_i/Z_i)\, h} = 0 \quad (1)$$

where A_i, Z_i and v_i are respectively the atomic mass, the charge and the concentration of the ion species i. Here $h = \omega/\Omega_{cH}$, linked to the radial coordinate X by $X/R_0 = h(X_{cH}/R_0 + 1) - 1$, with X_{cH} the position of the fundamental Hydrogen cyclotron resonance. Figure (1.a) shows the position and width of the cutoff-resonance layers (CRL) for different concentration of ^{3}He and D, with the background H guaranteeing charge neutrality. Since there are three ions with different values of Z_i/A_i, there are two CRLs, one between Ω_{cD} and Ω_{c^3He} and the other between Ω_{c^3He} and Ω_{cH}. To avoid detrimental absorption of waves launched from the low field side (LFS) at the second harmonic of Deuterium, one has either to move this resonance outside the plasma (increasing the ratio ω/Ω_{cH}) or to keep the H concentration high (thereby increasing the screening due to the majority ions). In addition, to avoid that the transmitted FWs are absorbed at the fundamental of ^{3}He it is necessary to keep the ^{3}He concentration high enough (roughly speaking larger than 3% [10]). Finally to maximize the power mode–transferred at the LFS-CRL, the HFS–CRL must be outside the plasma, which is achieved either by keeping the D concentration low or decreasing the ratio ω/Ω_{cH}. The different set of plasma parameters considered in Alcator C-Mod [4] are indicatively reported in this diagram together with the approximate position of the plasma column for the discharges at 5.8 T on-axis magnetic field and $f = 80$ MHz. As one would expect, these plasma parameters meet all these criteria. Figure (1.b) shows the operational diagram in the case of a reactor relevant mixture D–T–H where H enters as a contaminant and the concentrations of T and D are comparable. In this case for acceptably low H concentration the detrimental absorption at Ω_{cH} can be avoided only by moving this resonance outside the plasma on the LFS. This implies also that the LFS–CRL is likely to fall outside the plasma, and the only accessible CRL is that between Ω_{cT} and Ω_{cD}.

Since the D concentration is high the absorption at the fundamental Ω_{cD} is negligible. This scheme could be experimentally tested with a D–^{3}He–H mixture in which D and ^{3}He play the role respectively of T and D. In this regime, however, the antenna loading might suffer from a sensitive reduction due to the low single pass absorption [3].

ONE-DIMENSIONAL MODEL

In the following we use Cartesian coordinates in the poloidal cross-section, ($R = R_0 + X, Z, \varphi$), where R is the radial distance from toroidal axis, Z the vertical distance from the midplane and φ the toroidal angle. The key assumption to derive the 1-dim wave equation is to neglect the radial variation of the plasma parameters in comparison with horizontal variation of $\vec{B}_0$ [9]. This is well justified near ion–ion and cyclotron resonances. Then both k_φ and k_Z (the vertical component of the wavevector) are constant, and one can write a wave equation for the propagation in the horizontal direction. Shifting the origin of the horizontal coordinate X at the cyclotron resonance of the minority species we write $S = S_0 - S'/(k_0 X)$ and $L = L_0 - 2S'/(k_0 X)$ [9]. Note that for the ion mixture considered σ has no resonant contributions, and can thus be regarded as constant, like the other coefficients of the wave equation, which for the left polarized component of the wave electric field E_+ is:

$$\left[\hat{H}_0(\hat{n}_\perp, \hat{n}_\parallel) X + S' \hat{H}_1(\hat{n}_\perp, \hat{n}_\parallel)\right] \frac{E_+(X)}{X} = 0 \qquad (2)$$

where:

$$\hat{H}_0 = -\sigma \hat{n}_\perp^6 + [S_0 + \sigma P]\hat{n}_\perp^4 + [P(\hat{n}_\parallel^2 - S_0) + (S_0 \hat{n}_\parallel^2 - RL_0)]\hat{n}_\perp^2 + P(\hat{n}_\parallel^2 - R)(\hat{n}_\parallel^2 - L_0) \qquad (3)$$

$$\hat{H}_1 = (P - \hat{n}_\perp^2)[\hat{n}_\perp^2 + 2(\hat{n}_\parallel^2 - R)] + \hat{n}_\perp^2 \hat{n}_\parallel^2 \qquad (4)$$

The differential operators $\hat{n}_\perp$ and $\hat{n}_\parallel$ depend on the pitch of the magnetic field line $\tan\Theta = B_{pol}/B_{tor} \ll 1$ and on the generalized poloidal angle τ:

$$\hat{n}_\parallel^2 = \left(\frac{n_\varphi}{k_0 R_0}\right)^2 + \sin\Theta \left[2\frac{n_\varphi}{k_0 R_0}\left(i\sin\tau \frac{d}{dX} - \cos\tau\, n_Z\right) - \sin\Theta \sin^2\tau \frac{d^2}{dX^2}\right] \qquad (5)$$

$$\hat{n}_\perp^2 = -\frac{d^2}{dX^2} + n_Z^2 - \sin\Theta\left[2\frac{n_\varphi}{k_0 R_0}\left(i\sin\tau \frac{d}{dX} - \cos\tau\, n_Z\right) - \sin\Theta \sin^2\tau \frac{d^2}{dX^2}\right] \qquad (6)$$

To visualize the mode transformation we have solved the local dispersion relation corresponding to Eq. (2) in the WKB approximation and in the limit $\sigma = 0$, where only the fast and the slow waves are taken into account. Figure 2 shows the real (I) and the imaginary (II) parts of n_X^2 as function of X and for constant values Z. The mode transformation starts at the $S = n_\parallel^2$, where in toroidal geometry $n_\parallel = n_\varphi/(qR_0) + m/a$. Figure 2.III shows the values of the poloidal wave number m taken by these waves in toroidal geometry: they are considerable high, as observed in TORIC simulations [4].

In Ref. [9] this equation has been written in the limit $\sigma = 0$, and the optical thicknesses and the transmission and reflection coefficients have been derived by means of Laplace

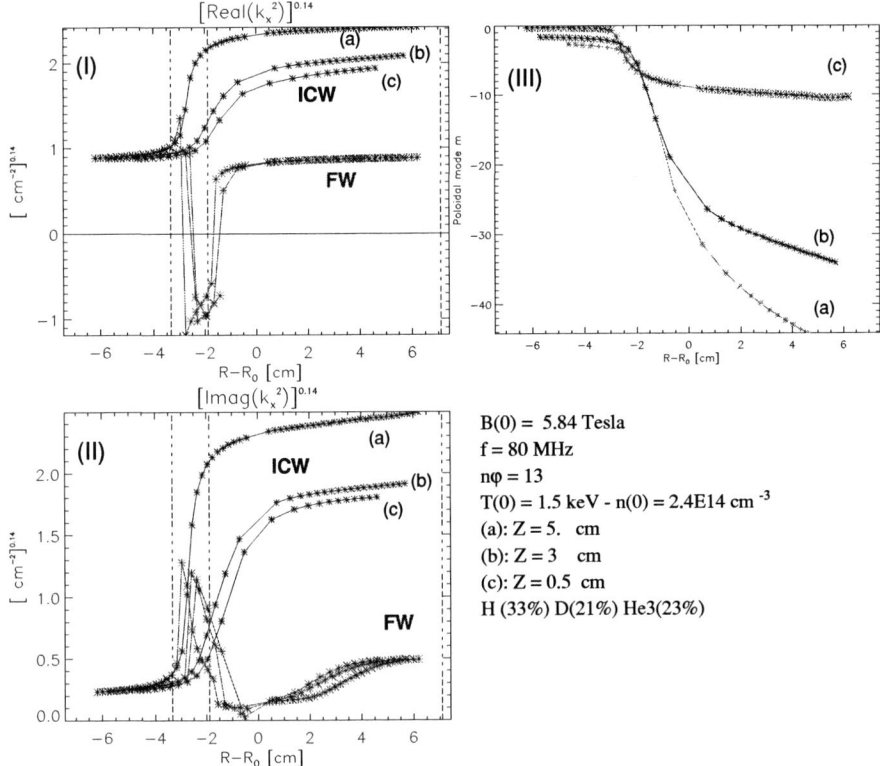

FIGURE 2. Dispersion relation $\hat{H}_0(n_X) = 0$ in the limit $\sigma = 0$, i.e. in presence of only fast and slow waves. To fit in the same plot the values of n_X^2 of the two waves, $\left(n_X^2\right)^{0.14}$ is mapped.

transform and an asymptotic analysis [11]. In progess it is the extension of this approach to the more general equation (2).

REFERENCES

1. PERKINS, F., Nucl. Fusion **17** (1977) 1197.
2. BRAMBILLA, M., *Kinetic Theory of Plasma Waves*, Oxford Science Publications, Oxford, 1998.
3. JACQUINOT, J. et al., Phys. Rev. Lett. **39** (1977) 88.
4. NELSON-MELBY, E. et al., Phys. Rev. Lett. **90** (2003) 155004.
5. LIN, Y. et al., Plasma Phys. Control. Fusion **45** (2003) 1013.
6. BRAMBILLA, M., Plasma Phys. Control. Fusion **41** (1999) 1.
7. PORKOLAB, M. et al., Proc. of the 20th IAEA Fusion Energy Conf. (Vienna: 2004) EXP/P4.32.
8. JAEGER, E. F. et al., Phys. Rev. Lett. **90** (2003) 195001.
9. BRAMBILLA, M. and OTTAVIANI, M., Plasma Phys. Control. Fusion **17** (1985) 1.
10. MAYORAL, M. et al., at this conderence .
11. GAMBIER, D. and SCHMITT, J., Phys. Fluids **26** (1983) 2200.

Numerical Studies of poloidal field effects on ICRF mode conversion

A. Parisot [1], S.J. Wukitch, P. Bonoli, Y. Lin, R. Parker, M. Porkolab, A.K. Ram and J.C. Wright*

Plasma Science and Fusion Center, MIT, Cambridge MA 02139, USA

Abstract. In multi-ion species plasmas with a moderate minority fraction, the fast wave is partially mode converted to shorter wavelength modes at the ion-ion hybrid layer. In addition to Ion Bernstein Waves (IBW), mode conversion to Ion Cyclotron Waves (ICW) in the presence of a poloidal field has been predicted by Perkins [1] and confirmed experimentally on Alcator C-Mod [2]. In this work, we take a similar approach as in [1] and study the effect of the poloidal field on ICRF mode conversion in a one dimensional geometry. Solving the local dispersion relation numerically in the vicinity of the ion ion hybrid layer, we show dispersion curves for the fast wave and mode-converted Ion Cyclotron Wave in the cold plasma limit and with finite electron temperature. Due to the rapid upshift in $k_\parallel$, the Ion Cyclotron Wave can enter the acoustic domain $\frac{\omega}{k_\parallel v_{the}} < 1$ and propagate towards to the minority ion cyclotron layer, although it will strongly damp on electrons.

In multi ions plasmas with a moderate minority fraction, the fast wave (FW) dispersion relation [3] $n_\perp^2 = \frac{(R-n_\parallel^2)(L-n_\parallel^2)}{S-n_\parallel^2}$ indicates possible mode conversion (MC) to shorter wavelength modes at the ion-ion hybrid (IIH) layer, defined by $S = n_\parallel^2$. This process has been extensively studied in slab geometries [4, 5] where $k_\parallel$ is conserved. In such conditions, the FW is converted into an electrostatic ion Bernstein mode (IBW) [6] propagating on the high field side (HFS) of the IIH layer. In toroidal geometries however, the poloidal field can lead to a significantly different situation. Perkins [1] showed theorically that the fast wave could be mode converted to an Ion Cyclotron Wave (ICW) propagating towards the low field side (LFS) of the IIH layer. Experimental and numerical studies on Alcator C-Mod [2] showed that this process could dominate over MC to IBWs.

In this paper, we use a simple one-dimensionnal model to determine how poloidal field effects make mode conversion to ICW possible in terms of the local dispersion relation. We show the key role played by the $k_\parallel$-upshift in this process, and how this model can be used to predict some general characteristics of the mode converted Ion Cyclotron Waves.

---

[1] Email contact : parisot@mit.edu

ONE DIMENSIONNAL MODEL

We consider a slab geometry in which the plasma is infinite and uniform in the y and z directions. The antenna excitation imposes $k_z =$ constant and $k_y = 0$. The static magnetic field has the following form $\vec{B} = B_t \vec{z} + B_p \vec{x} = B(x)(b_t \vec{z} + b_p \vec{x})$. For different values of $B(x)$, the local dispersion relation $D(k_\parallel, k_\perp) = 0$ can be used to determine $k_x(x)$, with:

$$k_\parallel = b_t k_z + b_p k_x$$

$$k_\perp^2 = k_z^2 + k_x^2 - k_\parallel^2$$

COLD PLASMAS RESULTS

Figure 1 shows the typical shape of the dispersion curves in the cold plasma limit for a $D(^3He)$ with 20 % helium, as obtained by solving the cold plasma dispersion relation. We considered $k_z = 0$ for simplicity.

The main effect of the poloidal field in this limit is a shift of the cold plasma ion-ion hybrid resonance (defined here as $k_x \to \infty$) towards the low field side. This situation corresponds to a mode transformation from the fast magnetosonic wave to the cold shear Alfvén wave. Stix [3] called this branch Ion Cyclotron Wave as the shear wave is strongly modified by the proximity of ion cyclotron resonances. An analytic study of the dispersion relation shows that the condition for resonance can be written $S + b_p^2 P = 0$. The shift increases as b_p is increased, but as $S \to \infty$ at the ion cyclotron layer, the resonance will always occur between the IIH and the ^3He cyclotron layer.

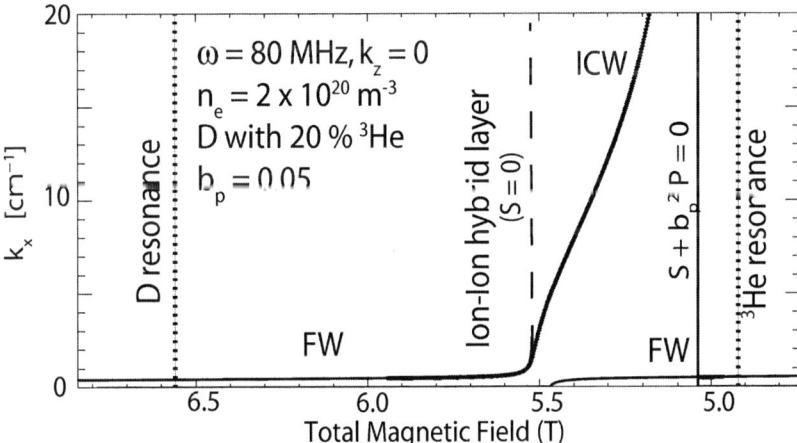

FIGURE 1. Typical dispersion curves in the cold plasma model show the cold ICW branch on the LFS of the ion-ion hybrid layer.

$K_\parallel$ UPSHIFT AND PARALLEL ELECTRON DISPERSION

Finite temperature effects are expected to resolve the cold plasma resonances. With a zero poloidal field, this resolution has been shown to involve mode conversion to the IBW on the HFS of the Ion-Ion hybrid layer at finite electron and ion temperature.

In presence of a poloidal field, the cold plasma resonance can be resolved at finite electron temperature but zero ion temperature. Due to the upshift in $k_\parallel$ as k_x increases for the shear wave, the parallel phase velocity becomes comparable or lower than the electron thermal speed $v_{the} = \sqrt{\frac{2T}{m}}$. The shear wave enters the acoustic regime for which $\frac{\omega}{k_\parallel v_{the}} < 1$ (figure 2) and becomes propagative beyond the shifted cold resonance layer, although it will be strongly damped by electron Landau damping. This situation is very analogous to the case of an Alfven resonance in the low frequency (MHD) regime, where the fast wave mode converts to a kinetic Alfven wave [7].

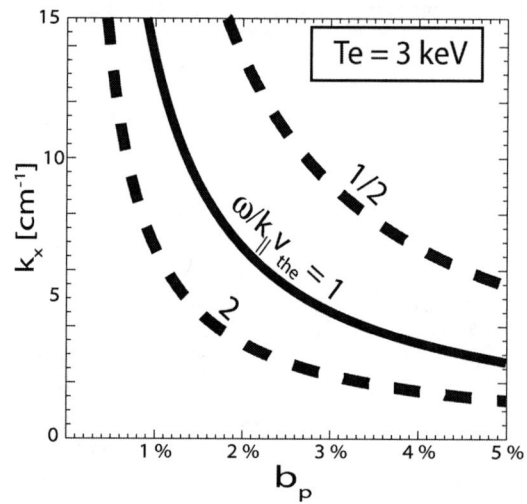

FIGURE 2. The upshift in $k_\parallel$ makes the parallel phase velocity comparable or larger than the electron thermal speed.

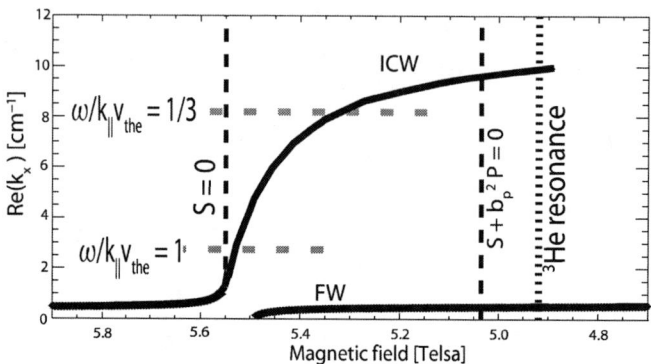

FIGURE 3. Electron temperature effects on the Ion Cyclotron branch for $T_e = 3keV$, $T_i = 0$. Other parameters are the same as in figure 1.

As $\frac{\omega}{k_\| v_{the}} \sim 1$, the plasma dispersion function [8] for electrons must be retained without approximation in the vicinity of the IIH. Approximate values for the roots of the dispersion relation are obtained using a test array of real values for k_x, covering the typical range for mode converted waves. The plasma dispersion function is computed for each test value, which reduces the dispersion relation to a four-order polynomial equation in k_x. Roots which have close proximity to the original test values for k_x are considered valid solutions and are retained as approximate solutions of the full dispersion relation. The result of this procedure is shown on figure 3.

EFFECTS OF ION TEMPERATURE AND CONCLUSIONS

The numerical approach can be extended to finite ion temperature. As T_i is increased, minority ion cyclotron damping competes with electron damping for the Ion Cyclotron Wave. The upshift in $k_\|$ Doppler-broadens the ion cyclotron resonance region significanty. Therefore ion cyclotron damping can dominate over electron damping for the mode converted ICW even in the vicinity of the IIH. Additionally, mode conversion from the fast wave to the IBW branch can dominate over MC to ICW as the ion temperature is increased and finite Larmor effects for ions become more important. The conditions for this transition can be studied in this one-dimensional configuration and will be reported in future work.

In summary, the poloidal field affects significantly wave propagation around the ion-ion hybrid layer in a multiple ion species plasma. In the cold plasma limit, the fast wave is mode-transformed into the Ion Cyclotron Wave. For typical tokamak electron temperatures, the ICW can transition into the acoustic regime $\frac{\omega}{k_\| v_{the}} < 1$ as $k_\|$ is upshifted. The wave propagates towards the Ion Cyclotron resonance layer, and is strongly damped on electrons and ions.

ACKNOWLEDGMENTS

This work is supported by USDOE Coop. Agree. No. DE-FC02-99-ER54512.

REFERENCES

1. Perkins, F., *Nucl. Fusion*, **17**, 1197 (1977).
2. Nelson-Melby, E., et al., *Phys. Rev. Lett.*, **90**, 155004 (2003).
3. Stix, T. H., *Waves in plasma*, American Institute of Physics, 1992.
4. Swanson, D., *Phys. Fluids*, **28**, 2645 (1985).
5. Brambilla, M., et al., *Plasma Phys. Control. Fusion*, **27**, 1 (1995).
6. Ono, M., *Phys. Fluids B*, **5**, 241 (1993).
7. Stix, H., "Alfvén Wave Heating," in *Proc. 2nd. Int. Symp. on Heating in Toroidal Plasmas*, Grenoble, 1980, vol. 2.
8. Fried, B., and Conte, S., *The Plasma Dispersion Function*, Academic Press, 1961.

Modeling of Ray Splitting in a Tokamak

E. R. Tracy *, A. Jaun,[†] and A. N. Kaufman **

*Physics Department, College of William and Mary, Williamsburg, VA, ertrac@wm.edu
[†] NADA, Royal Institute of Technology, Stockholm, Sweden, jaun@kth.se
** LBNL and Physics Department, UCB, Berkeley, CA, ankaufman@lbl.gov

Abstract. We summarize recent progress toward developing a practical ray-tracing algorithm that includes linear mode conversion, for realistic tokamak geometry. The goal of this code development project is to create new and easy-to-use numerical ray-tracing tools for the study of RF heating and current drive. After a brief summmary of our recent results, a series of figures showing mode conversion in tokamak geometry is presented, emphasizing the effect of poloidal magnetic field.

Keywords: mode conversion, ray splitting, numerical, ray tracing, tokamak
PACS: 52.35.Bj, 52.35.Lv

Ray-based methods for studying wave propagation in plasmas have several attractive features. They are much faster than full-wave calculations, and also give much-needed physical insight concerning energy propagation. They are not meant to replace full-wave methods, but to complement them. For example, because they run faster than full-wave simulations, they allow a wider range of parameter studies. This will lead to the identification of interesting parameter values for input to full wave studies. If both the ray-tracing and full-wave algorithms give similar physical predictions, this provides significantly greater confidence in the correctness of the calculations.

Ray methods have been used with success for many years, but until recently they have been unable to treat the phenomenon of *linear mode conversion* in nonuniform plasma. This is because an *incident* ray (see Figure 1) that enters a conversion region *splits* into two outgoing rays: a *transmitted* ray and a *converted* ray (which is a continuation of the incident ray). The *reflected* ray emerges after a second conversion, as described in our earlier papers. Our algorithm, described in [1], is the first practical method for dealing with ray splitting. (Our methods for treating the evolution of polarization and amplitude have also been developed theoretically, but have not yet been implemented.) Figure 1 also presents a conceptual summary of the conversion process in a tokamak cavity. A family of rays is launched from the low-field side of a tokamak. As each ray enters the ion-hybrid resonance region, it undergoes a sequence of *two* conversions. This double conversion is confirmed by these direct ray tracing results presented in Figure 3.

We are developing a multidimensional ray tracing code RAYcON (for *ray conversion*), using the MATLAB language to provide for flexibility during its development and to make it easy to modify the code. RAYcON performs ray tracing in realistic tokamak geometry including mode conversion, and is the first code of its kind. In its first version we are concentrating on the ICRH range of frequencies, and use the cold plasma model for the dispersion tensor. Hence the code will be useful for studying heating and current- and flow-drive using the ion-hybrid resonance. Future enhancements would include thermal effects and the extension to electron waves.

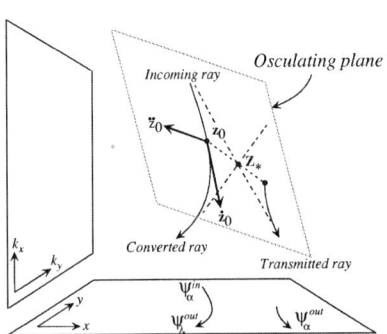 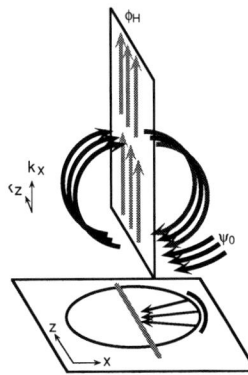

FIGURE 1. (Left figure) Phase space diagram of the mode conversion region. The simplest type of conversion in multidimensions involves an 'avoided crossing' type of geometry in phase space. The rays are locally confined to a two-dimensional osculating plane and are hyperbolic. It is important to note that this two-dimensional plane is embedded in four-dimensional phase space. (Right figure) A conceptual diagram showing double conversion in the poloidal plane of a tokamak. The rays are launched by an antenna on the low-field side of the tokamak. This family is denoted ψ_0. As each ray enters the conversion region, it produces a transmitted magnetosonic ray (shown in black on the high-field side of the resonance) and an ion-hybrid ray that remains in the resonance layer (shown as gray arrows). This ray propagates primarily in k-space, and enters a second conversion region, splitting off a reflected magnetosonic ray.

Our code is built for general tokamak equilibria using the same interface as developed for the PENN code [2]. The magnetic geometry uses a Soloviev equilibrium with density profiles modeled in terms of normalized poloidal magnetic flux $s^2 = \psi/\psi_s$: $n = n_0(1 - as^2)^b$ (a and b are parameters that fix the minor radius and steepness of the density profile). Figures 2-4 use a magnetic field equilibrium similar to those in JET: magnetic axis radius $R_0 = 3.0$ m; magnetic field strength $B_0 = 2.3$ T; aspect ratio = 3; elongation = 1.4; safety factor on axis $q_0 = 1.1$; deuterium-hydrogen plasma: $n_D = 5.5 \times 10^{18}$ m^{-3}, $n_H = 4.5 \times 10^{18}$ m^{-3}. The magnetic field geometry and density profiles are used to compute the dispersion tensor from a cold plasma model, either from a reduced 2×2 tensor (neglecting the parallel electric field component) or keeping all components. In the reduced model, we seek to find electric fields $\mathbf{E}$ that are solutions of $\mathbf{D}(\mathbf{k} \to -i\nabla, \mathbf{x}; n_\varphi, \omega) \cdot \mathbf{E}(\mathbf{x}) = 0$, with appropriate boundary conditions; ∇ is the gradient in the (R,Z)-plane ($\mathbf{k}$ is canonically conjugate to $\mathbf{x}$, $\mathbf{x} = (R,Z)$ are radial and vertical coordinates in the poloidal plane; $k_\varphi = n_\varphi/R$ is toroidal wavenumber. Poloidal magnetic field effects are included. Extension to fully 3–D equilibria (such as stellarators) is conceptually straightforward as far as the ray tracing aspects of the code are concerned.

The determinant of the dispersion tensor $\mathbf{D}(\mathbf{k}, \mathbf{x}; n_\varphi, \omega)$ plays the role of ray Hamiltonian [3, 4]: $\frac{d\mathbf{x}}{d\sigma} = -\frac{\partial H}{\partial \mathbf{k}}, \frac{d\mathbf{k}}{d\sigma} = \frac{\partial H}{\partial \mathbf{x}}$, with $H(\mathbf{x}, \mathbf{k}; n_\varphi, \omega) = det(\mathbf{D})$. The dispersion surface is defined by $H = 0$. The ray parameter σ is related to physical time via $\frac{dt}{d\sigma} = \frac{\partial H}{\partial \omega}$. These equations, along with transport equations for amplitude and Phase, yield WKB solutions to the wave equation outside mode conversion regions [3]. Following a ray, we detect where it enters a conversion region, assign initial conditions to the transmitted and converted rays, and apply connection formulas, using methods described in [1].

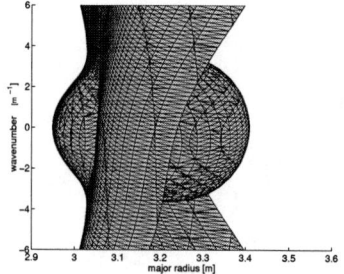
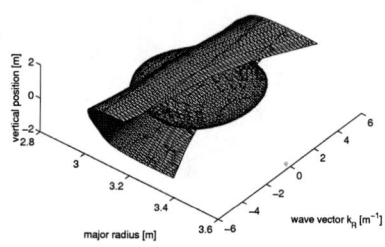

FIGURE 2. The dispersion manifold for the 2×2 cold plasma model is plotted as a three dimensional surface in (R, k_R, Z) with $k_Z = 0$ from two different perspectives. The left figure shows a view in (R, k_R) with Z perpendicular to the page, while the figure at right shows the same surface from a different angle.

The reduced 2×2 cold plasma dispersion tensor is used to model a D-H mode conversion scenario, using the JET-like parameters specified above. Figure 2 shows the dispersion manifold, *i.e.* the locus of points in the ray phase space (R, Z, k_R, k_Z) where $H = det(\mathbf{D}) = 0$. The frequency is $\omega/2\pi = 28$ MHz, and the rays are confined to the horizontal plane by setting $k_Z = 0$. The magnetosonic (fast) dispersion surface is the ellipsoidal shaped surface. If there were no ion-hybrid wave present (or if we could somehow 'turn off' the coupling between the fast wave and the ion-hybrid wave), the fast ray would be confined to this elliptically shaped surface. With no coupling, the ion-hybrid resonance would be a surface that is primarily vertical in the figure at left. The presence of coupling to the fast wave, and poloidal field effects that 'twist' the hybrid surface in an interesting manner, leads the hybrid-wave portion to develop a more complicated geometry.

Figure 3 shows results for a realistic tokamak geometry, with $\omega/2\pi = 28$ MHz, $n_\varphi = 0$. A single fast magnetosonic ray (1) is launched above the mid-plane from the low-field side of the tokamak, and splits into transmitted (2) and reflected magnetosonic rays (1) plus an ion-hybrid (3) ray. The MS rays subsequently propagate away from the IH resonance, are reflected by the low density cutoff, and return to the IH resonance below the mid-plane. The calculation is stopped soon after a second splitting takes place for the MS ray on the low-field side (LFS) of the resonance, while the MS ray on the HFS smoothly turns into an IH ray with an increasingly large wavenumber.

Figure 4 presents three calculations performed for a family of five MS rays. The three figures show how the poloidal magnetic field modifies the symmetry of the splitting, when rays with different toroidal mode numbers $n_\varphi = -9, 0, 9$, corresponding to wave vectors $k_\varphi = -3, 0, +3 \ m^{-1}$, are launched above and below the tokamak mid-plane.

Important future extensions of this code would include kinetic effects to treat absorption of the rays via Landau damping, and fully three-dimensional equilibria. It will be possible to treat ECRH by including electron dynamics.

This research was supported by the US DOE Office of Fusion Energy under contracts DE-FG02-96ER54344 and DE-AC03-76SFOO098.

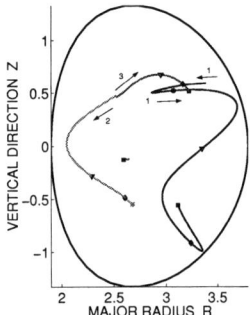 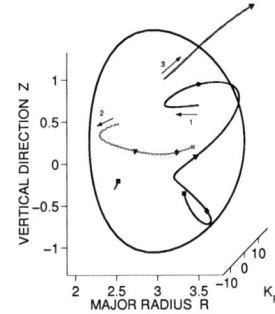

FIGURE 3. Ray splitting in a tokamak. The figure at left shows the splitting in the (R,Z)-plane, and the figure at right shows the same splitting in the projection of phase space (R,k_R,Z). (The symbols attached to each ray indicate points of equal ray parameter.) Note that the ion-hybrid ray (green) propagates to large k_R. Physically, this means that the disturbance develops short wavelength structure and will eventually Landau damp on the background plasma. See text for details.

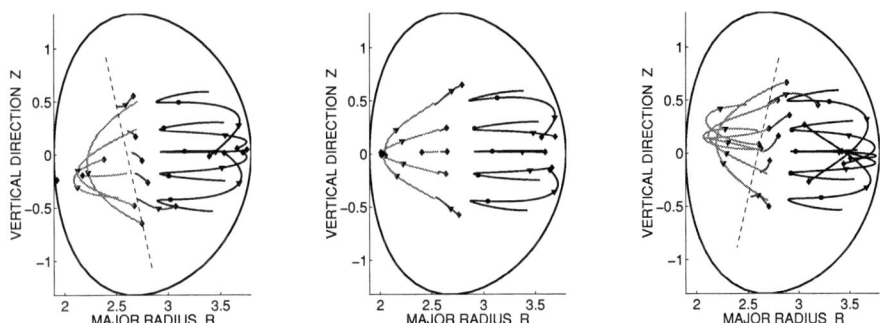

FIGURE 4. Ray splitting in a tokamak with poloidal field as a function of k_φ. Here, $k_\varphi = (-3, 0, +3) m^{-1}$ (left, center, right). Each figure shows ray splitting for a family of five rays, each of which undergoes a double conversion. The slanted dashed line guides the eye and suggests the orientation of the resonance layer. In these figures, the antenna wave vector k_φ is varied to study poloidal magnetic field effects, which modify the geometry of the dispersion manifold and the orientation of the resonance layer. Notice that the poloidal field effects cause the propagation characteristics of the ion-hybrid rays to be asymmetric above and below the midplane.

REFERENCES

1. E. R. Tracy, A. N. Kaufman, A. Jaun, Phys. Lett. A**290** (2001) 309-316.
2. A.Jaun, K.Appert, J.Vaclavik, L.Villard, *Comput. Phys. Commun.* **92** (1995) 153
3. A. N. Kaufman, in *Nonlinear and chaotic phenomena in plasmas, fluids and solids*, W. Rozmus and J. A. Tuszynski, eds. (World Scientific, New Jersey, 1991),
4. E. R. Tracy, A. N. Kaufman and A. J. Brizard, Phys. Plasmas **10**, 2147 (2003).

Triplicate Budden resonance in the presence of sheared flow

A. N. Kaufman*, A.J. Brizard[†] and E. R. Tracy**

Lawrence Berkeley National Laboratory, University of California, Berkeley, CA
[†]*Department of Physics, Saint Michael's College, Colchester, VT*
**Physics Department, College of William and Mary, Williamsburg, VA*

Abstract.
In order to reduce turbulent transport in tokamaks, several groups have introduced a layer of toroidal or poloidal sheared flow. We study the influence of such a layer on the process of linear wave resonance, also known as mode conversion. Because of the local Doppler shift caused by the flow, the dispersion relations of waves are modified, and a single Budden resonance may be replaced by a triple resonance. The occurence of such triplication depends on the layer's parameters: its location, maximum flow speed, and radial width. We use a standard one-dimensional slab model, with cold-plasma response in the ion-gyrofrequency range, with the example of conversion of a magnetosonic wave to ion-hybrid waves. When triplication occurs, wave energy is converted at three locations instead of one. This possibility must be taken into account in designing conversion scenarios for heating, for current-drive, or for flow-drive.

Keywords: mode conversion, shear flow, tokamak
PACS: 52.35.Bj, 52.35.Lv

In the standard Budden resonance, in which plasma properties vary linearly in one dimension x, resonant absorption of a non-dispersive wave a of frequency ω occurs at a single value of x. This absorption is interpreted as a linear conversion from mode a to a different mode b, whose dispersion relation is modeled as linearly x-dependent but k_x-independent:

$$\omega_b^0(x) = \omega_0 \cdot (1 - x/L),$$

where L is the scale length of the relevant plasma property, the x-component of the wave-vector is denoted $k_x \equiv k$, and $\omega_0 = \omega_b^0(0)$. Resonant conversion requires that its frequency function $\omega_b^0(x)$ match the frequency ω of mode a: $\omega = \omega_b^0(x)$; we choose the arbitrary zero of the x-scale so that resonant conversion (in the absence of Doppler shift) occurs at $x = 0$: $\omega_0 = \omega$.

In the slab model of a tokamak with background magnetic field $\mathbf{B}(x) = B_0(1-x/L)\hat{z}$, we take the radially-localized flow as $\mathbf{u}(x) = u(x)\hat{u}$, where $u(x)$ is the flow speed along a constant direction $\hat{u}$ in the (y,z)-plane. The frequency function is then Doppler-shifted as $\omega_b(x) = \omega_b^0(x) + K u(x)$, where $K \equiv \mathbf{k} \cdot \hat{u}$ is the (constant) component of the wave-vector in the direction of the plasma flow. Let us introduce a Gaussian model for the shear-flow layer:

$$u(x) = u_0 \exp\left(-\frac{(x-x_f)^2}{2\Delta^2}\right), \tag{1}$$

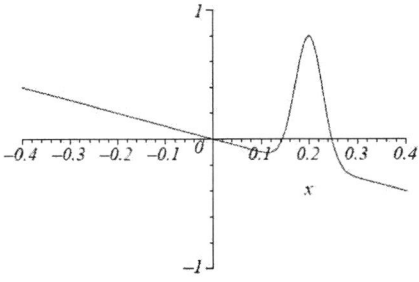
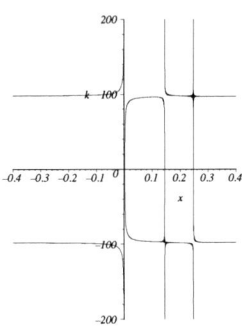

FIGURE 1. (Left) Plot of the ion-hybrid dispersion function $D_b(x)$ versus x. The zero-crossings are the resonances. (Right) Plot of the dispersion function $\det D = 0$ in the ray phase space (x,k). The vertical lines are the resonances x_1, x_2, and x_3.

where $u_0 = u(x_f)$ denotes the maximum velocity, while x_f and Δ denote the location and thickness of the shear-flow layer, respectively. The x-dependent frequency function of mode b is, therefore, modified to

$$\omega_b(x) = \omega \cdot (1 - x/L) + K u_0 \exp\left(-\frac{(x-x_f)^2}{2\Delta^2}\right),$$

and the dispersion function for mode b is

$$D_b(x) = \omega_b(x) - \omega = -\frac{x}{L}\omega + K u(x), \qquad (2)$$

where the plasma velocity $u(x)$ is given in Eq. (1). For a particular choice of parameters, this function is displayed in Fig.1(left).

We see that resonance can now occur at three locations: $x_1 = 0$ (the standard Budden resonance), x_2, and x_3. Note that the new resonances at x_2 and x_3 appear as a result of the existence of the shear-flow layer and require that the plasma flow satisfy the condition

$$\frac{K u_0}{\omega} > \frac{x_f}{L}, \qquad (3)$$

under the assumption that $K u_0 / \omega \gg \Delta / L$.

The wave equation for Budden resonance can be obtained from the normal form [1] of the dispersion matrix for linear mode conversion:

$$\mathsf{D}(x,k) = \begin{pmatrix} D_a(k) & \eta \\ \eta^* & D_b(x) \end{pmatrix}, \qquad (4)$$

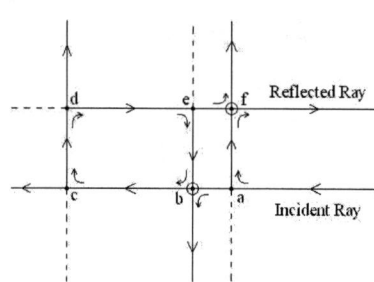 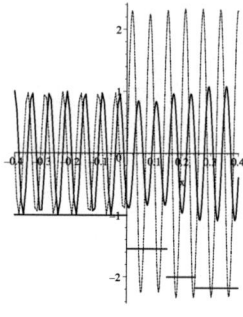

FIGURE 2. (Left) Plot of the uncoupled dispersion relations for low-field incidence. (Right) Plot of the real and imaginary parts of a solution for a wave launched from the right (low-field incidence, only transmitted wave at left). The horizontal line segments represent the flux.

where D_a, D_b are dispersion functions for modes a, b, and η is the (constant) coupling strength. For mode a, we take the expression appropriate to a magnetosonic wave:

$$D_a(k) = k^2 - k_a^2, \qquad (5)$$

where $k_a = \omega/c_A$, with c_A the (constant) Alfvén speed. Here, for simplicity, we ignore the Doppler-shift in the magnetosonic dispersion relation in order to demonstrate the resonance triplication from the Doppler-shifted ion-hybrid dispersion relation. We note that Doppler effects in the magnetosonic dispersion relation can also significantly enrich the resonance structure (for example, by introducing more cut-offs).

The WKB relation $k(x)$ is obtained by setting the determinant of D to zero:

$$k^2 = k_a^2 + \frac{|\eta|^2}{D_b(x)}. \qquad (6)$$

This relation is displayed in Fig. 1(right) for the same parameters used in Fig. 1(left). We note the three resonances (x_1, x_2, x_3), which appear as vertical lines in the phase-space plot. To interpret Fig. 1(right), we display the *uncoupled* dispersion relation in Fig. 2(left). The arrows indicate the direction of energy flow. Note that the ray of the middle resonance propagates to negative k; thus, its phase velocity in x opposes that of the outer two rays. Note also the ray crossings (indicated by circles) where two input channels interfere [2], as well as the recirculation structure [3].

The wave equation corresponding to the dispersion relation (6) is obtained by replacing k with $-i\,d/dx$:

$$\frac{d^2\psi(x)}{dx^2} + \left[k_a^2 + \frac{|\eta|^2}{D_b(x)}\right]\psi(x) = 0, \tag{7}$$

where $\psi(x)$ represents the magnetosonic wave field. If there is no plasma flow present, then $D_b(x)$ is linear in x and we have the standard Budden equation [4]. The singularities representing resonant absorption are resolved by the customary Plemelj prescription, from $\omega \to \omega + i\varepsilon$:

$$\frac{1}{D_b(x)} \to P\left(\frac{1}{D_b(x)}\right) + i\pi\,\delta(D_b(x)). \tag{8}$$

The numerical solution of Eq. (7) is shown in Fig. 2(right), for incidence from the right.

For interpretation, we examine the energy flux density, proportional to

$$F(x) = i\left[\psi^*(x)\frac{d\psi}{dx} - \psi(x)\frac{d\psi^*}{dx}\right]. \tag{9}$$

Multiplying Eq. (7) by ψ^*, and subtracting the complex conjugate, we obtain the flux jump at each resonance:

$$\Delta F = 2\pi \frac{|\psi|^2}{|dD_b/dx|}. \tag{10}$$

The flux is also shown in Fig. 2(right).

The conversion to the three (vertical) rays of mode b is given by the respective flux jumps. Thus we see that the shear-flow layer may significantly modify the locations of the wave energy transfer.

This research was supported by the US DoE Office of Fusion Energy under contracts DE-FG02-96ER54344 and DE-AC03-76SFOO098.

REFERENCES

1. A. N. Kaufman and L. Friedland, Phys. Lett. A **123**, 387 (1987).
2. E. R. Tracy and A. N. Kaufman, Phys. Rev. E **48**, 2196 (1993).
3. A. J. Brizard, J. J. Morehead, A. N. Kaufman, and E. R. Tracy, Phys. Rev. Lett. **77**, 1500 (1996); Phys. Plasmas **5**, 45 (1998).
4. A. N. Kaufman, E. R. Tracy, J. J. Morehead, and A. J. Brizard, Phys. Lett. A **252**, 43 (1999).

Theory and Practice in ICRF Antennas for Long Pulse Operation

L. Colas, E. Faudot*, S. Brémond, S. Heuraux*, R. Mitteau, M. Chantant, M. Goniche, V. Basiuk, G. Bosia, J.P. Gunn & the Tore Supra team

Association Euratom-CEA, CEA/DSM/DRFC, Centre de Cadarache,
13108 Saint-Paul lez Durance, France
*LPMIA, Unité du CNRS 7040, Université Henri Poincaré, Nancy 1
BP 239, 54506 Vandœuvre Cedex, France

Abstract. Long plasma discharges on the Tore Supra (TS) tokamak were extended in 2004 towards higher powers and plasma densities by combined Lower Hybrid (LH) and Ion Cyclotron Range of Frequencies (ICRF) waves. RF pulses of 20s×8MW and 60s×4MW were produced. TS is equipped with 3 ICRF antennas, whose front faces are ready for CW operation. This paper reports on their behaviour over high power long pulses, as observed with infrared (IR) thermography and calorimetric measurements. Edge parasitic losses, although modest, are concentrated on a small surface and can raise surface temperatures close to operational limits. A complex hot spot pattern was revealed with at least 3 physical processes involved : convected power, electron acceleration in the LH near field, and a RF-specific phenomenon compatible with RF sheaths. LH coupling was also perturbed in the antenna shadow. This was attributed to RF-induced DC $\mathbf{E} \times \mathbf{B_0}$ convection. This motivated sheath modelling in two directions. First, the 2D topology of RF potentials was investigated in relation with the RF current distribution over the antenna, *via* a Green's function formalism and full-wave calculation using the ICANT code [14]. In front of phased arrays of straps, convective cells were interpreted using the RF current profiles of strip line theory. Another class of convective cells, specific to antenna box corners, was evidenced for the first time. Within 1D sheath models assuming independent flux tubes, RF and rectified DC potentials are proportional. 2D fluid models couple nearby flux tubes *via* transverse polarisation currents. Unexpectedly this does not necessarily smooth RF potential maps. Peak DC potentials can even be enhanced. The experience gained on TS and the numerical tools are valuable for designing steady state high power antennas for next step devices. General rules to reduce RF potentials as well as concrete design options are discussed.

Keywords: ICRF, LHCD, high power long pulses, IR imaging, RF sheaths, $\mathbf{E} \times \mathbf{B_0}$ convection.
PACS: 52.50.–b, 52.40.Kh, 52.65.–y, 52.35.–g, 52.40.Fd

Long plasma discharges on the Tore Supra tokamak, first sustained by LHCD alone, were extended in 2004 towards higher powers (up to 11MW) and plasma densities (up to 80% of $n_{Greenwald}$) by combined LH and ICRF waves. For core physics programs such target plasmas are more ITER-relevant than previous ones [1]. For RF wave launchers, more than the mere extrapolation of shorter ones, such high power long pulses raise specific issues, combining technology, experiments and modeling. Any spurious phenomenon (parasitic power loss, impurity production, density rise), tolerable over transient phases because of its modest amplitude, can become dangerous in steady-state regime, due to spatial localization or long-time accumulation. Another specificity was the combined operation of several wave heating systems, whose cohabitation over long pulses needed assessment. Such phenomena need careful experimental characterization and subsequent modeling, in order to propose adequate operational rules and design options for next step devices.

SPECIFIC LONG PULSE EXPERIMENTAL SETUP

Tore Supra (TS) is equipped with 3 ICRF antennas including 2 straps in metallic boxes, shielded by a B_4C-coated Faraday Screen (FS), and surrounded by CFC guard limiters or bumpers, all actively cooled by pressurized water circulation. Cooling constraints at generators and vacuum transmission line fix the upper limit of pulse pulse energies to 3×120MJ, for a full RF power of 3×4MW. In 2004 RF pulses of 20s×8MW and 60s×4MW were produced, in combination with up to 3MW LHCD.

In addition to "standard" diagnostics, specific ones were used to study plasma facing components (PFCs). Surface temperatures are monitored with infrared (IR) cameras [2]. In the future real-time image processing will trigger safety feedbacks. IR films were analysed *a posteriori*, bringing valuable time-resolved information on hot spot spatial location. Within simplest thermal models, the steady state temperature difference between front faces and the water is proportional to heat fluxes from the plasma. Parametric dependencies can thus be documented from relative comparisons. Quantifying fluxes from IR temperatures requires antenna thermal properties. Cooling-down time constants suggest that these properties can evolve between experimental campaigns or antennas. Quantitative information is yet available from calorimetry, which compares inlet and outlet coolant temperatures for a given measured water flow in the pipes [3]. One energy balance per cooling loop is thus produced, with a time resolution limited by water transit time through the circuit.

HIGH POWER LONG PULSE EXPERIMENTS ON TORE SUPRA

Figure 1 shows the time history of record shot TS33613. 8MW RF power were coupled for 20s and combined with 1MW LHCD. The D(H) minority scheme was used, with dipole strap phasing. After 5 seconds all PFCs reached a constant temperature. Over the shot the impurity level remained low and showed no sign of accumulation. The energy evacuated by the FS as measured by calorimetry was only 2.5% of the injected ICRF energy. This fraction sets an upper bound on parasitic RF losses on the antenna front face. In spite of these encouraging results the steady state FS

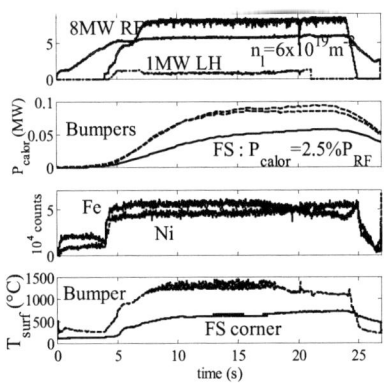

FIGURE 1. Time history of shot TS33613.

surface temperatures locally exceeded 700°C, i.e. near operational limits for this PFC. Inspection at shutdown revealed localised degradations on the front faces. The reason appears on IR images : losses were concentrated on a small surface of FS.

Figure 2 is an antenna IR picture after 60s operation at 4MW total RF power and 3MW LH power. It reveals a complex surface temperature pattern. A systematic parametric study identified at least three physical processes causing hot spots, depending on the location on the front face [4]. Convected power is visible on the lower part of the left guard limiter. It was verified that the local surface temperature scales linearly with the total injected power, whatever its origin. The right guard limiter was connected magnetically to the six waveguide rows of a LH grill. Fast electrons accelerated in the LH near field [8] caused the six localised beam impacts on this private limiter. It was checked that the local surface temperature reacts to the LH power, and that the beam strike points are sensitive to the edge safety factor. The FS and the upper part of the bumpers are sensitive to RF-specific mechanisms.

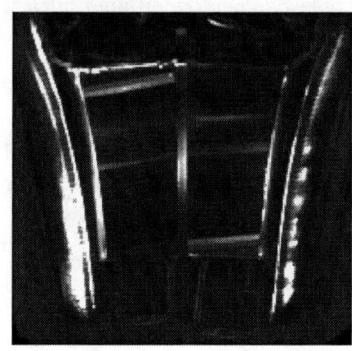

FIGURE 2. IR image (antenna Q1, shot TS33748)

A striking observation from figure 2 is the spatial asymmetry between the thermal behaviour of the different antenna box corners, in spite of the geometrical symmetry of the antenna itself. Upper-left antenna corners were systematically hotter than the lower left ones over all long shots and for all antennas, so that this can hardly be explained by a mechanical misalignment. A similar asymmetry was already observed on several machines [5-7], but previously on TS the lower right corners were hotter [7,12]. Since then the total confinement magnetic field was reversed, suggesting that this is the likely cause of the hot spot pattern asymmetry and reversal.

Figure 3 summarizes the parametric dependences of the surface temperature on one upper left corner in dipole strap phasing. Parametric behavior was similar for all three antennas, but the magnitudes differed, in part due to differences in antenna thermal properties. Good correlation was found with a scaling of the form $n_l P_{local}^{1/2} \exp(-d/d_0)$, where n_l is the line integrated density, P_{local} is the RF power from the observed antenna, d is the distance from the antenna to the last closed flux surface, and d_0 a characteristic radial length of the order of 14.5mm. To reduce the hot spots at given RF power, one could act on n_l or on d, but that deteriorates RF coupling, so that the RF antenna radial position for high power long shots resulted from a trade-off between

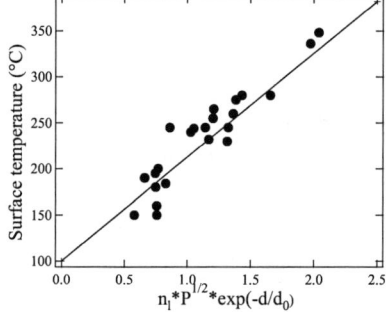

FIGURE 3. Parametric behavior of the surface temperature on upper left antenna corner.

these conflicting operational constraints. Heat loads on RF antennas were also sensitive to strap phasing. As soon as phasing departed significantly from dipole, the PFCs got hotter, the FS temperature pattern was changed, and the impurity content was raised. High power long shots were only attempted in dipole phasing.

Combined ICRF and LHCD operation also revealed delicate, due to interactions between magnetically connected objects. Electron acceleration in the LH near field was already illustrated on figure 1. Figure 4 compares over the 2004 campaign the LH reflection coefficients before and during ICRF power application. It shows quite often LH coupling degradation, which was correlated with measured RF-induced local density reduction [9]. Combined operation was yet achieved, by adjusting the relative radial positions of LH and ICRF launchers. Cohabitation was not as critical as on JET [9] because the target plasma was at high density.

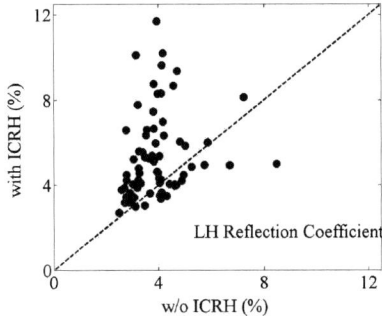

FIGURE 4. LH reflection coefficient after RF switch on versus its value before RF pulse.

RF SHEATH MODELLING

The observed phenomenology of FS upper left corners and of RF-LH interaction is reminiscent of sheath rectification processes [10]. In this framework ICRF antenna operation drives a pure RF oscillating potential V between the extremities of open magnetic flux tubes, given by $V = \int_L E_{//} \cdot dl$, where $E_{//}$ is the RF parallel electric field and integration is along the open field line. As a reaction to V, and due to the non-linear electrical characteristic of the sheaths at both flux tube extremities, the field line gets biased to a rectified DC potential, which at high RF power is well above typical floating potentials in the Scrape-Off Layer (SOL). Ions accelerated across the high DC potential cause enhanced sputtering and localised high heat fluxes at both field line ends. Within simplest models the energy gained by the ions scales as $P_{local}^{1/2}$, while the ion flux through the sheath is the Bohm flux $n_{local}c_s$, with $c_s = (k(T_e+T_i)/m_i)^{1/2}$ the local sound speed, and n_{local} the local density. Eventually parallel heat losses to the PFCs scale as $n_{local}c_s P_{local}^{1/2}$. Such parametric behaviour is close to that of figure 3.

In this model with independent flux tubes, each field line develops its own RF potential. Since the mapping of $E_{//}$ is spatially inhomogeneous, the differential biasing of nearby flux tubes creates RF-induced DC $\mathbf{E} \times \mathbf{B_0}$ particle convection transversally to the field lines [11]. It was shown that $\mathbf{E} \times \mathbf{B_0}$ convection could create an up-down density unbalance in the vicinity of TS antennas, and hence an asymmetrical hot spot pattern that reverses upon $\mathbf{B_0}$ reversal [12]. In the antenna shadow the local density is reduced, which could explain the RF-induced perturbations of LH coupling on waveguide rows connected to RF antennas [9].

In order to understand the observed hot spot pattern, and to reduce it by acting on the antenna front face design, one has to investigate how the spatial distribution of RF currents over the RF launching structure determines the topology of RF potentials radially in front of this structure. Linking these two quantities is not straightforward in the general case, but is considerably simplified on the particular category of "long" open field lines, extending toroidally far away on both sides of the antenna structure, and where convective cells are suspected to develop [13]. The geometry adopted is the slab development of the tokamak in figure 5. The radial, poloidal and toroidal directions are respectively x, y and z. The DC confinement magnetic field $\mathbf{B_0}$ is homogeneous and tilted by a pitch angle α with respect to the z axis. RF fields are described in

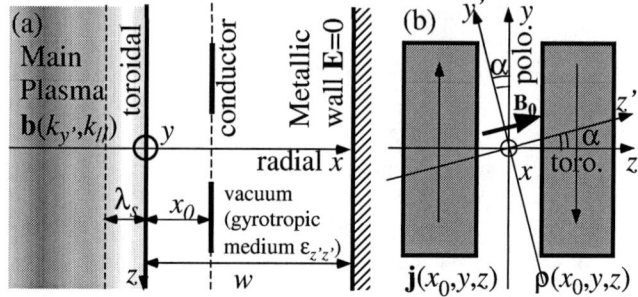

FIGURE 5. Geometry for convective cells modelling.

the tilted frame (x,y',z') whose axis z' is parallel to $\mathbf{B_0}$, and they are computed in a radial domain $0<x<w$, filled with an homogeneous dielectric medium. The metallic chamber wall is in $x=w$. $x=0$ constitutes the radial frontier with the "main plasma", whose dielectric properties for the slow wave are characterized by a skin depth λ_s. Harmonic oscillations at the RF pulsation ω_0 are assumed. The contribution of long field line (x_0,y') to the RF potential arises from the line integrated parallel RF current $\int_L j_{//}(x_0,y',z').dz'$. The 2D transverse map $V(x,y')$ is derived from this source term through convolution with a Green's function G. Physically, $G(x,x_0,y')$ can be regarded as the parallel electric field induced in (x,y') by an infinitely long and thin current wire placed in $(x_0,0)$ between two mirrors : one is the back wall, the other one lies radially λ_s inside the plasma. If all transverse dimensions are small with respect to the local wavelength for the slow wave $2\pi c \varepsilon_{z'z'}^{1/2}/\omega_0$, G takes the form :

$$G(x,x_0,y') = \frac{i\mu_0 \omega_0}{2\pi}\ldots$$

$$\ldots a \tanh\left[\frac{\sin(\pi(x+\lambda_s)/(w+\lambda_s))\sin(\pi(x_0+\lambda_s)/(w+\lambda_s))}{\cosh(\pi y'/(w+\lambda_s))-\cos(\pi(x+\lambda_s)/(w+\lambda_s))\cos(\pi(x_0+\lambda_s)/(w+\lambda_s))}\right] \quad (1)$$

Such formalism was first applied to a phased strap array. Strap s of the array is centered in (x_0,y'_s,z'_s) and is fed with an RF current of amplitude I_s and phase φ_s. It develops a sinusoidal poloidal current profile. Introducing the weighted average of y'_s, $\langle y'\rangle = \sum_s I_s y'_s / \sum_s I_s$, and assuming that $y'_s-\langle y'\rangle$, as well as all radial dimensions, are small with respect to the strap electric length, the convective cells adopt a simple pattern in front of the straps. Figure 6 illustrates the case of 4 straps. 3D spatial distributions of $\mathbf{j}$ and $\mathbf{E}$ were determined self-consistently around the launching structure with the antenna code ICANT [14], assuming $\lambda_s=0$. $E_{//}$ was subsequently integrated across the field maps along tilted field lines with $\alpha=7°$. In the radial direction V decreases linearly from the strap surface to λ_s inside the plasma. In the poloidal direction, within a shift by y'_s-$<y'>$, V follows approximately the RF current profile along one strap. The overall amplitude of V scales as $\tan\alpha \left|\sum_s I_s \exp(i\varphi_s)\right|$. Such result suggests two ways of reducing the sheath potentials : either tilt the straps perpendicular to field lines, so that α=0, or minimize the sum of complex RF currents $\left|\sum_s I_s \exp(i\varphi_s)\right|$. In the case of 2 straps, this latter criterion imposes dipole phasing. 4 straps offer more latitude: current drive with reduced convective cells seems possible with phasing $[0,\pi/2,\pi,3\pi/2]$.

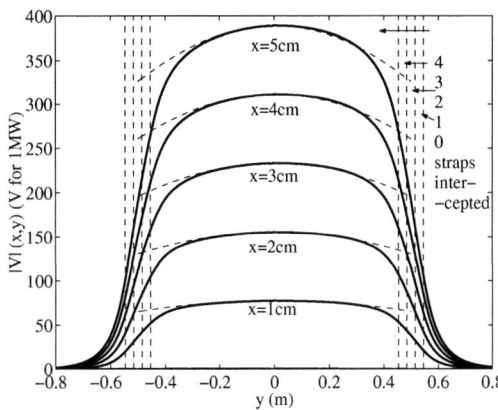

FIGURE 6. Mapping of $|V|$ in front of 4 identical 1m-long straps centered in (x_0=5.5cm,y_s=0), phased $[0,0,\pi/2,\pi/2]$. λ_s=0, w=16.5cm and α=7°. (--) : RF current profile ; (-) : $|V|$ profiles

Another class of convective cells, specific to antenna box corners, was evidenced for the first time Figure 7 describes the antenna structure used for simulation of the phenomenon with ICANT. It includes two straps phased in dipole, embedded in individual thick metallic boxes. Figure 7 also displays two tilted field lines limiting an (upper) antenna box corner region. Figure 8 shows a poloidal profile of $|V|$ in front of the antenna structure of figure 7. In the strap region $|V|$ is rather low, due to the "dipole" compensation of line-integrated $j_{//}$. In the box corner region of figure 8 specific high potential structures appear.

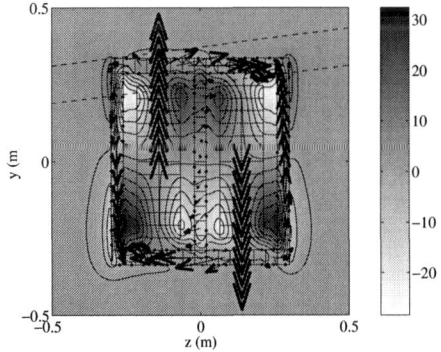

FIGURE 7. Mapping of Im($E_{//}$) in x=1.5cm, for 2 straps (π phasing) in boxes. Units : kV/m for 1MW coupled. Superimposed: antenna structure, RF current flow, and 2 tilted field lines (α=7°) limiting an antenna box corner zone.

FIGURE 8. $|V|(x=2.5cm,y)$ for : Dashed line : antenna in figure 7 ; solid line : same as figure 7, but toroidal conductivity is suppressed in the lower and upper part of the antenna box.

They are attributed to the parallel part of RF currents flowing in the lower and upper edges of the box. Figure 7 shows that these box RF currents are of moderate magnitude compared to those on the straps, but their projection in the parallel direction is larger. The above interpretation is confirmed on figure 8 by another simulation, in which the box toroidal conductivity was artificially suppressed in its lower and upper parts. $|V|$ was then drastically reduced in box corner zones, without affecting the antenna RF coupling properties. These results suggest to prevent RF currents from flowing toroidally in future front face designs.

All above results relied implicitly on a 1D sheath model assuming independent flux tubes, and figure 8 shows that within such framework large transverse gradients of V build-up locally. Due to the transverse plasma conductivity, these gradients can drive transverse currents between adjacent flux tubes. Influx of transverse current into a flux tube can in turn influence the sheath rectification process on this field line, i.e. the potential map. Self-consistent 2D fluid models of flux tubes coupled *via* transverse polarization currents were studied in [15]. Polarization current being a capacitive process, one could think that its average effect over a RF period is null. But the reaction of a flux tube to an influx of transverse current becomes highly non-linear when this influx approaches twice the ion saturation current j_{sat}. If such a limit is reached during a RF period, the transverse current is clamped to $2 j_{sat}$ and rectified DC potential maps get substantially modified. Such modifications were quantified on a 2D "test" RF map having initially a Gaussian shape in cylindrical geometry, of width r_0 and amplitude V_0. If V_0 is far above the floating potential, the simplest independent flux tubes model would produce a DC map with the same Gaussian shape and with a peak DC potential of V_0/π. At the top of the test RF map, the transverse current influx reaches the critical current if

$$(\omega_0/\Omega_{ci})(eV_0/kT_e)(L_{//}\rho_s/r_0^2) > 1 \quad (2)$$

where Ω_{ci} is the ion cyclotron frequency, T_e is the electron temperature, $L_{//}$ is the field line length and $\rho_s = c_s/\Omega_{ci}$ is the ion Larmor radius at the ion sound speed. For typical TS conditions ($kT_e \sim 20$eV, RF frequency 50MHz, 1MW coupled, $\omega_0 \sim \Omega_{ci}$, $L_{//} \sim 60$cm), eV_0/kT_e can reach 100 (see figure 8), $\rho_s \sim 0.1$mm so that structures of size $r_0 < 7.5$cm on the map are concerned. Figure 9 illustrates the test DC map above threshold (2) : peak DC potentials get enhanced, by up to $\pi/2$. Modifications are also observed in the shape of the DC map, which still need investigation.

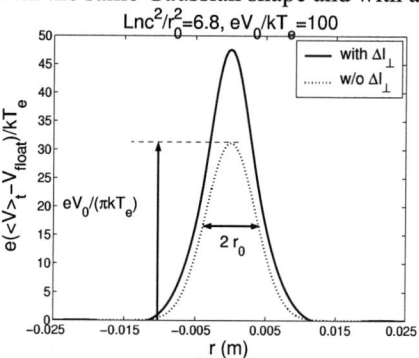

FIGURE 9. DC potential map in a transverse plane with and without transverse RF current. $eV_0/kT_e = 100$, $L_{nc}^2/r_0^2 = (\omega_0/\Omega_{ci})(L_{//}\rho_s/r_0^2) = 6.8$. DC potentials are normalized to kT_e/e.

LESSONS FOR ITER

The experience gained on Tore Supra and the numerical tools developed are valuable for designing steady state high power antennas for next step devices.

The first lesson drawn from high power long pulses is that any spurious power loss in the plasma edge, even modest, is a potential source of damage for PFCs, if it is concentrated over a small surface. Therefore no impurity rise in the plasma core does not guaranty quiet hot spots. IR monitoring thus proved to be mandatory, not only as a safety tool but also as a diagnostic: IR observation enabled to localise spatially the weak points, and to determine some parametric dependencies of the heat loads. In complement calorimetry offers quantitative estimates of the power losses.

Systematic front face observation revealed a complex hot spot pattern [4]. At least three different physical processes were identified: convected power, electrons accelerated in the LH near field, and a RF mechanism compatible with RF sheaths. Each process calls for a specific answer, but also for a global compatibility assessment of all solutions. Objects connected magnetically proved to be in close interaction, and also need to be tested together. This is a typical integration challenge, which often results in trade-offs between conflicting operational constraints, e.g. cooler FS *vs* better RF coupling, or adjusting the relative radial positions of RF and LH launchers. TS is a good test bed for this challenge, and should experiment in 2005 a prototype 2×2 strap array featuring the ITER-like matching scheme [16].

High power long RF pulses were achieved by finding operational solutions to limit hot spots, but another way to do so is to act on the antenna front face design to modify the RF current distribution. Hot spot minimization should be incorporated in the design process of future antennas, in the same way as the optimization of electrical properties, or the prevention of arcs. As an indispensable complement to experiments, edge RF physics modelling is the only alternative to empirical antenna design. A number of simulation tools were developed and could be further upgraded. Beyond interpreting some of the experimental observations, modelling already produced general rules to reduce RF potentials, and proposed concrete design options for the front face, e.g. reducing the antenna box toroidal RF currents in its horizontal parts.

REFERENCES

1. G.T. Hoang & al. , "Long pulse, multi-MW operation in Tore Supra", this conference.
2. D. Guilhem & al. "IR surface temperature measurements for long pulse operation, and real time feed-back control in Tore-Supra, an actively cooled Tokamak" , To be published in QIRT journal
3. J.C.Vallet & al., "Calorimetry in Tore Supra. An accurate tool and a benchmark for ITER", *Journal of Nuclear Materials* **313-316** (2003), pp 706-710.
4. S. Brémond & al., "Heat loads on Tore Supra ICRF Launchers PFCs", this conference.
5. R. Majeski & al. proc. 20th Eur. Physical Society Conf., Lisbon, July 26-30, 1993.
6. S.J. Wukitch & al., *Plasma Physics and Controlled Fusion* **46** (2004) pp. 1479-1491.
7. L. Colas, L. Costanzo, C. Desgranges, S. Brémond, & al, *Nucl. Fusion* **43** (2003) 1–15.
8. M. Goniche & al. *Nuclear Fusion* **38** (6) 1998, p. 919
9. A. Ekedahl & al., Proc. 15th Topical conf. On RF Power in Plasmas, Moran (Wy) 2003, AIP Conference Proceedings 694, pp. 259-262.
10. Perkins F.W., *Nucl. Fusion* **29** (4) 1989, p. 583.
11. D.A. D'Ippolito, J.R Myra & al. *Phys. Fluids* **B5** (10), 1993, p. 3603.
12. M. Bécoulet, & al. *Phys. Plasmas*, **9**, (6) 2002, pp. 2619-2632
13. L.Colas, S.Heuraux, S.Brémond, G.Bosia, "RF Current Distribution and Topology of RF Sheath Potentials in Front of ICRF Antennas" accepted in *Nuclear Fusion*
14. S. Pécoul, S. Heuraux, R. Koch, G. Leclert, *Comp. Phys. Com.* **146** (2) 2002, pp. 166-187.
15. Eric Faudot & al, this conference
16. G. Bosia & al., "First results of the Tore Supra ITER like ICRF antenna prototype", proc. 23rd Symposium on Fusion Technology, Venice (2004)

Investigation of "Conjugate T" Load-Resilient ICRF Antenna Systems - Application to the JET ITER-Like and to a Possible ITER ICRF System

P.U. Lamalle, A. M. Messiaen, P. Dumortier, F. Durodié, M. Evrard, F. Louche, M. Vervier and R. Weynants

Laboratory for Plasma Physics, Association EURATOM - Belgian State
Partner in the Trilateral Euregio Cluster
Royal Military Academy, 30 av. de la Renaissance, B-1000 Brussels, Belgium
e-mail : Philippe.Lamalle@rma.ac.be

Abstract. The paper reports on the radio-frequency (RF) analysis of multiple-short-strap load-resilient ICRF antenna systems, applied to the JET ITER-Like and to a proposed ITER ICRF system. The short radiating straps minimize the antenna voltage and the "conjugate T" load resilient matching circuit aims at reliable power delivery to ELMy H mode plasmas. The two designs mainly differ by the use of in-vessel matching capacitors for the JET array, whereas the proposed ITER design uses an optimized combination of straps in parallel and ex-vessel matching by means of line stretchers. Asymmetries and mutual coupling between straps strongly influence the performance of such load-resilient circuits and complicate their operation. These effects have been analyzed in detail along two parallel lines of investigation: *(i)* Detailed RF simulations, in which the input impedance matrix of the ICRF arrays has been computed with a three-dimensional electromagnetic code and incorporated in realistic models of the transmission and matching circuits. *(ii)* Comprehensive RF measurements on a scaled-down mockup of the proposed ITER antenna. Ongoing work to optimize array performance and to develop practical matching procedures and reliable automatic control of the matching elements is discussed. The main outstanding issues are reliable arc detection and demonstration of a robust array control algorithm.

Keywords: Ion Cyclotron Resonance Frequency, ICRF, antenna array, load-resilient impedance matching, JET, ITER.
PACS: 52.35.Hr, 52.50.-b, 52.50.Qt.

INTRODUCTION

The ITER ICRF launcher is requested to deliver 20MW through a single ITER port, at nominal distances between plasma last closed flux surface and machine wall on the order of 14cm (and more) [1]. To try and meet the resulting challenging power density and low antenna loading, the reference design was based on multiple short radiating straps minimizing voltage on plasma-facing conductors, connected in parallel in the resonant double loop (RDL) scheme [2]. More recently a variant of this circuit, the so-called "conjugate T" (CT) load resilient matching scheme has been proposed [3], aiming at reliable RF power delivery to ELMy H mode plasmas. (The basic operation of the CT is outlined in the following section.) An experimental validation of these novel concepts in ITER-relevant plasmas is under realization with the JET ITER-Like (JET-IL) ICRF system, to be installed in 2006 [4]. Its design uses in-vessel adjustable matching capacitors, see Fig.1. Studies are concurrently under way to improve the reference ITER ICRF design: the proposal promoted in [5, 6] increases the number of straps from 16 to 24 (6 poloidal by 4 toroidal), groups them by triplets optimally connected in parallel, and uses line stretchers located outside the vacuum vessel as adjustable CT matching elements. This design is known as ITER-EM (for "external matching"), see Fig.2. Another layout based on in-vessel capacitors is proposed in [7].

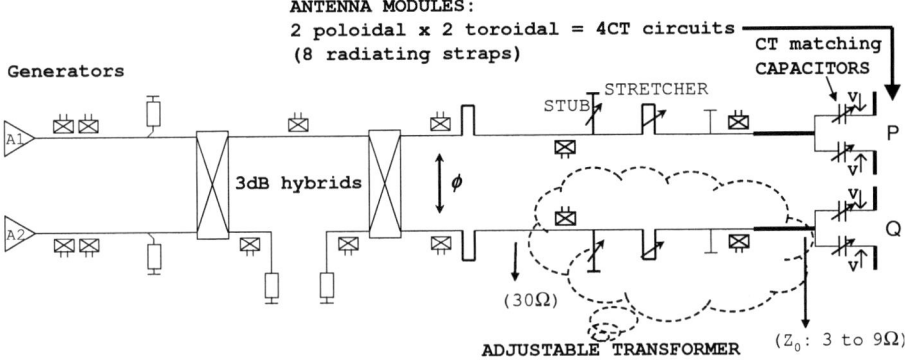

FIGURE 1. Schematic layout of one toroidal half of the JET-IL ICRF circuit (after [4]) showing two of the four CT circuits. The CT capacitors and the adjustable (stub-line stretcher) impedance transformer provide four effective matching parameters per CT.

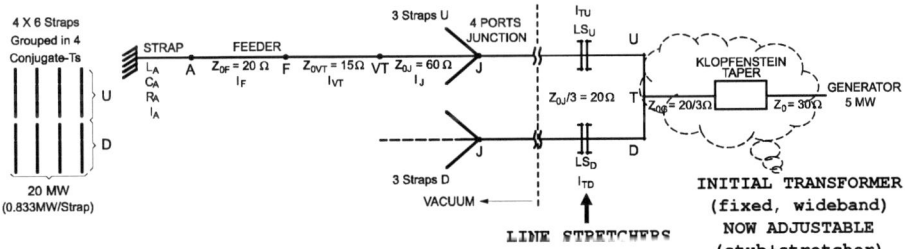

FIGURE 2. Schematic layout of the ITER-EM ICRF circuit [5, 6]. The CT line stretchers and the adjustable impedance transformer provide four effective matching parameters per CT circuit.

The paper reports on the RF analysis of multiple-short-strap load-resilient ICRF antenna systems, applied to the JET-IL and the ITER-EM ICRF systems. Asymmetries and mutual coupling between the radiating straps have been found to strongly influence the performance of such load-resilient circuits and complicate their operation [8, 9]. These effects have therefore been analyzed in depth, following two lines of investigation: Detailed RF simulations have been carried out by computing the input scattering matrix of the ICRF arrays with a three-dimensional electromagnetic code, and incorporating these equivalent parameters in realistic models of the transmission and matching circuits. In parallel with this activity comprehensive RF measurements have been made on a scaled-down mockup of the proposed ITER antenna, and are reported in the companion papers [10, 11]. Let us stress that good consistency is found between measurements and simulations. The complementary use of the two approaches has proved extremely fruitful and they jointly contribute to the results presented here.

Conjugate T (CT) Circuits

The conventional Resonant Double Loop (RDL) type of circuit [2] exploits susceptance cancellations between two branches connected in parallel. The CT circuit is a highly load-resilient variation on this concept, see Figs. 3 to 5. The generic "radiating units" of Fig.3 are the antenna straps with their feeders in the case of JET-IL, and triplets of straps connected in parallel through 4-port junctions in the case of ITER-EM. High load resilience is obtained by imposing a CT input impedance $Z_0 \sim 4$ times the ELM-free input resistance R_0 of these radiating units. This yields Z_0 between 3 and 6Ω for JET-IL and ≈7Ω for ITER-EM, much below the characteristic impedance of the main feeding lines (to which a conventional RDL would directly be connected). An impedance transformer between Z_0 and the main 30Ω line is thus required for these designs. In alternative layouts such as the CT with passive power repartition between 2 or 4 ITER-EM triplets considered in [11], a high level of prematching is achieved within the "radiating units", and with suitable design direct connection of the CT branches to the main lines (i.e. Z_0=30Ω) is even compatible with high load resilience.

The main characteristics of the CT circuit are summarized on Figs.4 and 5 in the ideal case of a single circuit with symmetric non-interacting antenna straps. The figures correspond to matching with series capacitors, but the case of line stretchers yields similar curves. Their most striking features are the following:

- The VSWR local maximum S completely determines the VSWR and branch current phase curves vs. the normalized branch resistance R/Z_0. In particular the branch reactances at the ELM-free matched point are $\pm X$ with $X = Z_0/S$.

- For the design value S=1.5, the matched loads are $R_0 = 0.25Z_0$ and $1.75Z_0$, with respective branch current phases 138° and 42°. The admissible (VSWR<1.5) resistance range spans 0.16 to $2.84Z_0$ (a factor of 18), and the ratio [maximum admissible R] / R_0 is 7.4 times that of a standard matching circuit. (The "classical" RDL antenna with 30Ω input would have $S \sim 3.2$ to 4.5 for the same R_0.)

- There are large excursions of branch current relative phase during the ELM, with quadrature at $R = X$.

- There is no resilience to reactive loading variations alone, nor to decreases in loading: a large increase of resistive loading is always necessary.

- The ideal CT circuit has two generic matched configurations for a given Z_0: "upper branch inductive + lower branch capacitive", and vice-versa.

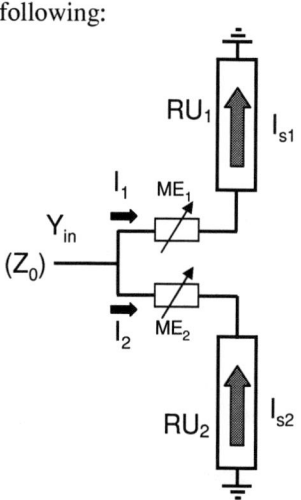

FIGURE 3. Generic CT circuit. "RU": radiating units, connected in parallel through "ME", adjustable matching elements; Z_0: reference input characteristic impedance. I_1, I_2: branch currents; I_{s1}, I_{s2}: direction of strap currents.

160

- The short radiating straps have a high quality factor: $X_{in}/R_{in} \approx \omega L'/R' \gg 1$ and typically ≥ 30 (ELM-free loading). (R_{in}, X_{in}: strap input resistance and reactance; R', L': distributed (lineic) resistance and reactance.) For S=1.5 this yields a CT branch input reactance $X \leq 0.1 X_{in}$, i.e. this matching scheme requires accurate reactance cancellations within each circuit branch and between branches.

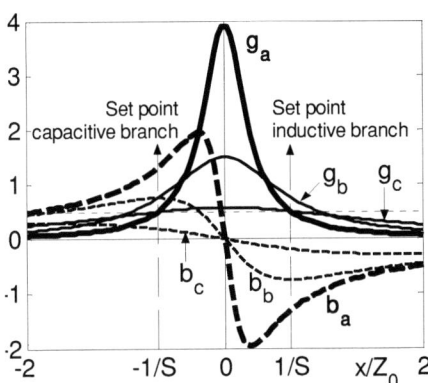

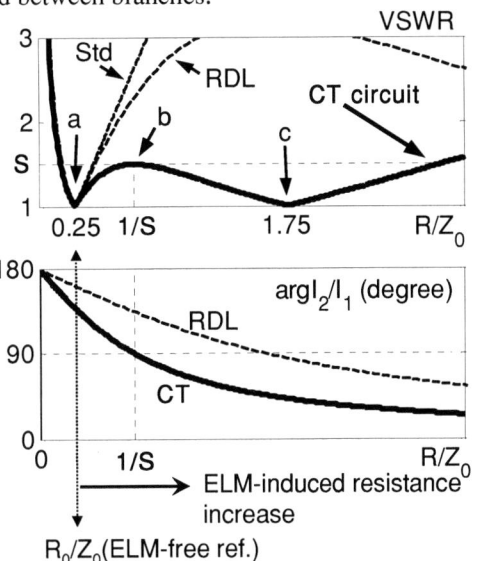

FIGURE 4. Normalized CT branch input conductance g and susceptance b (g+jb=Y Z_0) vs. normalized branch reactance x/Z_0. Subscripts a, b, c indicate different resistive loadings, see Fig.5: a, ELM-free reference; b & c during ELM, resp. at VSWR maximum and at second matched load. The branch reactances are opposite and the susceptances cancel throughout the ELM. Note the mild conductance variations.

FIGURE 5. Response of the ideal CT circuit to simultaneous resistive load variations in the branches: vswr and relative phase of branch currents. The cases of the conventional RDL and of standard matching of a single strap (Std) are also shown.

3D ELECTROMAGNETIC MODELLING OF ICRF ARRAYS

Models of the JET-IL and ITER-EM arrays have been developed with the commercial code CST MICROWAVE STUDIO® (MWS) [12-14]. This has allowed an accurate representation of the two launcher geometries and accurate evaluation of the RF near field, hence realistic estimates of the array input reactance matrix as a function of frequency. As MWS cannot model a magnetized plasma, loading of the antennas was simulated by means of a high permittivity dielectric slab. This approximation has been shown to capture the essential features of wave reflection and refraction at a steep plasma edge [15, 16]. It does obviously not account for plasma gyrotropy effects, which can be evaluated with plasma coupling codes and fed into the circuit analysis of the following section. In some simulations multiple stacked dielectric slabs were used to approximate the plasma scrape-off layer (SOL). However realistic quantitative predictions of resistive loading remain elusive due to the large uncertainties on the ITER SOL and far SOL density profiles, and we attach more significance to the relative magnitudes of the resistance matrix elements than to their absolute level which can a priori span a wide range. Important loading differences are found between straps, both toroidally and poloidally, due to their different surroundings within the array. Figure 7 shows sample strap impedance matrix elements for ITER-EM. Note the small value of the mutual impedances relative to the diagonal reactances. However their role is not negligible at all given the high degree of reactance cancellation inherent to the CT circuits.

The resistive mutual couplings reach 50% of the diagonal resistances, a feature which reflects the strong dependence of loading on strap phasing customary of ICRF arrays. The Faraday shield (FS) reduces the mutual impedances by 30 to 50% but also has an adverse effect on loading. Beside key thermo-mechanical considerations outside the scope of the present work the decision to install a FS on ITER or not will have to carefully ponder these antagonistic effects.

The MWS simulations have also guided the optimization of the JET-IL feeder design [13], leading to an important reduction of the estimated maximum electric field; a similar exercise is under way for the ITER-EM feeders [14].

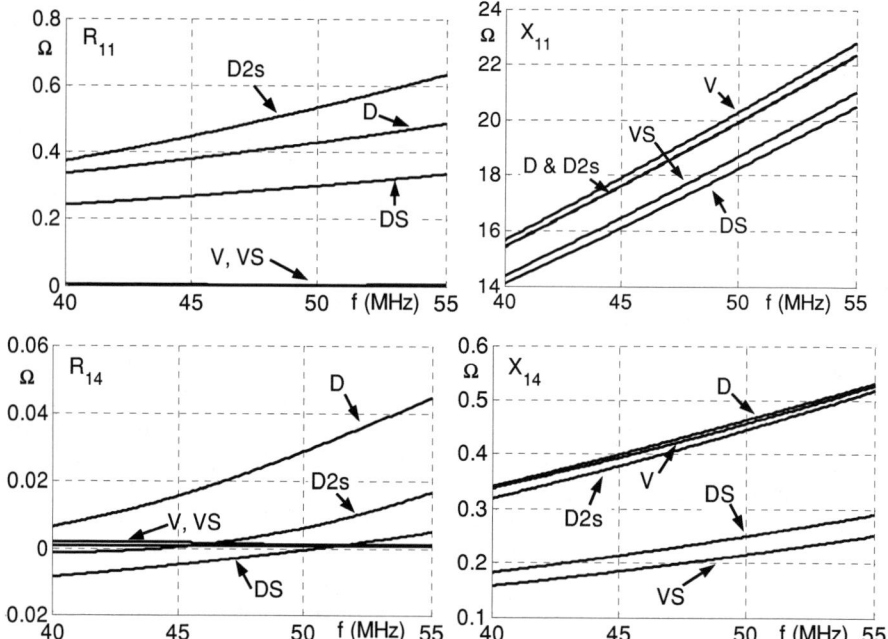

FIGURE 6. Partial view of the MWS model of the ITER-EM array, and frequency dependence of two impedance matrix elements $Z_{ij} = R_{ij} + j X_{ij}$ of the radiating sections. The subscripts refer to the strap indices. Z_{11}: proper impedance of the corner strap; Z_{14}: mutual coupling with its toroidal neighbour. V, VS: radiation in vacuum resp. without / with Faraday shield (FS); D, DS: dielectric slab loading at 14cm from straps without / with FS; D2s: dielectric loading with 2 additional SOL-like steps.

MATCHING THE COUPLED ICRF ARRAYS

The array input impedance matrix computed with MWS is incorporated in a model of the system transmission and matching circuit. This allows to realistically account for mutual coupling effects on array matching which had not been addressed in the ITER reference design [1]. In the general case of several (p) mutually coupled CT circuits the number of matched configurations can be very large for given input Z_0 and array feeding mode (i.e. relative amplitudes and phases of the CT input currents): *a priori* all 2^p combinations of inductive and capacitive CT branches can provide a solution. (This number rises to a theoretical maximum of $8^p/2$ for a fully asymmetric array matched with line stretchers, due to a lifting of degeneracy by mutual coupling.) Among these many solutions only two are normally of interest: e.g. for ITER-EM the combination with all upper branches inductive and all lower branches capacitive. The other branch combinations yield relative toroidal phasings between antenna strap currents quite different from the phasings imposed at the CT inputs, which hampers control of the radiated power spectrum. These undesired configurations also play an adverse role in creating spurious "basins of attraction" with a potential to lure matching procedures. Finally, note that the number of matched configurations varies with the various parameters of the system: magnitude of mutual coupling coefficients and of the reference Z_0, asymmetries in resistive loading, relative phase and amplitude of the RF voltages at the CT circuit inputs.

Mutual coupling and asymmetries have been found to strongly affect the antenna current relative amplitude and phase for given Z_0, as well as the quality of load resilience. This high sensitivity is due to the large reactance cancellations of the CT matching scheme: even though the mutual resistances and reactances at the strap inputs are only a few percent of the strap reactances, the mutual coefficients seen from the inputs of the CT branches are not negligible and have a strong influence on the circuit configurations. This also leads to strong interaction between the four CT matching systems and between the power sources feeding the array, making matching procedures much more complex than in the absence of mutual.

To assess load resilience we use phenomenological ELM perturbations in which the strap resistance matrix is scaled up by a factor of 4 to 5 with respect to the ELM-free reference; simultaneous reduction of the strap reactance matrix by ~10 to 25% is also taken into account. In the case of the single asymmetric CT circuit with coupled straps we have obtained general analytical results permitting optimization of load resilience for any type of matching elements, see the illustrations in [9]. For capacitor matching and purely resistive ELM perturbations the reference impedance Z_0 providing optimal load resilience (i.e. a minimal S on Fig.5) follows a simple locus (a parabola):

$$\left(\text{Im}\,Z_0 - X_{A12}\right)^2 = \left(R_{A22} - R_{A11}\right)^2 \left(\text{Re}\,Z_0 - R_{\min}\right) / \left(R_{A11} + R_{A22} - R_{A12} - R_{A21}\right) \quad (1)$$

Selecting the matched reference loading determines the remaining degree of freedom. Note that Z_0 is now generally complex. $R_{Aij}+jX_{Aij}$ are the antenna impedance matrix elements and R_{min} the theoretical minimum circuit input resistance achievable for a given R_A:

$$R_{\min} = \left(R_{A11}R_{A22} - R_{A12}R_{A21}\right) / \left(R_{A11} + R_{A22} - R_{A12} - R_{A21}\right) \quad (2)$$

In the case of symmetrically loaded straps ($R_{A11} = R_{A22}$) equation (1) takes the form $\mathrm{Im}\, Z_0 = X_{A12}$, i.e. the optimal imaginary part of Z_0 equals the mutual reactance between circuit branches. Homologous analytical expressions have been obtained for line stretchers. The theory (to be reported elsewhere) has been extended to ELMs with a reactive component, and generalization to several coupled CT circuits is in progress.

These analytical results, the numerical simulations in the case of several CT and the mockup measurements show the practical interest of an adjustable transformer stage between the CT and the main feed line, which allows operation with a complex impedance Z_0 at the T ensuring best load tolerance at each operating frequency. This new design feature (with respect to the ideal CT of [3]) is available and will soon be tested on the JET-A2, TEXTOR and JET-IL systems; it has also been implemented in the reference layout of the ITER-EM system.

The response of the JET-IL system to various load perturbations (ELMs, plasma L-to H-Mode transition, arcs at various locations in the circuit) and its closed-loop behaviour with the positions of the matching capacitors controlled by error signals derived from RF measurements have been studied with a dynamical model. Details are given in the companion paper [17]. The detection of arcs in the low impedance, low voltage vicinity of the T junctions is an outstanding question, because their signature on RF signals is hardly distinguishable from that of ELMs. In a first series of simulations we have experimented with feedback control of the capacitors based on admittance measurements at the CT circuit inputs. Feedback control is disabled during the ELMs to avoid drifting away from the matched configuration. This approach works well in the absence of mutual coupling. In the actual situation with mutual, there is convergence to a matched configuration for dipole toroidal phasing (0π), provided *(i)* that only one column of 4 capacitors is feedback controlled and their settings copied to the other column, and *(ii)* that initial errors are "sufficiently small". Divergent behaviour with capacitor(s) reaching an endstop is observed in more general configurations. This is attributed to the system lying in the attraction basin of an unwanted configuration, where the control error signals have the wrong polarity; theoretical analysis of these observations is under way. More robust control schemes using RF measurements in the CT branches are now under investigation, which should also apply to the nonsymmetric array feeding modes required for current drive.

Experimental work on the ITER-EM mockup [10] strikes a positive note, as matched configurations have been obtained iteratively in relatively straightforward fashion in the case of an array fed by two independent power sources; the case of four sources will soon be addressed. Experimental demonstration has also been made of alternative circuit layouts using passive power distribution between various combinations of strap triplets [11], which have the advantage of reducing the number of CT matching elements. However systems based on these layouts transmit larger powers per feed line (10MW in two lines or 20MW in a single line), and more work is required to fully appraise their feasibility.

CONCLUSIONS

The benefit of high load resilience brought by the Conjugate T matching circuits proposed for ICRF on ITER comes at the cost of a more complex array operation resulting from an intrinsically high sensitivity to mutual coupling. Demonstration of the successful operation of this matching scheme with the forthcoming JET ITER-Like antenna is therefore crucial, the main open issues being arc detection and reliable system control. We hope that the ongoing work summarized here, providing detailed RF analysis of two implementations of the Conjugate T circuit, can usefully contribute to the comparative assessment of the ITER External Matching design against other proposals and help the final design selection for ITER.

ACKNOWLEDGMENTS

We gratefully acknowledge fruitful discussions with the JET ITER-Like and U.S. High Power Prototype teams, Drs D. Swain, M. Carter, G. Bosia and Ing S. Brémond.

REFERENCES

1. ITER Detailed Design Description, Ion Cyclotron Heating and Current Drive System, WBS 5.1 (DDD), G 51 DDD 4 01-07-19 W 0.2, July 2001.
2. Hoffman, D., et al., in "Radio-frequency Power in Plasmas", 7th Top. Conf. on Radio-Frequency Power in Plasmas, USA, AIP Conf. Proceedings **159** p.302 (1987).
3. Bosia, G., *Fusion Science and Technology* **43** p.153 (2003).
4. Durodié, F., et al., "ITER-Like ICRF antenna for JET", to appear in Proc. 23rd SOFT, Venice (2004)
5. Dumortier, P., et al., Final report on Task FU05-CT 2002-00094 (EFDA/02-675), Laboratory Report n°121, Laboratory for Plasma Physics, Royal Military Academy, Brussels, Belgium (2004).
6. Messiaen, A.M., et al., in "Radio-frequency Power in Plasmas", 15th Top. Conf. on Radio-Frequency Power in Plasmas, Moran, USA, AIP Conf. Proceedings **694** p.142 (2003); Dumortier, P., et al., ibid. p.94.
7. Bosia, G., et al., these Proceedings.
8. Lamalle, P.U., Durodié, F., Goulding, R.H., Monakhov, I., et al., "Radio-frequency matching studies for the JET ITER-Like ICRF system", in "Radio-frequency Power in Plasmas", 15th Top. Conf. on Radio-Frequency Power in Plasmas, Moran, USA, AIP Conf. Proceedings **694** pp. 118-121 (2003).
9. Lamalle, P.U., Messiaen, A., et al., "Study of mutual coupling effects in the antenna array of the ICRH plug-in for ITER", to appear in Proc. 23rd SOFT, Venice, 2004.
10. Messiaen, A.M., Vervier, M., et al., "Tests of the ITER ICRH system with external matching by means of a mock-up loaded by a variable water load", these Proceedings.
11. Vervier, M., Messiaen, A., et al., "Experimental proof of a load resilient external matching solution for the ITER ICRH system", these Proceedings.
12. CST MICROWAVE STUDIO User Manual, Version 5.0, CST GmbH, Darmstadt, Germany, 2004.
13. Lamalle, P.U., Durodié, F., Whitehurst, A., Goulding, R.H. and Ryan, P.M., "Three-dimensional electromagnetic modelling of the JET ITER-Like ICRF antenna", in "Radio-frequency Power in Plasmas", 15th Top. Conf. on Radio-Frequency Power in Plasmas, Moran, USA, AIP Conf. Proceedings **694** pp. 122-125 (2003).
14. Louche, F., Lamalle, P.U., Dumortier, P. and Messiaen, A.M., "Three-Dimensional Electromagnetic Modeling of the ITER ICRF Antenna (External Matching Design)", these Proceedings.
15. Messiaen, A., Dumortier, P., Koch, R., Lamalle, P.U. and Vervier, M., "Use of dielectric or electrolyte to simulate the plasma loading of ICRH antennas", Laboratory Report n°123, Laboratory for Plasma Physics, Royal Military Academy, Brussels, Belgium (2004).
16. Lamalle, P.U., Messiaen, A., Dumortier, P. and Louche, F., "Recent developments in ICRH antenna modelling", in Proc. 20th IAEA Conference, Vilamoura, Portugal (2004), poster /FT/P7-18.
17. Evrard, M., Durodié, F. and Lamalle, P.U., "RF Circuit Simulation of the JET ITER-Like ICRH Antenna", these Proceedings.

Validation of a 3D/1D Simulation Tool for ICRF Antennas

R. Maggiora[1], V. Lancellotti[1], D. Milanesio[1], G. Vecchi[1], V. Kyrytsya[1], A. Parisot[2], S. J. Wukitch[2]

1. Dipartimento di Elettronica, Politecnico di Torino, 10129 Torino, Italy
2. Plasma Science and Fusion Center, M.I.T., Cambridge, MA 02139 USA

Abstract. TOPICA is an innovative tool for the simulation of the Ion Cyclotron Radio Frequency (ICRF) antenna systems that incorporates commercial-grade graphic interfaces into a fully 3D self-consistent description of the antenna geometry and an accurate description of the plasma; it can be considered as a "Virtual Prototyping Laboratory" to assist the detailed design phase of the antenna system. Recent theoretical and computational advances of the TOPICA code has allowed to incorporate a CAD drawing capability of the antenna geometry, with fully 3D geometrical modeling, and to combine it with a 1D accurate plasma description that takes into account density and temperature profiles, and FLR effects; the profiles are inserted directly from measured data (when available), or specified analytically by the user. The coaxial feeding line is modeled as such; computation and visualization of relevant parameters (input scattering parameters, current and field distributions, etc.) complete the suite. The approach to the problem is based on an integral-equation formulation for the self-consistent evaluation of the current distribution on the conductors. The environment has been subdivided in two coupled region: the plasma region and the vacuum region. The two problems are linked self-consistently by representing the field continuity in terms of equivalent (unknown) sources. In the vacuum region all the calculations are executed in the spatial (configuration) domain, and this allows triangular-facet description of the arbitrarily shaped conductors and associated currents; in the plasma region a spectral representation of the fields is used, which allows to enter the plasma effect via a surface impedance matrix; for this reason any plasma model can be used, and at present the FELICE code has been adopted; special techniques have been adopted to increase the numerical efficiency. The TOPICA suite has been previously tested against assessed codes and against measurements of mock-ups and existing antennas. This work is devoted to an extensive set of comparisons between measured and simulated reflection coefficients (magnitude and phase), both in vacuum and with plasma during ALCATOR C-MOD operation. The comparison demonstrates a very good agreement, leading to a validation of TOPICA as a predictive tool.

Keywords: Antenna, heating, plasma, loading, modeling.
PACS: 52.50.Qt

OBJECTIVES AND TOPICA ARCHITECTURE

The reliable prediction of antenna input parameters in realistic plasma-facing conditions is a key issue for a successful ICRF antenna design. Realistic plasma-facing conditions mean an accurate 3D modeling of the antenna geometry and an accurate description of plasma (including its density, temperature and static magnetic field profiles, FLR effects, etc.). The capability of obtaining affordable running times with affordable computing resources is another key issue in this direction.

TOPICA is a complete suite that computes the antenna input parameters and the antenna radiated fields and spectra starting from the CAD drawing of the antenna structure and from the plasma parameters. The main modules of TOPICA and their links are reported in Fig. 1. In TOPICA the antenna boundary-value problem is transformed into a set of integral equations, transformed into a linear system by a finite-element procedure; therefore, the antenna design or analysis phase begins with the antenna drawing and meshing by means of a CAD-mesher tool. The CAD drawing is meshed with triangular facets that constitute the support to the basis functions; mesh data are saved on a file in a format that may vary for different tools. Thus, the pre-processing unit is aimed at reading the geometry data, generating the basis functions and, then, saving all that to a new file. This is the input both to the plasma and to the vacuum modules: the former computes the entries of the system (interaction) matrix pertaining to the plasma contribution and couples to FELICE code to obtain the plasma surface matrix (the other capabilities of this code are not employed here), while the latter fills the remaining part of the matrix as well as the forcing term at the right hand side of the linear system. The solver module is charged with the task of solving the system and obtaining the currents, whence the antenna parameters are computed. Finally, a plotting tool has been coded to help in displaying the electric current distribution all over the conductors and the electric field.

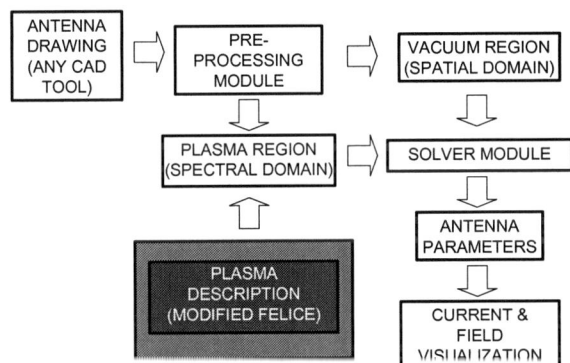

FIGURE 1. Main modules of TOPICA and their links.

THEORY OVERVIEW

The theoretical background of TOPICA has been described in [1] and will be only briefly summarized here. TOPICA is a self-consistent code: the distribution of electric current on the conducting parts is determined by the code solution, rather than from an Ansatz. A key point in achieving a high degree of detail in describing the antenna geometry, and yet keeping an accurate plasma description is to formulate the initial boundary-value problem so as to separate the antenna region from the plasma region by applying the Equivalence Theorem [3](ET) twice.

First the ET is applied in the plasma region; the curved surface that delimits the antenna recess from the toroidal vacuum chamber is identified, and (unknown) magnetic current densities are placed on both sides of this surface: this allows to close this aperture by a Perfect Electric Conductor (PEC), thus accomplishing the separation of the antenna region from the plasma region inside the vacuum chamber. In the antenna region the ET is applied again on all metal parts and also the aperture, now also a PEC; in this case the conductors are replaced by (unknown) surface electric currents that radiate in vacuo. As a result, the unknown current extends over all metal parts and over the aperture, while the magnetic current resides only on the latter. Upon enforcing the proper boundary and continuity conditions a set of two coupled integral equations ensues for the electric and magnetic currents.

The aperture is actually curve, while FELICE handles a slab and transversely invariant plasma; in order to accommodate the coupling, the magnetic current on the plasma side of the curved aperture is "stretched flat" by mapping the facets on the aperture to a plane by means of a suitable projection. The validity of this approximation is supported by the good agreement between simulation results and measured data presented in this work.

The obtained set of coupled Integral Equations are solved via finite-element discretization of the currents and a weighted-residual method, often called Method of Moments [7][8](MoM); the finite element basis functions are defined over the facets of the triangular mesh. In particular, the MoM is applied both in the spatial (antenna region) and in the spectral domain (plasma region), the latter being the natural domain in which the FELICE data are available; the overall procedure turns the integral equations into an algebraic system, whose unknowns are the projection coefficients of the currents on the set of basis functions; filling the system (interaction) matrix and solving it are the main tasks of TOPICA (see Fig. 2). The knowledge of the currents distribution allows direct calculation of radiated power and fields and the input admittance matrix. The antenna excitation is the TEM mode voltage in the coax feeding the antenna [8]. The admittance matrix can be converted to the scattering matrix that is the most useful set of antenna parameters for assisting the accurate and reliable design of the feeding and tuning-matching systems.

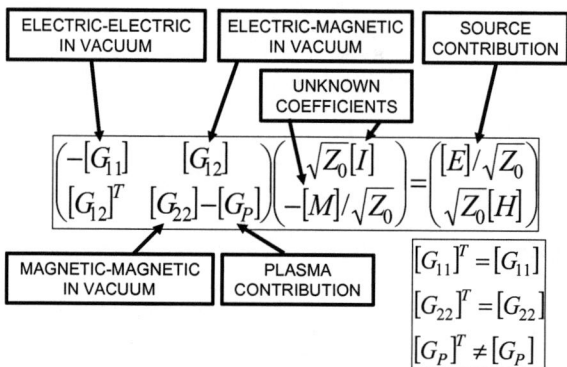

FIGURE 2. Main parts of the TOPICA interaction matrix.

EXPERIMETAL DATA

An experiment has been carried out in the ALCATOR C-MOD tokamak around the main objectives of obtaining ICRF coupling measurements and radial density profiles in the core, edge and scrape-off layer, in low ICRF power discharges. The broader goal is the benchmarking the predicting capability of the antenna-plasma computer code TOPICA.

Previous studies of ICRF loading on ALCATOR C-MOD, carried out with routine measurements, suggest that the density profile in the scrape-off layer can have a significant effect on loading, most notably during long EDA H-modes and for confinement mode transitions in general. The density in the scrape-off layer is monitored by Langmuir probe measurements, which are meaningful usually for non-ICRF or low power ICRF operations. To obtain a direct comparison between density profiles and coupling measurements, we dedicated low ICRF power discharges, in configurations which allow the access to a relevant range of scrape-off layer conditions.

The approach is to use minimum ICRF and ohmic power in order to obtain meaningful fast-scanning Langmuir probe density measurements, while varying the scrape-off conditions, namely with a density scan, current scan, and for L and H-mode. Using D and E-antennas and the A-port scanning probe, located on the other side of the vessel, will reduce the sheath rectification effects and should allow ICRF powers up to 1 MW. A wide scan of parameters can be done in L-mode with this range of ICRF power; to scan the density and current while triggering H-mode with relatively limited ICRF power, we will need then to target low ICRF-power threshold conditions. This can be done with lower magnetic field and ICRF deposition at the edge, and should allow a similar scan range as for L-mode (see Tab. 1).

TABLE 1. ALCATOR C-MOD Experiment Parameters.

Parameter	Value
Toroidal Field	4.3 – 5.4 T
Plasma Current	0.6 MA – 1.2 MA
Gas Species	Deuterium + Hydrogen minority < 5%
Density	$0.6 - 1.2 * 10^{20}$ m^{-3}
RF Power	500 kW (lowest required to trigger H-mode)

ANTENNA DRAWING

The first step for simulating an ICRF antenna, and arguably the most important for the code user, is to draw or import the antenna structure with an input CAD graphic tool (see Fig. 2a). The employed commercial graphic tool has been customized to ease antenna description (strap geometry, feeding, box, septa, etc.) and to permit the tagging of the various antenna model parts (mainly PEC or aperture, see Fig. 2).

The second step is the selection of the antenna feeding type (voltage gap or coax line excitation [8]) and the following identification of the corresponding feeding ports. An example is shown in Fig. 3.

The third step is to mesh the antenna by means of an appropriate triangular meshing and to export the mesh in an external file. The mesh can be structured for some surfaces (typically for the aperture): this allows for re-use of the plasma or antenna region results in subsequent simulations.

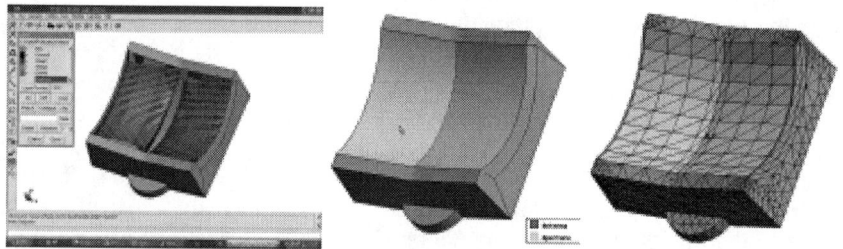

FIGURE 2. Drawing the antenna, identifying antenna parts and aperture, meshing of the structure.

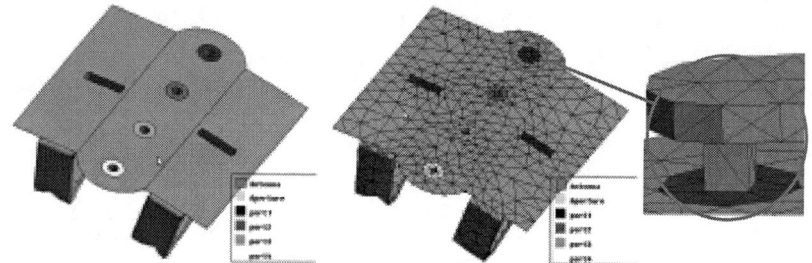

FIGURE 3. Feeding the antenna, identifying ports, meshing of the structure.

VALIDATION

The measurements have been realized by means of two directional coupler installed in the transmission line system feeding the E-antenna and have been provided as reflection coefficient values (magnitude and phase) with an error bar of about ±5% on magnitude. A sketch of the transmission line system showing the position of the direction couplers is reported in Fig. 4.

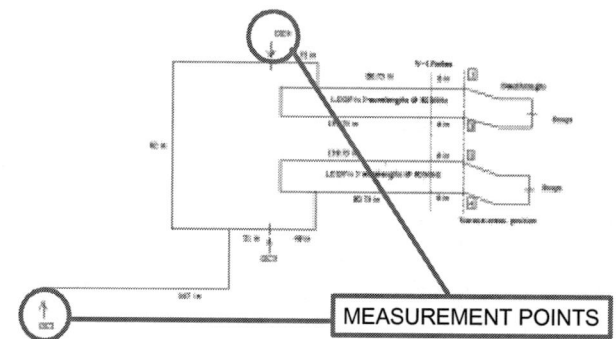

FIGURE 4. ALCATOR C-MOD E-antenna feeding system.

The antenna input scattering matrix computed by TOPICA are input in a circuit model that includes the (measured) feed-through characterization and all the transmission lines present in the system (losses are included). This yields the reflection coefficients at the locations where measurements have been taken.

Fig. 5 shows the comparison between computed and measured values at the directional coupler named DC4; both magnitude and phase of the reflection coefficient are reported, versus the plasma shot identifier. In Fig. 6 the same values are compared at the directional coupler named DC2. The phase shown here is the difference between the phase of the reflection coefficient with a plasma in front of the antenna and the corresponding phase of the reflection coefficient with the antenna in free space.

For each of the plasma shot numbered in Fig. 5 and 6 different plasma parameters have been considered in the simulation, i.e. different toroidal field, gas composition, density profile, and temperature profile (see Fig. 7 for an example).

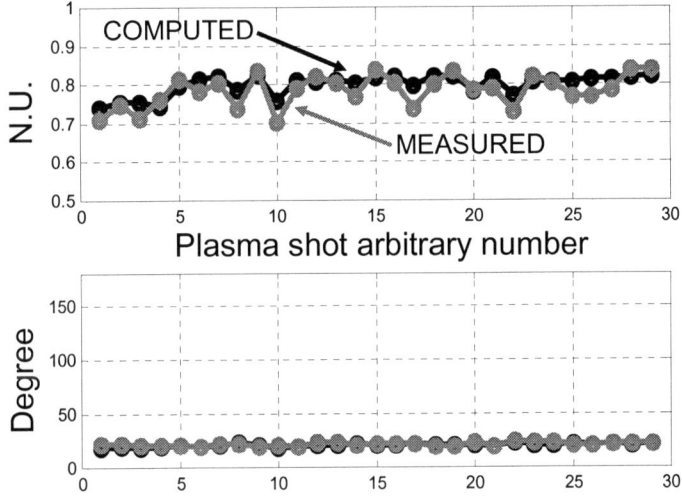

FIGURE 5. Computed vs. measured reflection coefficient at directional coupler named DC4.

CONCLUSIONS AND NEXT STEPS

In conclusion, a very good agreement (both in magnitude and in phase) has been demonstrated between computed and measured values of the antenna input reflection coefficients in presence of a real plasma in front of the antenna. Moreover, this is the only known case in literature of ICRF antenna input parameter comparison.

From the results of this work it is possible to state that with TOPICA a 1D plasma description with an extremely accurate density profile description is suitable for antenna analysis and for the prediction of the antenna behavior. TOPICA is capable of reliably predicting the behavior for plasma facing antennas.

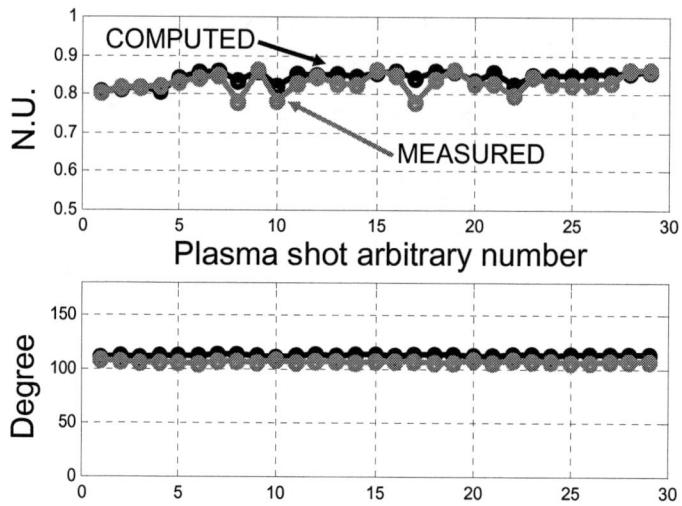

FIGURE 6. Computed vs. measured reflection coefficient at directional coupler named DC2.

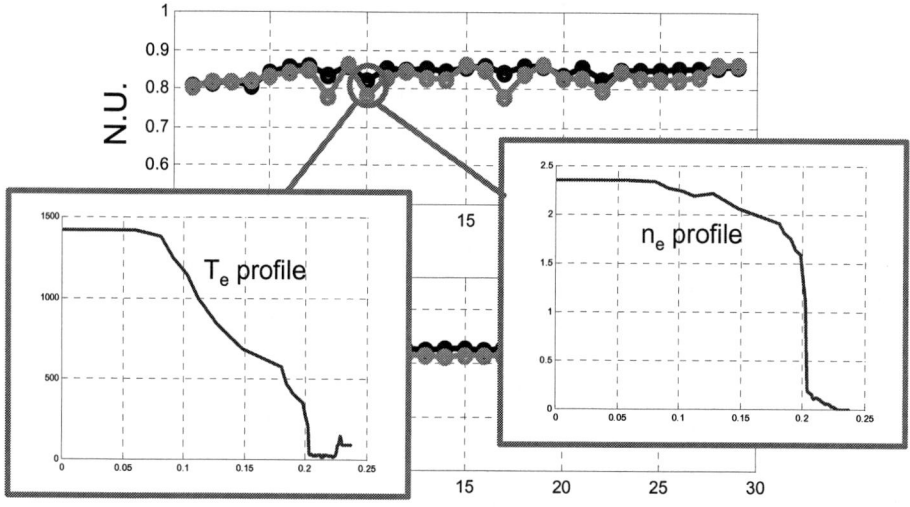

FIGURE 7. Example of plasma parameters available for each shot.

Moreover TOPICA, thanks to the implementation of numerous numerical algorithms, has demonstrated a very good efficiency and a controllable convergence. This yields to affordable CPU times for simulations: the ALCATOR C-MOD E-antenna has been meshed with about 10,000 surface unknowns (no symmetry assumed) and its analysis takes about 4 hours for each case on a LINUX based PC, single 2.66 GHz processor, 2 GB RAM, with a cost of about 1,000 USD.

The next steps in the TOPICA development include the coupling with a complete 3D plasma model (thus interfacing with TORIC instead of FELICE) and an extension of the antenna feeding schemes to permit the analysis of waveguide antennas used in the lower hybrid range of frequency.

REFERENCES

[1] M. Brambilla, *Nuclear Fusion* 35, pp. 1265-1280 (1995).
[2] V. Lancellotti, D. Milanesio, R. Maggiora, G. Vecchi, V. Kyrytsya, *Prediction of plasma-facing ICRH antenna behavior via a Finite-Element solution of coupled Integral Equations*, this book.
[3] R. Harrington, *Time-Harmonic Electromagnetic Fields*, New York: McGrawHill, 1961.
[4] S. M. Rao, D. R. Wilton and A. W. Glisson, *IEEE Transactions on Antennas and Propagation* AP-30, pp. 409-418 (1982).
[5] V. Lindell, *Methods for Electromagnetic Field Analysis*, Oxford: Clarendon Press, 1992.
[6] R. Harrington, *Field Computation by Moment Methods*, New York: Oxford, 1993
[7] A. F. Peterson, S. L. Ray, R. Mittra, *Computational Methods for Electromagnetics*, New York: IEEE Press, 1998.
[8] R. Maggiora, V. Lancellotti, G. Vecchi, V. Kyrytsya, *Nucear Fusion* 44, pp. 846-868 (2004).

Automatic control of ITER-like structures

G.Bosia, S. Bremond

Association EURATOM-CEA, CE Cucaracha, F-13108 ST-PAUL-LEZ-DURANCE

Abstract. Abstract In ITER Ion Cyclotron System a power transfer efficiency in excess of 90% from power source to plasma in quasi continuous operation. This implies the availability of a control system capable of optimizing the array radiation spectrum, automatically acquiring impedance match between the power source and the plasma loaded array at the beginning of the power pulse and maintaining it against load variations due to plasma position and plasma edge parameters fluctuations, rapidly detecting voltage breakdowns in the array and/or in the transmission system and reliably discriminating them from fast load variations. In this paper a proposal for a practical ITER control system, including power, phase, frequency and impedance matching is described

Keywords: Ion Cyclotron Heating.
PACS: 52-50Qt

INTRODUCTION

In ITER, the Ion Cyclotron Heating System should routinely operate close to the nominal RF power output (20 MW) in quasi-continuous operation, with infrequent power interruptions, due to voltage breakdowns in the array and/or in the transmission system, or because of equipment failure. A reliable operation requires a control system capable of:
1) controlling the array radiation spectrum during the power pulse, so as to optimize plasma coupling and adsorption and to minimize parasitic power losses in the plasma edge;
2) establishing and maintaining the RF power flow to the plasma. This implies for all the RF power sources to automatically acquire, at the beginning of each RF pulse an impedance match suitable for an efficient RF power transfer, starting from a preset condition, and to maintain it during the whole power pulse, against significant load variations..
3) detecting and suppressing RF voltage breakdowns in the array and in the transmission system, with a time response of few microseconds, to limit the energy deposited in the arc and to avoid local equipment damage. This is usually done by a fast power interruption, with automatic recovery to normal operation, when a suitable dielectric strength is restored in the torus vacuum vessel.
4). Providing an accurate real time measurement of the plasma load.
The control method proposed here relies on the active, real time, and vectorial control, of all currents in all array elements. Currents in each element are controlled

independent of those in the other ones. The method is therefore applicable to any array order.

The IC launcher described in the ITER Final Design Report is an array of eight (2 poloidal x 4 toroidal) elements, referred to as "ITER-like structure" - ILS -) powered by a 2.5 MW source, for a total of 20 MW power output. It can be shown [1]) that i) if the ILS currents are controlled to be complex conjugate and symmetric with respect to the input current and ii) the input impedance of the circuit is kept not too high compared to the circuit load resistance(s), the ILS exhibits a significant tolerance to the resistive load variations such as the ones due to fast ELMs fluctuations. In a more recent CEA proposal [2]) an increase of the array poloidal order (2 poloidal x 4 toroidal elements) is proposed.

CONDITIONS FOR LOAD RESILIENCE AND PERFECT MATCH

In an "ideal" ITER-like structure (i.e. with load symmetry and no inter strap coupling) the conditions for perfect match and load resilience coincide. In [3]) it is shown that the same level of load resilience can be obtained for a generic ILS, for reasonable assumptions about coupling and load asymmetries, relevant to ITER IC applications, provided that the match is performed to a complex impedance $Z_{in} = R_0 - iX_{sm}$, where X_{sm} is the reactive part of the non diagonal term of the ILS impedance matrix. In short, with the ITER parameters used in the FDR report, and the notations used in [3]), if $k_p = |X_{sm}|/X_s$ is of the order of percent and the ILS load asymmetries below 10%, a level of impedance matching acceptable to the specified power sources (VSWR < 2 at full power) is possible with the ILS tuning capacitors only.

In this paper however we shall derive the perfect match conditions *at* optimum load resilience for a generic ILS (i.e. with arbitrary load asymmetry and inter strap coupling). It is well known that an arbitrary complex load can be perfectly matched to a resistive input impedance R_{in} by a series/parallel combination of two reactances. Thus, for an arbitrarily loaded ILS four reactances are needed.

We consider the circuit in Figure 1, where the generic ILS, including the two tuning capacitive reactances X_{C1} and X_{C2} (Pretuner) is connected to the series/parallel combination of two reactances X_S and X_P (Trimmer). The system is described by the equations:

$$(R_{s1} + i \cdot X_1) \cdot \iota_1 + (R_{sm2} - i \cdot X_{sm2}) \cdot \iota_2 = (R_0 + i \cdot X_0)$$
$$(R_{sm1} - i \cdot X_{sm1}) \cdot \iota_1 + (R_{s2} + i \cdot X_2) \cdot \iota_2 = (R_0 + i \cdot X_0)$$

with $\iota_1 = \dfrac{I_1}{I_{in}}$, $\iota_2 = \dfrac{I_2}{I_{in}}$, $\iota_{in} = \iota_1 + \iota_2 = 1$ 1)

If we impose to the current ratios to be complex conjugate and:

$$\iota_1 = \iota_0 \cdot (\cos(\phi) + i \cdot \sin(\phi))$$
$$\iota_2 = \iota_0 \cdot (\cos(\phi) - i \cdot \sin(\phi))$$

or $\iota_1 + \iota_2 = 2 \cdot \iota_0 \cdot \cos(\phi) = 1$, i.e. $\iota_0 = \dfrac{1}{2 \cdot \cos(\phi)}$ 2) substitution in 1) leads to :

$$\Delta X_C = \Delta X_s \pm \sqrt{(2 \cdot R_0 - R_s - R_{sm}) \cdot (R_s - R_{sm}) + \Delta X_{sm}}$$

$$\tan \phi = \pm \sqrt{\frac{(2 \cdot R_0 - R_s - R_{sm}) \cdot (\Delta X_C - \Delta X_s - \Delta X_{sm})}{(R_s - R_{sm}) \cdot [(\Delta X_C - \Delta X_s) + \Delta X_{sm}]}} + \Delta X_{sm} \qquad 3)$$

$$X_C = (X_C + X_{sm}) - \cotan(\phi) \cdot (\Delta R_s - \Delta R_{sm})$$

$$X_0 = \frac{1}{2} \left[(\tan(\phi) + \cotan(\phi)) \cdot \Delta R_s + (\tan(\phi) - \cotan(\phi)) \cdot \Delta R_{sm} \right] - X_{sm}$$

where R_s, X_s, R_{sm}, X_{sm} are the arithmetic average of $R_{sk}, X_{sk}, R_{smk}, X_{smk}$ (k = 1,2) and $\Delta R_s, \Delta X_s, \Delta R_{sm}, \Delta X_{sm}$ the relative asymmetries. On the other hand, from Figure 1 it is deduced that:

$$X_0 = \sqrt{R_0 \cdot (R_{in} - R_0)} - X_s \qquad R_0 = R_{in} \cdot \frac{X_p^2}{(R_{in}^2 + X_p^2)} \qquad 4)$$

From 3) it is seen that, if current symmetry and input resistance R_0 are imposed to in a generic ILS for optimum load resilience, a reactive input mismatch (X_0) depending on both circuit asymmetry and coupling is generated.

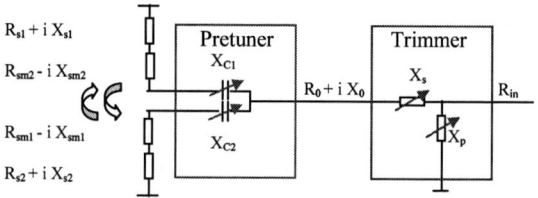

Figure 1 Layout of the tuning system

If unacceptable to the RF power source, the mismatch can however be easily corrected by the Trimmer circuit. In practice, for a moderate mismatch, it may be convenient to locate the trimmer at the generator end and use a wideband multi section transformer at the Pretuner input to bring the (low) R_0 value close to the transmission line characteristic impedance Z_c.

From the above considerations the matching algorithms for the Pretuner are:

$$[\arg(I_1) - \arg(I_{in})] + [\arg(I_2) - \arg(I_{in})] = 0 \quad \text{acting on } X_C \qquad 5)$$

and

$$\text{Re}(Z_p) = R_0 \qquad \text{acting on } \Delta X_C. \qquad 6)$$

where Z_p is the Pretuner input impedance. The Trimmer algorithms are straightforward from equations 4). In Figure 2 the two relations 5) and 6) are plotted against the tuning reactances X_{C1} and X_{C2}, for typical ITER values $R_s = 1\Omega$, $X_s = 30\Omega$, $R_{sm} = 0.5 \, \Omega$, $X_{sm} = 0$ and 0.3Ω mapped against the module of the reflection coefficient. In this case $\Delta R_s = \Delta X_s = \Delta R_{sm} = \Delta X_{sm} = 0 \, \Omega$ for simplicity. In the first case (Figure 2a) optimum resilience and perfect match conditions coincide, as expected. In the second case, they do not because of the term $X_0(R_s, X_s, R_{sm}, X_{sm}, \Delta R_s, \Delta X_s, \Delta R_{sm}, \Delta X_{sm})$.

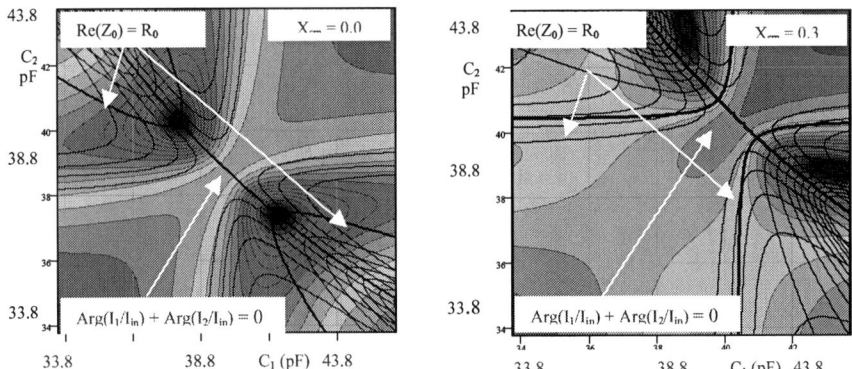

Figure 2 Conditions for perfect match and load resilience mapped against X_{C1} and X_{C2} : a) $X_{sm}= 0\Omega$; b) $X_{sm}= 0.3\Omega$.

The two control loops of the Pretuner should operate with different time constants so that the ΔX_C loop tracks the X_C loop. In the same way the Trimmer control loops (if any) should track the Pretuner loops. Finally, the tuning loops should track the phase and power control loops.

CONCLUSIONS

In conclusion, a simple matching algorithm for a generic ILS is proposed. The method requires the vectorial detection of the ILS currents. The design of a suitable vectorial current monitor cannot be included here for lack of space, but it is discussed in the companion poster. As the strap currents are symmetrically controlled with respect to the input current by the feedback system, a matched ILS behaves as a single matched strap. Therefore the extension of this method to a complex array can be performed as for an array of single straps. As all array currents are actively controlled to predictable values, phase instabilities described in [3] can be avoided.

REFERENCES

1) G. Bosia, Fusion Science & Technology, 43 pp. 153-159, (2003)
2) G. Bosia et al, "Proposals for ITER ICH design upgrades" Paper at this Conference
3) G. Bosia et al, "Effects of coupling and asymmetries on load resilience of ITER-like structures" Paper at this conference
4) G .Bosia et al, EFDA Design Task no 1131, Final Report, (2005

Effects of coupling and asymmetries on load resilience of IC ITER-like structures

G. Bosia, S. Brémond, L. Colas

Association EURATOM-CEA, CE Cadarache, F-13108 ST-PAUL-LEZ-DURANCE

Abstract. ITER-like structures feature an intrinsic resilience to load variations, which is related to the symmetry of the currents in the two branches of the structure. It has been suggested that the effects of coupling between the array elements would significantly impair the load resilience of the structure. In this paper the effect of inter strap coupling and of however induced electrical array asymmetries on the structure load resilience are quantitatively examined

Keywords: Ion Cyclotron Heating.
PACS: 52-50Qt

INTRODUCTION

In the ITER Ion Cyclotron (IC) System Reference Design (RD)[1], plasma loading is modelled by a "transmission line" estimate of the diagonal terms (Z_{ii}) of the coupling impedance matrix [2]. The resistance per unit length (R'- ohm/m) is obtained from the RANT3 2D code and a load inductance and capacitance per unit length, is deduced from the ARGUS 3D electrostatic code, as function of a simplified strap geometry, frequency and phasing. The analysis in [2] provides no estimate for the non diagonal terms of the matrix and therefore the effects of coupling between straps and possible matrix asymmetries induced by geometry effects and/or by plasma anisotropies [4], have been ignored in the launcher design. The basic element of the ITER array, hereafter referred to as ITER-like

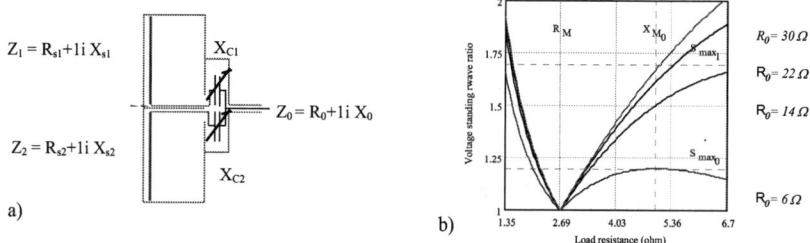

Figure 1. a) Layout of the ITER-like structure, b) Input VSWR as function of load resistance

structure, (ILS) is itself a poloidal array of two short-circuited current straps (Figure 1),each in series with to a tuning capacitive reactance, and connected in parallel to the output of a RF power source, via a step-up impedance transformer.

The circuit features a significant resilience to load variations[3] such as those caused by ELMs.. This arises, i) from the fact that the input admittances of the two sections are complex conjugate, and their imaginary parts cancel out when paralleled even in the case of resistive load variations and ii) because the loaded Q of the circuit is lowered by the imposed low input impedance.

In a relatively dense array, such as the ITER Ion Cyclotron array, a certain level of non conductive (i.e. inductive, and to less extent, capacitive) direct coupling between array elements is present at the plasma end, and an apparent inter element reactive and resistive coupling is reflected by the plasma load, back to the exciting array elements. Coupling adds non diagonal terms to the impedance matrix, which may not be symmetric, because a magnetized plasma is, by itself, an anisotropic and gyrotropic load [4].

It has been suggested [5] that coupling within the ITER array would impair the load resilience of the proposed scheme. In this paper the effects of plasma coupling and other asymmetries are analyzed for a generic ILS.

MATCH CONDITIONS OF A GENERIC ILS

For an "ideal" ITER like structure, such as the one considered the ITER RD, the match condition to an input impedance $Z_0 = R_0 + i\,X_0$ is described by the simple eigenvalue equation:

$$\begin{bmatrix} R + 1i \cdot X_1 & -(R_0 + 1i \cdot X_0) \\ -(R_0 + 1i \cdot X_0) & R + 1i \cdot X_2 \end{bmatrix} \cdot \begin{pmatrix} I_1 \\ I_2 \end{pmatrix} = 0 \qquad 1)$$

where $Z_s = R_s + i\,X_s$ is the (plasma loaded) impedances, $Z_0 = R_0 + i\,X_0$ the input impedance, $R = R_s - R_0$, $X_k = X_s - X_0 - X_{Ck}$, and X_{Ck} the variable capacitive tuning reactances. The solutions of 1) are:

$$X_{C1M} = X_s + X_0 - \left(\frac{R_0}{R} - 1\right) \pm \sqrt{\left(X_0^2 + R^2\right)\left[\left(\frac{R_0}{R}\right)^2 - 1\right]}$$

$$X_{C2M} = X_s + X_0 - \left(\frac{R_0}{R} - 1\right) \pm \sqrt{\left(X_0^2 + R^2\right)\left[\left(\frac{R_0}{R}\right)^2 - 1\right]} \qquad 2)$$

and

$$I_{M1} = \frac{V}{[(R+R_0) + i \cdot (X_{1M} + X_0)]} = \frac{V}{R_s + i \cdot (X_s - X_{C1M})}$$

$$I_{M2} = \frac{V}{(R+R_0) + i \cdot (X_{2M} + X_0)} = \frac{V}{R_s + i \cdot (X_s - X_{C2M})} \qquad 3)$$

It is seen that it is possible to find two reactances X_{C1} and X_{C2} capable of matching the load impedance Z_s to whatever input impedance Z_0. If $X_0 = \text{Im}(Z_0) = 0$, 2) and 3) reduce to:

$$X_{C1} = X_s + \sqrt{\left[R_0^2 - (R_0 - R_s)^2\right]} \qquad I_{1M} = \frac{R_0 \cdot I_0}{R_s + 1i \cdot \sqrt{\left[R_0^2 - (R_0 - R_s)^2\right]}}$$

$$X_{C2} = X_s - \sqrt{\left[R_0^2 - (R_0 - R_s)^2\right]} \qquad I_{2M} = \frac{R_0 \cdot I_0}{R_s - 1i \cdot \sqrt{\left[R_0^2 - (R_0 - R_s)^2\right]}} \qquad 4)$$

as obtained in [3].

We shall define the condition obtained for $X_0 = Im(Z_0) = 0$ the "optimum" load tolerance condition, characterized by the fact that the currents I_{1M} and I_{2M} are complex conjugated with respect to the input current. I_0. In this particular case it is coincident with the perfect match condition The match conditions for an abitrarly loaded and coupled ILS is :

$$\begin{bmatrix} (R_{s1} - R_0) + 1i \cdot (X_{s1} - X_0 - X_{C1}) & (R_{sm2} - R_0) + 1i \cdot (X_{sm2} - X_0) \\ (R_{sm1} - R_0) + 1i \cdot (X_{sm1} - X_0) & (R_{s2} - R_0) + 1i \cdot (X_{s2} - X_0 - X_{C2}) \end{bmatrix} \cdot \begin{pmatrix} I_1 \\ I_2 \end{pmatrix} = 0 \qquad 5)$$

where $Z_{sk} = R_{sk} + i X_{sk}$ ($k = 1,2$) are the (plasma loaded) impedances of the two current straps, $Z_{smk} = R_{smk} + i X_{smk}$ the apparent coupling impedances (assumed to be different) and $Z_0 = R_0 + i X_0$ the input impedance. X_{C1}, and X_{C2} are the variable capacitive tuning reactances. Equation 5) can be solved analytically with some more involved algebra [6]. More simply, if R_S, X_S, R_{sm}, X_{sm} are the arithmetic average of R_{sk}, X_{sk}, R_{smk}, X_{smk} ($k = 1,2$) and ΔR_S, ΔX_S ΔR_{sm}, ΔX_{sm} their relative asymmetry, it is in our case: $R_0 \sim 5\ \Omega$, $R_S \sim 0.5\ \Omega$, $X_S \sim 30\ \Omega$, $R_{sm} < R_S$, $\Delta R_S < R_S <<$ R0, $X_{sm} << X_S$, $\Delta X_S << X_S$, $\Delta R_{sm} << R_S < R_0$, $\Delta X_{sm} << X_{sm}$. If second order terms are neglected in respect to R_0, the equation 5) reduces to:

$$\begin{bmatrix} (R_0 - R_s) + 1i \cdot (X_0 - X_{s1} + X_{C1}) & (R_0 - R_{sm}) + 1i \cdot (X_0 - X_{sm}) \\ (R_0 - R_{sm}) + 1i \cdot (X_0 - X_{sm}) & (R_0 - R_s) + 1i \cdot (X_0 - X_{s2} + X_{C2}) \end{bmatrix} \cdot \begin{pmatrix} I_1 \\ I_2 \end{pmatrix} = 0 \qquad 6)$$

i.e. to the same equation as before, provided that the substitutions $R \to R_0 - R_s$, $R_0 \to R_{sm} - R_0$, $X_0 \to X_{sm} - X_0$ are made. The optimum load tolerance condition is now :
$$X_0 = X_m \qquad 7)$$
leading to the solutions:

$$I_{1M} = \frac{V}{R_0 + i \cdot X_{sm}} \cdot \frac{R_0 - R_{sm}}{(R_s - R_{sm}) - i \cdot \sqrt{(R_0 - R_{sm})^2 - (R_0 - R_s)^2}}$$

$$I_{2M} = \frac{V}{R_0 + i \cdot X_{sm}} \cdot \frac{R_0 - R_{sm}}{(R_s - R_{sm}) + i \cdot \sqrt{(R_0 - R_{sm})^2 - (R_0 - R_s)^2}} \qquad 8)$$

$$I_0 = \cdot \frac{V}{R_0 + i \cdot X_{sm}}$$

The currents I_{1M} and I_{2M} are again complex conjugate and symmetric with respect to the input current I_0 and load resilience is restored as in absence of interstrap coupling. However, the ILS input current is now not any longer in phase with the input voltage .In ITER relevant conditions, the effect od inter strap coupling is simply to add a (small) reactive term.to the desired input impedance at the perfect match, This term can be easily removed by addition of series reactance of opposite sign.

CONCLUSIONS

In conclusion it has been shown that any reasonable level of poloidal coupling and other electrical asymmetries do not affect load resilience of an ITER-like structures, provided that conjugate symmetry between half sections currents is maintained by a suitable control system.

In a large array, toroidal and poloidal coupling between array elements may cause parasitic power circulation between elements and prevent matching some array element at array phasing different from 0 or π.[7] For equal input powers, the phase angle between adjacent and symmetrically loaded array elements must be kept below a critical angle depending on coupling phasing and element input impedance. In Figure 2 the critical angle is plotted for a typical ITER parameters as function of apparent inter elements coupling and loading. It is impossible to match the array using reactive elements at current phasing angles exceeding the critical angle

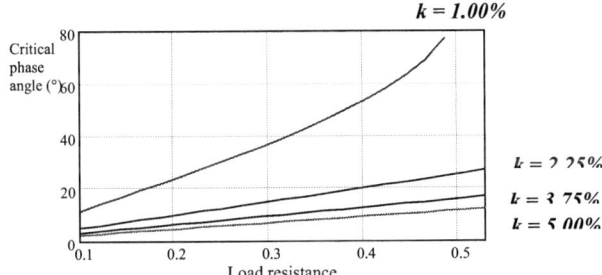

Figure 2. Critical angle as function of coupling coefficient (kp) and loading /Rs).

REFERENCES

1 ITER Detailed Design Description (DDD) No. 5.1 - Ion Cyclotron Heating and Current Drive System. IdoMS GAM MD 3 00-02-21 F1 (2000).
2 D. Swain et al. Final Report, ITER Task Agreement D320 (1996).
3 G. Bosia, Fusion Science & Technology, 43,153, (2003)
4 L.Colas et al, Proc EPS Conference (2004)
5 A. Messiaen, AIP CP 694, 142, (2003)
6 G. Bosia et al, EFDA Task 1137 report (2004)
7 G. Bosia, J.Jacquinot, Proc IAEA Technical Committee Meeting on Fast Wave Current Drive in Reactor Scale Tokamaks, pp 471- 495 Arles (1991).

Proposals for ITER Ion Cyclotron Reference Design Upgrades

G. Bosia, B. Beaumont, S. Brémond, P. Testoni [1], K.Vulliez

Association EURATOM-CEA, CE Cadarache, F-13108 ST-PAUL-LEZ-DURANCE
[1] *Università di Cagliari, Cagliari(Italy)*

Abstract A possible issue for the ITER Ion Cyclotron Heating and Current Drive system is related to the high electric field (2 kV/mm parallel to B_0) at which the array is planned to operate, which may be associated to a low dielectric rigidity of the of the vessel "vacuum". The combined effects are likely to set the upper limit to the RF power transfer efficiency to the plasma core. This paper addresses the problem of how to improve the power handling in the ITER IC launcher described in the ITER Final Design Report (FDR), to either operate at a significantly lower E-field in the torus vacuum or, alternatively, at a lower plasma coupling.

Keywords: Ion Cyclotron Heating.
PACS: 52-50Qt

INTRODUCTION

The Ion Cyclotron (IC) array described in the ITER Final Design Report (FDR) is designed for a RF output power of 20 MW, (approximately equivalent to a power density of 10 MW/m^2). A strap coupling impedance $Z_L \sim 1.2 + i\ 18\ \Omega$ is expected at 53 MHz [1]. At these load conditions, the strap maximum RF voltage is $V_{L\ max} = 30.7$ kV, and the maximum (peak) electric field in the torus vacuum, (reached in one of the tuning stubs) slightly exceeds 2 kV/mm, the maximum voltage reference limit, which is based on best experimental values obtained in current tokamaks operation.

However, there is now no guarantee that the ITER IC system will operate at these reference conditions, as they are predicted for the nominal antenna/plasma separatrix gap of 120 mm and $0,\pi, 0,\pi$ phasing. This value was found as a difficult compromise between conflicting constraints, including, on one hand, IC H&CD system requirements, asking for a narrow gap, and several others (such as radial control accuracy of the plasma position, plasma shaping, wall thermal loading at start up) all suggesting wider gaps. According to the FDR design estimates [1], the real part of the coupling impedance per unit length (R') decreases exponentially with the gap width, and the strap RF voltage doubles at a gap of 200 mm.

When the RF E-field temporarily exceeds the local dielectric rigidity R_d(volt/ mm) of the in-vessel "vacuum", a voltage break-down occurs. R_d is strongly affected by scrape-off plasma conditions, wall recycling and ELMs activity, which influence both local neutral pressure and "vacuum" dielectric conductivity. Accurate figures on the local neutral pressure are difficult to predict, as this is dependent of the molecular flow in the front end of the array, which depends on the details of the array geometry. In any case, voltage breakdowns should be infrequent in ITER IC array since RF arcs must be extinguished by interrupting the power flow for a time interval sufficient for the "vacuum" dielectric strength to be restored. This process complicates the operation of the IC system and often constitutes a major source of inefficiency and unreliability in power handling.

Operation at E- field substantially lower than the nominal value (such as half the value) would greatly improve the expected system reliability. In this paper a number of upgrades to the design of ITER IC array are proposed to achieve this goal.

PROPOSED UPGRADES

The IC launcher proposed in the ITER FDR (Figure 1a) is an array of eight (2 poloidal x 4 toroidal) structures - hereafter referred to as ITER-like structures (ILS) - having properties of load resilience, useful to maintain an efficient power flow in an ELMy plasmas, such as the one of ITER. Each IL structure is itself a poloidal array of two short circuited current straps, in series with two tuning capacitive reactances, connected in parallel to low characteristic impedance feeder. In the present proposal, the general ILS scheme is kept unchanged, but lay out changes are proposed which significantly improve the efficiency and the reliability of the array. These are: i) to increase the poloidal order of the array from eight to twelve ILS ii) to reduce dimensions and improve the power handling of the tuning reactances; iii) to change the lay out of the vacuum transmission line (VTL).

1) Increase of the array poloidal order. In a previous paper [2] it is shown that, in order to obtain a high power density in an IC array, it is of advantage to "segment" the current straps, i.e. to build it as a poloidal array of short straps.

The total strap + feeder length is about halved by increasing the array order (Figure 1b) and by choosing a more appropriate strap geometry and the input power is reduced by one third. The combined changes have the effect of approximately halving the maximum strap voltage.

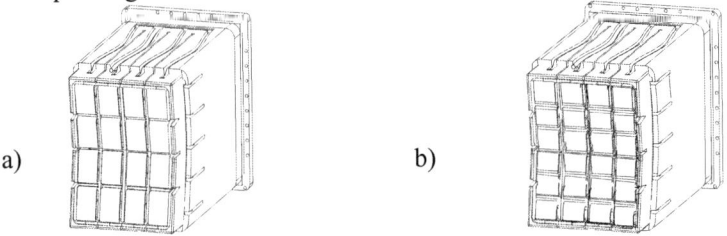

Figure 1 a) Layout of ITER IC array reference design b) proposed "segmented " upgrade

2) Changes in the lay out of the VTL. In the FDR, the capacitive tuning reactances are provided by sections of short-circuited coaxial transmission lines (Figure 2a). Coaxial components are not convenient for use in the environment of ITER vessel, because of their long length, large radial dimensions and weight

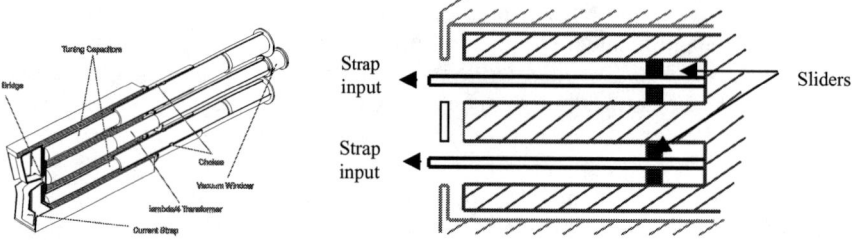

Figure 2. a) Section of a section of a ITER array, showing the two long short-circuited tuners. b) sketch of the location of the tuners in re-entrant cavities of the VTL

In the lay out of the vacuum transmission line adopted in the FDR design, the tuning stubs, are capacitively coupled to the surrounding neutron shield at ground potential. This distorts the input impedance of the tuners from a pure capacitive reactance and requires the use of λ/4 coaxial chokes, to decouple the tuners input. This increases the number of tuning adjustments and adds to the system a tri-axial geometry, which is difficult to implement, with components of that size and weight.

The use of the coaxial chokes becomes unnecessary if the tuners are located in re-entrant cavities within the VTL body, as shown in Figure 2b, as this geometry prevents coupling between tuner and ground potential. This implies an oblate geometry for the in-vessel section of the VTL (first stage of the λ/4 transformer), which can be easily interfaced to the following cylindrical sections.

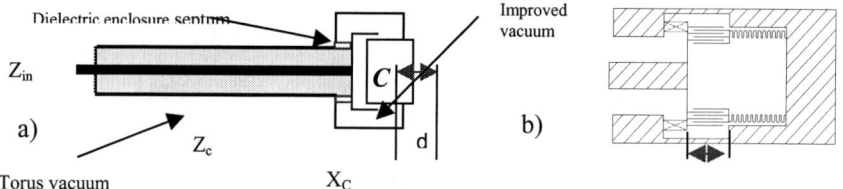

Figure 3. a) Sketch of a capacitively loaded line. b) End-of-line capacitor

3) Reduction of the dimensions and improvement of the dielectric strength of tuning elements A way of reducing length and weight of the tuners and of removing the need for sliding contacts is to use a capacitively loaded coaxial line, as shown in Figure 3a. The input reactance of the assembly in Figure 3a) is:

$$X_{in} = X_C \cdot \frac{1 - \frac{Z_c}{X_C} \cdot \tan(\beta \cdot l)}{\frac{X_C}{Z_c} \cdot \tan(\beta) + 1}$$, which remains capacitive as long as $\frac{Z_c}{X_C} \cdot \tan(\beta \cdot l) < 1$

where X_C is the load reactance, Z_C the coax characteristic impedance and l the line length. .

The load capacitance (X_C) of the assembly in Figure 2a) is plotted in Figure 4a for the ITER range of matching reactances. The coaxial multi-plate capacitor arrangement sketched in Figure 2b) is convenient in practice because it allows tuning the whole frequency range with small axial adjustments (Δl < 5cm), not requiring sliding but simply flexible components, as shown in the figure. The small volume where the electric field is high is isolated from the torus "vacuum" by means of a dielectric septum (shown in red in Figure 4b), and a high vacuum is kept in it by using coaxial vacuum capacitors technology. The ceramic material is shielded from neutron flux by a shielding thickness of 0.3 - 0.4 m, and kept well below a fluence causing dielectric and mechanical degradation..

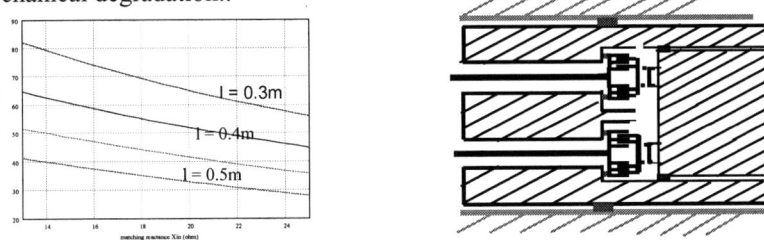

Figure 4 a) Matching reactance of the assembly for different values of line length and load capacitance. b) Layout of VTL, showing the removable capacitive tuning equipment.

The rear section of the tuner, which contains all moving parts, actuators, monitors, etc., requiring more frequent maintenance, is separable from the front section, which contains no moving components. In case of fault, they can be individually serviced and/ or replaced in situ with simple and possibly hands-on operations, without breaking the torus vacuum. In this sense this proposal has the same claimed advantages of the "external " matching.

CONCLUSIONS

The proposed upgrades significantly simplify the ITER IC array and improve its projected performances. This is obtained by minimizing the RF voltage in the unpredictable torus "vacuum" and by locating the high electri field points in restricted volumes where the a higher dielectric strength can be obtained, without in any way impairing operational aspects such equipment lifetime, neutron shielding, RH maintenance and torus vacuum confinement

REFERENCES

1   ITER DDD No. 5.1 - Ion C H & C D System, IdoMS GA MM D 3 00-02-21 F1 (2000).
2   G. Bosia, Proc.Topical Conf. on RF power application to plasma" Moran (2003).

RF Circuit Simulation of the JET ITER-like ICRH Antenna

M. Evrard, F. Durodié, P.U. Lamalle

*Laboratory for Plasma Physics, Association EURATOM - Belgian State
Partner in the Trilateral Euregio Cluster
Royal Military Academy, 30 av. de la Renaissance, B1000 Brussels, Belgium
e-mail : Michel.Evrard@rma.ac.be*

Abstract. In support of the development of a suitable matching algorithm, a RF circuit simulation of the complete JET ICRH antenna has been implemented in a Matlab/Simulink environment. The circuit model allows studying the time evolution of the capacitor positions under control of error signals computed from measurable RF quantities by proposed matching algorithms. The effect of arcs and their "visibility" on the measured RF signals is also examined.

Keywords: simulation,JET,ICRH,antenna.
PACS: 52.50.Qt

ARRAY MATCHING CIRCUIT DESCRIPTION

The JET ITER-like antenna consists of 4 <u>R</u>esonant <u>D</u>ouble <u>L</u>oop (RDL), disposed in a 2x2 square array, each RDL made of two straps connected to transmission lines through adjustable capacitors and a T-junction [Fig.1]. The capacitances must be chosen so that the input impedance at the junction is equal to a reference value Z_{ref} of the order of 3x (with possibly a small imaginary part) for each RDL. A λ/4 section of transmission line (eVTL) of characteristic impedance 9.5Ω is used to partially adapt this (low) impedance to standard 30Ω lines. Access to control the capacitors is permitted through service stubs, and the final adaptation to 30Ω is realised by a phase-shifter+stub matching system for each RDL.

Both the LHS and RHS pair of RDL are combined through 3dB hybrid couplers, their relative (poloidal) phase being controlled by additional phase-shifters, and connected to a generator. The (toroidal) phase between the LHS and RHS part of the array is controlled by the two generators.

Our first aim is to assess the influence of RDL mutual coupling on resilience and capacitors control. The full 8x8 impedance matrix $Z_{i,j}$ describing the array is obtained from a detailed 3D model using Microwave Studio, and the effect of the ELM is simulated by

$$Z_{i,j}(\lambda) = \lambda \Re e(Z_{i,j}) + i(41.-\lambda)/40.\Im m(Z_{i,j}) \qquad (1)$$

where λ is the 'elm-index' describing the importance of the ELM. Note that the imaginary part of the impedance is slightly reduced during the ELM. λ=1 correspond to L-mode, and we describe H-mode by setting $\lambda=0.75$. ELM themselves can be described by a top value of m between an optimistic $\lambda=3$ and a more pessimistic $\lambda=8$.

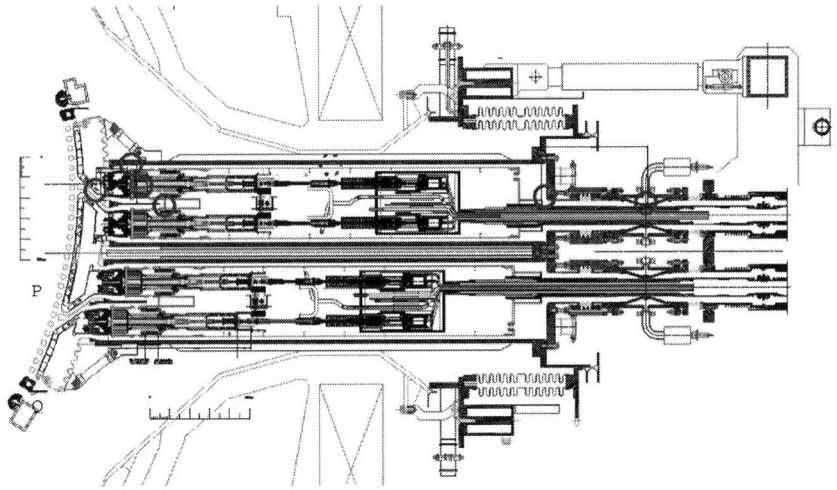

FIGURE 1. Half of the ITER-like JET array, showing 4 of the straps and of the capacitors, the eVTL and the beginning of transmission lines up to the ceramic windows for the RDL P and Q. The RDL R and S are adjacent and identical to P/Q. The red circles indicate locations where we have inserted 20nH parallel inductances to simulate arcs.

A Matlab model of the matching circuit is constructed, assembling the (numerical) scattering matrix of the array obtained from MWS and scattering matrices describing the various segment of lines, stubs, phase-shifters, hybrid couplers or capacitors obtained from standard discrete elements. This model then allows us to determine any desired quantity, like *f.i.* the RDL input impedance, corresponding to a given set of capacitors impedance, frequency, sources phasing etc....

MATCHING ALGORITHM IMPLEMENTATION IN SIMULINK

This Matlab model allows studying the dynamical evolution of the capacitors position in a Simulink environment, where the error signals $Y_{in}-Y_{ref}$ of each RDL are fed to actuators that compute the changes in capacitance in a realistic way, notably limiting the maximum capacitor speed to 200 pF/sec and including dead zone. The matching algorithm implemented in our simulink model translates the complex signals $Y_{in}-Y_{ref}$ in errors driving the capacitors movement by

$$\varepsilon(C_1) = \Re e(e^{i\alpha}(Y_{in}-Y_{ref})); \varepsilon(C2) = \Im m(e^{i\alpha}(Y_{in}-Y_{ref})) \qquad (2)$$

for each RDL. Finding the correct starting capacitances, a angles and Y_{ref} is crucial for the convergence of the simulation toward a match that is also resilient. We have used the natural symmetry between the LHS and RHS parts of the array to force identical capacitances in corresponding capacitors between the two parts of the array, hereby reducing the dimensionality of the problem from 8 to 4, a great simplification.

We have been able to find a valid set of parameters for all frequencies in the 30-55 MHz range foreseen for JET, although the resulting voltage differences across the capacitor plates are generally close to or above safety limits on some of the capacitors at the highest frequencies (above 50 MHz).

Typical results [Fig.2] for a mid-range frequency of 43 MHz shows that VSWR on the generator side of hybrid couplers remains below 1.2 and the maximum reflected power during ELM is 300kW for an incident power of 8MW. Z_{ref} was 3.0+i 0.6 Ω and the RDL phasing was [0ππ0]. This choice makes the antenna currents close to a poloidal monopole and a toroidal dipole when perfectly matched. Note that the capacitors actuation is stopped during the ELM phase.

In conclusion, the inter-RDL coupling, although it complicates the realisation of a resilient match, does not forbid it.

EFFECT OF ARCS ON RESILIENT MATCHING

Another important aspect that we have studied is the effect that arcs at different locations (see Fig.1) of our matching circuit had on the resilient match. In particular the 'visibility' of those arcs on the measured VSWR signals that are usually used to detect arcs. We simulated arcs by the sudden addition of a parallel inductance of 20nH at the chosen location. We have found that arcs occuring either on the straps or on the eVTL and further away produce very clear VSWR jump well above the ELM induced VSWR variations.

Arcs occuring at RDL junctions however cannot be seen and are below the level of variations due to ELM activity. We show an example of such an arc in Fig.3. For clarity, a single (slow) triangular ELM occurs first, followed by the arc at the junction of RDL P under exactly the same conditions (The capacitors are not actuated either during the ELM nor during the arc.) Arc detection based uniquely on VSWR signal would fail to notice such an arc.

Arcs occuring between the capacitor plates, within the capacitor, on the inductive strap of the RDL, could also be difficult to detect as they only produce a VSWR jump about twice those due to the ELM activity. On the other strap (capacitive strap), the VSWR jump produced by an arc within the capacitor is easily detectable.

Those simulation results are supported by experimental results obtained recently by the TORE SUPRA Team.

Obviously, this is a very important problem that we will need to address but for which we don't have yet a proven solution.

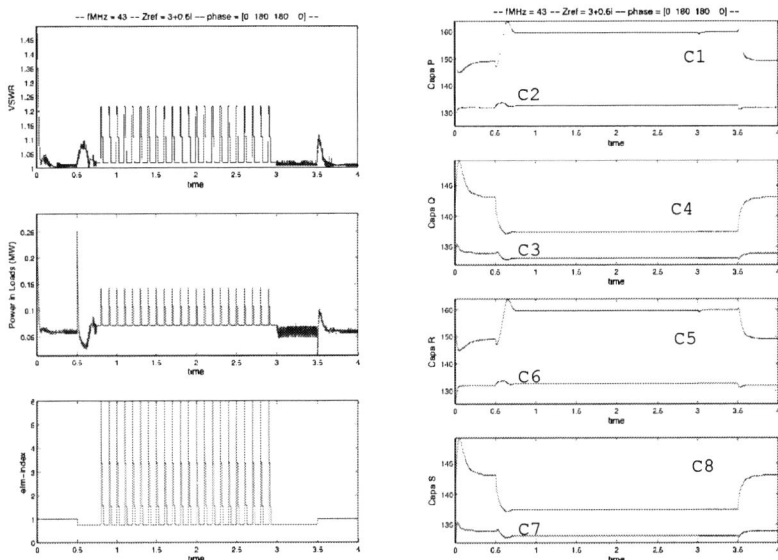

Figure 2: Typical simulink run, for f=43MHz, [0ππ0] phasing and $Zref = 3.0 + i\,0.6$ X. The RHS plot shows the capacitors evolution. From their starting values, the capacitances changes first to adapt to L-mode, then to H-mode. During H-mode, no adjustment are made, and the capacitances return to their L-mode values at the end of the shot. The VSWR measured at the generator and the power reflected on the load are shown on the LHS plot. Results for the other half are identical. The total incoming power is 8MW.

Figure 3: Simulation of an arc occuring at junction of RDL P, f=35MHz, [0ππ0] phasing and $Zref = 2.7 + i\,0.1$ X For clarity, a single (slow!) triangular ELM is represented. The arc occurs after the ELM is finished, under exactly the same conditions (The period during which the capacitors are fixed is indicated in grey) The effect of the arc on the VSWR at the 3dB couplers (antenna side) is well below the effect of ELMs. To check that the hybrid couplers played no particular role in the hiding of the arc by recirculating the power through the other RDL, we did our simulations both in presence of 3dB couplers and without them.

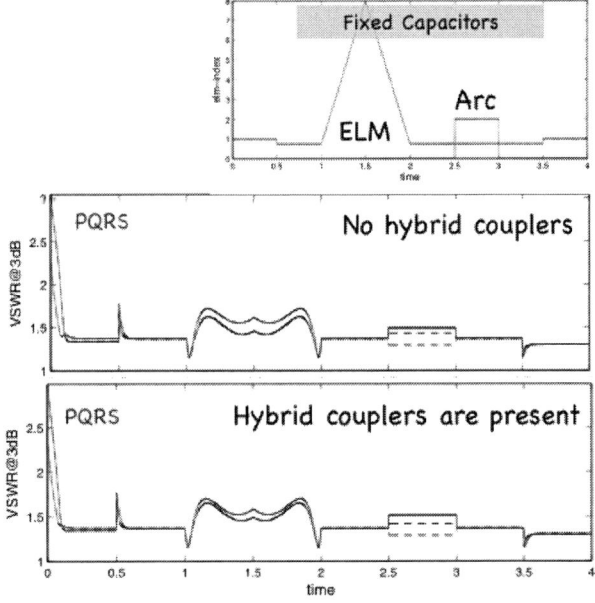

189

Three-Dimensional Electromagnetic Modeling of the ITER ICRF Antenna (External Matching Design)

F. Louche, P. U. Lamalle, P. Dumortier and A.M. Messiaen

*Laboratory for Plasma Physics, Association EURATOM - Belgian State
Royal Military Academy, 30 av. de la Renaissance, B-1000 Brussels - BELGIUM
Partner in the Trilateral Euregio Cluster*

Abstract. The present work reports on 3D radio-frequency (RF) analysis of a design for the ITER antenna with the CST Microwave Studio® software. The four-port junctions which connect the straps in triplets have been analyzed. Non-TEM effects do not play any significant role in the relevant frequency domain, and a well-balanced splitting of current between the straps inside a triplet is achieved. The scattering matrix has also been compared with RF measurements on a scaled antenna mockup, and the agreement is very good. Electric field patterns along the system have been obtained, and the RF optimization of the feeding sections is under way.

Keywords: ITER, external matching, modeling, ICRF, antenna array
PACS: 52.50.Qt

1. INTRODUCTION

An ICRH antenna array design has been proposed for ITER, in which 24 short radiating straps are combined in 8 triplets. A load-resilient "conjugate T" matching circuit located outside the tokamak vessel is considered, in which these triplets are connected in pairs through line stretchers [1]. Accurate simulations of this device are needed to estimate the effects of the mutual coupling between the straps. The commercial software CST Microwave Studio® (MWS) [2] has been used for this purpose. The main features of the antenna array are included, notably the non-planar geometry, the thickness of the radiating straps and all septa in the antenna box. As the software does not allow modelling a magnetized plasma, the antenna is located in front of a high-permittivity dielectric slab capturing the essential features of wave refraction by a steep plasma edge [3]. We address in this work important issues: non-TEM effects, current splitting among the straps of one triplet, loading properties and electric fields.

2. STUDY OF THE FOUR-PORTS JUNCTION

The 4-ports junction connects a set of three straps to one arm of the conjugate-T. It is constituted by a line of variable shape evolving from a stripline shape at one side to a circular coaxial line at the other side, with a characteristic impedance maintained at approximately 20 Ω. This stripline shape allows an easy connection to the three 60 Ω sections (see [1]). The junction has been basically designed to equally split the incoming

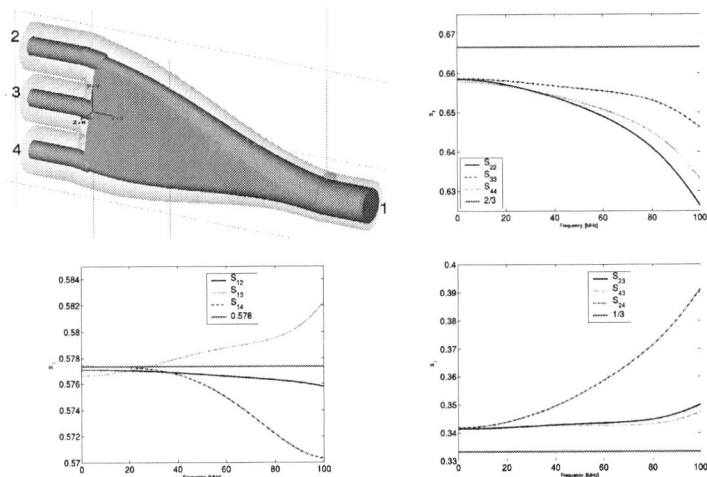

FIGURE 1. Magnitude of the scattering matrix elements in function of the frequency, obtained with MWS: (a) reflection coefficients at the output port 2,3 and 4 - (b) mutual scattering coefficients for the input port 1 - (c) mutual scattering coefficients for the output ports 2,3 and 4 - (d) Relative phases of S for the output ports. A π relative phasing is predicted by the theory. The value predicted by TLT is also given for each case.

RF current between each strap of the triplet, but a small curvature in the poloidal plane (8.5 degrees) is imposed by the constraint that the straps have to lie at approximately the same distance from the separatrix.

The scattering parameters of an ideal 4-ports junction, characterized by $3Z_1 = Z_2 = Z_3 = Z_4$ are given by (see for instance [4, section 4.3]): $S_{11} = 0$, $S_{1i} = \sqrt{1/3}(i \neq 1)$, $S_{ii} = 2/3 e^{i\alpha}$ ($i \neq 1$) & $S_{ij} = 1/3 e^{i(\pi+\alpha)}(i \neq j = 2,3,4)$, where α is the phase constant of the output diagonal terms, relative to the phase of S_{12} arbitrarily chosen equal to zero. If we compare this result with the results provided by a MWS run (see figure 1), in such a way to evaluate the non TEM effects of the junction, we see that these effects play a small role in the frequency domain of interest, and increase with frequency. Non-ideal effects are due to the geometrical complexity of the junction. There are fringing fields associated with the discontinuities at the junction between the sections at different impedances, leading to stored energy. The poloidal asymmetry is also visible, and its influence on the current distribution in one triplet has already been shown unsignificant [5].

3. MODELING OF THE ITER ANTENNA

A model of the 24 straps launcher with the 8 four-ports junction, inside an antenna box, is put in front of a piece of dielectric medium ($\varepsilon_r = 2000$). For the reference case the distance strap-dielectric is 14 cm. Radiating boundary conditions behind the dielectric simulates single pass damping of the wave in the tokamak center. Two approaches have been considered in this work for simulating the antenna: either the full system (24 straps and 8 junctions) is considered, or only the array of straps is simulated, the rest of the line being modeled by transmission lines.

We present the results obtained with the first model. The current repartition between the straps of one triplet is confirmed: our computations show that the current imbalance does not exceed 2 % in the frequency domain of interest. The electric fields allowed in the system must have an amplitude below 20 kV/vm. Our calculation show that peak values up to 30 kV/cm are obtained, while the fields are below the limit of 20 kV/cm along the rest of the line. The peak value are observed along the edges of the feeding part. The optimization the shape of the feeders is under way to reduce them as much as possible.

In the second approach, the scattering matrix of the 24 straps array has been computed over its frequency range for different loading cases: vacuum without and with Faraday shield, dielectric medium ($\varepsilon_r = 2000$) at 14 cm of the straps with and without shield, and dielectric medium ($\varepsilon_r = 2000$) at 14 cm of the straps with two intermediate SOL-like steps. The reference planes are very close to the straps. Figure 2 shows the values of the reactances for the different loading cases.

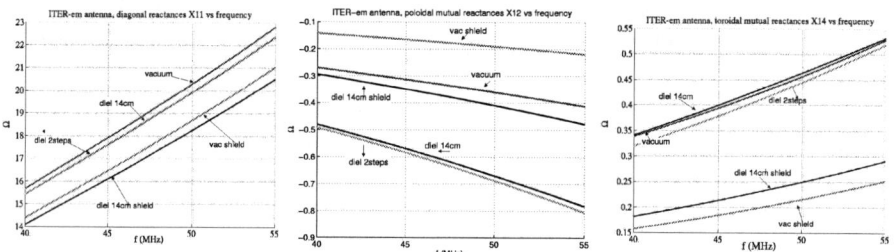

FIGURE 2. Computed straps reactances for different loading cases

We observe that the Faraday shield decreases the mutual reactances. It is to be noted that there are large uncertainties in the ITER edge profiles (SOL and far SOL). This results in large uncertainties in the antenna loading and array performance predictions. If the reference planes are put behind the feeder sections, unrealistically high reactances are obtained. This indicates the need to optimize feeder sections between the straps and the 4-port junctions. This work is under way.

We can estimate the equivalent parameters (resistance and inductance) for the second model. The following values of the equivalent parameters by unit length are obtained assuming poloidally uniform strap currents (poloidal π phasing of strap triplets):

TABLE 1. Equivalent loading resistance, [Ω/m]

frek, [Mhz]	$0\pi 0\pi$	0000	$00\pi\pi$	$0\pi\pi 0$	$\pi/2$
40	1.446	3.322	1.642	1.392	1.517
47.5	1.616	4.233	1.918	1.524	1.721
55	1.799	5.231	2.239	1.698	1.969 Ê

TABLE 2. Equivalent inductance, [nH/m]

frek, [Mhz]	$0\pi0\pi$	0000	$00\pi\pi$	$0\pi\pi0$	$\pi/2$
40	179.571	187.522	182.086	180.072	181.079
47.5	183.782	190.730	186.394	184.394	185.394
55	189.194	195.100	191.943	189.929	190.936 Ê

This underlines the importance of proper matrix analysis of transmission and matching circuit. In particular, array loading is significantly modified by the conjugate-T mode of operation [6].

4. COMPARISON WITH EXPERIMENTS ON MOCKUP

A scaled (1/5) mockup of the full ITER antenna has been developed and measurements of the S parameters have been performed [7] with a load made of salted water ($\varepsilon_r = 81$, $\sigma = 0.5$ S/m). The reflection coefficient for one triplet obtained with MWS is in good agreement with the measured result (see figure 3).

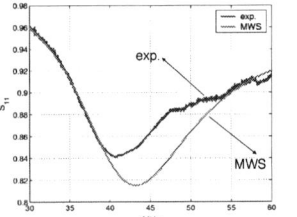
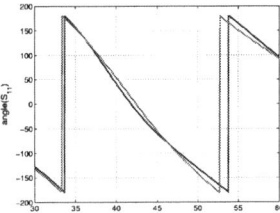

FIGURE 3. Comparison mockup in front of salted water/MWS model for the reflection coefficient of one triplet: amplitude (left) and phase (right)

REFERENCES

1. P. Dumortier *et al. LPP-ERM/KMS Int. Rep. 121*
2. CST MICROWAVE STUDIO® User Manual, Version 5.0, CST GmbH, Germany, 2004
3. A. Messiaen *et al.* (2004) *LPP-ERM/KMS Int. Rep. 123*
4. D. M. Pozar - Microwave Engineering, 2nd ed. John Wiley & Sons 1998
5. Louche F. *et al.* Europhysics Conf. Abstracts, 28B (2004) P-1.133
6. P. U. Lamalle *et al.* (2005), this conference
7. A. Messiaen *et al.* (2005), this conference

Status of the JET ITER-Like Antenna High-Power Prototype Test Program[*]

R.H. Goulding, F.W. Baity[1], F. Durodié[3], A. Fadnek[1], K. D. Freudenberg[1], J. C. Hosea[2], G. D. Loesser[2], B.E. Nelson[1], M. Nightingale[4], D.A. Rasmussen[1], D. O. Sparks[1], R. Walton[4]

[1]*Oak Ridge National Laboratory, Oak Ridge, Tennessee, USA*
[2]*Princeton Plasma Physics Laboratory, Princeton, New Jersey, USA*
[3]*Association Euratom-Belgian State, LPP-ERM/KMS, Brussels, Belgium*
[4]*Association Euratom-UKAEA, Culham Science Centre, Abingdon, UK*

Abstract. Previous tests of a High Power Prototype (HPP) comprising one quadrant of the JET ITER-Like ICRF Antenna have indicated the need for some design modifications in order to achieve 10 s pulses coupling the full design power (7.1 MW) into the reference plasma load ($R' = 4$ Ω/m). These modifications have now been made to the HPP, as well as to the design of the ITER-Like Antenna itself. In particular, maximum current densities have been reduced or otherwise accommodated in key areas. New current straps for the HPP have been fabricated from stereo-lithography-based investment castings. Design modifications to the antenna enclosure have also been implemented. This work has been materially assisted through the use of CST Microwave Studio (MWS), a commercially available 3-D electromagnetic modeling package. Essentially the full engineering CAD model of the HPP current straps and antenna enclosure has been ex-ported from ProE to MWS. Computed current density profiles have been introduced into an ANSYS thermal model. These activities will be discussed, as well as the current status of the HPP test program.

INTRODUCTION

The JET ITER-Like Ion Cyclotron Heating (ICH) Antenna High Power Prototype[1] closely replicates one quadrant of the actual JET antenna[2], which has been designed to couple 7.1 MW into a JET ELMy H-Mode plasma. The goal of HPP high power tests has been to achieve voltages and currents equal to those required to couple this power into a nominal resistive plasma load of 4 Ω / m. Several problems were encountered during the initial round of tests in 2003, which prevented the attainment of this goal at that time[3]. As a result, a series of

---

[*] Oak Ridge National Laboratory, managed by UT-Battelle, LLC, for the U.S. Department of Energy under contract number DE-AC05-00OR22725.

minor modifications were made to the HPP, which were also incorporated into the design of the actual antenna. In this paper we briefly review the previous test results, and describe the changes made to the HPP in response to them. The modeling leading to one of these changes is also discussed.

PREVIOUS HIGH POWER TEST RESULTS

There were several important results from the first HPP tests. The maximum voltage that could initially be obtained on the current straps was limited due to arcing across a small gap between graphite private limiter tiles and neighboring Faraday shield rods (drawings showing these and all other major antenna components can be found in ref. 3) Evidence of the arcing is shown in Figure 1, where discolored regions can be seen on the graphite tiles located on the left side of the photograph in the regions adjacent to the two shield rods. When the tiles were removed, the maximum voltage obtained at the junction between the current strap feedlines and internal matching capacitors increased from 24 kV to 40 kV (note: the operating frequency is 50 MHz for all tests discussed in this paper).

Another important result is that during long pulse operation, the achievable power and voltage was by limited by large pressure spikes (> 0.03 Pa) leading to plasma discharges both inside and outside the antenna. The pressure excursions are believed to be caused by excess heating in highly localized regions of the antenna. The heating also led to melting of a portion of the antenna "flexipivot". This structure consists of two thin (2mm thick) parallel Inconel plates whose purpose is to limit the stress on the Faraday shield rods and antenna box by deforming in response to differential thermal expansion of the rods, which are fabricated from beryllium on the actual antenna. On the HPP, they are made of

FIGURE 1. Evidence of arcs between private limiter tiles and Faraday shield rods.

FIGURE 2. Bottoms of flexipivot (left) and current strap (right). Arrow points to melted region of flexipivot.

nickel plated aluminum. Because of the excess heating, the maximum voltage at the capacitors achieved during a 10 s pulse was ~ 23 kV peak, approximately 60% of the value required to couple the required power into the specified JET plasma load. The corresponding input power was approximately 80 kW. The damaged portion of the flexipivot is shown in Figure 2.

In addition to the flexipivot, damage to the current straps was observed after 10 s pulses, with cracks in some of the welds, and nickel plating that flaked badly. A photograph of one of them can be seen in Figure 3. The plating damage may have contributed to the pressure spikes observed during these pulses.

ANTENNA MODIFICATIONS

As a result of these tests, several modifications were made to the HPP. These modifications have also been made to the design of the actual JET antenna. The main change was the addition of current redistribution wedges. These wedges have been designed with the assistance of CST Microwave Studio (MWS), a 3-D electromagnetic modeling package[4]. Their purpose is to reduce the current density in the region of the flexipivot near the grounded ends of the current straps, thus preventing the type of damage seen in Figure 2. In these regions there are horizontal plates with gaps between the plates and the flexipivots to allow the latter to deform. Image currents flowing in the flexipivots combine with currents from straps and become highly concentrated in these regions.

Figure 4. shows the results of MWS calculations using the actual HPP geometry input from the original 3-D CAD engineering model created with Pro/Engineer Wildfire[5]. Figure 4a shows current patterns before the wedge was added. The high current density is evidenced by the high concentration of arrows that can be seen in the same region in which the melting is shown in Figure 2. The case in which a wedge is added is shown in figure 4b. The current density is

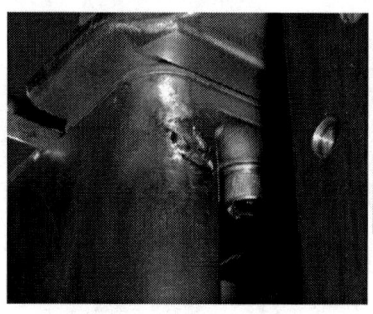  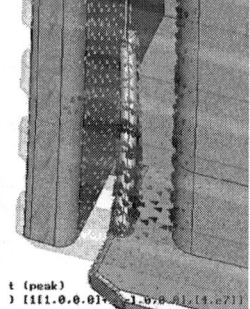

FIGURE 3. Top portion of upper strap showing crack in weld and peeled plating (arrow).

FIGURE 4a. Current pattern without wedge. Concentration seen at slot base.

FIGURE 4b. Current concentration is reduced when wedge is added

reduced substantially. Figure 5 shows the bottom of the actual antenna box with

FIGURE 5. Antenna box with wedge added   **FIGURE 6.** New top current strap casting before back plates attached.   **FIGURE 7.** Old and new (left) private limiter tiles

the wedge added. A similar wedge was added to the top of the box.

Another change that was made to the antenna is a modification to the current strap design. The sidewalls in the original straps, where the current density is the highest, are 2 mm thick. In the new straps the thickness in these regions is increased to 6 mm in order to increase the mass and reduce the temperature rise from rf dissipation. The original straps were fabricated by welding together many small parts. The weld depth was limited due to the need to minimize distortion. The shallowness of these welds likely led to cracking when the joints were subject to thermal stresses. In addition, the straps were plated by electroless nickel to a depth of 80 microns, sealed by a 40 micron thick layer of electroplated nickel. Phosphorus present in the electroless nickel plating may well have contributed to the cracking. The new straps were fabricated from investment castings, requiring less welds which could then be made deeper. Figure 6 is a picture of one of these castings before the back cover plate has been welded on. Electroless nickel plating was not used in the fabrication of the new straps.

Finally, figure 7 is a picture showing the original private limiter tile design subject to the arcing problem shown in figure 1, together with the modified version. In the new tile, the gap to the neighboring Faraday shield rods has been increased by ~ 6 mm. All of these modifications will be tested during further operation of the HPP, results of which will be reported in a future paper.

References

1. R. H. Goulding et al., "Initial operation of the JET ITER-like High Power Prototype Antenna", 15th Topical Conference on Radio Frequency Power in Plasmas, Moran, Wyoming, 2003, AIP Conf. Proc 694, pp. 102-105.
2. F. Durodié, et al., "The ITER-Like ICRH Launcher Project for JET", 15th Topical Conference on Radio Frequency Power in Plasmas, Moran, Wyoming, 2003, AIP Conf. Proc 694, pp. 98-101.
3. R. H. Goulding et al., "Results and Implications of the JET ITER-Like ICRF Antenna High Power Prototype Tests", Proceedings of the 20th IAEA Fusion Energy Conference, Vilamoura, Portugal, 1-6 November 2004.
4. http://www.cst.de
5. http://www.proe.com

Study of the ITER ICRH system with external matching by means of a mock-up loaded by a variable water load

A.Messiaen, M.Vervier, P.Dumortier, P.Lamalle, F.Louche

LPP-ERM/KMS, EURATOM-Belgian State Association,
Trilateral Euregio Cluster, B-1000 Brussels, Belgium

Abstract. A mock-up of the complete antenna array (24 straps grouped in 8 triplets) of the ICRH system with external matching for ITER has been constructed with a length reduction factor of 5. At a frequency increased by the same factor the electrical properties of the full-scale system can be measured. A movable water tank in front of the array simulates variable plasma loading. Measurements of the matching performances of various external circuit configurations and of the scattering matrix of the system show (i) the non-negligible effect of mutual coupling on load resilient matching by Conjugate-T (CT) or hybrid leading to coupling between the matching actuators and the generators and asymmetry in power distribution, (ii) good load resilience of a single CT for the right choice of configuration and number of matching parameters, (iii) the large number of matching solutions for coupled CT's, (iv) the benefit of passive power distribution to the straps. Keywords: ICRH, ITER, matching, mock-up, water load

PACS: 28.52.Cx, 52.50.Qt, 52.55.Fa

INTRODUCTION: MOCK-UP OF THE ITER RF ANTENNA

A conceptual design of a 20MW ICRH plug-in for ITER in the frequency band 40-55MHz with external matching has been developed [1]. The antenna is constituted by an array of 24 radiating straps (placed in a surface $1.9 \times 1.45 m^2$) in order to provide the large power density with an affordable antenna voltage. The main advantages of this design are the absence of in-vessel remotely operated components to achieve the matching and the use of 4-ports passive junctions that provide more uniform rf current distribution among the straps and minimize the number of matching circuits. Indeed these junctions combine the 24 straps in 8 triplets. Each strap, as shown in Fig.1, is put in an antenna box to decrease its mutual coupling with the others but they remain unavoidably coupled as they radiate in the same medium. The effect of this coupling is experimentally investigated in this paper.

Tests with realistic plasma-like load conditions can be obtained with a large dielectric constant medium facing the strap array. This is because for the chosen frequency range the antenna excites the fast Alfvèn wave in the magnetized plasma. This wave is characterized by a large propagation constant ($\sim \omega/V_A$) and therefore its antenna loading can be simulated within a good approximation by a dielectric medium with large dielectric constant facing the antenna array [2,3]. Water can advantageously be used as such a load. To avoid the spurious effect of wave reflection on the walls of the water tank salt can be added to the water to provide sufficient wave damping.

A full-scale model is not needed to measure the antenna characteristics: when decreasing the length and increasing the frequency by the same scale factor the impedance matrix of the array remains identical [2,3]. Therefore a mock-up of the complete antenna array has been constructed based on the Catia 3-D drawings of the ITER design [4]. A length reduction scale factor of 5 is used and therefore the frequency band corresponding to the full-scale system is 200-275MHz. A water tank is placed in front of the antenna array that is mounted on a sliding support for adjusting the distance array-water tank. An equivalent domain of distance load-antenna for ITER from 5 to 35cm is possible. Fig. 1 shows a front view of the antenna array and the inner part of one 4-ports junction feeding a triplet of straps through a pre-matching line system.

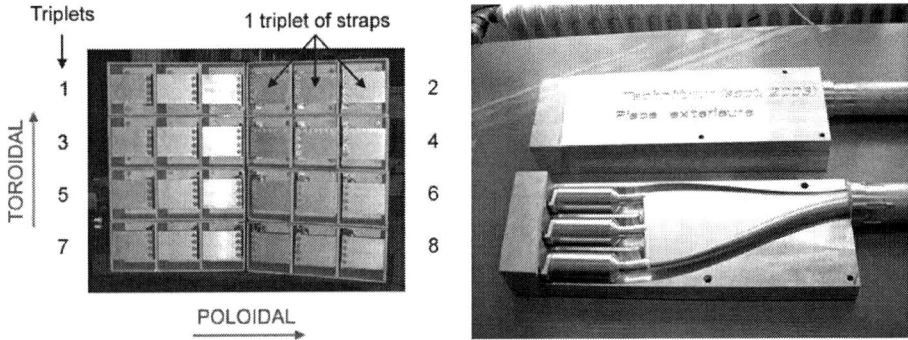

FIGURE 1. Mock-up: antenna box with straps (left) and inner view of a 4-ports junction feeding 1 triplet (right)

MEASUREMENTS ON THE MOCK-UP

The 8x8 scattering matrix S_8 of the 8 triplets is measured at the test point indicated in Fig.2a as a function of the frequency and as a function of the distance d_{sw} strap-water tank. The domain of variation of the S_{ii} parameters corresponds to an antenna loading resistance $\sim 1\Omega/m < R_A < \sim 8\Omega/m$. The S_{ij} parameters describe the mutual coupling which is not negligible as seen in Fig. 2b. The coupling between adjacent triplets in the toroidal direction is larger than in the poloidal direction. Their evolution versus the loading is different. From the S_8 matrix the S, Z or Y matrices in every location of the antenna plug or of the matching circuit can be derived from transmission line theory and compared with direct measurements. For instance such measurements of the rf input voltage of the straps by active rf probes has shown quasi-equal excitation in phase and amplitude of one triplet by the 4 ports junction in agreement with its derivation from S_8.

External matching circuits ($Z_0=50\Omega$ hardware) are connected to the output test points of the $Z_0=20\Omega$ lines from the mockup (see Fig. 2a). As the test points are chosen near a voltage anti-node there is a decrease of VSWR in the 50Ω lines. The properties of various matching schemes using the "conjugate-T" circuit or an hybrid junction to achieve the needed load resilience for operation with an ELMy plasma are investigated. In the present paper only the 4 central triplets are put into operation, the others being terminated on 50Ω loads. In [5] results with the 8 triplets fed are reported.

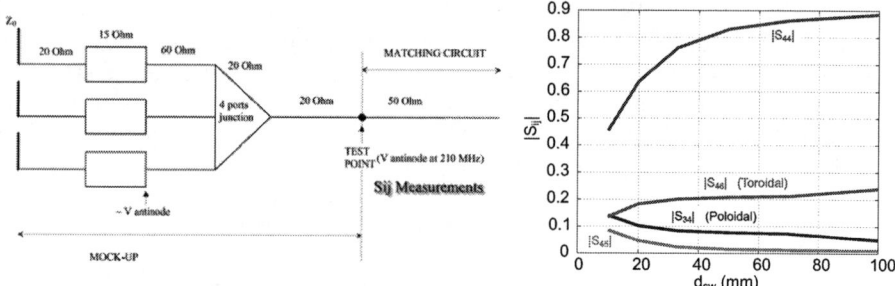

FIGURE 2. (a) Measurement set-up and (b) Amplitude of some measured S_{ij} parameters

The mutual impedances lead to the following effects: (i) Coupling between the matching actuators and the power sources of different CT's or hybrids, (ii) For the case of several CT's they induce a very large number of simultaneous tuning solutions: for 1 CT we have 4 distinct matching solutions [6], for 2 coupled CT's 32 and for n coupled CT's $8^n/2$; each set of solutions are only valid for given values of the forward voltages E_{i+} of the different power sources. This is illustrated in Fig.3 for the cases of 4 triplets connected in 2 CT's (circuit of Fig. 3a) where the reflection coefficients $|\Gamma_I|$ and $|\Gamma_{II}|$ are shown versus d_{sw} for half of the 32 matching solutions (adjusted by the lengths l_3, l_4, l_5 and l_6) and for $E_{2+}/E_{1+}=1$. We have $\Gamma_I=S_{T11}+S_{T12}E_{2+}/E_{1+}$, $\Gamma_{II}=S_{T22}+S_{T21}E_{1+}/E_{2+}$ where S_T is the scattering matrix of the network T_I-T_{II} seen from the generators. (iii) Power transfer between one matching circuit and the others resulting in asymmetry of the radiated power by the triplets [6]: if Va_i, Ia_i are respectively the input strap voltage and current and if Za is their input impedance matrix, the input active power to the straps is $P_i=Re(Va_i.Ia_i^*)/2$ which is the sum of the power effectively radiated $P_{ri}=(Ia_iRe(Za_{ij})).Ia_i^*/2$ and the exchanged active power between the straps is given by $P_{exch}=P_i-P_{ri}$. This is illustrated in Fig.4 where 4 triplets are fed by one 90° hybrid. In this case there is a large unbalance of the P_i's and a moderate one of the P_{ri}'s the difference being given by P_{exch}. There is also some input strap voltage dissymmetry. (iv) Possible reduction of the load resilience due to CT's or power dump by hybrids, depending of the circuit configuration. Examples of good resilience are shown in Fig.4 for the hybrid case and on Fig. 5 for the CT case. In this last case 2 of the 4 matching solutions have much better load resilience. The 4 parameters tuning of the CT is used [5,6]. In the circuits of Figs. 4 and 5 passive power share to the triplets is used to avoid coupling between several matching circuits. The triplets are put in parallel near voltage anti-nodes on the lines: this reduces the VSWR without increase of line voltage. The case of 8 triplets fed by 1 CT or 1 hybrid is discussed in detail in [5].

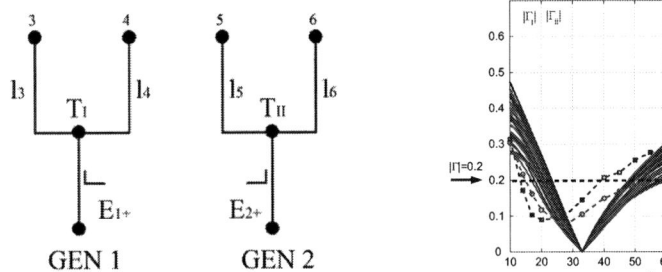

FIGURE 3. (a) Circuit of 4 triplets connected in 2 CTs and (b) Load resilience of the different matching solutions.

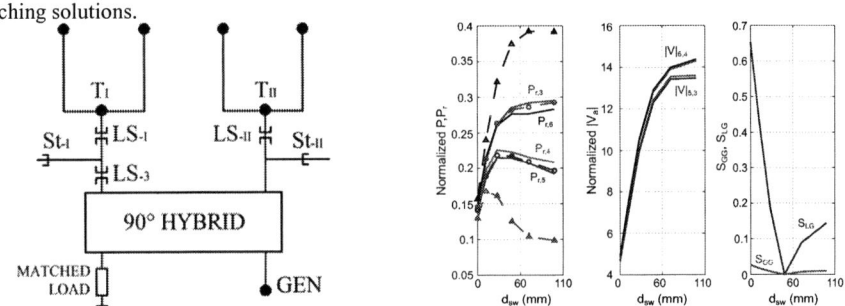

FIGURE 4. (a) Circuit of 4 triplets fed by 1 90° hybrid and (b) Measured power imbalance, voltages and load resilience. Measured |V| ratio = 1.28 for d_{sw} = 100 mm.

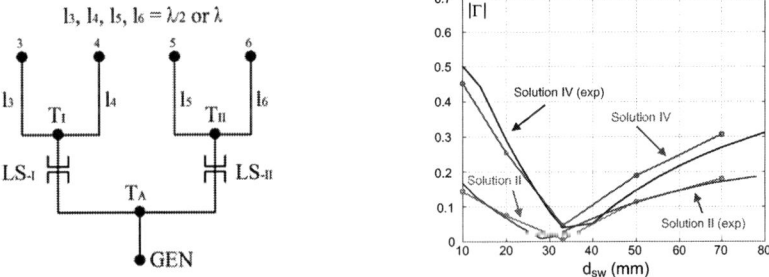

FIGURE 5. (a) 4 triplets connected in 1 CT and (b) Comparison of measured load resilience and its computation for measured S matrix for solution I (≈II) and IV (≈III).

References

[1] P.Dumortier et al., Final report on Task FU05-CT 2002-00094 (EFDA/02-675), LPP-ERM/KMS Int. Rep. 121, A.Messiaen et al., Proc. 15th Top. Conf. On Radio-Frequency Power in Plasmas, Moran, Wyoming, May2003, AIP conf proceedings volume 694 p.142; P.Dumortier et al., ibid. p.94; A.Messiaen et al.30th EPS conf. Contr. Fus. Plasma Phys., Europhysics Conf. Abstracts, 27A (2003)P-3.216.
[2] A.Messiaen et al. LPP-ERM/KMS Int. Rep. 123, May 2004.
[3] P.Lamalle et al. 20th IAEA Fusion Energy Conference (Vilamoura) 2004, to be published.
[4] A.Messiaen et al., SOFT conference (Venice) 2004, to be published.
[5] M.Vervier et al., this conference
[6] P.Lamalle et al., SOFT conference (Venice) 2004, to be published.

Experimental proof of a load resilient external matching solution for the ITER ICRH system

M.Vervier, A.Messiaen, P.Dumortier, P.Lamalle

*LPP-ERM/KMS,EURATOM-Belgian State Association,
Trilateral Euregio Cluster, B-1000 Brussels, Belgium*

Abstract. A reliable load resilient external matching scheme for the ITER ICRH system has been successfully tested on the mock-up of the external matching system with variable plasma load simulation. To avoid the deleterious mutual coupling effects the power has been passively distributed among the upper half and the bottom half of the 24 radiating straps of the antenna plug. In this plug the straps are grouped in 8 triplets by 4-ports junctions. The 4 top and 4 bottom triplets are respectively put in parallel outside the antenna plug near a voltage anti-node by means of T junctions. The load resilient matching is then obtained by a 4 parameters single "conjugate T" (CT) configuration. For an antenna loading variation of about 1 to 8 Ω/m the VSWR at the power source remains below 1.3. The maximum voltage along the line remains equal to the one in the antenna plug and there is a fair power share between the straps. A $\pi 0 \pi 0$ toroidal phasing is easily obtained. The poloidal phasing between the top and bottom triplets is determined by the loading. A straightforward matching procedure is described. Good load resilience is also obtained by replacing the CT by one hybrid.

Keywords: ICRH, ITER, conjugate T, hybrid, matching
PACS: 28.52.Cx, 52.50.Qt, 52.55.Fa

INTRODUCTION

The antenna system for the 20MW ICRH of ITER with external matching consists in an array of 24 short straps (see Fig.1 and Refs. of [1]) grouped in 8 triplets by means of 4-ports junctions. Due to its compactness mutual coupling effects between the radiating triplets are not at all negligible. These effects have been analyzed theoretically [2] and experimentally by a mock-up with variable load [1]. They lead to (i) a coupling between all matching actuators and power sources, (ii) a very large number of matching solutions and (iii) asymmetry in the radiated power distribution among the straps. Power dump by means of hybrid junctions can be used instead of CT's to obtain the needed load resilience (for operation on ELMy plasmas) but the mutual coupling effects are also important in this case as will be shown below. The experimental matching studies with the mock-up indicate the interest of decreasing the number of matching circuits by passive power distribution among the triplets. As the mock-up has a length reduction factor of 5 the electrical properties of the full-scale model are measured by working at 5 times the frequency planned for ITER.

FULL ARRAY MATCHING LAYOUT AND PROPERTIES

Conjugate T matching

Fig.1 shows the circuit used to group the 8 triplets in one CT. The different triplets inputs are labeled 1 to 8 and their positions are chosen at (or near) a voltage anti-node at the 20Ω outputs of the mock-up (called test point on Fig. 2a of Ref. [1]). At this point the line characteristic impedance Z_0 is increased from 20 to 50Ω. Then the four top triplets and the four bottom triplets are put in parallel also near a voltage anti-node by 50Ω lines and T's respectively at positions A, B, I and C, D, II. In this way the maximum voltage on the different lines remain approximately the same and the VSWR decreases towards the generator after the 20/50Ω transition and after each T. The transmitted power doubles after these T's. The line lengths between the tests points are adjusted to obtain a toroidal phasing $0\pi0\pi$. Finally I and II are connected together in CT by means of the line-stretchers LS-I and LS-II which are adjusted to perform a 2-parameters conjugate T (CT) matching. The tuner between the CT and the generator provides 2 additional parameters to maximize the load resilience (we refer to this optimization as 4-parameters CT matching). The passive power distribution avoids the problem of multiple interacting matching circuits but necessitates combining the power of several generators in one feeding line.

Fig. 1b shows the resulting load resilience vs. the distance d_{sw} between the straps and a water load simulating the plasma loading [3]. It corresponds to an equivalent antenna loading resistance variation from ~1 to ~8Ω/m and the 2 and 4 parameters matching results are shown. In the last case the VSWR seen by the power source remains always below 1.5.

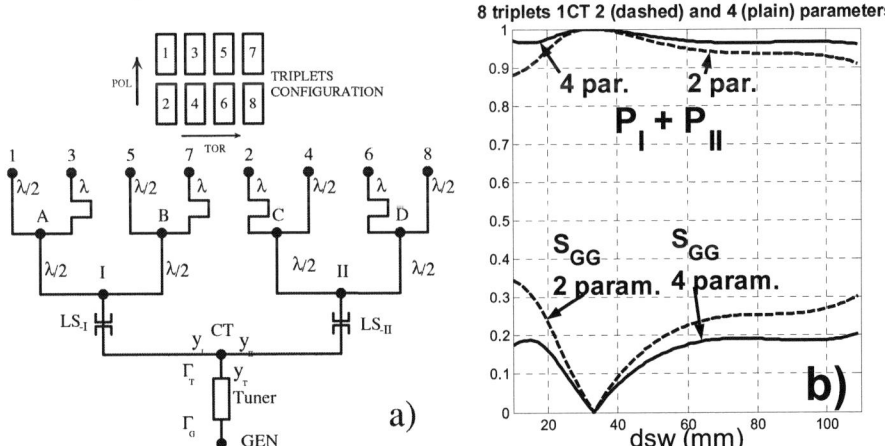

FIGURE 1. a) Configuration of the connections between the different triplets to obtain the desired phasing between the straps using the conjugate T. The junction points A, B, C, D, I and II are located at anti-node voltage; b) (bottom) coefficient of reflection at the generator (S_{GG}) in function of the distance strap-water (d_{sw}) corresponding roughly to a coupling resistance from ~1 to ~8 Ohm/m for the case with 2 (dashed line) and 4 (plain line) parameters; and (top) ratio of transmitted power to the antennae on forward power at the generator for the 2 and 4 parameters matching.

Direct measurements of input voltage distribution among the straps show good voltage repartition and toroidal phasing (see table 1 for d_{sw}=100mm). The poloidal phasing depends on loading with here $\Delta\Phi\to 0$ when the loading$\to 0$.

TABLE 1. Direct measurement of voltage and phase at each triplet relative to the forward voltage.

triplet	1	2	3	4	5	6	7	8
|V| (V)	1.1	0.92	1.1	0.91	1.13	0.91	1.13	0.93
φ(°)	159	-171	-20	8	161	-171	-24	4

Quadrature hybrid matching

Fig.2 shows a circuit where the CT is replaced by one 90° hybrid and 2 matching circuits (LS+ST) and an additional LS for phase adjustment to dump the reflected power in matched load. Note that good load resilience is also obtained with the hybrid.

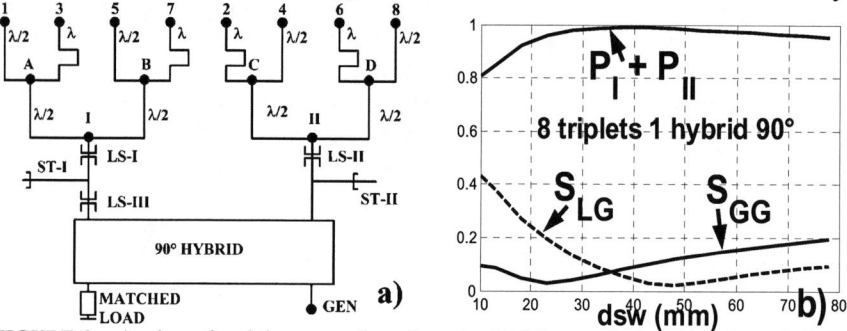

FIGURE 2. a) schematic of the connections from the 8 triplets to the hybrid to have 0π0π toroidal phasing between the straps ; b) (bottom) reflection coefficient at the generator (S_{GG}) and transfer coefficient to the load (S_{LG}) in function of the distance strap-water (d_{sw}). (top) ratio of power transmitted to the antennae on the forward power at the generator.

OPTIMIZATION OF CT LOAD RESILIENCE AND POWER DISTRIBUTION

One CT circuit has 4 distinct 2-parameters matching solutions. The solutions corresponding to the minima of the reflection coefficient |Γ| seen in CT from the generator side are shown in the LS-I and LS-II lengths plane for the conditions of Fig.1a with low loading (d_{sw}=100mm:~vacuum radiation) as shown in Fig. 3a. The 4 parameters matching of Fig. 1b is obtained (i) by adjusting LS-I and LS-II for one of the solutions of Fig. 3a (low loading) and (ii) by adjusting the tuner (LS+ stub) for minimum reflection at higher loading (d_{sw} =33mm). A minimization procedure using the measured S parameters vs. d_{sw} shows that the optimal load resilience is obtained with LS-I and LS-II adjusted on the path between a pair of solutions as indicated on Fig. 3a. In this condition the locus versus d_{sw} of the admittance $y_G=y_I+y_{II}$ seen in CT from the generator is folded in the Smith chart and concentrated in a narrow domain (y_T path in Fig. 3b). The tuner brings this path around |Γ|=0 as appears on the Fig.3b. Fig.4 shows the resulting measured resilience compared with the one deduced from the measured S parameters and compared with the one obtained in Fig.1b.

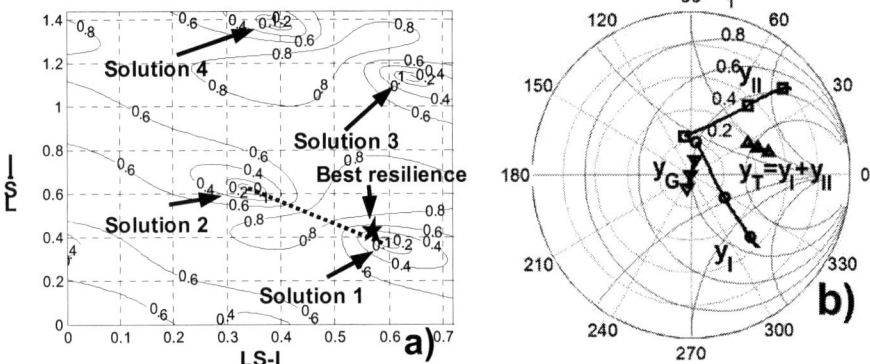

FIGURE 3. a) Contour plot of the absolute value of the reflection coefficient seen by the generator in function of LS-I and LS-II at a distance strap-water (dsw) of 100mm (~vacuum radiation) without tuner (stub length equal to λ/4). Note the 4 distinct solutions and the best resilience solution located on the way between the group of 2 solutions; b) Locus of y_I, y_{II}, y_T and y_G in the Smith chart (for $Z_0=50\Omega$) for the best resilience case.

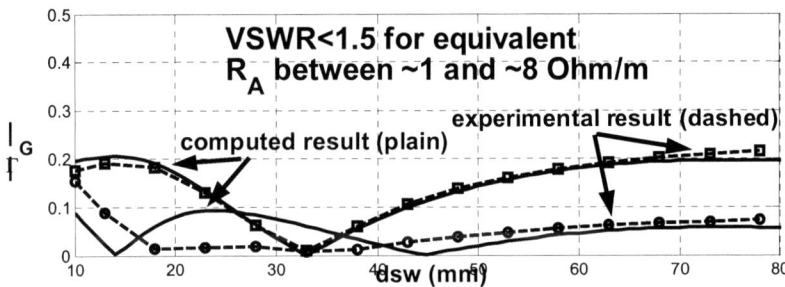

FIGURE 4. Measured (dashed line) and computed (plain line) resilience from the experimental S matrix. The solutions shown in Fig.1b (4 parameters) and 3b are compared.

CONCLUSIONS

A solution combining good load resilience and good power repartition has been experimentally demonstrated for the external matching option. The tuning procedure is greatly simplified by the reduction of the number of matching circuits. One power source delivering the total power is needed but due to the VSWR reduction after each T junction the maximum voltage on the lines does not exceed the one appearing for several power sources and matching circuits. Use of 14" 50Ω external lines would provide an excellent voltage standoff safety margin.

REFERENCES

[1] A. Messiaen et al., this conference.
[2] P. Lamalle et al., SOFT conference (Venice) 2004, to be published.
[3] A. Messiaen et al., ibidem

Initial Operation of the Alcator C-Mod ICRF Antennas with High-Z Metal Antenna Guards

G. Schilling[1], S. J. Wukitch[2], Y. Lin[2], A. Parisot[2], M. Porkolab[2], and the Alcator C-Mod Team

[1]*Princeton University Plasma Physics Laboratory, Princeton NJ 08543,* [2]*MIT Plasma Science and Fusion Center, Cambridge, MA 02139.*

Abstract. The Alcator C-Mod ICRF antennas have been operated with BN antenna guards since 2000. This modification had followed the observation that metallic impurities in the plasma increased with increasing ICRF power. Systematic improvements to the antenna structure have allowed the launched power to be raised to the 6 MW level, with good heating efficiency and few deleterious effects on the plasma. BN is inherently fragile, and disruption mechanical shocks have resulted in fracturing of the antenna tiles, exposing the supporting metal structure. Since the antennas are now electrically in reasonable shape, and one of C-Mod's goals is to study the behavior of plasmas with all-metal plasma facing components, it was decided to replace the BN tiles with molybdenum. High power operation was resumed in March, 2005, and the behavior of the antennas during the present run campaign will be reported.

INTRODUCTION

The C-Mod antenna complement includes two 2-strap antennas in a dipole configuration and a 4-strap design that allows efficient heating as well as providing a directed launched wave spectrum for current drive by changes in current strap phasing. An antenna's ability to deliver useful power to the plasma may be limited by the injection of impurities into the plasma or by arcing at high voltage limits. The 2-strap antennas operate at a power capability of ~10 MW/m^2. The 4-strap antenna power capability has increased from an initial value of 5 MW/m^2 to ~11 MW/m^2 by eliminating impurity generation and improving high voltage handling.[1,2,3]

INITIAL IMPURITY GENERATION BY PLASMA-FACING SURFACE INTERACTIONS

All antennas were first installed and operated with molybdenum protection tiles. Initial 4-strap antenna operation in 1999 resulted in high levels of metallic impurity influx at heating power levels above ~1.3 MW. The impurity source was identified as RF-induced arcing across tile gaps under the local edge plasma conditions. The gaps were short-circuited with stainless steel straps installed underneath the plasma protection tiles, eliminating this problem.

Operation with the metal plasma-facing components was satisfactory, but the level of Mo impurity at the plasma core was found to scale with the rf power. Although the source rate was low, plasma screening was poor.[4] All of the antennas' plasma protection tiles were therefore changed from the original molybdenum to boron nitride.

In addition to arcing to unprotected tile fasteners, a new front surface interaction limit appeared later in 4-strap antenna operation above 2.5 MW. An analysis of the hot spot observations suggests that the tokamak's field line pitch in front of the antenna results in nonzero rf magnetic flux linkage to tokamak field lines connecting antenna surfaces. The resulting rf electric field expels electrons, and plasma neutrality results in ion acceleration leading to an enhanced sheath potential.[5] All front protection tiles were realigned with side tiles, all remaining exposed metal surfaces were covered with boron nitride or removed, and a central boron nitride septum was installed to reduce the tokamak field line connection length.

ARCING IN ANTENNA INTERNAL STRUCTURE

Extensive arc damage was observed in the 4-strap antenna in 2000 between the striplines feeding rf current to the antenna straps, in a direction along the tokamak edge magnetic field. An effective stripline voltage limit of ~15-20 kV in plasma (45 kV in vacuum) limited the antenna heating power to ~2.5 MW. This corresponded to an empirical electric field limit of ~15 kV/cm under the local conditions, i.e. $E \| B$, and plasma edge neutral gas pressure up to ~0.5 mTorr. The striplines had been designed with $E \| B$ in order to achieve maximum compactness, but a redesign was performed in 2001 to reorient the striplines to an $E \perp B$ configuration.[6] High voltage gaps were increased to reduce electric fields, and in the case of arcing at the current strap crossover, electrodes were reshaped to reorient the region of highest field.

Series arcing was observed in 2002 in bolted contacts both in the current feeds and the antenna mounting plate. These have been redesigned with more bolts, improved mating surfaces, and copper plating where needed to improve electrical contact.

ANTENNA PERFORMANCE WITH BN TILES

The front-surface and internal antenna modifications have allowed operation up to 1.5 MW for each of the two 2-strap antennas and up to 3 MW for the 4-strap antenna. Good heating efficiency has been obtained, and no deleterious effects were observed on plasma operation resulting from the boron nitride.[7,8]

CHANGE TO METALLIC TILES

Boron nitride is inherently mechanically fragile, and in spite of mounting scheme redesign, the mechanical shocks resulting from C-Mod plasma disruptions led to repeated tile fracture, exposing the metal surface underneath to arcing. The evolutionary modifications to the antennas' internal structure had led to an ongoing reduction in internal arcing, possibly reducing metallic impurity generation as well. This led to the decision to revisit antenna operation with molybdenum protection tiles, and the changes were carried out in 2004.

FIGURE 1. 4-strap antenna with BN tiles (left) and molybdenum tiles (right). 2-strap antennas are similar.

PERFORMANCE WITH METALLIC TILES

At the time of this conference, the 2-strap antennas have been brought up to a power level of ~1.25 MW each, and the 4-strap antenna has been brought up to ~2.5 MW, into plasma. Achievable antenna voltages range from ~20-40 kV, as before.

A systematic study of plasma response to ICRF power applied with the modified antennas has just been started, but initial measurements with an unboronized wall at a power level of ~2.5 MW (4-strap at 2.5 MW followed by 2-strap at 1.25 MW each) indicate heating evident in stored energy, neutrons, and radiated power signals. There is no large density increase, and no large edge neutral pressure rise. The hydrogen minority fraction H/(H+D) stays below 8%, and Z_{eff} stays flat, near 1. This behavior is similar to the plasma response with the unmodified antennas.

A comparison has also been made of plasma response to the same power applied through each of the three modified antennas in sequence in the same discharge. The density rise is observed to be small and ~equal, stored energy is ~equal, neutron rate is ~equal, and radiated power is ~equal for each antenna.

No obvious plasma-facing surface interactions have been observed up to the 2.5 MW (2-strap at 1.25 MW each and 4-strap at 2.5 MW) power level, as is shown on the two images of Figure 2.

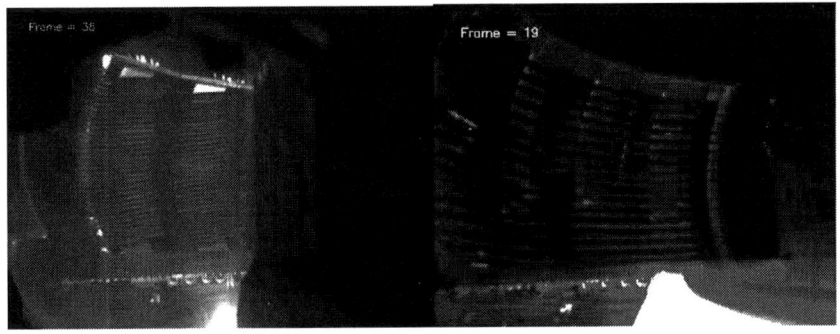

FIGURE 2. 2-strap antenna at ~1.25 MW (left), 4-strap antenna at ~2.5 MW (right).

SUMMARY

Initial performance of the C-Mod ICRF antennas with high-Z metal antenna guards is similar to the antennas with BN guards. Power and voltage capability are about the same. Plasma response with unboronized walls, comparable power levels, and 4-strap antenna heating phasing is also similar.

ACKNOWLEDGEMENTS

Work supported by US DoE Contract DE-AC02-76-CH0-3073 and Cooperative Agreement DE-FC02-99ER54512.

REFERENCES

[1] G. Schilling et al., "Upgrades to the 4-strap ICRF Antenna in Alcator C-Mod," Proceedings of the 14th Topical Conference on Radio Frequency Power in Plasmas, Oxnard CA, May 2001, 186-189.
[2] S. J. Wukitch et al., "Results and Status of the Alcator C Mod Tokamak," Proceedings of the 19th IEEE/NPSS Symposium on Fusion Engineering, Atlantic City NJ, January 2002, 290-295.
[3] S. J. Wukitch et al., "Performance of a Compact Four-Strap Fast Wave ICRF Antenna," presented at the19th IAEA Fusion Energy Conference, Lyon, France, 14 - 19 October 2002, FT/P1-14.
[4] B. Lipschultz et al., "A Study of Molybdenum Influxes and Transport in Alcator C-Mod," Nuclear Fusion **41**, (2001) 585.
[5] J. W. Myra and D. A. D'Ippolito, "Far Field ICRF Sheath Formation on Walls and Limiters," Proceedings of the Tenth Topical Conference on Radio Frequency Power in Plasmas, Boston MA, April 1993, 421-424.
[6] G. Schilling et al., "Analysis of 4-strap ICRF Antenna Performance in Alcator C-Mod," Proceedings of the 15th Topical Conference on Radio Frequency Power in Plasmas, Moran WY, May 19-21, 2003, 166-169.
[7] S. J. Wukitch et al., "Investigation of performance limiting phenomena in a variable phase ICRF antenna in Alcator C-Mod," Plasma Phys. Control. Fusion **46** (2004) 1479-1491.
[8] G. Schilling et al., "Assessment of ICRF Antenna Performance in Alcator C-Mod," Presented at the 31st European Physical Society Conference on Plasma Physics, 28th June to 2nd July 2004, Imperial College, London.

Heat Loads On Tore Supra ICRF Launchers Plasma Facing Components

S. Brémond, L. Colas, M. Chantant
B. Beaumont, A. Ekedahl, M. Goniche, P. Moreau, R. Mitteau

Association EURATOM-CEA, CEA/Cadarache, 13108 ST PAUL-LEZ-DURANCE, France

Abstract. Understanding the heat loads on Ion Cyclotron Range of Frequency launchers plasma facing components is a crucial task both for operating present tokamaks and for designing ITER ICRF launchers as these loads may limit the RF power coupling capability. Tore Supra facility is particularly well suited to take this issue. Parametric studies have been performed which enables to get an overall detailed picture of the different heat loads on several areas, pointing to different mechanisms at the origin of the heat power fluxes. Lessons are drawned both with regards to Tore Supra possible operational limits and to ITER ICRF launcher design.

Keywords: Ion cyclotron resonant heating, heat loads, edge plasma physics.
PACS: 52.35.–g , 52.35.Hr, , 52.40.Fd, 52.40.Kh , 52.50.–b

1. INTRODUCTION

Limitation in operation due to hot spots and possibly erosion of IC Launchers (ICL) parts were reported in several present tokamaks. On Tore Supra (TS), such events occurred in 1998[1,2], which was overcome with some modification of the Faraday shield electrical connections, but still, very high temperature were observed on recent high power medium duration pulses (8 MW 20 s), very close to the PFCs limits. The issue may be even more critical on ITER due to foreseen 10 MW/m^2 ICRF launcher power density, typically a factor of 2 or 3 higher than the current operating values, and to long pulse duration. The critical issue is not in the mean heat loads to be withstand by ICRF PFCs, but in the peaking of these loads on some small specific areas.

Tore Supra is particularly well suited to tackle this issue because i) the three TS ICRF launchers are now monitored with a new infrared imaging system with 1cm x 1 cm spatial resolution and 20 ms time resolution[3], ii)the TS ICRF launchers PFCs are water cooled. As a consequence, power heat fluxes can be inferred from steady state temperature and energy cross-checked with calorimetric data, iii) the TS ICRF launchers can reach high power density over the Faraday screen surface (8 MW/m^2 very routinely and up to 11 MW/m2 quiet currently)[4], have long pulse capabilities (60 s at 8 MW/m^2), and are radially movable. The TS ICRF launchers PFCs consists of CFC private limiters on both side of the launchers and of a B_4C coated Faraday screen (see Figure 1).

 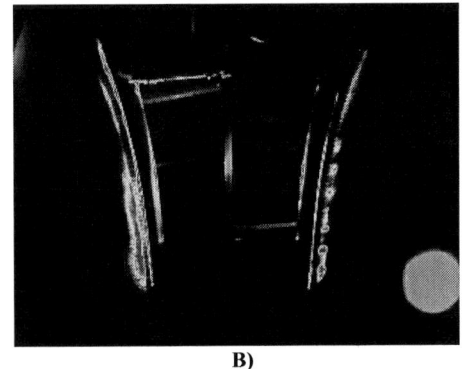

FIGURE 1. A) Picture of one ICRF launcher in Tore Supra vessel **B)** IR typical picture

A general analysis of both IR and calorimetry data over TS 2004 experimental campaign was performed, focusing on two high performance pulses (8 MW 20s and 4 MW 60 s of coupled ICRF power) and a series of dedicated pulses where parametric studies were carried out varying parameters such as rate of private/ total ICRF power and LH power, plasma density and current, and plasma - start-up limiter -launcher gaps. Whenever possible, change on parameters was performed during the pulse to avoid any possible biasing of the analysis. Several heat loads areas on the ICRF launcher PFCs were defined and systematically studied.

A brief inventory of expected heat loads from past experimental results and associated modeling can be made: apart from convected and radiative heat loads, two types of heat loads due to fast particles generated by IC and LH launcher near fields play a role: Fast ions produced by RF voltage rectification (sheath effect) near IC launcher were already shown to explain hot spots on the IC launcher FS corners[1]. In particular, the move of related hot spots from bottom right FS corner to top left FS corner when the plasma current and magnetic field was reversed in the framework of the CIEL upgrade of TS was explained by the development of convective cells due to radial gradient of the rectified voltage cross with the magnetic toroidal field. Fast electrons produced by stochastic acceleration near LH launchers were also evidenced to be responsible of hot spots on IC launcher magnetically connected to LH grill waveguides (with the same poloidal periodicity as the LH waveguides)[5]. The radial extension of the electron beam was found to be of the order of 10 mm. To end the list of possible heat loads, RF ohmic losses, fast particles from the core plasma and acceleration of particles on parasitic resonant layers can also be mentioned.

2. HEAT LOADS CHARACTERISATION

Many data were analyzed and only highlights can be given in this paper (see for instance Figure 2). From an overview of calorimetric data, it can be said that: i) heat fluxes on each ICL guard limiter range from 0.5 to 2.5 % of the total injected power, which would results in up to 15 MW/m^2 at present maximum total power

(10 MW) if all the fluxes were to be focused on the leading edge (but it is not the case). ii) heat fluxes on each ICL Faraday screen range from 0.3 to 1.5 %, which results in up to 1 MW/ m^2 mean heat flux over the FS surface at present maximum total power. Some correlation are found between IC private power and FS mean heat flux, but not systematically and less clearly on one launcher. The RF ohmic losses are checked to be negligible by comparison of similar shots with/without IC power on one launcher. Radiating power which can be estimated around 0.05 MW/ m^2 at present maximum power (assuming a typical 20% of radiating power) is also negligible with regards to other heat loads.

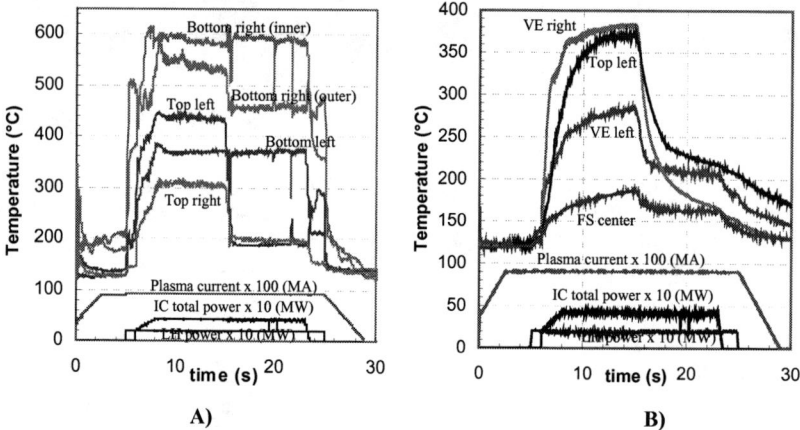

FIGURE 2. EFFECT OF IC TOTAL/PRIVATE/LH POWER A) Guard limiters time traces **B)** Faraday screen time traces (IC launcher switched off at 15 s at constant total IC power)

From parametric studies, it can be deduced that bottom areas of ICL guard limiters are mainly sensitive to total convected power. The warming up is found to be lower at high density and large plasma-launcher gap, as expected from convected power scaling, but i) differences between ICL is observed on IR data maximum warming up that are not consistent with calorimetry, and variable discrepancies between left/right side, confirmed in this case by calorimetry, are observed not due to geometrical misalignment. These observations are considered to be related to the effect of low thermal conductivity carbon type deposits –whose thickness could be related to the guard limiter lifetime in the vessel – and will be cross-checked in the next campaign after cleaning up of parts of these areas. Yet, the preferential collection of convected power on bottom areas is still to be clearly understood. ii) some specific areas are sensitive to LH power as expected when magnetic connection exist between ICL and LH launchers, but it is observed a much lower effect on the IC launcher located on the electron side of the LH launchers.

Top areas of ICL guard limiters are mainly sensitive to private IC power, especially on the left side. The warming up is found to be lower at large plasma-launcher gap. This is consistent to fast ions generated near IC launcher (maximum effect in the top left corner) but the same differences between ICL as observed on bottom areas is found.

Moving to the FS, it is confirmed that top left corner of FS is mainly sensitive to private IC power, with warming up lowering at lower density and larger plasma-launcher gap as already found in previous scalings. Differences between ICL are once again observed.

Vertical edges of the FS are found to be mainly sensitive to private IC power, but also with (total) convected power and LH power in case of magnetic connection between IC and LH launchers. Warming up is sensitive to start-up limiter radial position for the launcher that is shadowed and the effect of LH power is found to be very sensitive to IC to LH radial gap.

3. CONCLUSION

As an outcome of this study, in particular of the analysis of high performance pulses, it is found that the most critical items for TS operation are localized heat loads on the FS screen top left corner and vertical edges. Warming up close to maximum temperature limit originally set for protection of the PFCs is found on high power pulses, but no erosion was observed after detailed inspection of the launcher in TS vessel. Yet, the associated heat loads could be limiting for TS operation in the future, and some dedicated work is under progress to improve the understanding of these power fluxes, pointing out the importance of getting a better knowledge of particle flows in the scrape of layer. This work is also relevant to prepare the design of ITER IC launcher: Further experimental and modeling work is believed to be crucial in order to draw practical ways to optimize for instance the Faraday screen (and possibly the RF configuration) design, in order to reduce heat loads due to fast ions generated near ICL[6]. The result of this work could be tested on next TS IC launchers that are under discussion

REFERENCES

1. Becoulet, M., and al., "Edge plasma density convection during ion cyclotron resonance heating on Tore Supra, *Physics of Plasmas*, vol.9 n°6 (2002) p.1-14 (2002).
2. Colas, L., and al., "Hot Spot Phenomena on Tore Supra ICRF Antennas Investigated by Optical Diagnostics", *Nuclear Fusion*, vol.43 (2003) p.1-15 (2003).
3. Guilhem, D., « Infrared surface temperature measurements for long pulse operation, and real time feed-back control in Tore-Supra, an actively cooled Tokamak", submitted to Journal of Quantitative Infrared Thermography.
4. Brémond, S., and al., "High power density and long pulse operation with Tore Supra ICRF facility", *Fusion Engineering and Design*, vol.66-68 (2003) p.453-460 (2003).
5. Goniche, M., and al., "Enhanced heat flux in the scrape-off layer due to electrons accelerated in the near field of lower hybrid grills", *Nuclear Fusion*, vol.38, n°6 (1998) p.919-937 (1998).
6. Colas, L., and al., "Theory and Practice in ICRF Antennas for Long Pulse Operation", this Conference.

2D modeling of DC potential structures induced by RF sheaths with transverse currents in front of ICRF antenna

E. Faudot*, S. Heuraux* and L. Colas†

*LPMIA, UMR CNRS 7040, Univ. Henri Poincare, Nancy 1, BP 239, 54506 Vandoeuvre Cedex
†Association Euratom-CEA-DRFC, CEA Cadarache, 13108 St Paul les Durance

Abstract. Understanding DC potential generation in front of ICRF antennas is crucial for long pulse high RF power systems. DC potentials are produced by sheath rectification of these RF potentials. To reach this goal, near RF parallel electric fields have to be computed in 3D and integrated along open magnetic field lines to yield a 2D RF potential map in a transverse plane. DC potentials are produced by sheath rectification of these RF potentials. As RF potentials are spatially inhomogeneous, transverse polarization currents are created, modifying RF and DC maps. Such modifications are quantified on a 'test map' having initially a Gaussian shape and assuming that the map remains Gaussian near its summit, the time behavior of the peak can be estimated analytically in presence of polarization current as a function of its width r_0 and amplitude ϕ_0 (normalized to a characteristic length for transverse transport and to the local temperature). A 'peaking factor' is built from the DC peak potential normalized to ϕ_0, and validated with a 2D fluid code and a 2D PIC code (XOOPIC). In an unexpected way transverse currents can increase this factor. Realistic situations of a Tore Supra antenna are also studied, with self-consistent near fields provided by ICANT code. Basic processes will be detailed and an evaluation of the 'peaking factor' for ITER will be presented for a given configuration.

Keywords: open magnetic field line, RF sheath, rectified potential, transverse polarization current, DC potential map, peaking criterion
PACS: 52.40.Kh, 52.50.Qt, 52.55.Fa

INTRODUCTION

The study of 2D potential structures in front of ICRF antennas has been firstly motivated by the appearance, on antenna's structure, of hot spots [1] [2] in the upper left hand corner or in the lower right hand corner with respect to magnetic field direction (see fig. 1). The main component of RF electric field radiated by the antenna straps is along y (poloidal direction). But the mismatch orientation (and other effects) between the Faraday screen and the magnetic field lines creates a parallel electric field by projection of the poloidal component [3] [4]. The integration of this parallel electric field along the magnetic lines between the 2 bumpers (see fig 1) gives a non null RF electrostatic potential [5] able to drive RF sheaths. Magnetic lines can be seen as flux tubes ended by RF sheaths which rectify the RF potential and are able to create high DC potential structures in front of the Faraday screen of antenna. These structures can then accelerate parallel ion fluxes and convective fluxes on to the antenna's surface and dangerously heat the materials to finally create hot spots. Because they are not homogeneous, potential gradients induce transverse currents that will modify the driven RF potential in the plasma. At first, a 2D fluid model is elaborated taking into account only transverse polarization currents and neglecting other terms. From a simple assumption, an analytical expression for the rectified Gaussian potential is obtained [7]. The second part deals with the comparison between the numerical resolution of the fluid model and analytical expression, which show a good agreement. Next, the fluid code permits to deduce a peaking potential criterion to characterize the way the potential structure will be modified (peaking or smoothing) for a typical antenna (Tore Supra or ITER). Finally, some possible enhancements of the antenna structure are described to minimize high RF potential occurring in the upper and lower part of the antenna.

FLUID MODELING OF A RECTIFIED GAUSSIAN POTENTIAL WITH TRANSVERSE POLARIZATION CURRENT

To obtain this simple fluid model, the following assumptions are made : the model is electrostatic and $\omega < \omega_{pi}$, so that we can neglect the parallel dynamic of the sheaths. Moreover, plasma density is supposed to be constant, and electrons follow Boltzmann law. One can now apply current conservation (Eq. 1) with a double probe model to which we have added transverse currents to obtain the equation 2.

$$\nabla \cdot (j_\parallel + j_\perp) = 0 \tag{1}$$

Here $j_\parallel$ is the parallel density current and $j_\perp$ is the transverse density current. The Boltzmann electron density distribution given by $n_0.exp(\frac{eV}{kT_e})$, with n_0 the plasma density, e the electron charge, V the electrostatic potential, k the Boltzmann constant and T_e the electron temperature, associated to $j_{isat} = e.n_0.C_s$ with C_s the ion sound speed, leads to the I-V characteristic for a double probe and after integration over one flux tube, one obtains the equation 2. j_{isat} and j_{esat} are respectively ion and electron saturation currents, ϕ the potential of the plasma compared to the ground, ϕ_{RF} the RF potential resulting from the integration of the parallel electric field E_{RF} along an OMFL (open magnetic field line). For all calculations, ϕ and ϕ_{RF} are normalized to $\frac{kT_e}{e}$, and depend on the time t and the spatial coordinate r in a cylindrical geometry. The $\parallel$ direction is along B_0 and the $\perp$ direction is perpendicular to B_0.

$$\frac{j_{esat}}{2j_{isat}}\left(\exp\frac{\phi_{RF}}{2} + \exp-\frac{\phi_{RF}}{2}\right) = \left(1 + \frac{L_\parallel \nabla \cdot j_\perp}{2j_{isat}}\right)\exp\phi \tag{2}$$

Only the polarization current component is conserved. All other components as the collisional current, viscous current, convective current are ignored while $\omega\tau_\perp < 1$ (see eq. 6). According to the 'flute hypothesis', the expression of the perpendicular current normalized to the ion saturation current is given by equation 3:

$$\frac{L_\parallel \nabla \cdot j_\perp}{2j_{isat}} = -\frac{\Delta I}{2j_{isat}} = -\frac{n_0 m_i}{2j_{isat}B^2}\frac{\partial \nabla^2 \phi}{\partial t} = -\frac{L_\parallel \rho_s}{2\Omega_{ci}}\frac{\partial \nabla^2 \phi}{\partial t} \tag{3}$$

where $\rho_s = C_s/\Omega_{ci}$ is the ion Larmor radius at the sound speed $C_s = \sqrt{kT_e/m_i}$, and $\Omega_{ci} = eB/m_i$ is the ion cyclotron frequency. For the following calculations we work on a spatial Gaussian potential structure which oscillates with a pulsation ω in a cylindrical geometry. r_0 is half the width of the Gaussian potential structure given by equation 4.

$$\phi_{RF}(r,t) = \phi_0.cos(\omega t)exp(-\frac{r^2}{r_0^2}) \tag{4}$$

We deduce the expression of the rectified potential structure from equation 2 to which the current term of Eq. 3 is added:

$$\boxed{\phi(r,t) = \phi_{float} + \ln(\cosh(\frac{\phi_{RF}(r,t)}{2})) - \ln(1 - A\frac{\partial \nabla^2 \phi(r,t)}{\partial t})} \tag{5}$$

with $\phi_{float} = \ln(\frac{j_{esat}}{j_{isat}})$ is the floating potential of the plasma and $A = \frac{L_\parallel \rho_s}{2\Omega_{ci}}$.
Now, assuming that the top part of the structure remains gaussian during one period, one obtains the following expression for the amplitude of the rectified potential structure as a function of time.

$$\phi(0,t) = \phi_{float} + \ln(\cosh(\frac{\phi_{RF}(r,t)}{2})) - \ln(1 + \omega\tau_\perp \frac{\partial \phi(0,t)}{\partial \omega t}) \tag{6}$$

This expression can be linearized for $\omega\tau_\perp \phi_0/2 \ll 1$, that permits to define in which regime the polarization currents occur on the potential structure. For $\omega\tau_\perp \phi_0/2 \ll 1$, the correction due to transverse current term is weak, and the structure is slightly the same. On the contrary, for $\omega\tau_\perp \phi_0/2 \gg 1$, the current term being strongly non linear, becomes of the same order than the RF potential. The next section shows that the more $\omega\tau_\perp \phi_0/2$ is high, the more the time average amplitude of the structure tends to $\phi_0/2$. This effect is shown on fig. 2 on which the temporal evolution of a square shaped potential structure appears in a linear and in a weakly non linear case. These figures justify the good agreement between fluid code and PIC (XOOPIC [8]) computations (see fig. 2), and show the decreasing of the oscillation amplitude when $\omega\tau_\perp$ grows, that is to say for an intensification of polarization currents.

PEAKING CRITERION

The parametric study defines how the time averaged potential amplitude is rectified with respect to the 2 parameters ϕ_0, the initial RF potential amplitude and $\omega\tau_\perp$ the relaxing normalized time of the potential structure ($\omega\tau_\perp = -4\omega A/r_0^2$).

By spreading ϕ_0 (normalized to $T_e = 20eV$) in a range from 1 to 100 and $\omega\tau_\perp$ from .01 to 10, the obtained iso curves give the peaking ratio equal to the time averaged amplitude of the potential structure without currents over the maximum amplitude of RF potential.

This plot corresponds to a set of simulations given by a 2D fluid code which solves the equation 5 in a plane perpendicular to the magnetic field for a typical Gaussian potential structure. These contours permit to evaluate the domain representing the best the regime (linear/non linear) in which transverse currents occur on the structure for real antenna (Tore Supra) et antenna to come (ITER). Now, from the left figure 3, one can deduce that the characteristic domain for ICRF antennas is mainly non linear, so that the DC amplitude of the potential peaks saturates to $\phi_0/2$ while in a linear regime, this DC amplitude would be ϕ_0/π. The consequence is an increase of the DC amplitude due to polarization currents, which add some inertia to the temporal dynamic of ϕ and reduce the amplitude of oscillations, which shifts up the time average value of the potential.

NUMERICAL RESULTS OF A CHARACTERISTIC POTENTIAL MAP IN FRONT OF AN ICRF ANTENNA

The RF potential map deduced from ICANT code [6] computations reveals the high peaking potential area at the top and at the bottom of the antenna structure (see fig. 3 on the right). ICANT code is a code computing self-consistently the near fields in front of the antenna structure. The successive poloidal potential peaks and potential holes correspond to the successive strips of the Faraday screen. Only small scale structures (1 cm) can be smoothed by polarization currents, when $L_{nc}^2 > r_0^2$ (see legend of figure 3 on the right), L_{nc} is a characteristic diffusion length for potential structure.

In upper and lower parts of the antenna, RF potentials are largely rectified due to non linear response of RF sheaths and DC structures are of the order of half the initial amplitude of ϕ_{RF} in the range of 0.5 and 2 kV. The strong convective fluxes around these DC potential structures tend to deviate plasma fluxes on to the upper part or the lower part of antenna's structure with respect to magnetic field direction. Moreover for different reasons (magnetic lines angle, parallel current in the antenna structure, non homogeneous plasma, see [1]) the distribution deposition flux on the antenna structure is confined near the corner. This could explain why the heating takes place only in the corner of antenna.

CONCLUSION

The 2D fluid model giving the rectified potential along magnetic lines between the bumpers in front of ICRF antennas and including perpendicular polarization currents is able to predict the DC amplitude of these structures as a function of 2 parameters ϕ_0 and $\omega\tau_\perp$. It is shown that for a typical antenna (Tore Supra, ITER), the time averaged potential peaks appearing on the antenna boundaries are conserved because their amplitude saturates to approximately half the the maximum RF potential value $\phi_0/2$. To minimize antenna's corner heating, it is necessary to reduce RF parallel electric fields which induces RF sheaths. In this purpose, the design of the antenna could be modified by aligning the straps and the antenna's structure along B_0 to offset the RF potential along a magnetic line.

REFERENCES

1. L. Colas et al: Nuclear Fusion 43 (2003) 1-15
2. M. Becoulet et al. : Plasma Physics 9 (2002) 2619
3. D.A. D'Ippolito, J.R. Myra and al : Nucl. Fusion 42 (2002) 1357-1365
4. FW Perkins, Nucl. Fusion 29 (1989) 583
5. J-M Noterdaeme and G. Van Oost : Plasma Phys. Control. Fusion 35 (1993) 1481-1511
6. S. Pecoul, S. Heuraux, R. Koch and G. Leclert : Comp. Phys. Comm. 146 (2002), 166-187
7. E. Faudot, S. Heuraux, L. Colas : Czech. Journal of Physics (to be published)
8. J.P. Verboncoeur, A.B. Langdon and N.T. Gladd : Comp. Phys. Comm. 87 (1995) 199

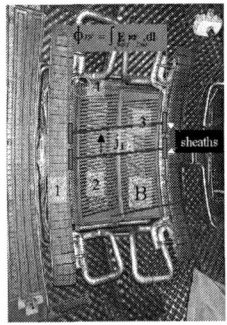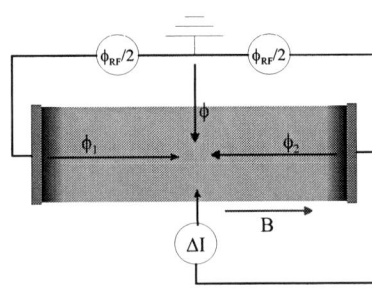

FIGURE 1. Left figure : The labels (1) and (2) are respectively the bumpers protecting the antenna and the straps radiating the waves. Magnetic lines (3) and sheaths are represented in front of the antenna. The hot spot appears (4) on the upper left hand corner of the Faraday screen [1]. Right figure : Scheme of 1 flux tube model with transverse current corresponding to the open magnetic line (3) on the left figure.

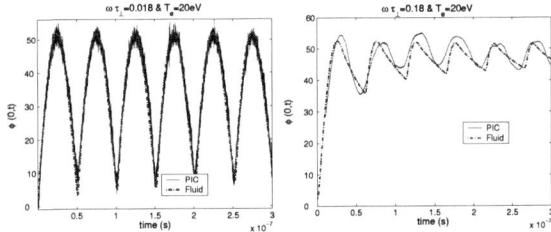

FIGURE 2. These 2 figures represent the amplitude of square shaped structure with respect to time for 2 values of $\omega\tau_\perp$. RF potential is normalized to $T_e = 20eV$ and half its amplitude is equal to 50. The RF square shaped potential is located in the middle of the simulation box and its width measures 5mm. For figure (a), $\omega\tau_\perp \phi_0/2 = 0.9$ (linear behavior), and for figure (b) $\omega\tau_\perp \phi_0/2 = 9$ (non linear behavior).

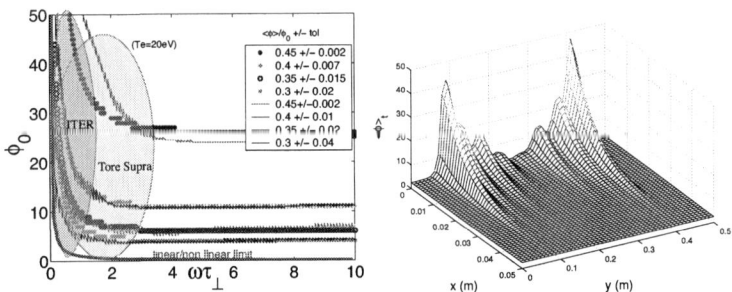

FIGURE 3. Left figure gives the typical domain for Tore Supra's antenna compared to iso curves corresponding to the ratio ϕ_{DC}/ϕ_0 with respect to the parameter $\omega\tau_\perp = 4L_{nc}^2/r_0^2$ corresponding to the temporal inertia of the potential structure or to its width compared to L_{nc} a characteristic 'diffusion' length. Right figure is the DC rectified potential map with transverse current effects. The potential decreases exponentially when the radial coordinate ('x') increases because the wave is evanescent. The poloidal direction is along "y".

Fusion Antenna Analysis Using the Modular Oak Ridge RF Integration Code (MORRFIC)

M. D. Carter*, D. A. D'Ippolito[†], J. R. Myra[†] and D. A. Russell[†]

*Oak Ridge National Laboratory, Oak Ridge, Tennessee
[†]Lodestar Research Corporation

Abstract.
Nonlinear and small scale effects occur in the RF near-field region of fusion antennas, including sheaths, ponderomotive effects, and gas build-up to self-generate plasma. Integrated modules in MORRFIC can estimate the importance of, and interplay between, these various effects. Modules include a linear RF Maxwell solver with plasma variations across and along field lines. Solutions from this module can be used to estimate ponderomotive effects, sheath driving terms, and collisional dissipation. Other modules can be iterated with these solutions to study plasma transport. Transport models include SOLT, which models the effect of RF fields on turbulent radial plasma transport, and a weakly-ionized two-dimensional model of collisional transport in RF self-generated plasma. The geometry allows azimuthally symmetric perfectly conducting boundary conditions and complex dielectrics, including a local plasma dielectric tensor. 3-D antenna structures can be modeled using Fourier analysis. A sheath mask is implemented to allow different non-linear sheath models to be included, and diagnostics are available to estimate non-linear effects.

INTRODUCTION

Models of nonlinear RF effects in the plasma edge must allow power to flow throughout the system, while simultaneously resolving very small sheath layers near solid surfaces. In lieu of a fully non–linear three–dimensional model of an entire tokamak with plasma, we describe progress that we have made using a system that allows three–dimensional antenna geometries with two–dimensional variations for the plasma and other dielectric materials. Two cases are considered. The first is the formation of RF driven sheaths on lateral protection in the near–field region of the antenna. The second is the effect of density perturbations in the edge plasma caused by turbulence or convective cells that do not conform to the flux structure of the equilibrium magnetic field. The Modular Oak Ridge RF Integration Code (MORRFIC) is used to model both cases.

In this paper, we demonstrate the feasibility of performing these calculations in realistic 2D geometries. We show the importance of modeling a sheath gap to consider local damage caused by rectified RF voltage. We also show the effects of RF, or turbulence–driven, convective cells on the RF propagation. Throughout the paper, we consider an RF frequency of 80 MHz and attempt to use parameters that that are relevant to the C–Mod experiment.

MORRFIC is a FORTRAN 95 code that integrates various physics–based models. The main program sets up geometry features in cylindrical coordinates including ceramic or metallic structures, the computational grid, the magnetic field, and various options for determining the plasma density and composition. The code allows iteration between the

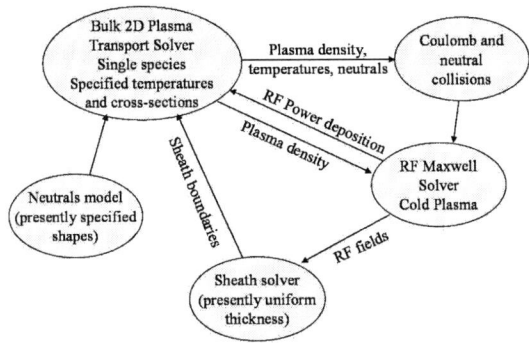

FIGURE 1. MORRFIC integrates physics–based FORTRAN95 modules allowing allows easy nonlinear iteration.

modules to consider nonlinear effects as shown schematically in Figure 1.

Present modules include a cold plasma dielectric for RF solutions on a staggered mesh[1], a 2D diffusive transport model, a rudimentary model using the RF power deposition with neutral gas to provide a plasma source term for the 2D transport equations, and a sheath masking utility to analyze the boundaries between plasma and solid objects.

RADIAL-TOROIDAL MODELING OF LATERAL PROTECTION

To model field lines intercepting the conducting lateral protection of an RF antenna, the MORRFIC code is run in a "screw pinch" geometry. For this case, the first step has been taken toward integrating the RF calculation with a radial density profile obtained from the SOLT[2] code. The SOLT code computes a radial density profile from RF voltages in the scrape–off region. MORRFIC then uses this profile to calculate RF sheath voltages suitable for feedback to SOLT, as shown in the two left–most pictures of Figure 2.

If the linear plasma dielectric is continued all the way to the conducting surface, the RF sheath is shorted out, as shown in the center–right picture of Figure 2. However, surrounding conductors by a uniform vacuum layer allows the fields in the RF sheath region to be enhanced by nearly three orders of magnitude, as shown in the right–most picture of Figure 2.

The lower hybrid resonance also plays a role in the near–field structure as shown in the two left–most pictures of Figure 3. These effects appear in the ponderomotive potential, which was calculated for with no static poloidal magnetic field, as shown in the center–right picture of Figure 3. These potentials are much lower than the electron temperature on the core–plasma side of the lower hybrid resonance, but can become large in the low–density regions where the plasma does not short out the parallel electric field.

A static poloidal magnetic field was also added, indicated by the penetration of E_z^2 deep into the plasma as shown in the right–most picture of Figure 3. Note that the color–map in the right–most picture should not be interpreted as the ponderomotive potential because the proper use of $E_\parallel$ for finite poloidal field is not yet operational in MORRFIC.

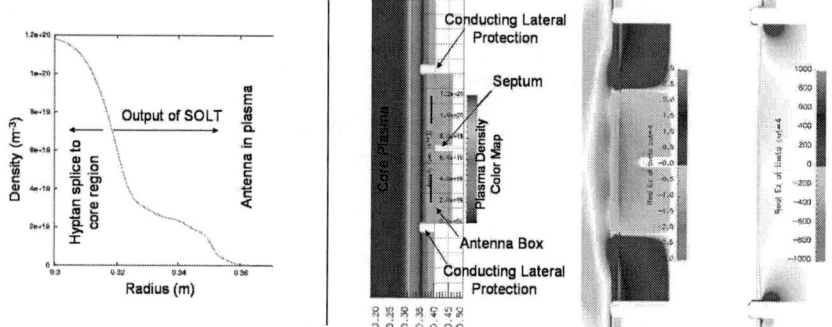

FIGURE 2. MORRFIC can be used in a "screw-pinch" geometry to model RF lateral protection. The radial density profile in the scrape–off–layer is obtained from the SOLT code (left), and the sheath is modeled by a small vacuum layer around solid materials (cross–section at center–left). Allowing linear plasma dielectric to reach the conductors shorts out the sheath fields (expanded cross–section at center–right). However, using a small vacuum gap to model the sheath, causes a large electrostatic enhancement of the fields in the sheath gap (expanded cross–section at right). Note the scales in the two right–most pictures are different by almost 3 orders of magnitude.

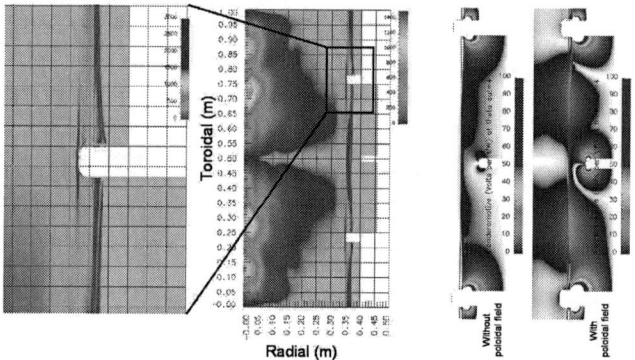

FIGURE 3. The lower hybrid resonance leads to field peaking, especially where it intercepts the conducting limiters, at low density for "C–Mod–like" parameters. The left–center picture shows the linear RF power density in the screw-pinch geometry. The left is a close–up of the lateral protection. The right–center shows the Ponderomotive potential with no static poloidal field for 1 MW of power. The far–right shows a similar calculation with a poloidal static field, but note that the diagnostic used E_z rather than $E_\parallel$, which is required for the ponderomotive potential calculation.

RADIAL-POLOIDAL MODELING OF "BLOBS"

Localized cells of plasma that do not conform to flux surfaces, sometimes known as blobs, can also affect the RF propagation through the edge and the RF sheaths. In Figure 4, we show the effect of adding Gaussian–shaped density perturbations in the edge region, on the RF propagation and sheaths. To model this tokamak geometry, MORRFIC used poloidal field coils and plasma currents to generate an equilibrium with a toroidal

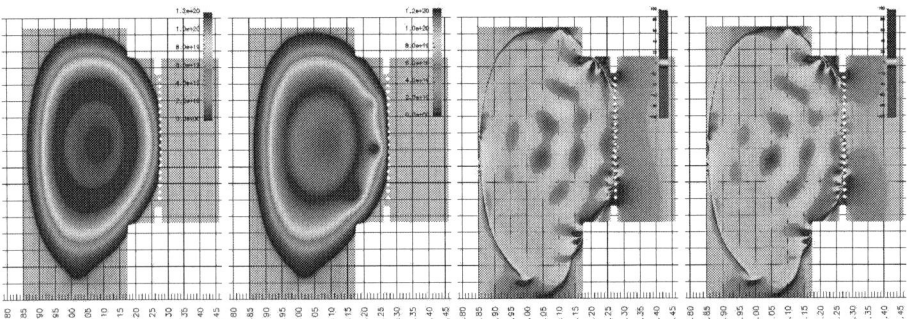

FIGURE 4. The presence of blobs of plasma that do not conform to flux surfaces can change the near-field pattern and scatter waves. From left to right: density profiles without and with blobs, followed by the Real part of the "Z" component of the RF electric field without and with blobs.

(azimuthal) magnetic field having $1/R$ scaling. The two–dimensional diffusive transport module was then used with ad-hoc sources and transport coefficients to generate a density profile having "C–Mod–like" parameters. The two left–most pictures in Figure 4 show the density profiles that were used, and the two right–most show the results of the real part of the "Z" component of the electric fields for the zeroth azimuthal mode number, with and without blobs.

CONCLUSIONS

MORRFIC can be used to study nonlinear issues of importance to the RF–edge problem including two dimensions plus Fourier analysis for the third dimension. The basic infrastructure is in place to calculate sheath voltages at locations where field lines intercept RF conducting surfaces, such as the lateral protection of the antenna, in a "screw pinch" geometry. The cylindrical coordinate system used in MORRFIC can also be reoriented to allow studies in conventional tokamak geometry. Small sheath layers can be analyzed near conducting surfaces, including Faraday screens. An open–loop coupling has been made between MORRFIC and the radial density profiles generated by the RF–dependent transport code, SOLT.

In the future we hope to replace the vacuum sheath layer with an appropriate non–linear boundary condition to speed the code and allow convergence of the sheath fields. We also intend to use of the 2–D diffusive transport module in MORRFIC to more consistently analyze flows along private flux tubes in the edge, and simulate RF self–generated plasma with gas recycling.

ACKNOWLEDGMENTS

Research was sponsored in part by the USDOE Grant DE-FG02-04ER86216 and by Oak Ridge National Laboratory, managed by UT-Battelle, LLC, for the USDOE under contract DE-AC05-00OR22725.

REFERENCES

1. M. D. Carter, et.al., Physics of Plasmas **9** 5097 (2002)
2. D .A. D'Ippolito, this conference.

Integrated Codes for ICRF-Edge Plasma Interactions

D.A. D'Ippolito,[1] J. R. Myra,[1] D. A. Russell,[1] and M. D. Carter,[2]

[1]*Lodestar Research Corporation, Boulder, Colorado;* [2]*ORNL, Oak Ridge, Tennessee*

Abstract. Progress towards a suite of integrated codes for computing the mutual interaction of ICRF antennas with the turbulent scrape-off-layer (SOL) plasma is described. The rf waves are calculated by the 2D MORRFIC antenna code, modified to include a vacuum sheath boundary condition; the SOL profiles are evolved using the SOLT 2D turbulence code, modified to include nonlinear sheath and ponderomotive physics. The SOLT code includes the physics of 2D turbulence, blob transport, rf convection, and ponderomotive density depletion. Iteration of these codes to convergence would provide self-consistent solutions for the density and the rf waves. Finally, an rf sheath boundary condition is described, which permits an iterative solution for rf fields at the boundary and the associated sheath potential and sheath power dissipation. Physics results and future plans for code integration are discussed.

Keywords: ICRF, rf sheath convection, ponderomotive effects, SOL turbulence and transport
PACS: 52.50.Qt, 52.40.Fd, 52.40.Kh, 52.35.Mw, 52.35.Ra

INTRODUCTION

An outstanding problem in ICRF modeling is to make quantitative predictions of nonlinear ICRF antenna-plasma interactions, including rf sheath, ponderomotive force (PF) and parametric decay effects. (A review of these effects and their importance is given in Ref. [1].) Accurate estimates of these effects require a self-consistent treatment of the rf waves and the plasma. For example, predicting antenna loading, local rf fields, and sheath interactions requires a knowledge of the SOL density profile and the particle flux to the antenna, but the density profile is strongly influenced by nonlinear rf effects such as rf-sheath-induced convection [2] and ponderomotive density expulsion [3]. The density profile can also be strongly affected by turbulence and blob transport (as reviewed in Ref. [4]). Thus, a quantitative study should include the mutual interactions of linear and nonlinear ICRF physics with SOL turbulence, including scattering of ICRF waves off turbulent density fluctuations (blobs), and the effect of the rf-induced sheared flow layers and currents on the underlying turbulence.

The goal of the present work is to develop an integrated suite of 2D codes for making quantitative prediction of ICRF antenna-plasma interactions and for studying the interplay between rf waves and SOL turbulence. The rf fields launched by the antenna are computed by the 2D antenna code MORRFIC [5]. This code can be run in (x,z) geometry (here, x,y,z refer to the radial, poloidal, and toroidal directions) to compute the sheath voltage $V_{sh}(x,y)$ at the boundaries or the ponderomotive potential $\Psi(x,y,z)$ along each field line. The poloidal dependence is included by summing over poloidal harmonics of the antenna spectrum. In (x,y) geometry, the code can investigate wave scattering off a turbulent density distribution n(x,y) [5]. The SOL profiles are computed by the Scrape-Off-Layer Turbulence (SOLT) code, originally developed to study turbulent blob transport [6], but now modified to include the nonlinear rf physics. The density n(x,y) from SOLT, and the rf fields and the nonlinear rf terms from MORRFIC, can be iterated to obtain a self-consistent solution. This has been tested by hand and will be automated in the next stage of code development.

SOL PHYSICS

The physics of the SOLT code is illustrated by the following vorticity equation:

$$\frac{d}{dt}\left(\frac{nmc^2}{B^2}\nabla_\perp^2 \Phi\right) = \nabla_\| J_\| + \frac{2c}{B}\mathbf{b}\times\mathbf{\kappa}\cdot\nabla p + \frac{c}{B}\mathbf{b}\cdot\nabla\times\mathbf{F}, \qquad (1)$$

where $d/dt = \partial/\partial t + \mathbf{v}\cdot\nabla$. (In our 2D SOL model, this equation must be averaged along the field lines.) The first term is the divergence of the ion polarization current, important in describing rf-sheath-driven convection [2]. The next term is the divergence of the parallel current. Integrated along the field line and matched to the dc sheath boundary condition (BC) ensuring quasineutrality, this term becomes [7]

$$\frac{\langle J_\| \rangle}{nec_s} = 1 - \upsilon\, e^{-e\Phi/T} I_0(\xi) \qquad (2)$$

where $\langle Q \rangle$ denotes the field-line average of Q, $\upsilon = (m_i/2\pi m_e)^{1/2}$, $\xi = ZeV_{sh}/T$, and $T \equiv T_e$. The "rectified potential" Φ_0 of 1D sheath theory is obtained by setting $\langle J_\| \rangle = 0$ to obtain $\exp(\Phi_0/T) = \upsilon I_0(\xi)$. In our SOL simulations, the rf convection is driven by setting $\Phi = \Phi_0$ at the antenna. Here, $V_{sh}(x,y)$ can be an analytic function (see [2]) or the MORRFIC code result. Equation (2) allows a smooth transition from the rf sheath regime ($\xi \gg 1$) near the antenna to the Bohm sheath regime ($\xi \sim 1$) in the turbulent zone. The third term in Eq. (1) is the curvature drive for the instabilities that drive the turbulence. The saturated turbulence produces 2D "blobs" (or filaments along **B**) of enhanced density, which convect outwards at a velocity $\mathbf{v} = c\mathbf{b}\times\nabla\Phi/B$ determined by the vorticity equation [4]. The final term in Eq. (1) describes the drift current due to an external force $\mathbf{F} = \mathbf{F}_i + \mathbf{F}_e$ (e.g. the PF and neutral frictional forces).

In general, the charge balance in Eq. (1) must be supplemented by equations for conservation of particles, energy, parallel current and momentum. In the simplest system (considered here), only conservation of charge, particles and energy are considered, but more complete models can be treated by the same algorithms. To include PF density expulsion, one can solve the parallel momentum equation analytically in the limit of isothermal temperatures and rapid parallel electron motion past the antenna (compared to other time scales) to obtain

$$n = n_0 \exp[-\Psi/(T_e + T_i)], \qquad (3)$$

where $\Psi = m_e \left(e|E_\|\|/2m_e\omega\right)^2$ is the ponderomotive potential calculated in the MORRFIC code. PF physics has not yet been implemented in the SOLT code.

The SOLT-MORRFIC code suite was tested for a model of the C-Mod antenna, including the poloidal limiters (PL), bumper tiles (BT) and Faraday screen (FS). The

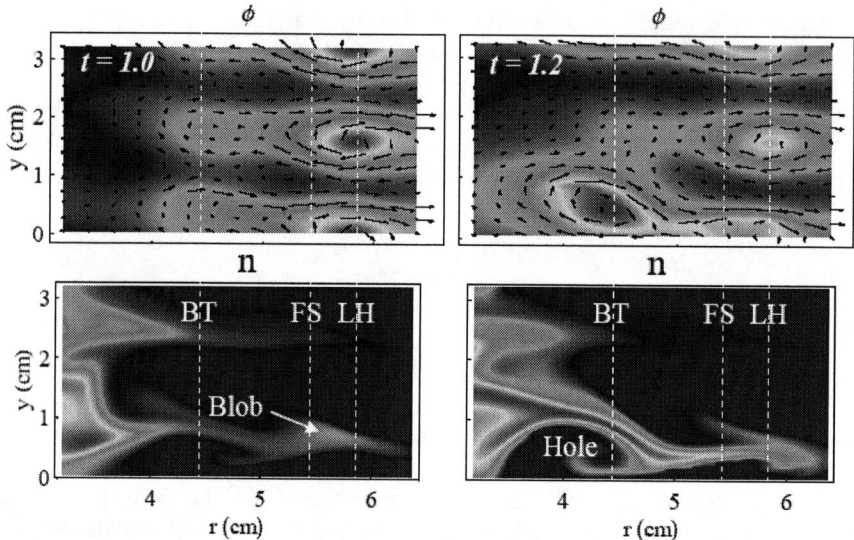

FIGURE 1. Snap-shots of electrostatic potential ϕ and plasma density n as functions of radial and poloidal variables, r and y respectively, from the SOLT simulation based on the MORRFIC sheath potential calculated after one iteration. LH denotes the lower hybrid resonance, based on the initial density profile. Arrows denoting the convecting E×B velocity overlay the ϕ plot, where $E = -\nabla\phi$. The notation in the figure differs slightly from the text: $\phi \equiv \Phi$ and $r \equiv x$.

edge and SOL plasmas were modeled with a particle source inside the separatrix and a particle sink BC ($n \to 0$) well inside the antenna box. The numerical algorithm and detailed BCs will be discussed elsewhere. A single iteration of the codes by hand showed that (i) the equilibrium density and sheath voltage profiles changed significantly in one iteration step, and (ii) the rf convection significantly flattened the density profile in front of the antenna. Shown in Fig. 1 is a blow-up of the SOLT solution for the instantaneous $n(x,y,t)$ and $\Phi(x,y,t)$ in the region near the antenna. The density plot shows the existence of coherent propagating density maxima (blobs) and minima (holes), and the potential plots show the rf convective cell structure near the antenna. Notice the pronounced convection of the density blob into canals between the isolated vorticity cells associated with the lower hybrid resonance. Also note the large vorticity cell shed by the antenna, correlated with a density depression or hole. This figure illustrates the complex interplay between rf and turbulence that can be studied with this suite of codes.

RF SHEATH BC

Rf sheaths can substantially modify the rf fields near the boundaries. Most antenna codes or full-wave codes do not include this physics and use simple conducting wall BCs to compute the rf fields. The MORRFIC code models the sheaths as thin vacuum regions near the boundaries [5]. In MORRFIC, the sheath voltage V_{sh} is calculated by integrating $E_\parallel$ across the vacuum layer. However, this approach requires good grid point resolution near the boundaries and can lead to numerical instabilities. To solve this problem, we have recently derived an analytic rf sheath BC generalizing the Appendix of Ref. [8]:

$$\nabla_t \cdot \mathbf{E}_t - \frac{\Delta}{1+i\nu} \nabla_t^2 D_z = 0 \ , \quad B_z = 0 \ . \tag{4}$$

The first relation is a jump condition expressing $\nabla \cdot \mathbf{D} = 0$ across a thin electrostatic sheath layer with dielectric constant $\varepsilon_{zz} = 1 + i\nu$, where z is the direction normal to the sheath and t denotes the tangential component. Here, the fields $\mathbf{E}$, $\mathbf{B}$ and $\mathbf{D}$ are defined on the plasma side of the sheath-plasma interface, and Δ is the time-averaged sheath width (which is related to the sheath capacitance). In the limit $\Delta \to 0$ the usual conducting wall BC ($E_t = 0$) is recovered. The sheath voltage and width are related by

$$V_{sh} \equiv \int_{z=0}^{\Delta} dz \, E_z^{(sh)} \approx \frac{D_z \Delta}{1+i\nu} \ , \quad \Delta = \lambda_D \left(\frac{eV_{sh}}{T_e}\right)^{3/4} , \tag{5}$$

where $E_z^{(sh)}$ and D_z are the defined at the sheath and plasma sides of the boundary, respectively. These relations imply the nonlinear scaling $\Delta \sim E_z^3$. Finally, the sheath dissipation parameter $\nu \equiv \mathrm{Im}(\varepsilon_{zz})$ is defined as

$$\nu = \frac{4\pi\Delta}{\omega V_{sh}^2} \left(\frac{P_{sh}}{A}\right) \ , \quad \frac{P_{sh}}{A} \equiv n_e c_s \, T_e \xi h(\xi) \frac{I_1(\xi)}{I_0(\xi)} \ , \tag{6}$$

where $\xi = ZeV/T_e$, $h(\xi) = (0.5 + 0.3\,\xi)/(1+\xi)$ is a form factor connecting known results in the limits $\xi \ll 1$ and $\xi \gg 1$, and P_{sh} is the sheath power dissipation [7] (the flux of ions into the sheath × the ion energy gain in the sheath). An rf field solution incorporating Eqs. (4) - (6) can be iterated to obtain a self-consistent solution for the rf fields, sheath voltage, and power dissipation. Note that this procedure gives the local distribution of the sheath effects on material surfaces, so that problem locations (e.g. "hot spots") can be identified, as well as global quantities computed (e.g. total sheath power dissipation). A related BC has been considered for plasma processing [9]. It is planned to test our sheath iteration procedure in both antenna and full-wave codes.

ACKNOWLEDGMENTS

This work was supported by US DOE grants DE-FG02-04ER86216 and DE-FG03-97ER54392.

REFERENCES

1. Myra, J.R., D'Ippolito, D.A., Russell, D.A., Berry, L.A., *et al.*, this conference.
2. D'Ippolito, D.A., Myra, J.R., Jacquinot, J., and Bures, M., *Phys. Fluids B* **5**, 3603 (1993).
3. Motz, H., and Watson, C.J.H., *Adv. Electron.* **23**, 153 (1967).
4. D'Ippolito, D.A., Myra, J.R., Krasheninnikov, S.I., et al., *Contrib. Plasma Phys.* **44**, 205 (2004).
5. Carter, M.D., D'Ippolito, D.A., Myra, J.R., Russell, D.A., this conference.
6. Russell, D.A., D'Ippolito, D.A., and Myra, J.R., Bull. APS **49** (2004), paper CP1.066.
7. D'Ippolito, D.A., and Myra, J.R., *Phys. Plasmas* **3**, 420 (1996).
8. Myra, J.R., D'Ippolito, D.A., and Bures, M., *Phys. Plasmas* **1**, 2890 (1994).
9. Jaeger, E.F., Berry, L.A., Tolliver, J.S., and Batchelor, D.B., *Phys. Plasmas* **2**, 2597 (1995).

A Particle-in-Cell Approach to Time Domain Simulations of the ICRF Edge Regions*

D. N. Smithe[1] and C.K Phillips[2]

[1]*ATK Mission Research, Newington, VA*
[2]*Princeton Plasma Physics Laboratory, Princeton, NJ*

Abstract. Traditional particle-in-cell / time-domain methods excel in situations of complex geometry, but are difficult to use in fusion plasma type problems, primarily because of explicit time-step algorithms that become numerically unstable when confronted with temporally or spatially unresolved physics, at high frequency and short scale-length, such as plasma oscillations and sheaths. We discuss an approach for a progressive treatment of these difficulties with carefully planned intermediate steps, each fully functional and useful in its own right. In our first year we plan to implement, in an implicit algorithm, the standard linear magnetized cold-plasma dielectric tensor in time-domain to treat the RF propagation. Particles will be pushed, but initially only for comparison purposes with the currents of the linear dielectric medium. This will permit cataloging of kinetic and non-linear departures from the linear model, while at the same time assuring numerical stability. The linear equations are identical to the fluid equations, and so subsequent upgrades involve relaxation of fluid linearity, and eventual feedback from particle information into the fluid quantities, and incorporation of new sheath boundary physics models. *This work supported by Princeton Subcontract S-04267-F.

This paper is an overview of a planned approach to implementing a Particle-in-Cell (PIC) Approach to Time Domain Simulations of the ICRF Edge Regions. We plan to use an existing 3-D particle-in-cell code, ATK-Mission Research's MAGIC Tool Suite.[1] The initial emphasis, called Stage 1, is on developing a time-domain tool, and acquiring experience in its use. In this first stage, a Linear Cold Plasma Dielectric will be implemented as a type of dielectric "material", rather than as particles. The PIC code will still compute the particle orbits, and the implied plasma currents, as it normally would, however, these particle currents will not feed back into the Maxwell's equations, since the dielectric material will already be used for this purpose. Thus, there will be two sets of plasma currents, one from the dielectric material, and one for particle orbits. We will compare the test particle current to that of the plasma material currents in hopes that the differences will show where non-linear and non-cold (kinetic) effects are important. While not a complete or consistent model at this point, this tool should still be very useful in its own right, able to show wave propagation patterns around complex geometry and able to determine basic power-level scaling of particle orbits, e.g., ponderomotive effects.

In the second stage we plan to relax the linearity and kinetic restrictions in gradual fashion, addressing issues of numerical stability which are expected to turn up. We also hope to have more sophisticated edge physics models from the SciDAC RF effort which might address specific issues raised in the Stage 1 studies. At first, the focus will be on the incorporation of such edge models. We then hope to relax to hybrid fluid-PIC model of Stage 1 and eventually towards full time-domain electromagnetic PIC, in as much as algorithm stability permits.

DISCUSSION

There are several reasons to desire a time-domain simulation capability for looking at edge physics phenomenon. Time-domain PIC methods are traditionally used for study of complex geometry, where the details of the geometry are as important as the physics itself. The geometry of the RF antennas is indeed, quite complex, and inherently three dimensional, see Figure 1 for example. Time-domain PIC has also been used for its inherent non-linear analysis. All frequencies are present all the time, within the temporal resolution of the time-step, of course. Hence all non-linear interactions are in play, and can develop naturally, without concern that one is building-in the expected non-linear modes, in-lieu of observing them in a natural setting. Even in Stage 1, which will use a linear plasma dielectric, one would expect to see significant power-related effects in the test particle trajectories. Finally, a third reason for a time-domain capability is the desire to view temporal behavior in "movie" format. Time essentially provides another axis to any visualization, and that can aid in understanding complex phenomenon.

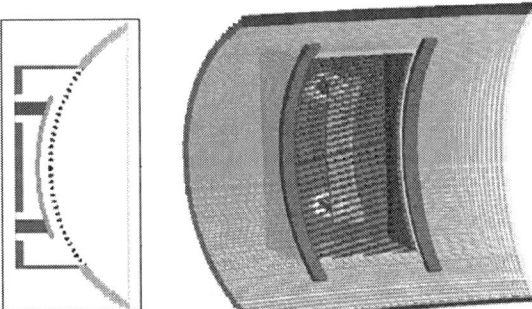

FIGURE 1. Illustration of representative RF antenna geometry, modeled with a PIC code

We expect to start with the MAGIC Tool Suite Software, of which one of the authors (DNS) has been a developer for the last 15 years. This software has both commercial and government-sponsored users. It is used in industry, government labs, and universities.2 This provides us with several built-in advantages, primarily the availability of mature algorithms and diagnostics. This software has also recently been endowed with MPI-based parallelization in a collaborative effort with Stanford Linear Accelerator Center, 2002-2004. It is currently running on a 40-node linux cluster at SLAC, and an 11-node linux cluster at ATK-MR, and an 8-node Windows cluster at ATK-MR. The existing PIC software has several other pre-existing capabilities which may benefit the study of edge effects. One such capability is secondary emission off metallic and dielectric materials, with secondary yield a function of incident energy and angle. Also available are accurate drag, energy spreading and straggling based upon the Integrated–TIGER series of codes.3 These particle features are available in Stage 1, with the test particles.

There are, however, several limitations in a typical time-domain approach, and this drives our cautious multi-stage approach. A typical restriction is that physical wavelengths must be resolved. This is the same restriction as any full-wave code. Likewise, in a time-domain code, physical frequencies must be resolved in explicit (leap-frog) time-step algorithms. However, implicit algorithms are not restricted. We intend to use a combination of an explicit algorithm for the vacuum operators, and implicit algorithms for the plasma currents. Boundary conditions often carry some of the physics load, and are usually not idealizable to the extent that analysis would like. Typical ones already implemented are outgoing waves at specific phase velocity, general wide-band absorbers, particle creation, and particle-with-field import. We expect that in Stage 1, with test particles, we will rely on existing technology, primarily wide-band RF absorbers. It is expected that Stage 2 may require creation of specific boundary conditions for self-consistency of particles and hybrid plasma representation.

The linear cold plasma medium will be implemented in terms of the unperturbed plasma current of each of the species. The linear dielectric tensor in frequency domain,

$$\varepsilon = 1 - \Sigma_{species} \begin{bmatrix} \frac{\omega_p^2}{\omega^2 - \Omega^2} & \left(\frac{\Omega}{-i\omega}\right)\frac{\omega_p^2}{\omega^2 - \Omega^2} & 0 \\ \left(\frac{\Omega}{i\omega}\right)\frac{\omega_p^2}{\omega^2 - \Omega^2} & \frac{\omega_p^2}{\omega^2 - \Omega^2} & 0 \\ 0 & 0 & \frac{\omega_p^2}{\omega^2} \end{bmatrix} \quad (1)$$

is equivalent to a cold plasma fluid current in time-domain, when making the substitution, $-i\omega \rightarrow \partial_t$,

$$\{\partial_t + v_s\}\mathbf{J}_s = \varepsilon_0 \omega_{ps}^2 \mathbf{E} - \Omega_s \times \mathbf{J}_s \quad (2)$$

where for *linear* plasmas, ω_{ps} and Ω_s are fixed in time at the equilibrium values. Hence, we will implement the linear plasma dielectric by introducing new fields over the spatial grid, corresponding to the species current, $\mathbf{J}_s$, which is time-updated according to this equation. Note that this cold plasma update equation contains no spatial derivatives. Hence $\mathbf{J}_s$ can be at the same time-step as $\mathbf{E}$. Then the above equations, together with Ampere's Law,

$$\partial_t \varepsilon_0 \mathbf{E} = -\Sigma_s \mathbf{J}_s + \nabla \times \mathbf{B}/\mu_0 \quad (3)$$

can be solved implicitly for $\mathbf{E}$ and $\mathbf{J}_s$ on the same time step, with a small matrix inversion for each point, on each time-step. The implicit solve will provide numerical stability for temporally unresolved plasma oscillations, for example.

It also sets up some easy non-linear investigations for Stage 2, because simply relaxing the fixed-nature of ω_{ps} and Ω_s, to include the time variations of the plasma density and RF magnetic field, immediately results in a nonlinear-cold plasma model, with the ability to investigate parametric interactions,

$$\{\partial_t + v_s\}\mathbf{J}_s = \varepsilon_0 \omega_{ps}(t)^2 \mathbf{E} - \Omega_s(t) \times \mathbf{J}_s \quad (4)$$

The great advantage of time-domain at this point is that all frequencies are present and available, hence all non-linear modes can compete naturally, so non-linear interactions happen naturally.

In the case of coupling to Bernstein waves, we need warm plasma effects. An All-Orders method is not considered to be feasible in this effort. Instead, a simpler approach is recommended. The usual pressure term, added to the fluid equation gives

$$\{\partial_t + v_s\}\mathbf{J}_s = \varepsilon_0 \omega_{ps}^2 \mathbf{E} - \mathbf{\Omega}_s \times \mathbf{J}_s - \tfrac{1}{2} v_{th}^2 \nabla \rho_s \qquad (5)$$

where ρ_s is the fluid density, which is time-updated via the continuity equation according to $\partial_t \rho_s = -\nabla \cdot \mathbf{J}_s$. Note that this introduces a $\partial_t^2 \mathbf{J}_s \sim \nabla \nabla \cdot \mathbf{J}_s$ wave which is the characteristic form of electrostatic waves. Normally v_{th} is the effective thermal velocity for the pressure effect, e.g., $\tfrac{1}{2} m_s v_{th}^2 = T_s$, but here the quantity $(\tfrac{1}{2} m_s v_{th}^2)$ would be selected to approximate the Bernstein wave dispersion relation, giving what would essentially be a 1st-order warm-plasma Larmor approximation in time-domain. Happily the $\nabla \nabla \cdot \mathbf{J}_s$ operations are already implemented in the existing software, due to past use of this operation in a different context. This feature can be implemented at both the linear and non-linear stages. One difficulty with such a model, though, will be numerical stability issues. The spatial derivatives do not participate in the implicit method, hence numerical instability is possible. Thus, in the case of this warm plasma add-on, some basic algorithm testing and validation will be required to insure the algorithm's viability.

Stage 2 will focus on improvements for better non-linear analysis and kinetic effects. There are several ways in which this can be done, each with its own set of merits and difficulties Some of the physical phenomenon it may make sense to look at with these models are parametric interactions, dynamics of magnetized sheaths, scrape-off layers including DC electric fields, and $\mathbf{E} \times \mathbf{B}$ poloidal drifts, and Trivelpiece-Gould modes.

SUMMARY

A two stage process is envisioned for developing time-domain particle-in-cell simulations of ICRF edge regions. The first stage will be a Linear Cold Plasma model, formulated in terms of a linear cold fluid, together with test particles. The second stage will generalize easily to non-linear cold fluid, and eventually will be made to include self-consistent kinetic effects as much as possible. It is expected that the simulations will utilize existing electromagnetic PIC software with a large user base (MAGIC Tool Suite), and take advantage of existing algorithms, diagnostics, and user experience, in order to speed-up results and concentrate development time on edge physics issues in fusion energy devices.

REFERENCES

1. "User-Configurable MAGIC for electromagnetic PIC calculations," Computer Physics Communications 87 (1995) 54-86.
2. http://www.mrcwdc.com/magic/.
3. "ITS Version 3.0: The Integrated TIGER Series of Coupled Electron/Photon Monte Carlo Transport Codes," by J. A. Halbleib, R. P. Kensek, T. A. Mehlhorn, G. D. Valdez, S. M. Seltzer, and M. J. Berger, SAND91-1634 (March 1992). Also Oak Ridge National Laboratory document CCC-467.

Prediction of plasma-facing ICRH antenna behavior via a Finite-Element solution of coupled Integral Equations

V. Lancellotti, D. Milanesio, R. Maggiora, G. Vecchi, V. Kyrytsya

Dipartimento di Elettronica, Politecnico di Torino, Torino, Italy

Abstract. The demand for a predictive tool to help designing ICRH antennas for fusion experiments has driven the development of codes like ICANT, RANT3D, and the early developments and further upgrades of TOPICA code. Currently, TOPICA handles the actual geometry of ICRH antennas (with their housing, etc.) as well as a realistic plasma model, including density and temperature profiles and FLR effects. Both goals have been attained by formally splitting the problem into two parts: the vacuum region around the antenna, and the plasma region inside the toroidal chamber. Field continuity and boundary conditions allow writing a set of coupled integral equations for the unknown equivalent (current) sources; finite elements are used on a triangular-cell mesh and a linear system is obtained on application of the weighted-residual solution scheme. In the vacuum region calculations are done in the spatial domain, whereas in the plasma region a spectral (wavenumber) representation of fields and currents is adopted, thus allowing a description of the plasma by a surface impedance matrix. Thanks to this approach, any plasma model can be used in principle, and at present Brambilla's FELICE code has been employed. The natural outputs of TOPICA are the induced currents on the conductors and the electric field in front of the plasma, whence the antenna circuit parameters (impedance/scattering matrices), the radiated power and the fields (at locations other than the chamber aperture) are then obtained. An accurate model of the feeding coaxial lines is also included. This paper is precisely devoted to the description of TOPICA, whereas examples of results for real-life antennas are reported in a companion paper [1] in this proceedings.

BASIC THEORY AND FORMULATION

The design of an ICRH system requires evaluating the $[S]$ matrix and the radiated fields of a realistic antenna (i.e. of arbitrary shape and finite thickness) that operates inside a cavity with curved conducting walls and faces the plasma as well as the Faraday screen rods. From a practical standpoint the cavity is the toroidal chamber of a *tokamak* machine, wherein the fusion plasma column is generated and confined, plus a number of lateral recesses (*ports*), which host the antennae: a radial cross section of a typical structure is depicted schematically in Fig. 1a. Also shown therein is the source M_P, i.e. an equivalent magnetic current density extending over a surface S_P; simply put, M_P is the transverse electric field (rotated by $90°$) over the aperture of the coax lines feeding the antenna. Without losing generality, all the metallic parts and the cavity walls are considered perfect electric conductors (PEC), since in practice the conductivity σ is high, albeit not infinite; on the other hand, a finite value of σ may be accounted for successively as a perturbing parameter in the solution process.

Thanks to the procedure outlined below, TOPICA [2] is a full-wave and self-consistent code, as no distribution of electric current on the conducting parts is assumed known,

CP787, *Radio Frequency Power in Plasmas:16th Topical Conference on Radio Frequency Power in Plasmas*
edited by Stephen J. Wukitch and Paul T. Bonoli
© 2005 American Institute of Physics 0-7354-0276-0/05/$22.50

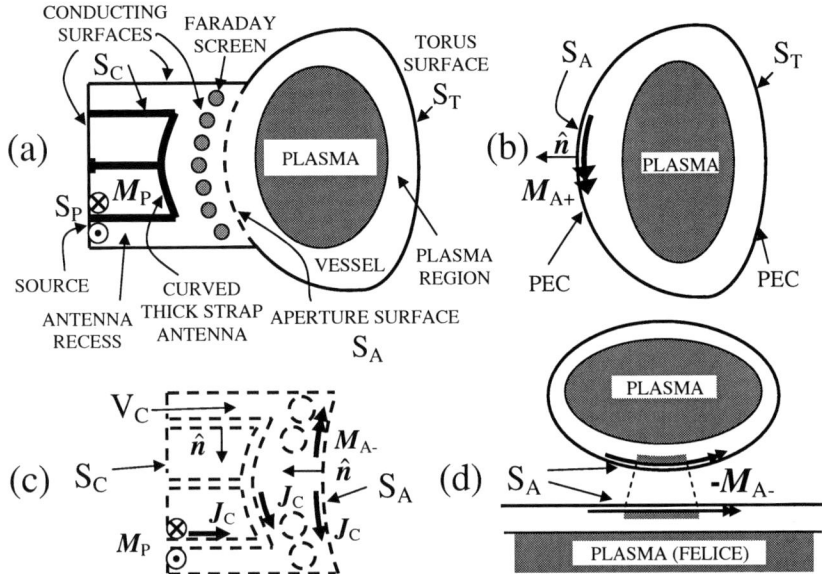

FIGURE 1. (a) Schematic radial cross section of a typical ICRH antenna setup, (b) application of the ET in the torus region to separate it from the antenna, (c) application of the ET in the antenna region to reduce it to the radiation of currents in free space, (d) pictorial representation of $-M_{A-}$ stretching procedure to help couple TOPICA to FELICE.

and it can be coupled with available assessed codes, such as FELICE [3] or TORIC [4], to include the description of a real plasma (i.e. with FLR effects, inhomogeneous profile of density and temperature). A key point in achieving the above features is to conceptually separate the antenna region from the plasma one by applying the Equivalence Theorem (ET) [5, 6] twice. This approach should be contrasted with the one followed in developing the early version of TOPICA [7], wherein a much simpler antenna model was addressed.

To proceed, first the ET is applied to the torus region; namely, a curved surface S_T coinciding with the torus wall is introduced and magnetic and electric current densities $M_T = E \times (-\hat{n}), J_T = (-\hat{n}) \times H$ are placed thereon. Hence, the aperture can be closed by a PEC patch to accomplish separation from the antenna, as shown in Fig. 1b. Owing to the boundary condition at a PEC interface, it is seen that J_T does not radiate any field, while the magnetic current exists on S_A only, so more precisely $M_T = M_{A+} = E \times (-\hat{n})$.

In the recess the ET is applied on a surface S_C that enfolds all conducting parts *and* the aperture S_A, then the unbounded volume outside V_C is entirely substituted with *free space*, so that the problem reduces to the radiation of surface current densities in vacuum (see Fig. 1c). As a result, the unknown current J_C extends over the whole S_C, while the magnetic current is non-null only on S_A and is given by $M_C = M_{A-} = E \times \hat{n}$.

Upon enforcing the proper boundary and continuity conditions the fields must obey on S_β, $\beta = A,C,P$, a set of two coupled integral equations ensues:

$$\chi_C \left(E_1^p\{M_P\} + E_1^s\{J_C, M_{A-}\}\right)\big|_{\tan} = \hat{n} \times (\chi_P M_P + \chi_A M_{A-}), \qquad (1)$$

$$\chi_A \left(H_1^p\{M_P\} + H_1^s\{J_C, M_{A-}\}\right)\big|_{\tan} = H_2^s\{-M_{A-}\}\big|_{\tan}, \qquad (2)$$

where 1 (2) denotes recess (plasma) region, p (s) means *primary* (*scattered*), χ_β are characteristic functions equal to 1 on S_β and 0 elsewhere, and the identity $M_{A-} = -M_{A+}$ has been used. The dependence of the fields on the currents, implicit in (1), (2), is linear and represented by suitable surface integrals with appropriate kernels (i.e. Green's functions) [8]. Concerning this, note that the calculation of $H_2^s\{-M_{A+}\}$ requires knowing the dyadic Green's function G_P in the plasma region. At present, TOPICA relies on FELICE [3] to obtain G_P in the spectral (wavenumber) domain, wherein it is simpler. Besides, since the aperture S_A is actually curve, while FELICE handles a transversely invariant plasma, a strategy has been conceived to accommodate the coupling, which consists of formally *flattening* the current M_{A-} when it enters (1), (2) through H_2^s and leaving it unchanged when it contributes to E_1^s and H_1^s. More precisely, the stretching of $M_{A-}(r)$ is accomplished by mapping the points r on S_A to a plane by means of a suitable projection, as qualitatively depicted in Fig. 1d. The validity of this assumption is widely supported by the good agreement between simulation results and measured data presented in [1] in this conference.

Equations (1), (2) are solved via the Moment Method (MoM) [9], also known as weighted-residual method, upon representing J_C and M_{A-} by a finite set of subdomain basis functions (finite elements), defined over a triangular mesh [10]. In particular, the MoM is applied both in the spatial and in the spectral domain [2], the latter being the natural domain wherein the G_P is available; the overall procedure turns the integral equations into an algebraic system, whose unknowns are the projection coefficients of J_C and M_{A-} on the set of basis functions; filling the system (interaction) matrix and solving it are the main tasks of TOPICA. The knowledge of J_C and M_{A-} allows direct calculation of radiated power and fields and the admittance matrix $[Y]$ through:

$$Y_{ij} = -\langle M_{Pj}, H_{1i}^s\{J_C, M_{A-}\}\rangle / V_{Cj}^2 \big|_{V_{Ci}=0, i\neq j}, \qquad (3)$$

wherein $\langle \cdot, \cdot \rangle$ indicates a suitable inner product, i, j run over all the antenna terminal pairs (i.e. coax apertures), V_{Cj} is the TEM mode voltage in the j-th coax and the magnetic field is evaluated at the i-th coax location. $[Y]$ can be converted to the scattering matrix $[S]$ that is the only set of antenna parameters capable of assisting the accurate and reliable design of the feeding and tuning-matching systems.

TOPICA CODE STRUCTURE

The main modules of TOPICA and their links are reported in Fig. 2a. The design or analysis phase begins with the antenna drawing and meshing by means of a CAD tool: an example of a complicated antenna realized by a commercial tool is shown in Fig. 2b. The CAD generally yields a file containing information on the triangular mesh, whose

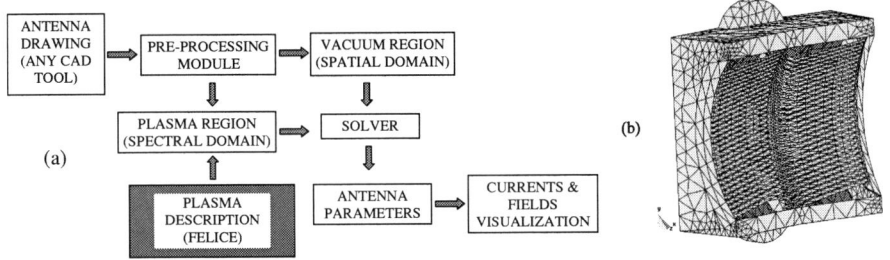

FIGURE 2. (a) Block diagram of TOPICA code structure, (b) a sample ICRH antenna model obtained by a commercial CAD tool.

facets are support to the basis functions, in a format that may vary for different tools. Thus, the pre-processing unit is aimed at reading the geometry data, generating the basis functions and, then, saving all that to a new file. This is input both to the plasma and the vacuum modules: the former computes the entries of the interaction matrix pertaining to the plasma contribution and couples to FELICE just to obtain the plasma surface matrix [3], while the latter fills the remaining part of the matrix as well as the forcing (RHS) term. The solver module is charged with the task of solving the system and getting the unknown currents, whence the antenna parameters ($[Y]$, via Eqn. (3), along with the radiated power and fields) can be computed. Finally, a plotting tool has been coded in MATLAB® to help display the electric current distribution all over the conductors and the electric field over the aperture S_A.

Numerical results obtained with TOPICA are presented in [1] along with their experimental validation, and will not be repeated here for the sake of conciseness.

REFERENCES

1. R. Maggiora, V. Lancellotti, D. Milanesio, G. Vecchi, V. Kyrytsya, A. Parisot, S. J. Wukitch 2005 *Proc. 16th RFPP*, Park City, UT
2. D. Milanesio 2003 *Analysis of plasma-facing antennas via coupled Integral Equations and a hybrid Method of Moments MS Thesis* University of Illinois at Chigaco and Politecnico di Torino
3. M. Brambilla 1992 *IPP Report* 5/45
4. M. Brambilla 1996 *IPP Report* 5/66
5. I. V. Lindell 1992 *Methods for Electromagnetic Field Analysis* Oxford Engineering Science Series (Oxford: Clarendon Press)
6. A. F. Peterson, S. L. Ray, R. Mittra 1998 *Computational Methods for Electromagnetics* (New York: IEEE Press)
7. R. Maggiora, V. Lancellotti, G. Vecchi, V. Kyrytsya 2004 *Nucl. Fusion* **44** 846
8. L. B. Felsen, N. Marcuvitz 1973 *Radiation and Scattering of Waves* (Englewood Cliffs: Prentice Hall)
9. R. Harrington 1993 *Field Computation by Moment Methods* (New York: Oxford)
10. S. M. Rao, D. R. Wilton and A. W. Glisson 1982 *IEEE Trans. Antennas Propagat.* **AP-30** 409-418

An alternative method for calculating the RF plasma dielectric response in ICRH simulations

E.A. Lerche and P.U. Lamalle

Laboratory for Plasma Physics, Association "EURATOM – Belgian State", Trilateral Euregio Cluster, Royal Military Academy, Brussels

Abstract. A new procedure for computing the plasma dielectric response in the ion-cyclotron range of frequencies (ICRF) is proposed. It is based on a coordinate system in which the parallel component of the wave-vector $k_{//}(m,n)$ of each wave mode is constant on magnetic surfaces. This representation allows drastic simplifications in the evaluation of the plasma dielectric response, and is very attractive both analytically and numerically. In the case of Maxwellian plasma distributions, the computations are significantly faster than in standard coordinates whereas for non-Maxwellian plasmas, further analytical developments become possible. We illustrate the new approach by simulating IC minority heating with the CYRANO full-wave code in the case of Maxwellian particle distributions. The results obtained with constant-$k_{//}$ coordinates are benchmarked against the standard coordinates results, and speed and performance comparisons are presented. Besides performing substantially faster than the standard coordinates, the new coordinates allow the easy tabulation of the plasma dielectric response at the beginning of the computations and its subsequent interpolation according to the $k_{//}(m,n)$ spectrum required by each toroidal mode (n) in 3D simulations.

Keywords: ICRH, tokamak, full-wave simulations, constant $k_{//}$ coordinates

PACS: 52.55.Fa, 52.50.Qt, 52.25.Mq

INTRODUCTION – THE NEW METHOD

Perhaps the tightest bottleneck in full-wave ICRH simulations is the calculation of the kinetic plasma dielectric response. Even for Maxwellian plasma distributions, where the plasma response can be expressed in terms of the well known Fried-Conte dispersion function Z_{FC} [1], the repeated Fast Fourier Transforms (FFT) of the Z_{FC} function at each magnetic surface for the complete $k_{//}(m,n)$ spectra usually considered is rather time consuming. For non-Maxwellian plasmas, the calculation of the plasma response involves numerical integrals over the individual guiding centre orbits and the practical application requires excessively cumbersome computations.

We propose a new method for calculating the RF dielectric plasma response, based on a coordinate system where the parallel wave-vector $k_{//}$ is constant on magnetic surfaces [2,3]. This formalism avoids performing the time-demanding FFT's of the Z_{FC} function for Maxwellian plasmas and permits further analytical development of the otherwise strictly numerical terms in the non-Maxwellian plasma response, through a semi-analytical evaluation of the trajectory integrals. This method is being implemented in the full-wave code CYRANO [4], and has shown a significant reduction in the computing time needed for calculating the plasma response.

Since the Fourier basis used in the new formalism is different from the standard (θ,φ) one, it is clear that the Fourier harmonics of the RF fields are also different and only the total field components computed with a sufficiently wide poloidal mode interval will exhibit convergence. This fact is illustrated in Fig.1, where we show the lowest order poloidal Fourier components of the poloidal electric field obtained with the standard (a) and with the new (b) coordinate systems for a typical (H)D heating scenario in the TEXTOR tokamak, with $B_0 = 2.25T$, $f_{RF} = 32.5MHz$ and $n = +13$. The thick curves represent the total poloidal electric field in each case.

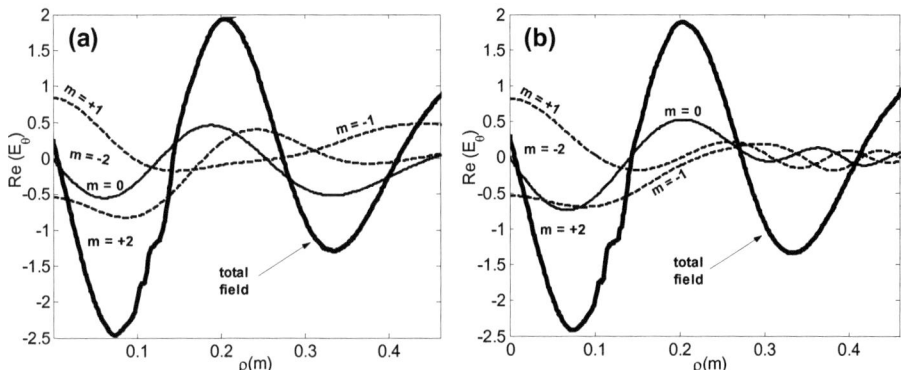

FIGURE 1. Lowest order poloidal Fourier harmonics (m) of the total (m = -30...+30) poloidal electric field component (thick curve) obtained with the standard (a) and with the new (b) coordinate systems with the full-wave code CYRANO, illustrating the different Fourier expansions used in each method.

As expected, despite the individual Fourier harmonic contributions being quite different, the total electric field component E_θ (summed over all M = 61 modes considered) agrees very well. Also note that the difference between the Fourier harmonics is more evident for larger minor radii (higher safety-factor values), where the new coordinates representation differs considerably from the standard one.

The new method has thus been successfully implemented in the CYRANO code for general axisymmetric toroidal geometry and Maxwellian plasma distributions. Its benchmarking against the standard coordinates calculations and the exploitation of the advantages of the new formalism are currently under way.

COMPARISON WITH STANDARD COORDINATES

As a first application of the new procedure, we compare the results obtained with the CYRANO code for a regular (H)D ICRF minority heating scenario in the JET tokamak with the new (constant $k_{//}$) and with the standard coordinate systems. In Fig.2 we show the radial RF electric field component (a) and the averaged power deposition profile (b) obtained with the new (solid) and with the standard (dashed) methods. The main parameters were $B_0 = 3.4T$, $f_{RF} = 52.0MHz$ and $n = +14$ (+π/2 antenna phasing).

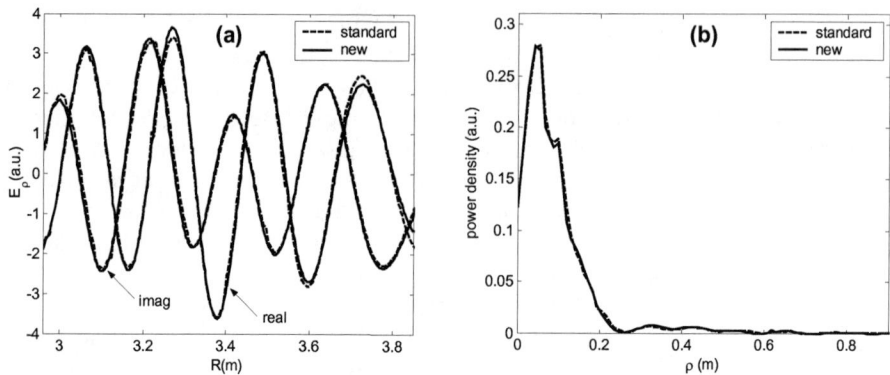

FIGURE 2. Results of (2%H)D minority heating simulations in the JET tokamak obtained with the new (solid) and the standard (dashed) coordinate systems with the CYRANO code: (a) radial RF electric field; (b) average power deposition profiles. The parameters were $B_0 = 3.4T$, $f_{RF} = 52MHz$ and $n = +14$.

First note the very good agreement between the RF fields calculated with the two methods. The small discrepancies in the electric field profiles are due to the common truncation of the poloidal spectrum considered in the two simulations (M = 81). If a broader spectrum is considered, the fields agree completely. The power deposition profiles are also in excellent agreement, again confirming the correct implementation of the new method. Other heating scenarios such as (T)D and (^{3}He)H have also been investigated and showed similar agreement.

Finally, let us comment on the computational speed improvement achieved with the new formalism. In the previous example (Fig.2), the overall simulation with the new coordinates was approximately 30% faster than the standard calculations, mainly due to the accelerated calculation of the plasma response. The results of a more systematic investigation of the speed enhancement achieved with the new formalism is shown in Fig.3, where the time needed for computing the dielectric plasma response (at one radial position with 3 Maxwellian plasma species) with each method is plotted against the number of poloidal harmonics M considered in the simulations.

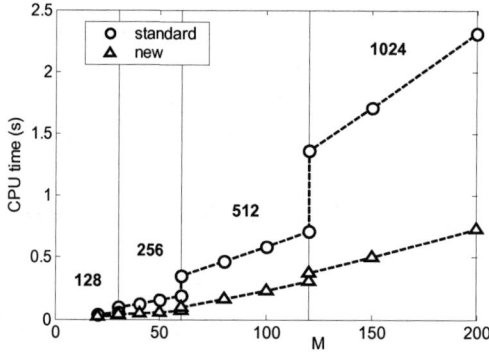

Figure 3. Time needed to compute the dielectric plasma response (at one radial position for 3 Maxwellian species) as function of the number of poloidal harmonics M with the standard (circles) and the new (triangles) methods. The represented numbers indicate the number of FFT points used in each M domain, progressively increased to correctly describe the Z_{FC} function in the standard coordinates.

From Fig.3 we clearly see the significant advantage of computing the plasma response in the constant $k_{//}$ coordinates, particularly for a high number of poloidal harmonics, where the new method performs up to 3 times faster than the standard one. In addition, it can be seen that the computations based on the new method have very low dependence on the number of FFT points considered in the simulations, in contrast to the standard calculations, where the time needed to compute the plasma response roughly doubles when increasing the FFT points to the next power of two.

In addition to the evident speed-up of the computations for the large number of poloidal harmonics typically required in tokamak geometry, we expect an even larger improvement in the case of 3D simulations, in which several toroidal Fourier field modes are superposed. In this case, with the new coordinates, the plasma response can be tabulated once and for all as a function of the one-dimensional variable $k_{//}(\rho)$ during code initialization, and simply be 'read from file' according to the $k_{//}(m,n)$ spectrum required by each toroidal mode.

CONCLUSIONS

A new procedure for computing the RF dielectric plasma response in ICRH simulations has been presented. It is under study in the full-wave code CYRANO and the initial benchmarking with the standard calculations showed a considerable speed increase in the numerical computations for Maxwellian plasma distributions. This performance enhancement is currently being explored to perform 3D ICRH simulations (with the full antenna toroidal spectrum), where the new method has the additional advantage of allowing the 1D tabulation of the plasma dielectric response at the beginning of the computations for subsequent use during the code execution. Second order Finite Larmor radius corrections and the description of non-Maxwellian plasmas will be addressed hereafter. The final goal is to take advantage of the new formalism to couple the CYRANO wave code to a quasi-linear Fokker-Planck diffusion code [5], in order to self-consistently calculate the RF fields and the plasma species distributions in large size tokamak experiments.

ACKNOWLEDGMENTS

The authors gratefully thank Dr. D. Van Eester for very fertile discussions. This work is supported by an EURATOM Intra-European Fellowship under contract number FU06-CT-2003-00331.

REFERENCES

1. T. H. Stix, *Waves in Plasmas*, AIP, New York, 1992, p.202.
2. P. U. Lamalle, ECA **25A** (2001), p.1145, 28th EPS Conference on Plasma Physics, Madeira.
3. E. A. Lerche and P. U. Lamalle, in *Theory of Fusion Plasmas*, Joint Varenna-Lausanne International Workshop, 2004, p.359.
4. P. U. Lamalle, *Internal Laboratory Report* **101**, LPP - ERM/KMS, Brussels (1994).
5. D. Van Eester, *J. Plasma Physics* **65**, p.407-452 (2001).

Antenna Optimization By Using Finite Element Programs

F. Braun, ICRF Group

Max Planck Institut für Plasmaphysik, Garching Germany

Abstract. Ion Cyclotron Frequency Heating and Current Drive play an important role in fusion experiments. The recent availability of powerful commercial finite element programs for PC's [1], now allow detailed optimization of critical components as antennas to improve the power handling capability. Present ASDEX Upgrade antenna simulations indicate that the design can be optimized to symmetry the input power. Simulations also show the influence of Faraday shields and current straps with various geometries on different antenna arrays.

Keywords: Modeling, Simulation, Antennas.
PACS: 52.35.Mw, 52.40.Fd, 52.50.Qt

INFLUENCE OF SLOTS ONTO S- PARAMETERS

Based on the present design of the ASDEX Upgrade antenna (Fig. 1 left) with slightly different resonance frequencies, which narrows the operation regime, calculations show an improvement of the S- parameters (Fig. 2) by implementation of slots, which gives a wider operation range of the system and also increases the pumping and therefore the voltage handling capability. The new design is shown in Fig.1 right.

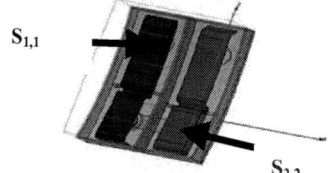
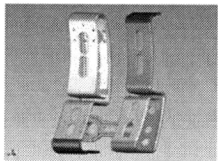

FIGURE 1. Simulation model (left) and design drawing (right) of the ASDEX Upgrade antenna.

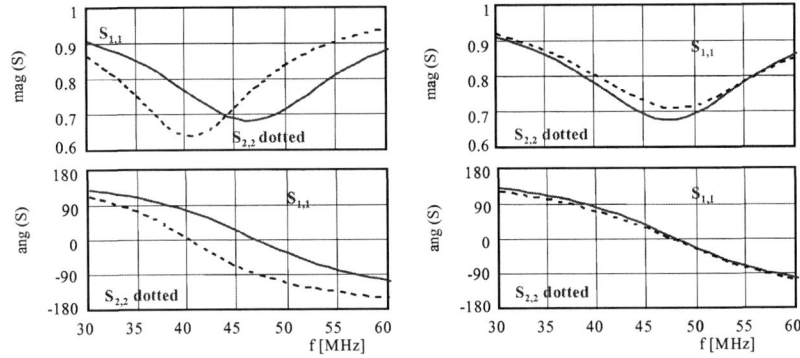

FIGURE 2. S- Parameters of the actual AUG-antenna (left) and modified antenna (right)

INFLUENCE OF THE FARADAY SCREEN ONTO S-PARAMETERS

On a simplified model different spacing of the Faraday rods have been calculated and compared under air and water radiation conditions [2, 3].

Distance Current strap (CS)- Faraday screen:	1.5cm
Distance CS- ground:	18 cm
Width Current strap:	20 cm
Faraday screen spacing:	1*1 cm
Ground : L*W*H:	98*74*23.5 cm

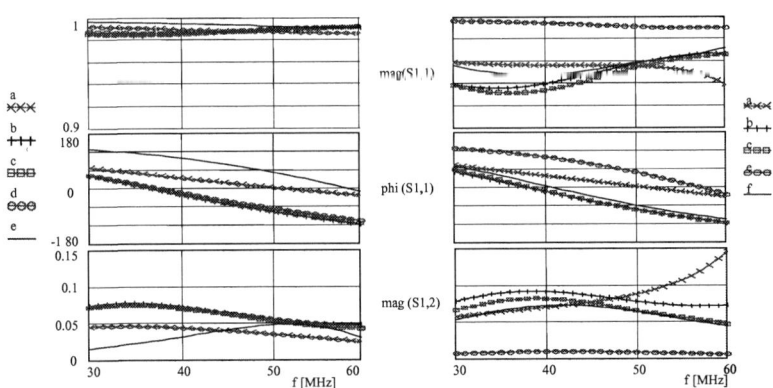

FIGURE 3. Influence of different Faraday configurations onto the S-parameters at (left: air absorption, right: water absorption both in 10 cm distance of Faraday screen.
a: without Faraday screen (FS), b: horizontal FS, c: FS angle 15deg ratio rod/space= 1,
d: FS angle 15deg ratio rod/space= 2, e: double FS, angle 15deg ratio rod/space= 1, f: double FS, angle 15deg ratio rod/space= 2

An antenna with only one layer of the Faraday screen (Fig. 3 trace c), rod diameter = gap, shows the minimum value of the scattering parameters into the absorptive medium and therefore the lowest maximum voltage in the transmission line.

COMPARISON OF ANTENNAS WITH DIFFERENT WIDTH, AT FREE SPACE ABSORPTION

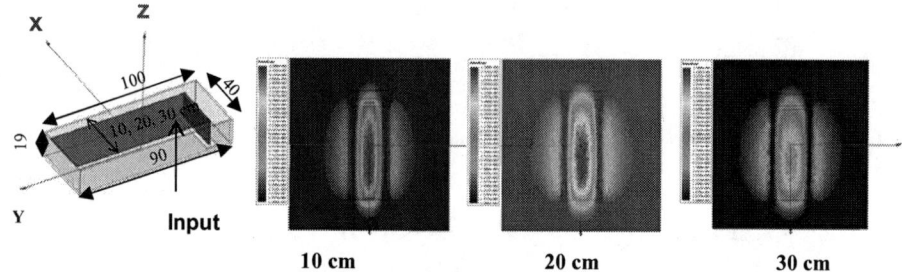

FIGURE 4. Bx- field 15 cm above current strap (CS) with different widths for a single antenna. Boundary conditions for simulation: Components are made of stainless steel, Distance backplane-current strap constant (19 cm), antenna radiates into a box of size 1*1*.5 m, net power is kept constant 1 Watt

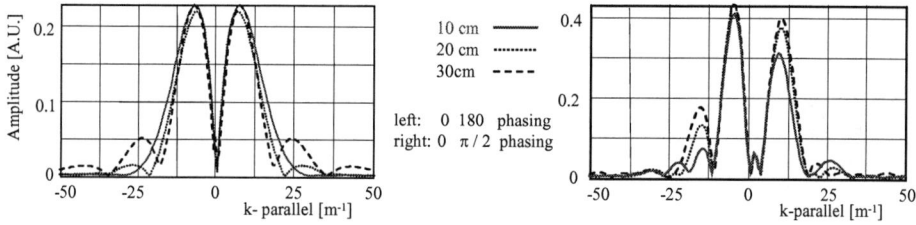

FIGURE 5. Spectrum of two adjacent antennas, located side by side, width of straps varied

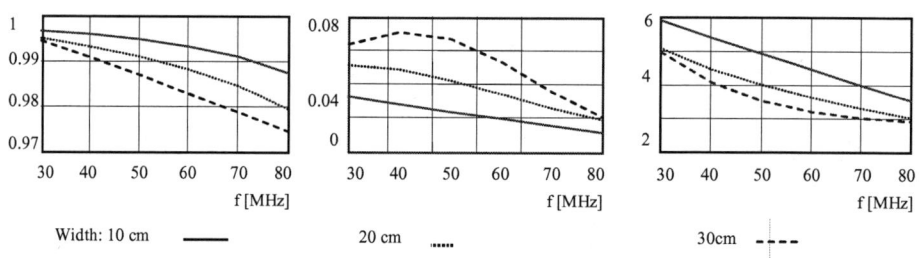

FIGURE 6. Scattering parameters for different width of the CS. Left: mag (S1,1) center: mag (S1,2) right: max. voltage for $Z_o = 30\ \Omega$, input power 1 kW

A variation of the width of the current straps shows that a wider CS results in a broader radiation area (Fig. 4), but influences the k-parallel spectrum by adding sidelobes in both, heating (Fig. 5 left) and current drive case (Fig. 5 right), the radiated

240

power and mutual coupling increases and the maximum voltage in the transmission line is reduced (Fig.6.). A good compromise can be achieved, if the width of the CS is half of the antenna box size.

COMPARISON OF 4x4 AND 4x6 ANTENNA ARRAY FOR ITER

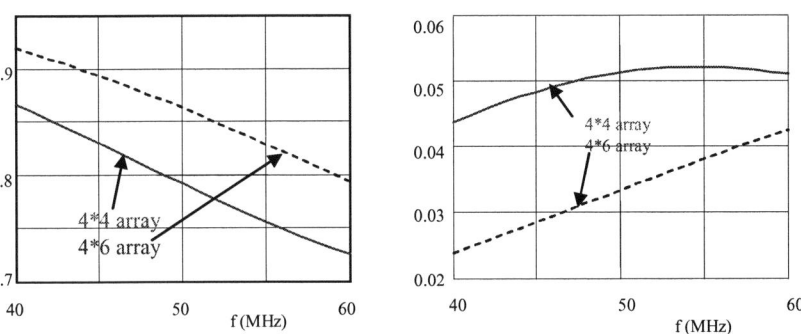

FIGURE 7. Scattering parameters for different ITER configurations: left: mag (S1,1) right: mag (S1,2)

Two different cases of a possible ITER antenna configuration, a four times four and a four times six array within the same box have been compared, the distance between CS / backplane has been kept constant. Fig. 7 shows better coupling (mag (S1,1) but slightly more mutual coupling (mag (S1,2) for the 4*4 array, therefore this type could be preferable due to cheaper manufacturing costs and lower voltages within the adjacent transmission lines. Assuming transmission lines with Zo = 30 Ω and the same total input power in both cases, the maximum voltages are comparable.

CONCLUSION

It has been demonstrated, that powerful Finite Element Programs for PC's are very helpful to support the improvement or redesign of antennas. A simplified absorbing model can be used to check the field strength and voltages within the geometry.

REFERENCES

1. "High Frequency Structure Simulator", v.9.2, Ansoft Corp., Pittsburg, PA.
2. P. U. Lamalle, A. M. Messiaen, P. Durodier and F. Louche, "Recent developments in ICRF modelling", 20th IAEA Fusion Energy Conference, Portugal, November 2004
3. D. A. Hartmann et al., " Measurements and Calculations of Electrical Properties of ICRF Antennas", 15 th Topical Conference on Radio Frequency Power in Plasmas , Moran, May 2003

Measurements and Calculations of CW Transmission Lines for ICRF in W7-X

D. Birus, D. A. Hartmann, W. Becker and W7-X Team

Max-Planck-Institut für Plasmaphysik, Garching / Greifswald, Germany

Abstract. Continuous wave capable transmission line components for use in the ion cyclotron range of frequencies are required for forthcoming high power fusion experiments like Wendelstein 7-X and ITER. For example, ITER requires 30 Ω transmission line components for transmitting 2.5 MW at a VSWR of 1.5. This corresponds to maximum voltages and currents of about 12 kVrms and 360 Arms, respectively. Since commercially available transmission lines do not meet these requirements, a 9" coaxial transmission line resonator was built with a resonance frequency of 60 MHz and a length of 5 m, pressurized with dry air at 3 Bar. Since its quality factor is about 2500 the ITER requirements can be locally obtained using an available 50 kW CW RF source. This setup was used to test the suitability of electrical spring contacts and ceramic spacers made of AlN and Al_2O_3. The temperature increase in the ceramics was found to be in good agreement with calculations based on measured values of tan δ and ε. In a modified resonator with a total length of 7.5 m, water cooled outer conductor but no cooling of the inner conductor it was found that heat transport from the inner to the outer conductor via convection and radiation was sufficient to keep the temperature of the inner conductor at the current maximum below 120°C.

Keywords: Transmission Line, .Resonator, Ceramic, AlN, Al2O3, Spring Contact, W7X, ITER
PACS: 84.40Az:

EXPERIMENTAL SETUP

Presently no ICRH plant has generators with sufficient CW power to test components at the specified values or current and voltage. However, since the components contact elements, conductors or ceramic spacers experience the highest load either due to the current or due to the voltage; it is possible to use a resonator setup where locally the specified values of current or voltage can be achieved.

Such a transmission line resonator was built at the ICRH test stand at IPP site in Garching. The schematic setup of the resonator is shown in Figure 1. The resonator consist of two sections of 30 ohm transmission line (Spinner RL 140-230 parts mostly), whose total length was λ – later 3λ/2 – at the resonance frequency of 60 MHz. Both sections terminated with a short; one section, if of fixed length, the other section has a continuously movable short. They are joined at a T-Connector at the 50 ohm location that is fed with a 50 ohm transmission line connected to an ASDEX Upgrade generator. This generator is capable of supplying 50 kW power CW. Fine tuning and matching of the resonator is done by changing the operation frequency and by adjusting the length of the variable short section. Typically the whole arrangement is pressurized to 3 Bar with dry air. Both shorts - aluminium discs that provide electrical contact to the inner and outer conductor of the transmission line via MULTILAM contacts – are water cooled. A voltage probe is installed near the voltage maximum and a current probe is installed near the current maximum at the fixed short.

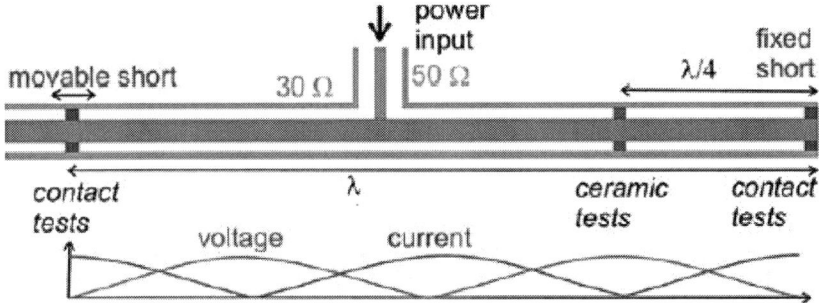

FIGURE 1: Schematic of a 1λ-transmission line resonator to test contact elements at the current maximum and ceramic elements in the voltage maximum

The quality factor of the unloaded resonator (i.e. without a ceramic at the voltage maximum) was estimated from the materials properties and measured and found to be about 2650. With this quality factor the value of the current and the voltage at their respective maxima for an input of 50 kW are about V(max) = 63kV, I(max) = 2.1kA. These values are well above what is needed for the tests, thus it is possible to operate the generator at less power.

THERMAL MEASUREMENTS OF CONTACTS

MULTILAM contacts of type LA-CU were used on the fixed short and on the movable short. The MULTILAM contacts were installed on the inner conductor. Since direct measurement of the temperatures is difficult an observation port with a Sapphire window as added to the resonator outer conductor which was used for observation with an IR camera.

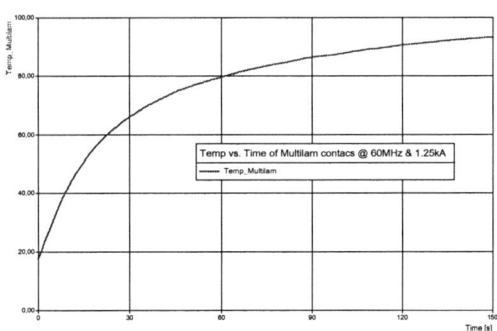

FIGURE 2: Temporal Evolution of the tip of the MULTILAM contacts.

The MULTILAM contacts were cooled via heat conduction through the water cooled copper inner conductor. The temporal evolution of the hottest spots close to the upper tip of the MULTILAM contacts stops below 100°C, while the allowed maximum operating temperature of the MULTILAM contacts is 150°C. After the operation no apparent damage was found of the MULTILAM contacts. Thus the contacts are suitable to be used in joints, e.g. for the connectors used for inner conductors.

THERMAL MEASUREMENT OF CERAMIC

Identical discs of AlN and Al2O3, both with a thickness of 11mm, were manufactured and shrink fitted into CuCrZr rings. These rings were installed in the resonator at the approximate location of the voltage maximum. A special section of the outer conductor was also manufactured that had a port with a Sapphire window for oblique observation with the IR camera.

In both cases the quality factor of the resonator was not much affected by the installation of the ceramic. Thus the available maximum CW power of the generator continued to be sufficient for the test. For the experiments the inner conductor of the transmission lines and both electric shorts were water-cooled throughout, the outer conductor was water-cooled near the locations of the current maximum.

The ceramics were operated with an input power of 25 kW (corresponding to a peak voltage at the ceramic of 63 kV) the temperature rise was measured with both the Al2O3 and the ALN disc and is shown in Fig. 3. The term "cooling" refers to the operation of the water cooling channel at the outside edge of the ceramic disc.

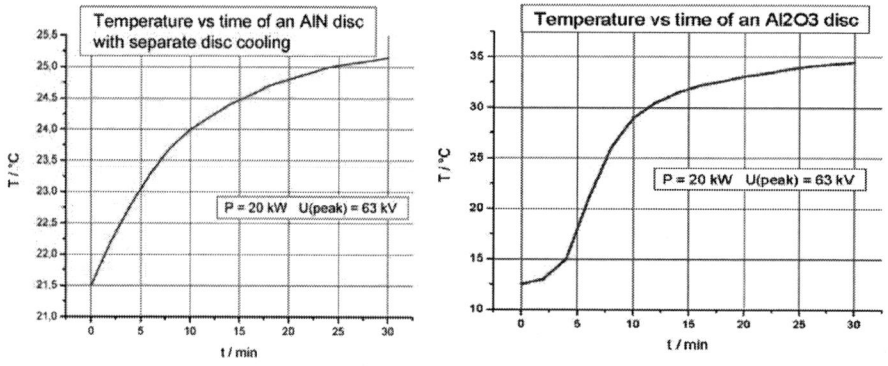

FIGURE 3: Temperature evolution of the Al2O3 and the AlN ceramic discs

Thus both, AlN and Al$_2$O$_3$ are ceramics suitable for matched and unmatched transmission lines. The loss factor of the ceramics has been measured about 100ppm at Al2O3 and about 50ppm at AlN. The Al2O3 ceramic supplied from WESGO and the AlN 180 ceramic supplied from ANCERAM, both from Germany.

COOLING OF INNER CONDUCTOR

Fig. 4a shows the temperature distribution along the inner conductor of the resonators every 5 minutes during 45 minutes of 12 kW RF power that is dissipated on the inner and outer conductor. The inner conductor is water cooled on the left side of the ceramic at 3.75m. The thermal heat conduction of the inner conductor via the MULTILAM contacts to the short at 7.5m is neglected. The temperature rise within the 45 minutes of RF and approximately reaches saturation where the power dissipated on the inner conductor is being conducted to the outer conductor. The temperature rise at the locations where there is a current minimum is due to thermal heat conduction along the inner conductor. In the regions where the inner conductor is water cooled there is hardly any temperature increase to the good heat transfer to water.

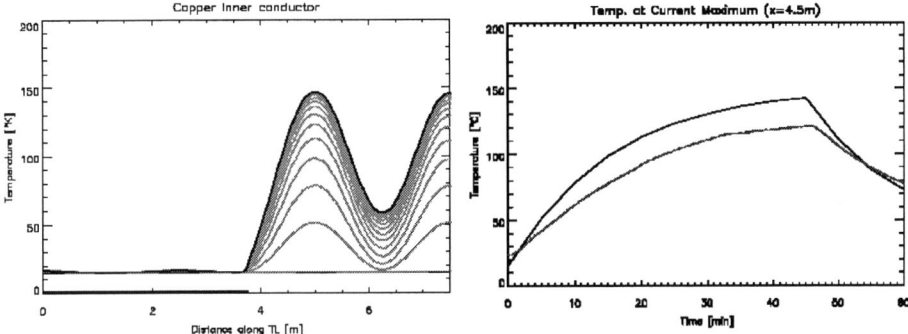

FIGURE 4: Temperature Distribution (a) and Evolution (b) of inner Conductor

The calculated temperature rise at the location of the current maximum (x= 5m) is compared with the measured temperature rise in Fig. 4b. The calculation is shown in black and the measurement is shown in red. The agreement is fairly good and indirectly states that the model for the heat transfer via convection is appropriate. In addition, the temperature rise was measured and calculated for an absolute pressure of 3.8 Bar. This led to an increase in the final temperature by about 10° in both experiment and calculation. The temperature rise was also calculated for transmission lines filled with helium at a pressure of 3.8 Bar. This calculation predicted a final temperature of about 110°C.

The inner conductor does not have to be actively cooled. If the outer conductor is actively cooled then the heat conduction via radiation and convection in pressurized air is sufficient to keep the inner conductor of temperatures of about 100°C. At these temperatures safe operation is possible.

REFERENCES

1. TW3 – THHI – GTFDS – 1, IPP Final Report, Dec. 2004
2. Multi Contact brochures. (www.multi-contact.com)
3. D. M. Pozar, Microwave Engineering, New York 1998

The ICRH System Planned for Wendelstein 7-X

D. A. Hartmann, D. Birus, J. Wendorf, F. Wesner

Max-Planck Institute for Plasma Physics, 17493 Greifswald, Germany

Abstract. The Max-Planck Institute for Plasma Physics is presently building the stellarator experiment Wendelstein 7-X (W7-X) at Greifswald, Germany. First plasma operation in planned after 2010. The plasma will be heated primarily using electron cyclotron resonance heating (ECRH) at a magnetic field of B=2.5 T, but also using ion cyclotron resonance heating (ICRH) and neutral beam injection heating (NBI). The latter heating methods are necessary to study high beta discharges and fast ion confinement. The ICRF system is planned to consist of two radially movable antennas powered by two generators with an output power of 2 MW each. The antenna plasma load will be matched in real time to the generator impedance using ferrite loaded transmission line elements. In addition, 3dB couplers will passively protect the generators from RF power reflections during rapid changes of the antenna plasma load that are too fast for the ferrite matching system.

Keywords: ICRH, antenna, transmission line, generator
PACS: 52.50.Qt, 52.55.Hc

INTRODUCTION

W7-X is a five fold modular super-conducting stellarator (R=5.5 m, <a>=0.55 m, B≤3T) designed for long pulse (30 minutes) operation at reactor-relevant densities. The confining magnetic field is generated by seven types of independently powered magnetic field coils. By varying the coil currents a wide range of different magnetic field configurations can be obtained. These configurations are optimized with respect to classical stellarators, e.g. for good fast particle confinement at high plasma beta, a reduction of the Shafranov shift and MHD plasma stability up to a plasma beta of 5 %.

The initial heating system consists of 10 MW cw ECRH, 4 MW pulsed ICRH and 5 MW pulsed NBI and is designed to test all the optimization criteria of the magnetic field configurations of W7-X. In this respect, ICRH is needed for studying fast particle confinement and generating high beta plasmas at densities higher than the cutoff density of ECRH of $1.2 \cdot 10^{20}$ m^{-3} (2nd harmonic X-mode at 140 GHz at 2.5 T). An additional RF system using a separate antenna is planned for torus conditioning in the presence of magnetic field and to generate target plasmas for NBI high beta discharges at magnetic fields non-resonant with ECRH.

REQUIREMENTS

A flexible ICRH system requires that heating scenarios with sufficiently strong single pass absorption exist for magnetic fields between 0.8 T and 3 T and for varying concentrations of hydrogen and deuterium. Latter is necessary since stellarators – in contrast to tokamaks – do not show an isotope dependence of the energy confinement time, thus stellarator plasmas not necessarily required low hydrogen concentration. A frequency range of 25 to 80 MHz allows such flexibility. In this range of magnetic field and hydrogen concentrations calculations with the slab geometry full wave code FELICE [1] predict that at least one heating scenario can be chosen with single pass absorption values in excess of 70 %. These heating scenarios include second harmonic hydrogen heating, hydrogen or helium 3 minority heating and hydrogen in deuterium mode conversion. Calculations with a three-dimensional full wave ICRF heating stellarator code [2] confirm the predicted good absorption of the different heating schemes.

The midplane distance of the last closed flux surface (LCFS) to the vessel wall varies by more than 10 cm for different magnetic field configurations. In addition, the density decay outside of the LCFS also depends on the configuration. Therefore the antennas have to be radially movable in order to obtain sufficient antenna-plasma coupling for all configurations. Two large ports (40 cm x 100 cm) facilitate the installation of the antennas from outside and accommodate a mechanism to move the antennas radially.

On W7-AS good heating efficiency of ICRF heating was obtained in stellarator geometry using antennas that closely followed the plasma contour. On W7-X, 30 minutes long plasma operation imposes the additional requirement to actively cool all plasma facing components of the antennas and most of the elements exposed to RF losses.

ANTENNAS

Initially, two different antennas will be developed for W7-X: one double-strap antenna and one single-strap antenna. The double strap antenna excites the fast wave at wave numbers of $k_\parallel \approx \pm 6m^{-1}$ with both good coupling and good ion absorption. For this, however, the antenna has to be toroidally wider than the port size.

Radially and poloidally, the antennas have to be sufficiently far from the plasma so as not to intercept the particle and heat flow to the divertor modules. If all plasma configurations are considered, the remaining maximum antenna depth, i.e. the distance between the strap and the torus wall, is only about 7 cm. The estimated antenna loading of such an antenna is too low to couple all of the installed generator power.

Therefore in the beginning only the single-strap antenna will be installed. This antenna can be pulled back into the feeding port almost completely and thus can be moved to positions where it does not interfere with the edge plasma in all plasma configurations. During experimental campaigns with reduced plasma configurations that do not extend too far outward, also a double strap antenna will be installed with sufficient depth to ensure sufficient coupling.

Fig. 1 shows a vertical cut view of different plasma cross sections, of the antenna shape and of the amount of radial motion required to achieve sufficient antenna plasma coupling for different plasmas.

Both antennas have an interface to the "liner". This liner is a boron nitrite coated actively cooled stainless steel structure installed inside of the vacuum vessel to protect it against the plasma Its operational temperature can be adjusted between room temperature and 150 °C. Sliding finger contacts between antenna and liner avoid coupling RF power into the region between liner and vessel.

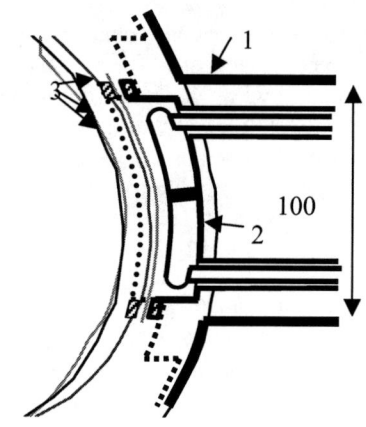

FIGURE 0. Vertical cut of W7-X torus (1) including the antenna (2) in the port at two radial positions with the corresponding plasma profiles (3).

Both antennas consist of poloidal current straps fed from both ends and short-circuited in the horizontal midplane. For a given maximum RF voltage on the strap the maximum power to the plasma by such a set-up can be almost twice as large compared to a strap short-circuited at its end. In W7-AS a similar antenna without a central short was installed. By adjusting the phase in the feeding lines the current maximum was resonantly adjusted in the horizontal midplane creating a virtual short [4] and the external high voltage stand off was found to reach up to about 50 kV for several seconds. On W7-X a real short is required to provide water connections to the straps and the inner conductor of the antenna feeding lines. In addition, ceramic rods are being tested in the feeding lines in order to be able to omit the central short.

The S-matrix of the antenna and the electric field distribution inside of the antenna are calculated with commercially available codes.

Both, single- and double-strap antennas will be installed by moving them through the feeding port on a cart which later will be used for the radial motion. Due to the restricted space the straps of the double strap antenna have individual housings with a hinge in between. This way the antenna can be moved through the port sideways.

All plasma facing components of the antenna (the antenna limiters, the Faraday screen and the antenna current straps) have to be water-cooled to dissipate RF losses and an average heat load from the plasma of 200 kWm^{-2} during full power plasma operation. This heat load is expected due to radiation and plasma particle loss for the outside plasma region that is not intercepted by the divertor modules. First orbit particle losses of ICRF heated plasma particles are expected to increase the heat load on the ICRF antenna but have not been calculated in detail.

The current feeders inside of the port and the vacuum feedthroughs will also be fully water cooled. This is not required for the initial phase but is useful for the baking of the antenna and for a later upgrade for cw operation. To this extend a test campaign has been started at the ICRH teststand in Garching to test all the required components for the suitability for cw operation: ceramic discs, ceramic rods, electric spring contacts, cooling water feedthroughs.

POWER TRANSMISSION, MATCHING, GENERATORS

The straps of each antenna are fed in parallel by pressurized coaxial transmission lines. The transmission lines leading to both antenna straps are connected at the voltage maximum and matched to the generator impedance. With a transmission line model, that uses antenna S-parameter matrix data obtained from a scaled-down mock-up antenna (for cases with and without loading), the behavior of the set-up is being investigated and optimized for the desired frequency range of the matching system. In particular, the influence of the mid-plane short and the characteristic impedance of the feeding lines in the ports is analyzed.

Remotely movable trombones are used to change the length of the transmission lines in the unmatched section in order to allow operation over the whole frequency range of the generators without requiring downtime of the system for mechanical alterations.

Matching of the antenna impedance to the generator impedance was planned to be done using ferrite loaded transmission lines that were supposed to be developed commercially [5]. Even though the first tests were able to match any antenna load of the design range within less than 1 ms, increasing the bandwidth proved too difficult. Thus the matching system is presently under revision. In addition, 3dB couplers [3, 6] are planned to decouple the generators from the antennas also under fast load changes.

In the beginning the transmission lines will not be cw capable and consist of the standard 6 and 9" pressurized transmission line elements. It is planned to use available generators at the Max-Planck Institute in Garching that were formerly used on the predecessor experiment Wendelstein 7-AS. These generators can each operate in a frequency range from 34 to 110 MHz at a nominal power of 2 MW for 10 seconds every 3 minutes. A later upgrade to cw capable generators and transmission lines is envisioned.

REFERENCES

1. M. Brambilla, *Plas. Phys. Contr. Fusion* 31, 723 (1989).
2. V. Vdovin, *28th EPS Conf. Contr. Fusion Plasma Physics*, Madeira, P4.058 (2002).
3. F. Wesner et al., *2nd Europhysics Conf. on Radio Frequency Heating and Current Drive of Fusion Devices*, 22A, 13, (1998).
4. D. A. Hartmann et al., *12th Topical Conf. Radio Frequency Power in Plasmas*, AIP Conf. Proc. 403, 49 (1997).
5. F. Braun et al., *Proc. 18th IEEE/NPSS Symposium on Fusion Engineering*, 395 (2001).
6. F. Wesner et al., *12th Topical Conf. Radio Frequency Power in Plasmas*, AIP Conf. Proc. 403, 101 (1997).

Development of 2 MW Dummy Load for KSTAR ICH System

Jong-Gu Kwak, Son Jong Wang, Young Dug Bae, Jae Sung Yoon, and Bong Guen Hong

Korea Atomic Energy Research Institute, 150 Dukjin-Dong, Yusong-Ku, Daejon 305-353, Korea

Abstract. A 2 MW dummy load with a frequency ranging from 30 to 60 MHz is developed for the KSTAR ICH transmitter and the cold test shows that VSWR is less than 1.35 for the temperature variation of 17℃. DC test also shows that overall temperature increase is less than 3℃ for 400 kW. RF test is done for the RF power of 140 kW and Max. VSWR is about 1.4 at the temperature of 40 ℃.

INTRODUCTION

RF power source with the frequency ranging around 50 MHz is widely used in broadcasting and plasma heating applications. Especially, in the thermo-nuclear fusion experiments, the ICRF(Ion Cyclotron Range of Frequency) is located around HF and VHF and the high power transmitter is required to achieve effective fusion yields. For the ITER(International Thermonuclear Experimental Reactor) case, the RF power of the unit transmitter is 2 MW and 10 sets of 2 MW transmitters will be used. It is necessary to equip it with a dummy load for the power test of the transmitter. Generally, the resistive film type dummy load is widely used, however, it has a limit for an application above 300 kW because there is an adhesion problem between the resistive film material and ceramic substrate at the high temperature. Therefore, the most promising candidate above 2 MW is the soda water dummy load where the soda water is used for both a material for absorbing rf power and a cooling medium. In addition, there is no power density limit for the soda dummy load if it is operated below the boiling point of soda water.

In this work, the electrical and mechanical design requirements of the 2 MW soda dummy load for the KSTAR(Korea Superconducting Tokamak Advanced Research) ICH(Ion Cyclotron Heating) system are introduced[1] and the results of the power test using DC and RF is discussed.

GENERAL DESCRIPTION OF SODA DUMMY LOAD

Fig. 1(a) shows the resistive part of the soda dummy load. It is composed of three sections. One is the first inner cylinder(I) where the main soda coolant is introduced. Another is the second inner cylinder(II) for the coolant path to the outlet. The last one is the outer enclosure(III) with the shape of a conical as described in Fig 1(b) and it is made of metal conductor. The rf current from the top is shorted to the ground at the bottom. The cooling water flows into the bottom of the first inner cylinder and then it flows out from the bottom between the first and second cylinder. Insulators such as Polyethylene are typically used for the material of inner cylinders. The profile of inner surface for the outer conductor(III) is important because it mainly determines the VSWR(Voltage Standing Wave Ratio) characteristics of the dummy load. Another critical factor for VSWR is the skin depth effect. Compared with the nominal resistive dummy load where the resistive medium is only coated on outer cylinder, soda water inside the inner cylinder is also the resistive medium for the soda dummy load so the skin depth affects on the VSWR with the frequency. The skin depth effect also makes it difficult to deduce the VSWR of the dummy load via the analytical solution and VSWR is calculated using the commercial software. The size of the dummy load is determined by the cooling capability of the water for the 2 MW heat generated on the water. As the resistivity of the soda water has a negative temperature coefficient, it is crucial to maintain the temperature difference of the soda water between the inlet and the outlet within 20 ℃ for low VSWR of the dummy load and this is usually done by the flow control valve at the secondary loop as shown in fig.2. The overall control is done by PID controller based on the EPICS(Experimental Physics and Industrial Control System).

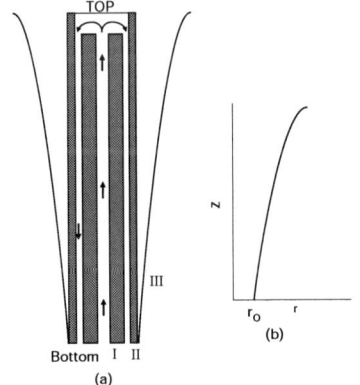
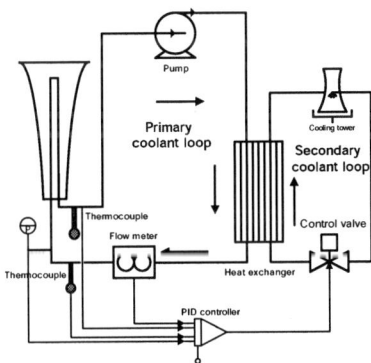

Fig. 1. Schematic diagram of the dummy load column(a) and profile of the outer conductor(b).

Fig. 2. Schematic diagram of cooling line and the control circuit of the soda dummy load.

The outer enclosure is made by the aluminum and the size of the input port is 9-3/16 EIA. The axial length of the soda column at the dummy load is 1m and the flow rate is nominally 100 m^3/h. The pressure drop at the heat exchanger and dummy load is less than 3 atm.

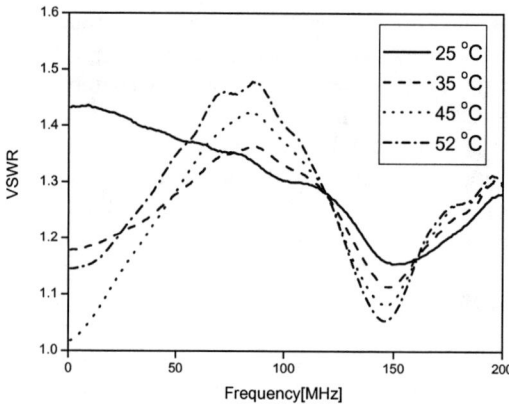

Fig. 3. VSWR vs. frequency for various temperatures.

The cold test for the dummy load is done for various temperatures as shown in fig 3. At the temperature of 45 ℃ VSWR is less than 1.25 up to 50 MHz and it is less than 1.35 for the temperature variations of 17 ℃. The conductivity is 10 mS at 40℃. The peak around 80 MHz for 35, 45 and 52 ℃ comes from the axial resonance of the dummy load column. When the axial length of the soda column is increased, its resonance frequency moves to the lower frequency

POWER TEST OF THE DUMMY LOAD

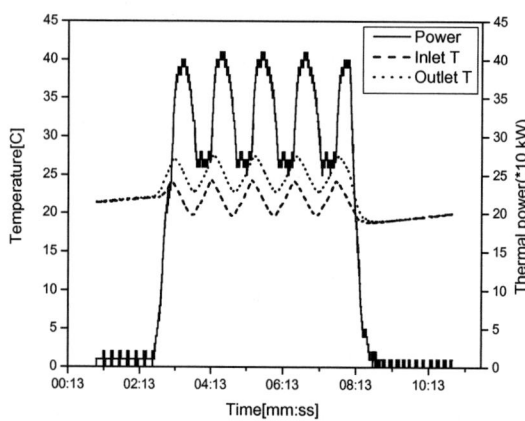

Fig. 4. Time evolutions of the coolant temperature and the thermal power at DC power test.

The DC power test is done to calibrate the control circuit and to investigate the thermal characteristics of the coolant system including the dummy load column. The soda solution is diluted and adjusts its resistance value as 150 ohm. It takes approximately 10 s for the soda water to travel from the outlet of the dummy column to the inlet. So the thermocouple at the inlet responses 10 sec later after DC power is applied. To reduce the response time, the high flow rate and very short path of the primary coolant loop is required. The temperature is regulated within 3℃. The difference of the temperature between the inlet and outlet is about 20 ℃ for 2 MW rf power and the permissible VSWR would be less than 1.4 at 50 MHz.

RF test is done for the RF power of 140 kW as shown in fig. 5. The frequency is 45 MHz. Max. VSWR is about 1.4 at the temperature of 40 ℃. The full power test will be done at the factory acceptance test for the 2 MW transmitter.

SUMMARY

The 2 MW soda dummy load is designed and tested. DC and RF tests show that it can be used for 2 MW dummy load.

ACKNOWLEDGMENTS

This work was supported by the Korean Ministry of Science and Technology under KSTAR Project Contract.

REFERENCES

1. B. G. Hong et al., 18th IAEA Fusion Energy Conference, Sorrento, Italy, 2000, p. 96.

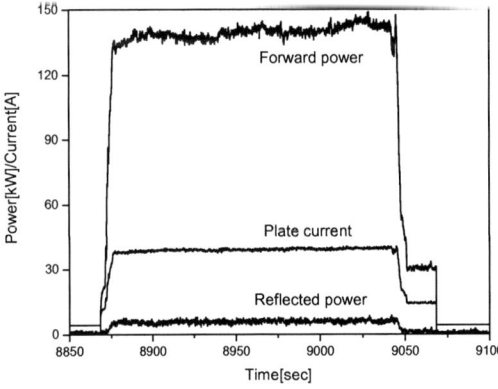

Fig. 5. Time evolutions of P_{FWD}, P_{REF} and the plate current at RF power test

CURRENT DRIVE

Synergy in RF Current Drive

R. J. Dumont* and G. Giruzzi*

*Association Euratom-CEA, CEA/DSM/DRFC, Centre de Cadarache,
13108 Saint-Paul-lez-Durance, France

Abstract. Auxiliary methods for efficient non-inductive current drive in tokamaks generally involve the interaction of externally driven waves with superthermal electrons. Among the possible schemes, Lower Hybrid (LH) and Electron Cyclotron (EC) current drive have been so far the most successful. An interesting aspect of their combined use is the fact that since they involve possibly overlapping domains in velocity and configuration spaces, a synergy between them is expected for appropriate parameters. The signature of this effect, significant improvement of the EC current drive efficiency, results from a favorable interplay of the quasilinear diffusions induced by both waves. Recently, improvements of the EC current drive efficiency in the range of 2-4 have been measured in fully non-inductive discharges in the Tore Supra tokamak, providing the first clear evidence of this effect in steady-state conditions. We present here the experimental aspects of these discharges. The associated kinetic modeling and current state of understanding of the LH-EC synergy phenomenon are also discussed.

Keywords: Radiofrequency waves, current drive, efficiency, LHCD, ECCD, synergy
PACS: 52.35.Hr, 52.35.Mw, 52.55.Wq

INTRODUCTION

In tokamaks, most modern modes of operation involve one or several sources of non-inductive current drive, both to sustain a large fraction of the toroidal current responsible for plasma confinement[1] and to control the plasma stability and transport properties by an appropriate choice of the control parameters[2]. To reach these two objectives, radiofrequency (rf) waves launched with an asymmetric toroidal spectrum have been successfully employed to drive fast electron tails, providing reliable sources of current with a certain amount of flexibility determined by the chosen scheme. Lower Hybrid Current Drive (LHCD) is a well-tested and efficient method[3], based on Landau damping of the wave power. Its main drawback is that the current profile remains difficult to control. On the other hand, Electron Cyclotron (EC) waves are absorbed through a resonant interaction with the short-scale electron gyro-motion and are characterized by a narrow deposition, thus providing a means to induce a local modification of the current profile. Moreover, by an adequate choice of magnetic field, frequency and launching angle, it is possible to control this location. The efficiency, however, is known to be significantly lower than for LH waves[4], owing to the absence of direct transfer of parallel momentum, the influence of trapped particles, and the fact that the interaction usually involves less energetic electrons. The combination of LH and EC waves in the same discharge therefore constitutes an appealing solution: their features are largely complementary and the possibility of a synergy between them implies that the total current in the presence of both waves can exceed the sums of separate currents, as was shown either in purely kinetic [5, 6, 7, 8] or in kinetic+transport[9] simulations. Experimental efforts have been devoted to the associated use of LH and EC waves with varying outcomes, depending on the operating conditions[10, 11, 12, 13, 14]. More recently, a direct proof of the synergy was obtained in steady-state conditions in the Tore Supra tokamak[15]. In this paper, the results obtained in Tore Supra and their kinetic interpretation are presented. Also addressed is the current state of understanding of the LH-EC synergy mechanism.

LH-EC EXPERIMENTS ON TORE SUPRA

The requirements to provide an experimental demonstration of the LH-EC synergy are 1) Stationary conditions, with no significant electric field effects, 2) Good confinement of the fast electrons, accountable for most of the non-inductive current, 3) Large optical depth for the EC waves. Specific experiments meeting these requirements have been performed on the Tore Supra tokamak (Major radius $R_0 = 2.40$m, Minor radius $a_0 = 0.72$m, Magnetic field $B_0 \approx 3.8$T, circular cross section). Deuterium plasmas lasting 30s have been realized with plasma current $I_p = 0.58$MA, central electron density $n_{e0} \approx 1.8 \times 10^{19}$m^{-3}, central electron temperature $T_{e0} \sim 6 - 8$keV and effective ion charge $Z_{eff} = 4$. These experiments were all based on the same scheme: after the initial Ohmic phase, the transformer flux was kept constant and the plasma current was sustained by $P_{LH} = 3$MW of LH power at frequency $f_{LH} = 3.7$GHz. The three criteria listed above were satisfied by resorting to the following actuators: 1) gas puff to maintain a constant density, 2) LH power to ensure a constant plasma current, and 3) transformer flux to keep the loop voltage exactly equal to zero. EC waves at power $P_{EC} = 0.7$MA and frequency $f_{EC} = 118$GHz have been injected in this target plasma in fundamental ordinary mode polarization, by means of a set of steerable mirrors. During the ECCD phase, the LH power is found to drop by an amount ΔP_{LH}, which can easily be translated into a corresponding EC current ΔI given by

$$\Delta I = (I_p - I_{bs})\frac{\Delta P_{LH}}{P_{LH}}, \qquad (1)$$

where I_{bs} is the calculated bootstrap current, using the NCLASS code[17]. This formula is valid provided that the loop voltage is zero, and that both I_{bs} and the LH efficiency vary little during the relevant phases of the discharges, which is the case in these experiments. The current ΔI deduced from Eq. 1 can be compared to the current I_{EC} that would have been driven by EC waves alone in the same plasma conditions. I_{EC} can be evaluated either by a Fokker-Planck code to obtain a quasilinear estimate of the driven current, or by linear formulas for the current drive efficiency. These simplified expressions have been extensively tested against experimental measurements on DIII-D for a variety of plasma conditions (see, e.g., Ref. [16]) and are found to be in close agreement with the quasilinear results for the parameters under study. Therefore, they can be used as references to evaluate the ECCD efficiency in good single-pass absorption situations. On Fig. 1(a) and (b), the time traces corresponding to one of the LH-EC discharges (#31463) performed on Tore Supra are shown.

On Fig 1(a), it can be seen that the temperature for $r/a_0 \leq 0.4$ vary very little between the ECCD phase, and after the ECCD has been switched off. This implies that 1) the core confinement has been durably affected by the EC power application, 2) an improvement

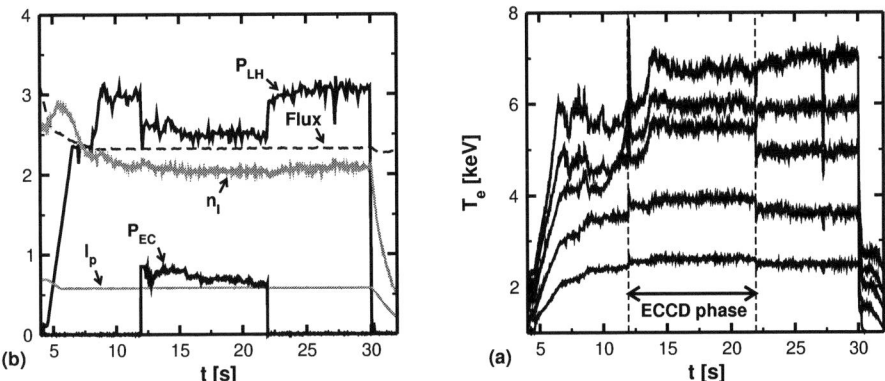

FIGURE 1. Time traces for Tore Supra shot #31463. (a) From top to bottom: LH Power [MW], transformer flux [Wb], line integrated density [$\times 10^{19} \mathrm{m}^{-3}$], EC Power [MW], plasma current [MA]. (b) Electron temperature from ECE measurement at various positions with $0 \leq r/a_0 \leq 0.4$.

of the LHCD efficiency caused by an increase of the electron temperature can be ruled out. In order to quantify a potential synergy effect, the synergy factor[8], defined as $F_{syn} \equiv \Delta I/I_{EC}$ is introduced. In the discharge shown on Fig. 1, the LH power drops by $\Delta P_{LH} \approx 0.5$ MW during the ECCD pulse. The computed bootstrap current is $I_{bs} \approx 80$ kA, so that Eq. 1 gives $\Delta I \approx 90$ kA. On the other hand, kinetic simulations, as well as linear formulas, predict $I_{EC} \approx 24$ kA, yielding $F_{syn} \approx 3.8$. On Fig. 2(a), ΔI is compared to I_{EC} for various discharges versus ρ_{EC}, the location of maximum EC power deposition, determined by Ray-Tracing calculations. The corresponding values of F_{syn} are shown on Fig. 2(b).

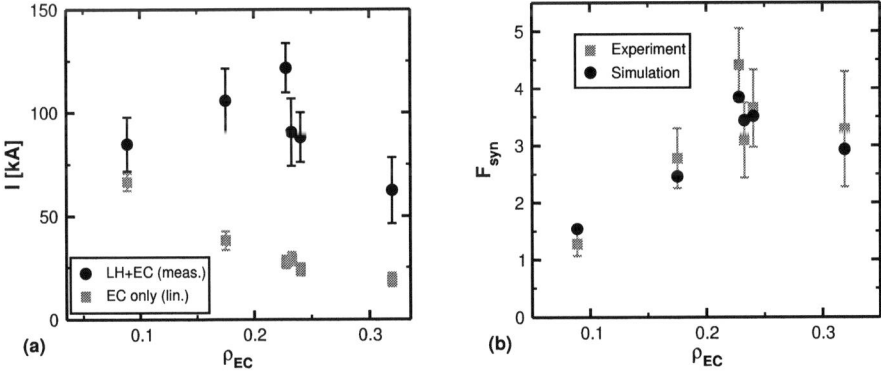

FIGURE 2. (a) Measured additional current driven by EC waves in the presence of LH waves (dots) and computed linear EC current (squares) vs ρ_{EC}, location of maximum EC power deposition. (b) Synergy factor from experiment (squares) and from kinetic simulations (dots) vs ρ_{EC}.

The difference between these discharges is essentially the result of variations of the EC wave injection angles. The error bars are obtained from standard deviation of the various parameters entering Eq. 1 and experimental uncertainties on the injection angles. Despite these corrections, it can be seen that ΔI always exceed I_{EC}, meaning that the EC current drive efficiency has been improved by the presence of LH driven fast electrons. On Fig. 2(b), F_{syn} is plotted versus ρ_{EC}, and compared to kinetic predictions, performed using a 3-D Fokker-Planck code[18]. The plasma parameters are used as input to the Fokker-Planck code. Each EC beam is modeled by 150 rays and the individual contributions to the quasilinear diffusion coefficient are summed to deduce the driven current. The results, shown on Fig. 2(b), are in good agreement with experimental measurements.

KINETIC DESCRIPTION

Although the experimental measurements and kinetic simulations are in good agreement, the mechanisms underlying the LH-EC synergy phenomenon still need to be identified. Simpler kinetic models may help to gain insight into these mechanisms and constitute tools to assess the CD efficiency for given plasma and waves parameters. Such two methods are the adjoint technique and the Langevin equations (see Ref. [1] and references therein). The starting point is the Fokker-Planck equation, which can be written as

$$\frac{\partial f}{\partial t} - \hat{C}f = \hat{D}_{LH}f + \hat{D}_{EC}f, \qquad (2)$$

with $\hat{C}$ the collision operator, $\hat{D}_{LH}$ and $\hat{D}_{EC}$ the quasilinear operators for LH and EC waves, respectively. It is assumed that the static electric field is zero, which is consistent with the experiments described previously. In order to describe the current drive efficiency in the presence of two waves, the distribution function can conveniently be written as $f \equiv f_m(1 + \phi_{LH} + \phi_{syn})$ with

$$\frac{\partial f_m \phi_{LH}}{\partial t} - \hat{C}(f_m \phi_{LH}) = \hat{D}_{LH}\left(f_m(1 + \phi_{LH})\right) \equiv -\frac{\partial}{\partial \mathbf{u}} \cdot \mathbf{S}_{LH}, \qquad (3)$$

where f_m is the Maxwellian, $\mathbf{u} \equiv \mathbf{v}/v_{th,e}$ is the velocity in terms of the electron thermal velocity and $\mathbf{S}_{LH}$ is the quasilinear flux associated with LH waves.
Subtracting Eq. 3 from Eq. 2 yields

$$\frac{\partial f_m \phi_{syn}}{\partial t} - \hat{C}(f_m \phi_{syn}) - \hat{D}_{LH}(f_m \phi_{syn}) = \hat{D}_{EC}\left(f_m(1 + \phi_{LH} + \phi_{syn})\right) \equiv -\frac{\partial}{\partial \mathbf{u}} \cdot \mathbf{S}_{EC}, \quad (4)$$

with $\mathbf{S}_{EC}$ the EC quasilinear flux. It should be noted that up to this point, the only approximation consists in neglecting ϕ_{syn} in the definition of $\mathbf{S}_{LH}$ used in Eq. 3. Physically, this means that we implicitly suppose that the velocity space direction of the interaction of LH waves with the plasma is not modified by the EC power, which is a reasonable assumption for the parameters of most present experiments. The cross-effect between the waves is essentially contained in the term $\hat{D}_{LH}(f_m \phi_{syn})$ on the left-hand side of Eq. 4, and in the magnitude of $\mathbf{S}_{EC}$. A convenient way to estimate the current drive efficiency of a given scheme is to separate the relaxation from the rf drive, by introducing a response function χ to describe the former, while the latter is determined by the quasilinear flux[8]. For a given rf quasilinear flux $\mathbf{S}_w$, the normalized current drive efficiency η can indeed be expressed as

$$\eta = \frac{\int d^3\mathbf{u}\, \mathbf{S}_w \cdot \frac{\partial \chi}{\partial \mathbf{u}}}{\int d^3\mathbf{u}\, \mathbf{S}_w \cdot \frac{\partial}{\partial \mathbf{u}}\left(\frac{u^2}{2}\right)} \approx \frac{\hat{\mathbf{e}}_s \cdot \frac{\partial \chi}{\partial \mathbf{u}}}{\hat{\mathbf{e}}_s \cdot \frac{\partial}{\partial \mathbf{u}}\left(\frac{u^2}{2}\right)}, \quad (5)$$

where it is assumed that the plasma-wave interaction is sufficiently well localized in velocity space. The major advantage of this formalism is that the magnitude of $\mathbf{S}_w \equiv S_w \hat{\mathbf{e}}_s$ cancels out. Therefore, to evaluate the CD efficiency, the only information needed is 1) The direction of the quasilinear flux, which is well-known: parallel for the LH-induced diffusion, essentially perpendicular for the EC-induced diffusion; 2) The response function χ. To estimate this quantity, one can resort either to the adjoint method, or solve the Langevin equations[1]. While both methods are known to be formally equivalent, the latter is more cumbersome than the former, but has the advantage of being physically transparent. The response function associated to Eq. 3, χ_{FB}, can be evaluated analytically and is known as the Fisch-Boozer response function. It describes the collisional relaxation of the wave-driven electrons. The same procedure can be applied to Eq. 4 and leads to a modified response function χ_{LH}, which takes into account the quasilinear influence of the LH wave on the electron relaxation[8]. On Fig. 3(a), the response function modified by the presence of LH waves is shown as contours in velocity space. The LH quasilinear domain is assumed to be a vertical band approximately bounded by two parallel velocities, $u_{\|,1} = 4$ and $u_{\|,2} = 8$. Also shown as dashed contours is the Fisch-Boozer response function.

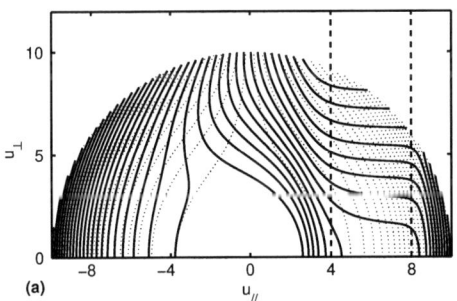
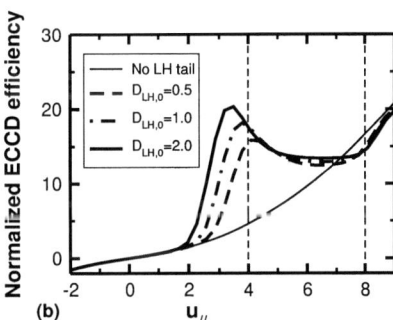

FIGURE 3. (a) Contours of a typical response function in the presence of LH power in velocity space. The dashed contours represent the collisional (Fisch-Boozer) response function. (b) Local ECCD efficiency improvement versus normalized parallel velocity in the presence of LH power, for various values of the normalized LH diffusion coefficient and $u_\perp = 2$. In both (a) and (b), the vertical lines denote the boundaries of the LH quasilinear domain.

It appears that the presence of LH waves affects the response function, mostly in the region $u_{\|,1} \lesssim u_\| \lesssim u_{\|,2}$, but also for velocities $u < u_{\|,1}$ due to pitch-angle scattering. In order to evaluate the consequences of this deformation in terms of ECCD efficiency, it is necessary to compute the gradient of χ_{LH} in velocity space along the perpendicular direction, as is readily seen from Eq. 5, letting $\mathbf{S}_w \equiv \mathbf{S}_{EC}$. The result obtained for various values of the normalized LH quasilinear diffusion coefficient $D_{LH,0}$ is shown on Fig. 3(b), for $u_\perp = 2$. The ECCD efficiency appears to be generally enhanced in the LH quasilinear region, with a peak appearing in the vicinity of $u \sim u_{\|,1}$.

LANGEVIN EQUATIONS

The Langevin equations provide a useful tool to confirm the results obtained by direct solution of the adjoint equation and can help to gain some insight into the mechanisms responsible for the ECCD efficiency improvement. The application of this formalism to the synergy problem is extensively discussed in Ref. [8]. Practically, the idea is to transform the Fokker-Planck equation in a set of stochastic equations tracking the electron trajectories on their relaxation paths. The response function is then obtained by considering a large number of electrons and performing ensemble-averages to obtain the desired quantities. For instance, the response function is given by

$$\chi(\mathbf{u}) = \int_0^\infty d\tau \langle u_\| \rangle(\tau), \qquad (6)$$

where $\langle \cdot \rangle$ denotes the ensemble average and the time-integral is carried out from the initial position $\mathbf{u}$ at $\tau = 0$ to complete relaxation of the considered set of electrons. In order to deduce the efficiency from Eq. 5, the gradient in velocity space is deduced from the difference between the response function at positions $(u_\|, u_\perp)$ and $(u_\|, u_\perp + \Delta u_\perp)$, with $\Delta u_\perp$ sufficiently small. Firstly, the efficiency obtained for $D_{LH,0} = 2$ and $u_\perp = 2$ is shown on Fig. 4(a), computed by the adjoint method and by numerical solution of the Langevin equations. Both method are in good agreement. The slight scattering of Langevin points is due to statistical errors and to the approximate method employed to estimate $\partial \chi / \partial u_\perp$.

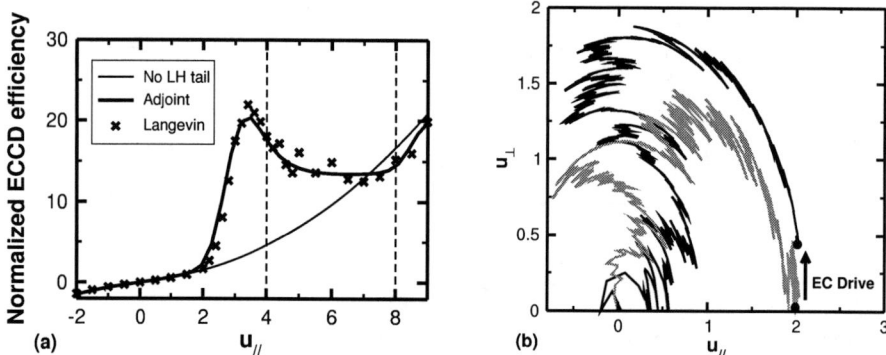

FIGURE 4. (a) Normalized ECCD efficiency computed by solving the adjoint equations (solid), and by numerical integration of the Langevin equations (crosses). Also shown is the Fisch-Boozer efficiency (thin line). (b) Electron relaxation trajectories in velocity space without (grayed) and in the presence of (black) EC drive for $u_\| < u_{\|,1}$.

On Fig. 4(b), two sample trajectories are shown. Several kinetic processes clearly influence the relaxation process: collisional friction slows down the electron until thermalization, and pitch-angle scattering causes it to explore a large part of velocity space during its relaxation. Collisional energy diffusion is found to be insignificant. In the case shown on Fig. 4(b), the EC driven electrons do not experience any influence of the LH wave, since they are pushed at parallel velocities below the LH quasilinear domain boundary $u_{\parallel,1}$. Thus, in this case, the collisional efficiency is recovered, as can be seen on Fig. 4(a) for $|u_\parallel| \lesssim 2$. On Fig. 5(a), two such trajectories are shown when an electron having $u_{\parallel,1} < u_\parallel < u_{\parallel,2}$ is driven by EC waves.

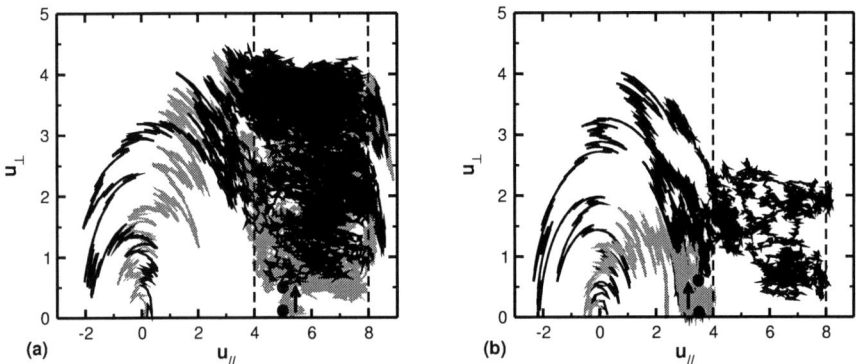

FIGURE 5. Same as Fig. 4(b), but for initial parallel velocity $u_\parallel$ such as (a) $u_{\parallel,1} < u_\parallel < u_{\parallel,2}$, (b) $u_\parallel \lesssim u_{\parallel,1}$. The arrows denote the EC drive.

It can be seen that both trajectories are now strongly influenced by the LH-induced parallel diffusion and friction. This has the consequence that the precise value of $u_\parallel$ has only a limited influence, in terms of CD efficiency, as long as the driven electrons belong to the LH quasilinear domain. This explains the rather flat behavior of the ECCD efficiency for $u_{\parallel,1} \lesssim u_\parallel \lesssim u_{\parallel,2}$. However, the most striking feature of the ECCD enhancement appears to be the peak obtained for $u_\parallel \lesssim u_{\parallel,1}$. The cause of this enhancement is a subtle interplay between pitch-angle scattering and parallel quasilinear diffusion. This situation occurs when the EC power drives an electron having an initial velocity too low to undergo the influence of the LH diffusion to a position where pitch-angle scattering will cause it to be "trapped" in the LH domain. This is illustrated on Fig. 5(b), where it is readily seen that the difference between the two trajectories is very large, causing the CD efficiency to be significantly enhanced.

CONCLUSIONS

In this paper, the experimental aspects and kinetic understanding of the LH-EC synergy in fully non-inductive CD experiments were presented.

On Tore Supra, fully non-inductive discharges were performed, where the LH power was adjusted to ensure that the plasma current was constant with exactly zero loop voltage. By switching on the EC system and measuring the drop in LH power, the amount of current driven by EC waves has been deduced. Values of the ECCD efficiency as high as four times the predicted values in the absence of LH waves were obtained.

These results are consistent with numerical solutions of the 3-D Fokker-Planck equation accounting for both waves.

In order to investigate these features and identify the underlying mechanism, a kinetic model has been used to evaluate the ECCD efficiency in the presence of LH waves. This is done by solving either the adjoint equation directly, or the stochastic Langevin equations to obtain the response function of a plasma in which most of the current is driven by LH waves. The obtained results confirm that an enhancement of the ECCD efficiency can be expected in a LHCD plasma. Future work in this area will consist in using the obtained adjoint solution in a Ray-Tracing code for the EC wave in order to provide a fast estimate of the ECCD efficiency for given plasma conditions.

In next step devices such as ITER, the kinetic simulations performed so far seem to indicate that only a weak enhancement of the ECCD efficiency is to be expected in the presence of LH power. The main reason is that for the considered parameters, quasilinear effects play only a minor role, due to the large dilution of the LH power, so that the LH driven fast electron tail is not as significant as in present machines. Improvement of about 10% are nevertheless obtained. Moreover, the fact that LH and EC waves do not exhibit a large synergy effect for these parameters does not rule out their combined use: they still possess complementary features. In this domain, only integrated simulations accounting for kinetic and transport effects can help to reach a definite conclusion.

REFERENCES

1. N. J. Fisch, Rev. Mod. Physics **59**, 175 (1987).
2. T. S. Taylor, Plasma Phys. Controlled Fusion, **39**, B47 (1997).
3. E. Barbato, Plasma Phys. Controlled Fusion **40**, A63 (1998).
4. V. Erckmann and U. Gasparino, Plasma Phys. Controlled Fusion **36**, 1869 (1994).
5. I. Fidone, G. Giruzzi, G. Granata, and R. L. Meyer, Phys. Fluids **27**, 2468 (1984).
6. I. Fidone, G. Giruzzi, V. Krivenski, E. Mazzucato, and L. F. Ziebell, Nucl. Fusion **27**, 579 (1987).
7. G. Giruzzi, I. Fidone, and R. L. Meyer, Nucl. Fusion **29**, 1381 (1989).
8. R. J. Dumont and G. Giruzzi, Phys. Plasmas **11**, 3449 (2004).
9. R. Dumont, G. Giruzzi, and E. Barbato, Phys. Plasmas **7**, 4972 (2000).
10. A. Ando, K. Ogura, H. Tanaka *et al.*, Nucl. Fusion **26**, 107 (1986).
11. Y. Yamamoto, K. Hoshino, H. Kawashima *et al.*, Phys. Rev. Lett. **58**, 2220 (1987).
12. T. Maekawa, T. Maehara, T. Minami *et al.*, Phys. Rev. Lett. **70**, 2561 (1993).
13. J. A. Colborn, J. P. Squire, M. Porkolab, and J. Villaseñor, Nucl. Fusion **38**, 783 (1998).
14. G. Granucci *et al.*, in *Proceedings of the 12th Joint Workshop on ECE and ECRH*, Aix-en-Provence, France, 2002, edited by G. Giruzzi (World Scientific, Singapore, 2003), p. 341.
15. G. Giruzzi *et al.*, Phys. Rev. Lett. **93**, 255002 (2004).
16. C. C. Petty *et al.*, Nucl. Fusion **42**, 1366 (2002).
17. W. Houlberg *et al.*, Phys. Plasmas **4**, 3230 (1997).
18. G. Giruzzi, Plasma Phys. Controlled Fusion **35**, A123 (1993).

Modulated Current Drive Measurements

C.C. Petty*, W.A. Cox[†], C.B. Forest[†], R.J. Jayakumar[‡], J. Lohr*,
T.C. Luce*, M.A. Makowski[‡], and R. Prater*

*General Atomics, P.O. Box 85608, San Diego, California, 92186-5608 USA
[†]University of Wisconsin, Madison, Wisconsin, USA
[‡]Lawrence Livermore National Laboratory, Livermore, California, USA

Abstract. A new measurement approach is presented which directly determines the noninductive current profile from the periodic response of the motional Stark effect (MSE) signals to the slow modulation of the external current drive source. A Fourier transform of the poloidal magnetic flux diffusion equation is used to analyze the MSE data. An example of this measurement technique is shown using modulated electron cyclotron current drive (ECCD) discharges from the DIII-D tokamak.

Several methods are available for determining the noninductive current profile from the measured internal magnetic field structure (e.g. motional Stark effect (MSE) polarimetry [1,2]). In the deductive approach, the noninductive current drive is found from the evolution of the poloidal magnetic flux per radian (ψ) obtained from a time series of magnetic equilibrium reconstructions [3]. In the inductive approach, the measured MSE signals are compared to realistic simulations of the MSE evolution using a coupled transport-equilibrium code that contains a model for the location, width, and magnitude of the current drive source [4]. In this paper, a new modulation approach is presented, where the current drive profile is found directly from the periodic response of the MSE signals to a slow modulation of the external current drive source using the poloidal flux diffusion equation. This yields a local measurement of the noninductive current profile in analogy to measuring the rf power deposition profile from the oscillations in the electron temperature (T_e).

DERIVATION OF MODULATION METHOD

The evolution of ψ is governed by a diffusion equation that is arrived at by combining the flux-surface-average Ampere's, Faraday's and Ohm's laws. In general toroidal geometry, this equation can be written as

$$\frac{\partial \psi}{\partial t} = \frac{\eta}{\mu_0 \rho_b^2 \hat{F}^2} \frac{1}{\rho} \frac{\partial}{\partial \rho} \left(\hat{F}\hat{G}\hat{H}\rho \frac{\partial \psi}{\partial \rho} \right) + \frac{R_0 \eta \hat{H}}{B_{\phi,0}} \left\langle \vec{J}_{NI} \cdot \vec{B} \right\rangle \quad , \tag{1}$$

where η is the plasma resistivity, ρ is the normalized toroidal flux coordinate, ρ_b is the effective boundary radius, J_{NI} is the noninductive current density, and $B_{\phi,0}$ is the vacuum field at the center of the plasma boundary (at major radius R_0). The symbol $\langle ... \rangle$ denotes a flux surface average. The three dimensionless geometry factors, defined as

$$\hat{F} \equiv \frac{R_0 B_{\phi,0}}{R B_\phi} \quad , \quad \hat{G} \equiv \left\langle \frac{R_0^2}{R^2} \rho_b^2 \left|\vec{\nabla}\rho\right|^2 \right\rangle \quad , \quad \hat{H} \equiv \frac{\hat{F}}{\left\langle R_0^2/R^2 \right\rangle} \quad , \tag{2}$$

tend towards unity in the infinite aspect ratio, low β, circular equilibrium limit. Considering for the moment the simplified situation where the plasma flux surfaces are fixed in space, the Fourier transform of Eq. (1) yields an ordinary differential equation that relates the modulated current drive source ($\tilde{J}_{NI}$) to the modulation in the poloidal magnetic flux ($\tilde{\psi}$),

$$\frac{\partial^2 \tilde{\psi}}{\partial \rho^2} + \left[\frac{1}{\rho} + \frac{\partial}{\partial \rho} \ln\left(\hat{F}\hat{G}\hat{H}\right)\right] \frac{\partial \tilde{\psi}}{\partial \rho} - \frac{i}{\hat{D}} \tilde{\psi} = -\mu_0 R_0 \rho_b^2 \frac{\hat{F}}{\hat{G}} \left[\langle \tilde{J}_{NI} \rangle + \frac{\tilde{\eta}}{\eta} \langle J_{oh} \rangle\right] , \tag{3}$$

$$\hat{D} = \frac{\eta}{\mu_0 \rho_b^2 \omega} \frac{\hat{G}\hat{H}}{\hat{F}} \quad , \tag{4}$$

where $\tilde{\eta}$ is the modulated resistivity, J_{oh} is the ohmic current density and $\langle J \rangle = \langle \vec{J} \cdot \vec{B} \rangle / B_{\phi,0}$. Utilizing Eq. (3), the measured modulation in the vertical magnetic field from the MSE signals, $\tilde{B}_z = (1/R)\partial\tilde{\psi}/\partial R$, can be used to experimentally determine the modulated current drive profile. A useful feature of the modulation method is that fiducial comparison discharges are not needed to separate the modulated current drive from the unmodulated noninductive currents since the act of detrending the MSE data to remove the non-oscillating component serves this purpose.

There are several practical complications in using Eq. (3) to measure the current drive profile. First, modulating T_e along with the driven current adds a phantom source to the right hand side of Eq. (3) through the $\tilde{\eta}$ term. This may be unavoidable if one wishes to measure the bootstrap current profile by modulating the pressure profile, but for measurements of an external current source it is preferred to minimize $\tilde{\eta}$ by using a push/pull setup where co- and counter-injecting current sources at the same deposition location alternate during each cycle so that the total heating power remains constant with time. Second, the MSE diagnostic actually measures $\tilde{B}_z$ at fixed (R,z) coordinates rather than at a fixed toroidal flux. Therefore, for moving flux surfaces the contribution of the electric field to the left hand side of Eq. (1) should be written as

$$\frac{\partial \psi}{\partial t} = \left.\frac{\partial \psi}{\partial t}\right|_{R,z} + \frac{\rho_b^2 B_{\phi,0}}{q} \rho \left.\frac{\partial \rho}{\partial t}\right|_{R,z} - \rho \frac{\partial \psi}{\partial \rho}\left(\frac{1}{\rho_b}\right)\frac{\partial \rho_b}{\partial t} \approx \left.\frac{\partial \psi}{\partial t}\right|_{R,z} + R\tilde{B}_z \frac{\partial R_0}{\partial t} \quad , \tag{6}$$

where the final relation is derived assuming concentric elliptical flux surfaces. Oscillations in the $\hat{F}$, $\hat{G}$, and $\hat{H}$ factors can give rise to additional phantom source terms in Eq. (3), but these are generally negligible.

To experimentally determine the modulated current drive, the Fourier transform is first taken of the oscillating MSE signals, and the resulting $\tilde{B}_z$ profile is integrated over major radius to convert it to $\tilde{\psi}$. Next $\tilde{\psi}$ is corrected for any periodic movement of the flux surfaces using Eq. (6). The constant of integration for $\tilde{\psi}$ can be taken from the measured oscillation in the surface loop voltage ($\tilde{V}_{\phi,b}$). Alternatively, for off-axis current drive the boundary condition $\tilde{\psi}(\rho=0) = 0$ works well. In the high frequency limit, which is usually approached since $\hat{D}$ is small except near the plasma boundary, the modulated current drive profile is found from

$$\langle \tilde{J}_{NI} \rangle = \frac{i\omega}{R_0 \eta \hat{H}} \tilde{\psi} - \frac{\tilde{\eta}}{\eta} \langle J_{oh} \rangle = \frac{i\omega}{R_0 \eta \hat{H}} \left(-\int_{R_b}^{R} R \tilde{B}_z \, dR + \frac{i}{2\pi\omega} \tilde{V}_{\phi,b} \right) - \frac{\tilde{\eta}}{\eta} \langle J_{oh} \rangle \quad , \quad (7)$$

where R_b is the major radius of the plasma boundary on the low field side. Since the current profile does not have time to change significantly during the modulation period in the high frequency limit, the driven current profile is being determined in effect by the measured back emf. The above equation shows that $\tilde{\psi}$ mirrors the modulated current drive profile (with a 90 deg phase lag); interestingly, the MSE diagnostic actually records a null measurement ($\tilde{B}_z = 0$) at the peak current drive location.

MODULATED ECCD EXAMPLE

The direct measurement of the modulated current drive is demonstrated using the DIII-D discharge shown in Fig. 1, which utilized 3.0 MW of alternating co/counter electron cyclotron current drive (ECCD) at 10 Hz. There was a small residual modulation in T_e because the co and counter heating powers were not matched exactly. This caused a small undesired modulation in η that is neglected in the analysis presented here. The expected MSE response, determined by numerically solving Eq. (3), is shown in Fig. 2 where the amplitude and phase of $\tilde{B}_z$ at the fundamental modulation frequency are plotted as a function of the major radius. The modulated ECCD source had a peak deposition at $\rho = 0.22$, which corresponds to $R = 1.59$ m on the plasma inboard midplane and $R = 1.92$ m on the plasma outboard midplane. The largest value of $\tilde{B}_z$ should occur at locations that have a strong spatial gradient in the ECCD source; it also should be noted that the phase of $\tilde{B}_z$ changes by 180 deg across the major radius of the ECCD peak. The measured MSE response to the modulated ECCD was in qualitative agreement with the numerical simulation, as shown in Fig. 2.

The first direct measurement of the flux-surface-average ECCD profile ($\langle J_{EC} \rangle$) using the modulation method is shown in Fig. 3. Since $\langle J_{NI} \rangle$ found from Eq. (7) actually represents the 0-to-peak amplitude for the sinusoidal component at a single frequency, the magnitude needs to be multiplied by $\pi/2$ to yield the peak-to-peak

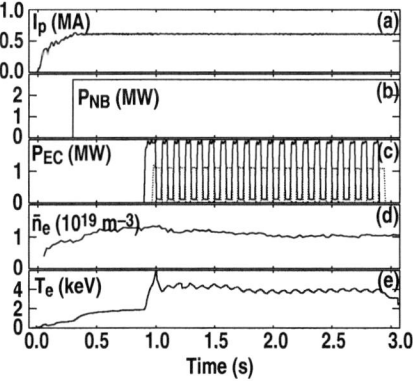

Figure 1. Time history of DIII-D discharge 115425: (a) plasma current, (b) neutral beam power, (c) alternating co-ECCD (solid lines) and counter-ECCD (dotted lines) powers, (d) line average electron density, and (e) central electron temperature.

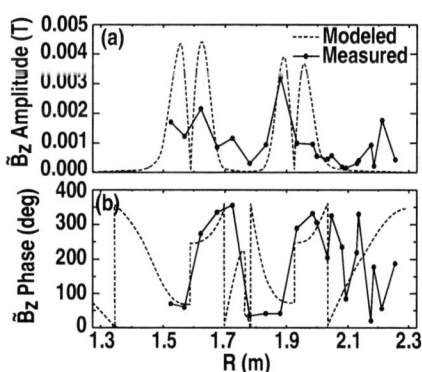

Figure 2. (a) Amplitude and (b) phase (relative to ECCD) of the oscillating vertical magnetic field measured by MSE polarimetry as a function of major radius for DIII-D discharge 115425. Both the modeled (dashed curves) and measured (solid curves and symbols) MSE responses are shown.

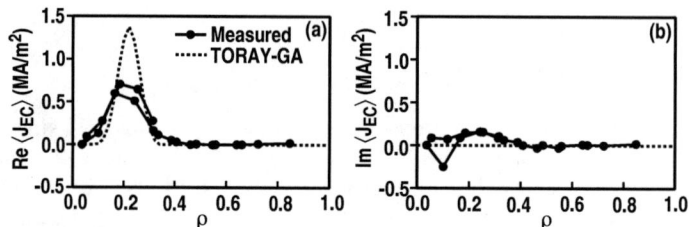

Figure 3. (a) Real and (b) imaginary components of the flux-surface-average ECCD profile as a function of normalized toroidal flux coordinate for DIII-D discharge 115425. Both the experimental (solid curves and symbols) and theoretical (dashed curves) ECCD profiles are plotted.

swing in the driven current density for square wave modulation. The real part of $\langle J_{EC} \rangle$ shown in Fig. 3(a) is the component that is in phase with the modulated ECCD source, whereas the imaginary part of $\langle J_{EC} \rangle$ plotted in Fig. 3(b) is the (90 deg) out of phase component that ideally should be zero. Figure 3(a) shows that localized current drive was measured at the predicted ECCD location, and there was good overlap between the MSE points on the inboard and outboard sides of the axis. Integrating the experimental ECCD profile gives a total driven current (co minus counter) of 0.12 MA, which is in agreement with the predicted ECCD of 0.14 MA from the TORAY-GA ray tracing code [5,6]. However, the TORAY-GA code predicts a more narrow ECCD profile than measured. This could be due to a steering misalignment between the six gyrotrons, or radial diffusion of the current carrying electrons.

CONCLUSIONS

A new method has been developed to directly measure the local current drive profile from internal magnetic measurements such as those from MSE polarimetry. In this approach, the noninductive current profile is determined from the periodic response of the MSE signals to a slow modulation of the current drive source using the poloidal flux diffusion equation. (To measure the bootstrap current profile, the plasma pressure can be slowly modulated.) Corrections to the raw MSE data may be needed to account for oscillations in the plasma flux surfaces and resistivity. The first direct measurement of localized current drive was successfully demonstrated using 10 Hz modulation of ECCD on the DIII-D tokamak.

ACKNOWLEDGMENT

Work supported by the U.S. Department of Energy under DE-FC02-04ER54698, DE-FG03-99ER54541, DE-FG02-86ER53223, W-7405-ENG-48, and the National Undergraduate Fellowships in Plasma Physics and Fusion Energy Sciences.

REFERENCES

1. F.M. Levinton, et al., Phys. Rev. Lett. **63**, 2060 (1989).
2. B.W. Rice, et al., Phys. Rev. Lett. **79**, 2694 (1997).
3. C.B. Forest, et al., Phys. Rev. Lett. **73**, 2444 (1994).
4. C.C. Petty, et al., Nucl. Fusion **41**, 551 (2001).
5. K. Matsuda, IEEE Trans. Plasma Sci. **17**, 6 (1989).
6. Y.R. Lin-Liu, et al., in Controlled Fusion and Plasma Physics (Proc. 26th Euro. Conf., Maastricht, 1999), Vol. 23J (European Physical Society, Geneva, 1999) p. 1245.

RF Current Drive in Internal Transport Barrier

Y. Peysson, J. Decker[*], V. Basiuk, A. Bers[*],
G. Huysmans, and A. K. Ram[*]

*Association Euratom – CEA, CEA/DSM/DRFC
CEA-Cadarache, 13108 St Paul-lez-Durance, France*
[*]*Plasma Science and Fusion Center, Massachusetts Institute of Technology,
Cambridge, MA 02139 U.S.A.*

Abstract The current drive (CD) problem in regimes with internal transport barrier (ITB) is addressed using a fast solver of the electron drift kinetic equation which may be used for arbitrary tokamak plasma magnetic equilibrium and any type of electron radio-frequency (RF) wave. Parametric studies are performed for the Lower Hybrid (LH) and Electron Cyclotron (EC) waves.

1. INTRODUCTION

Preliminary studies have shown that distortions of the electron momentum distribution function by radial drifts due to magnetic field gradient and curvature could interplay with RF quasilinear diffusion, thus leading to expect a significant gain of plasma current by comparison to usual Fokker-Planck (FP) calculations for similar wave parameters [1]. A favorable scaling with the local electron pressure gradient suggests to investigate this effect in the vicinity of strong transport barrier, a region where current profile is known to play a crucial role [2]. Besides the possible additional benefits for the CD efficiency, it is necessary to evaluate this current profile accurately. The recent development of a fast relativistic solver for the electron drift kinetic equation dedicated to the CD problem for arbitrary magnetic equilibrium and LH or EC type of RF waves has open the possibility to investigate in detail this effect which is specific to advanced scenarios now routinely achieved in many tokamaks [3].

2. CODE DESCRIPTION

A self-consistent description of the RF driven current with the effect of radial drifts is obtained from the steady-state drift kinetic equation (DKE),

$$v_s \frac{\partial f}{\partial s} + \frac{v_\parallel}{\Omega} I(\psi) \frac{|\nabla \psi|}{R} \frac{\partial}{\partial s}\left(\frac{v_\parallel}{B}\right) \frac{\partial f}{\partial \psi} = C(f) + Q(f) + E(f)$$

where f is the electron distribution function, (s,ψ) are the generalized poloidal and radial (poloidal magnetic flux) coordinates, v_s the poloidal component of the electron velocity along the magnetic field line and R is the toroidal major radius. The effects of collisions, RF quasilinear diffusion and Ohmic electric field acceleration are described respectively by the operators $C(f)$, $Q(f)$ and $E(f)$. In the small drift approximation, f is expanded as $f \simeq f_0 + \delta f_1 = f_0 + \tilde{f} + g$, where the small parameter δ is the ratio of the drift velocity v_D across magnetic field lines given by $v_D \cdot \nabla \psi = -\frac{v_\parallel}{\Omega} I(\psi) B \cdot \nabla\left(\frac{v_\parallel}{B}\right)$ to v_s. Here f_0 is the RF-modified distribution unperturbed by radial drift given by the usual bounce-averaged FP equation. Since in the banana regime, the collision detrapping time τ_{dt} is much shorter than the bounce time τ_b of trapped electrons, a subordering $\delta \ll \tau_b/\tau_{dt} \ll 1$ may be used to further expand and solve DKE in three steps as in [1],

$\{C(f_0)\}+\{Q(f_0)\}+\{E(f_0)\}=0$, then calculate $\tilde{f}=\frac{v_\parallel}{\Omega}I(\psi)\frac{\partial f_0}{\partial \psi}$ and finally $\{C(g)\}+\{Q(g)\}+\{E(g)\}=-\{C(\tilde{f})\}-\{Q(\tilde{f})\}-\{E(\tilde{f})\}$. Here, $\{...\}$ denotes the bounce-averaging, $v_\parallel$ is the electron velocity along the magnetic field line and Ω the electron gyrofrequency. The self-consistent flux surface-averaged current $\langle J \rangle$ and power $\langle P \rangle$ are then calculated. Without external RF perturbation, f_0 is Maxwellian and the thermal electron bootstrap current $\langle J_B \rangle$ deduced from neoclassical theory is well recovered [4]. In order to evaluate the role of radial drifts on CD, the gain in current is calculated by the relation $\langle \Delta J \rangle = \langle J \rangle - \langle J_{RF} \rangle - \langle J_B \rangle$, where $\langle J_{RF} \rangle$ is the solution of the FP equation without drift corrections. The figure of merit for the CD given by $\eta = (\langle J \rangle - \langle J_B \rangle)/\langle P \rangle$ is compared to the value $\eta_{RF} = \langle J_{RF} \rangle/\langle P_{RF} \rangle$ deduced from FP calculations.

3. RESULTS

Calculations are performed for a plasma configuration of the tokamak C-MOD [5], with electron densities, temperatures profiles given in Fig. 1, using numerical magnetic equilibrium solver HELENA [6]. The effective charge profile is taken flat, at $Z_{eff}=1$. The ITB is located at a normalized minor radius $\rho=0.7$, and by varying the local pressure gradient ∇P_e, the electron contribution to the bootstrap current, which represents 70% of the total value varies from $\langle J_B^e \rangle = 9.4$ to $17.5 MA.m^{-2}$.

A simplified model is considered for the LH wave CD problem, with a constant power spectrum in $n_\parallel$ between two limits usually fixed by strong Landau Damping and accessibility. In momentum space, they correspond roughly to $v_{\parallel min}=3.5 v_{th0}$ and $v_{\parallel max}=6 v_{th0}$ respectively, where v_{th0} is the thermal velocity in the center of the plasma. The normalized LH quasilinear diffusion coefficient is chosen $D_{LH}=0.5 v_{e0} p_{th0}^2$, where v_{e0} is the collision frequency and p_{th0} the thermal momentum taken at $\psi=0$. A parametric study shows in Fig. 2 a linear dependence of the ratio $\langle \Delta J \rangle/\langle J_{LH} \rangle$, which predominantly arises from temperature gradient, as observed for weak pressure gradient calculations [7]. As expected, $\langle \Delta J \rangle/\langle J_{LH} \rangle$ may reach large values, when the ITB is strong, up to nearly 60% in the examples here considered, thus confirming previous tendencies. Interestingly, $\langle \Delta J \rangle/\langle J_{LH} \rangle$ is found to be independent of D_{LH} ranging from 0.01 to 2 (saturated quasilinear regime with a flat tail), which means that usual FP calculations neglecting drift corrections may strongly underestimate the driven current at high input power level when $\langle J_{LH} \rangle$ is large. Regimes where $\langle \Delta J \rangle + \langle J_{LH} \rangle$ exceeds $\langle J_B^e \rangle$ may be easily found while $\langle J_{LH} \rangle \leq \langle J_B^e \rangle$ for similar wave parameters. This result has favorable consequence for the ability to control the current density profile in the region of strong ∇P_e where the bootstrap current is large. It emphasizes the importance of drift kinetic calculations for an accurate description of the current density profile in ITB. While synergistic effects vanish at the top and foot of the ITB, as shown in Fig. 2, the effect of impurity is found less deleterious than for the standard Fisch-Boozer CD characterized by the well known $4/(5+Z_{eff})$ scaling. The drop of $\langle J_{RF} \rangle$ is almost fully compensated by the increase of $\langle \Delta J \rangle$ which results from the improvement of the electron density in the vicinity of the trapped-passing boundary in momentum space, because of pitch-angle scattering. Current generation in ITB is therefore a robust process regarding impurity level for fixed wave parameter, an effect which can be only described by DKE. The CD efficiency η is found to scale like $\langle \Delta J \rangle$, though the relative gain is lower, as shown in Figs. 2 and 3. Since the LH power deposition is usually broad, the benefit on η is expected to be almost negligible on the total input power required to drive then same amount of current in the plasma. Indeed, the plasma volume concerned by the improvement of η is small, due the strong localization of the ITB.

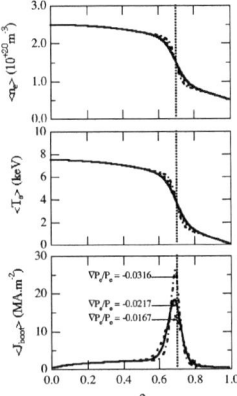

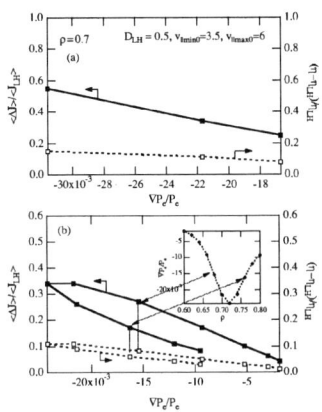

Fig. 1. Electron density (top), temperature (middle) and total (electron + ion, $T_i = T_e/2$) thermal bootstrap current (bottom) profiles. The ITB is located at $\rho = 0.7$ for three different normalized electron pressure profiles.

Fig. 2. (a) Ratio of the synergistic to the LH current and relative increase of the CD efficiency as function of the electron pressure gradient at $\rho=0.7$. (b) Localization of the synergistic effect for the LH wave in the vicinity of the ITB. Profiles correspond to full lines in Fig.1.

The situation is completely different for the Electron Cyclotron (EC) wave, since the width of the region of strong absorption may be comparable or even narrower than the ITB itself. A possible gain of η has therefore a direct impact on the required power level for driving the current. For the study, here just addressed to investigate the difference with the LH wave since the EC wave is not planned to be used on C-MOD, a simplified model is also considered, assuming a Gaussian power spectrum centered around $n_{\parallel 0}$ with a width $\Delta n_\parallel = 0.02$. A quasi-X mode for which wave-particle interaction takes place near the second harmonic of the cyclotron frequency, $\omega \approx 2\Omega$ is investigated, since it corresponds to usual CD conditions encountered in most tokamak experiments. When the beam is assumed to be launched horizontally in the mid-plane from the low field side (poloidal angle $\theta_b = 0$), Ohkawa CD predominates at $\rho = 0.7$, because the quasilinear diffusion lies closely to the trapped-passing boundary in momentum space [8]. Consequently, the parallel index is taken negative, $n_{\parallel 0} = -0.3$, so that EC and bootstrap currents have the same positive sign. Since $n_\parallel$ remains approximately constant when the wave propagates across the resonance region, the resonance in momentum space depends mostly upon the variations of the ratio $2\Omega/\omega$. For $2\Omega/\omega = 0.98$, the power deposition is found maximum around the ITB, but $\langle \Delta J \rangle / J_{EC}$ never exceeds 10% for the largest value of ∇P_e, as $(\eta - \eta_{EC})/\eta_{EC}$ (see Fig. 4). The smallness of the relative gain may be explained by the fact that EC wave in Ohkawa regime and the radial drift equally contribute to reduce the electron density in

271

the region $p_\parallel \leq 0$ close to the trapped-passing. Consequently even if $\langle J_{EC} \rangle$ and $\langle J_B^e \rangle$ have the same sign, kinetic effects of the two mechanisms nearly cancel, and beneficial effects are only marginal for the ECCD from LFS in transport barrier closed to the edge [8].

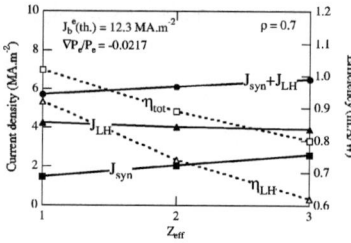 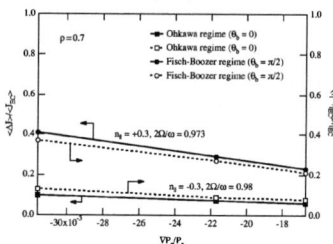

Fig. 3. Effect of impurity content on plasma current in ITB for the LH wave.

Fig. 4. Ratio of the synergistic to the EC current and relative increase of the CD efficiency as function of the electron pressure gradient at $\rho=0.7$ for the Fisch-Boozer and Ohkawa regime.

Possible significant improvements of $\langle \Delta J \rangle$ may only be foreseen when the Fisch-Boozer CD mechanism predominates. This situation may be obtained for rather central ITB and if the barrier is located in the outer-half of the plasma section, by launching the EC wave vertically ($\theta_b = \pi/2$) or from the high field side region ($\theta_b = \pi$) ($2\Omega/\omega = 0.973$ at the maximum of absorption). For $\theta_b = \pi$, the resonance domain becomes usually too far from the region where an excess of electrons is produced in momentum space by the radial drift. Even if η_{EC} is large (0.12 m.A/W), $(\eta - \eta_{EC})/\eta_{EC}$ is small whatever the strength of the ITB (less than 15%). For $\theta_b = \pi/2$, an intermediate situation may be found, with an lower efficiency η_{EC} (0.05 m.A/W) but $(\eta - \eta_{EC})/\eta_{EC}$ may be much larger, reaching 40%. Conversely to the case of the LH wave, $(\eta - \eta_{EC})/\eta_{EC}$ and $\langle \Delta J \rangle / J_{EC}$ are of the same order of magnitude (Fig. 4). This effect may be ascribed to the perpendicular diffusion of electrons in momentum space by resonant interaction with the EC wave. This mechanism contributes to enhance interaction with the region closed to the trapped-passing boundary, as done by enhanced pitch-angle scattering when $Z_{eff} \geq 1$, a mechanism which is also found very favorable for the synergy between bootstrap and EC wave. Finally, $\langle \Delta J \rangle$ is always found positive, which means that counterbalancing bootstrap current by external means leads always to a decrease of the CD efficiency.

REFERENCES

1. S.D.Schultz, et al., in AIP Conf. Proc. 485, N.Y. (1999) p.317
2. Y. Peysson et al., in Advances in Plasma Phys. Research, Vol.4, Nova Eds. (2003) p.1
3. J. Decker and Y. Peysson, Euratom-CEA report EUR-CEA-FC-1736 (2004).
4. O. Sauter et al., Phys. Plasma, 6 (7) 2834 (1999).
5. J. Decker et al. In Proc. EC-12 Conf., Aix-en-Provence, France (2002)
6. G.T.A. Huysmans et al. in Proc. CP90 Conf., Singapore Word Scientific (1991) p.371
7. J. Decker et al., in 29th EPS Conf., Montreux, Switzerland (2002).
8. J. Decker in AIP Conf. Proc. 694, N.Y. (2003) p.447, Y. Peysson et al., in Proc. 19th IAEA Fusion Energy Conf., Lyon, France (2002).

Fast Wave Current Drive in JET ITB-Plasma

T. Hellsten[1], M. Laxåback[1], T. Bergkvist[1], T. Johnson[1], M. Mantsinen[5], G. Matthews[6], F. Meo[2], F. Nguyen[3], J.-M. Noterdaeme[8,12], C. C. Petty[4], T. Tala[6], D. Van Eester[9], P. Andrew[7], P. Beaumont[7], V. Bobkov[8], M. Brix[8], J. Brzozowski[1], L.-G. Eriksson[3], C. Giroud[7], E. Joffrin[3], V. Kiptily[7], J. Mailloux[7], M.-L. Mayoral[7], I. Monakhov[7], R. Sartori[10], A. Staebler[8], E. Rachlew[1], E. Tennfors[1], A. Tuccillo[11], A. Walden[7], K.-D. Zastrow[7] and JET-EFDA contributors[13]

[1]Association VR-Euratom, Sweden, [2]Association Euratom-Risø, National Laboratory, Denmark, [3]Association Euratom-CEA, CEA-Cadarache, France, [4]General Atomics, San Diego, USA. [5]HUT, Association Euratom-Tekes, Finland, [6]VTT Processes, Association Euratom-Tekes, Finland, [7]UKAEA Fusion Association, Culham Science Centre, U.K., [8]Max-Planck-Institut für Plasmaphysik, EURATOM Association, Garching, Germany, [9]LPP-ERM/KMS, Association Euratom-Belgian State, Belgium, [10]EFDA CSU-Garching, Germany, [11]ENEA, Association Euratom-ENEA, Italy, [12]EESA Department, University of Gent, Belgium, [13]see appendix of J. Pamela et al., Proc. 20th IAEA Fusion Energy Conf. 2004 Villamoura.

Abstract. Fast wave current drive has been performed in JET plasmas with internal transport barriers, ITBs, and strongly reversed magnetic shear. Although the current drive efficiency of the power absorbed on the electrons is fairly high, only small effects are seen in the central current density. The main reasons are the parasitic absorption of RF power, the strongly inductive nature of the plasma and the interplay between the fast wave driven current and bootstrap current. The direct electron heating in the FWCD experiments is found to be strongly degraded compared to that with the dipole phasing.

Keywords: Fast Wave Current Drive, ICRH, ITB.
PACS: 52.55.Wq, 52.40.Fd, 52.40.Kh, 52.50.Qt, 52.50.Sw, 52.55.Fa.

EXPERIMENTAL RESULTS

Fast wave current drive experiments have been carried out in JET ITB plasmas. The magnetic field at the axis was 3.45T, plasma current 2MA, elongation 1.65, minor radius $a\approx0.9$m and magnetic axis $R_0\approx3$m. Hot low density plasmas, $n_e\approx1.2\times10^{19}$m^{-3}, with strongly reversed magnetic shear were produced with 2-2.5MW LHCD, which was switched off when around 13MW of NBI and up to 6MW of ICRF power at 37MHz were applied. During the NBI and RF-heating, the plasma current and toroidal field were kept constant. The time sequence of the coupled power and some plasma parameters for a typical discharge can be seen in Fig. 1 starting at 40s. The antennas were phased at +90° and −90° producing asymmetric toroidal mode spectra peaked at $n_\phi\approx+15$ and $n_\phi\approx-15$, respectively. During the initial main heating phase the ion temperature increased to 12keV and the electron temperature to around 8keV. Before

47.4s the outermost part of the internal transport barrier was located at about $R=3.25$m, $r/a=0.28$. At 47.4s q_{min} reached 2 and the barrier expanded, following the outer $q=2$ surface to 3.45m, $r/a=0.5$. In this improved confinement regime the central ion temperature increased to over 20keV and the neutron yield triggered a step down of the NBI power to avoid a disruption. The heating performance was compared to reference discharges with dipole phasing at 37MHz and H-minority heating with +90° at 51MHz and $n_H/n_D=0.06$. The reference scenarios had better heating.

Square-wave modulations of the RF-power were performed to measure the direct electron heating. The fraction of RF power directly absorbed on electrons varied strongly with plasma conditions and antenna phasing. The difference in heating efficiency between different phasings is clearly seen when comparing measured direct electron heating power deposition profiles for a triple of discharges that had similar electron temperatures and densities despite different levels of coupled RF power, Fig. 2. The heating efficiency is reduced to about 50% for -90° phasing and to about 60% for +90° compared to the dipole phasing. Parts of the discrepancies were due to parasitic damping by residual ^{3}He ions, whose presence was confirmed by several methods. A significant fraction of the RF power was not absorbed and transferred to the bulk plasma for ±90°. The time integration of this fraction of power was obtained from the balance between the energy delivered by the heating systems, including ohmic heating and excluding beam shine-through, and the energy delivered to the divertor and radiated from the plasma. The energy delivered to the divertor was measured with 6 thermocouplers and the energy radiated from the plasma with a bolometer camera. For discharges dominated by ICRH the method offers a good measure of the lost power, as for #58680 with +90° at 37MHz without NBI; 14MJ of energy could not be accounted for, corresponding to 48% of the injected RF energy and being well above the error bars of ±13%. Likely causes of the lost RF power are losses of RF-heated high-energy ions intercepted by the limiters and energy dissipated in rectified RF-sheaths. For the reference discharge with hydrogen minority heating at 51MHz with +90° phasing only 8MJ (7%) of the 109MJ delivered by the heating systems could not be accounted for.

Visible light spectrometers were used to study the intensity in the CIII, CIV and BeII lines in order to assess the level of sputtering or arcing taking place at the antenna Faraday screens and limiters. For the reference discharges with dipole phasing or hydrogen minority heating a steady BeII line radiation intensity of the order of 3×10^{12} photons/s/cm^2/sr was observed with the spectrometers viewing the divertor. For the FWCD experiments with ±90° phasings the radiation intensity varied strongly during the pulses with large spikes, Fig. 1d.

Studies of the effects by FWCD on the central current and heating could be done by comparing a pair of discharges #60664 and #60663 that had different RF-power, but similar temperatures and densities in the early part of the main heating phase, 44-46s. When the LHCD is switched off and the NBI and RF are applied the density increases and the electron temperature initially decreases. The central current density also increases due to increased current diffusion and possibly also due to small tearing mode activity in the reversed magnetic shear region. When the density barrier is established at 44.3s the central current density decreases as the bootstrap current in the barrier builds up. After 45s the central current starts to increase due to the inward diffusion of poloidal flux and due to tearing mode activity in the reversed magnetic shear region. A small but clear difference in the central current density, measured from eight lines-of-sight of the far-infrared polarimeter [1], could be seen, Fig. 3, which because of the similar electron temperatures is not expected to be caused by different current diffusion rates.

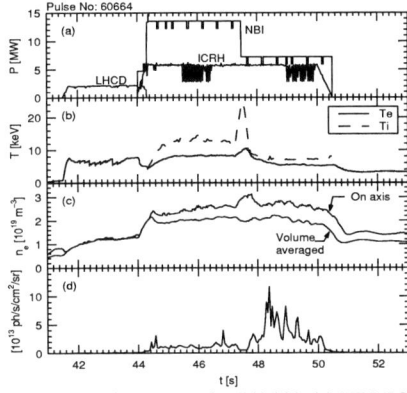

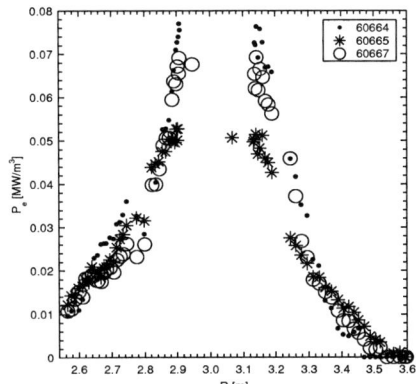

Figure 1. Time traces for #60664. (a) NBI, ICRF and LHCD power, (b) central electron, T_e, ion T_i, temperature, (c) on axis and volume averaged electron density, n_e, and (d) BeII line radiation.

Figure 2. Power deposition by direct electron heating between 45.5 and 46.5s for #60664 with 5.3MW $-90°$, #60665 with 4.8MW $+90°$ and #60667 with 2.9MW dipole.

MODELING OF HEATING AND CURRENT DRIVE

Due to the difficulty in determining the ^{3}He concentration, crucial for modeling the power deposition, the driven currents were calculated with the LION code [2] for a reconstructed experimental equilibrium with the power normalized so that the calculated power absorbed by electrons agreed with that measured. The plasma current response could then be calculated and compared to the measured difference in central current density between discharges with co and counter current drive. For the discharges #60664 and #60665 9-10% of the total power was absorbed on electrons in the centre of the plasma, which according to LION corresponds to a total electron absorption of around 17%. The power absorbed by TTMP/ELD was for $+90°$ 0.8MW and for $-90°$ 1MW, yielding driven currents of -55kA and 70 kA respectively, and current drive efficiencies of the order 0.07A per watt absorbed on the electrons. The effect of the current drive on the evolution of the central plasma current was simulated with the JETTO code [3]. To quantify the effects of the current drive the same discharge was simulated for 2.6s with both co and counter current drive $+70$kA and -55kA, respectively, and without current drive for reference. The plasma current profiles due to RF, NBI and bootstrap currents and poloidal flux diffusion are shown in Fig. 4. The total plasma current in the centre changes only with a small fraction of the RF driven current. After 2.6s the differences in current density inside $r/a = 0.3$ compared to the reference simulation without current drive are of the order $+10$kA/m^2 to $+30$kA/m^2 for co current drive and -10kA/m^2 to -15kA/m^2 for counter current drive. This should be compared to the driven current densities calculated with LION, which are around

+80kA/m^2 and -70kA/m^2 respectively at $r/a = 0.3$. Owing to the inductive nature of the plasma current the effect of the driven current is not seen until the back EMF diffuses away. In addition, the RF driven current is also partly compensated for by an opposite change in the bootstrap current due to the dependence of the bootstrap current on the poloidal field. Halfway into the simulation, 46s, the difference in the calculated central current densities is only between 10kA/m^2 and 40kA/m^2 between co and counter current drive. Measurements of the central current densities however showed a difference in current density of around 100kA/m^2, Fig. 3, indicating that poloidal flux diffusion takes place on a faster time scale than that by neo-classical resistivity.

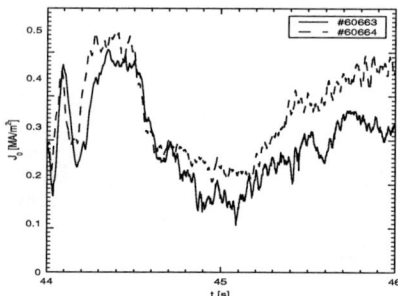

Figure 3. Central current density derived from for discharges #60663 with +90° and #60664 with −90° phasing.

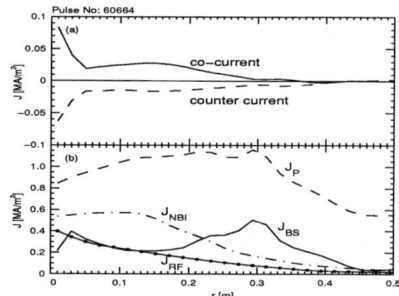

Figure 4. Calculated plasma current density response (a) and the total plasma, RF- and NBI-driven and bootstrap current density (b) after 2.6s of co current drive.

CONCLUSIONS AND DISCUSSIONS

A clear difference in the heating efficiency with respect to the antenna phasing was seen. The dipole phasing had the highest efficiency, comparable to hydrogen minority heating. Smaller differences in direct electron heating, lost power, production of impurities, fast ions and gamma-rays were seen with respect to ±90° phasings for 37MHz, and are consistent with the effect of the RF-induced spatial transport of ^{3}He ions. The modelling with the JETTO code demonstrates the difficulty in affecting the current in the transport barrier because of the interplay between the driven current and the bootstrap current and because of the inductive nature of plasmas with high electron temperatures and the resulting long diffusion times for the poloidal flux. The measured difference in central current density during co and counter current drive is much larger than the difference in the response of the plasma current as calculated with the JETTO code suggesting a current penetration faster than that given by neo-classical resistivity.

REFERENCES

1. Mazon, D., *et al Plasma Physics and Controlled Fusion* **44** 1087 (2002).
2. Villard, L., *et al Computer Physics Reports*, **4** 95 (1986) and *Nuclear Fusion* **35** 1173 (1995).
3. Tala, T., *et al* 2000 *Nuclear Fusion* **40** 1635 (2000).

LOWER HYBRID RANGE OF FREQUENCIES

Recent RF Experiments and Application of RF Waves to Real-Time Control of Safety Factor Profile in JT-60U

T. Suzuki, A. Isayama, S. Ide, T. Fujita, T. Oikawa, S. Sakata, M. Sueoka, H. Hosoyama, M. Seki, and the JT-60 Team

Japan Atomic Energy Research Institute, Naka Fusion Research Establishment, Japan

Abstract. Two topics of applications of RF waves to current profile control in JT-60U are presented; application of lower-hybrid (LH) waves to safety factor profile control and electron cyclotron (EC) waves to neo-classical tearing mode (NTM) control. A real-time control system of safety factor (q) profile was developed. This system, for the first time, enables 1) real time evaluation of q profile using local magnetic pitch angle measurement by motional Stark effect (MSE) diagnostic and 2) control of current drive (CD) location (ρ_{CD}) by controlling the parallel refractive index $N_{//}$ of LH waves through control of phase difference ($\Delta\phi$) of LH waves between multi-junction launcher modules. The method for real-time q profile evaluation was newly developed, without time-consuming reconstruction of equilibrium, so that the method requires less computational time. Safety factor profile by the real-time calculation agrees well with that by equilibrium reconstruction with MSE. The control system controls ρ_{CD} through $\Delta\phi$ in such a way to decrease the largest residual between the real-time evaluated q profile q(r) and its reference profile $q_{ref}(r)$. The real-time control system was applied to a positive shear plasma (q(0)~1). The reference q profile was set to monotonic positive shear profile having $q_{ref}(0)=1.3$. The real-time q profile approached to the $q_{ref}(r)$ during application of real-time control, and was sustained for 3s, which was limited by the duration of the injected LH power. Temporal evolution of current profile was consistent with relaxation of inductive electric field induced by theoretical LH driven current. An m/n=3/2 NTM that appeared at β_N~3 was completely stabilized by ECCD applied to a fully-developed NTM. Precise ECCD at NTM island was essential for the stabilization. ECCD that was applied to resonant rational surface (q=3/2) before an NTM onset suppressed appearance of NTM. In order to keep NTM intensity below a level, ECCD before the mode onset was more effective than that after mode saturation.

Keywords: real time control, safety factor profile, motional Stark effect (MSE) diagnostic, lower-hybrid current drive (LHCD), neo-classical tearing mode (NTM), electron cyclotron current drive (ECCD)
PACS: 52.55.Fa, 52.55.Wq, 52.35.Hr, 52.50.Sw, 52.55.Tn, 52.35.Py

INTRODUCTION

Recent progress of performance of tokamak plasma has been achieved utilizing current profile control/optimization [1]. In the studies, emphasis is placed on steady current profile optimized for stability and confinement [2]. Various current drivers are used to establish the steady current profile. RF waves in tokamaks are applied not only for heating but also, and more importantly, for non-inductive current drive (CD). JT-60U [3] has two different RF heating/CD systems at different frequency ranges: lower-hybrid and electron-cyclotron ranges of frequency (LHRF, ECRF). Both of them were proved to be effective current drivers. Their driven current profiles were studied in detail [4,5]. While ECCD drives spatially localized current in a scale smaller than minor radius [4], LHCD current profile has a scale structure as large as minor radius. We can easily control ECCD location, as desires, where ray trajectory crosses to the EC resonance. Absorption location of LH waves can be controlled by control of refractive index parallel to equilibrium magnetic field ($N_{//}$) in front of wave launcher. Advantage of the LHCD is its quite high CD efficiency (up to 3.6×10^{19} A/W/m^2 recorded in JT-60U), while ECCD have its strong controllability of local current density. Considering the advantages of the current drivers, we employed LHCD for large-scale safety factor profile (q(r)) control. ECCD is used to control neoclassical tearing mode (NTM) that limits attainable beta value ($\beta_N \equiv \beta_t a B_t / I_p$). The NTM can be stabilized/suppressed by compensating missing bootstrap current due to flattening of pressure gradient inside magnetic island. Thus, spatially localized current drive is necessary.

This paper handles two topics; the real-time control of safety factor profile by LHRF and NTM suppression/stabilization by ECRF. Takase presents one of other topics, plasma initiation by ECRF without use of central solenoid (or Ohmic heating) coils for reduction of reactor size, in this conference [6]. We must note that ECRF is also used to investigate thermal/magnetic property of plasma, such as "current hole" [7] and internal transport barrier structures.

Active feedback control of current profile is essential to obtain an optimized current profile for higher β_N. The optimization of the current profile could raise β_N limited by instabilities, such as NTM and sawtooth [2], expelling resonant rational surfaces to the modes from plasma. For this purpose, real-time feedback control system of current profile has been developed. We developed a new technique to quickly evaluate safety factor profile in real-time, using internal magnetic pitch angle measurement by MSE diagnostic [8]. Control of LH driven current profile controls q profile. The controllability of LHCD profile is examined. We propose a unique control logic that does not strongly depend on the choice of current driver. We examined the whole system in a low β plasma here. Final goal for the q(r) control system is to steadily sustain a high β plasma, where large bootstrap current closely links pressure and current profiles.

REAL-TIME SAFETY FACTOR PROFILE CONTROL

Diagnostics And Current Drivers

The real-time q(r) control system enables real time evaluation of q(r) using MSE and control of CD location by adjusting $N_{//}$ of LH. Real-time evaluation method of q(r) and controllability of LHCD location by $N_{//}$ are described here.

For the first time, q(r) has been evaluated in real time from the local measurement of pitch angle by MSE. The equilibrium was not reconstructed in real-time with MSE, but q(r) was estimated under an assumption that the shape of the last closed flux surface (LCFS) represents the shapes of internal magnetic surfaces. The following shape of magnetic surface was assumed, taking account of elongation κ, triangularity δ, and Shafranov shift profile $\Lambda(\rho) \equiv (R_p - R_{ax})\rho^2$;

$$\begin{cases} R = R_{ax} + \Lambda(\rho) + a\rho\cos(\theta + \delta\sin\theta), \\ Z = Z_{ax} + a\rho\kappa\sin\theta. \end{cases} \quad (1)$$

Cauchy condition surface (CCS) method [9] calculates the LCFS and global parameters, such as major radius of geometric center R_p, major radius (vertical position) of the magnetic axis R_{ax} (Z_{ax}), a, κ, and δ, every 1ms. The CCS method uses acquired data from external magnetics in real-time, that is, poloidal magnetic fields, poloidal magnetic fluxes, and external coil currents. The magnetic axis is a weighted center of the plasma current in the CCS method. Safety factor at each MSE channel location (R_{MSE}, Z_{MSE}) is calculated on the nested magnetic surfaces (eq. (1)), using local pitch angle $\gamma(R_{MSE}, Z_{MSE})$ measured by MSE diagnostic every 10ms. Currently, real-time processor can handle 16 MSE channels. Since the MSE measurement points in JT-60U (and in most tokamak devices) are near the equatorial plane, eq. (1) well approximates normalized minor radius ρ_{MSE} at (R_{MSE}, Z_{MSE}), and we have $\gamma(\rho_{MSE})$. The safety factor is calculated neglecting divertor configuration at separatrix, as in the following formula,

$$q(\rho) = \sqrt{\frac{1+\kappa^2}{2}} \frac{a\rho}{R_{ax}} \frac{1}{\tan(\gamma(\rho))}. \quad (2)$$

Whole of the above calculation finishes within 10ms, which is cycle of real-time control system and much shorter than time scale for global current profile change (O~1s). Figure 1 shows safety factor by real-time evaluation, in comparison with that by equilibrium reconstruction, at all used channels in various positive magnetic shear discharges. They show a good agreement in a wide range of plasma parameters (I_p=0.6-1.0MA, B_t=1.7-2.4T, β_p=0.3-0.8). Since the technique does not require time-consuming solution of Grad-Shafranov equation, it is applicable to various computational environments. Since the safety factor is inversely proportional to the poloidal magnetic field, which is an area-integral of current density, we can also evaluate current density profile in real-time. In case there is an external current driver, we can simply estimate the current drive location from the temporal evolution of the real-time current profile. The CD location ρ_{CD} is defined as a radius where temporal increment of current becomes the largest, under a quasi-steady state of global Ohmic current relaxation. This method to detect the CD location is similar to what we proposed to detect ECCD location in references [10], except that the detection of ρ_{CD} is performed in real-time. Thus, we can obtain q(r) and ρ_{CD} in real-time.

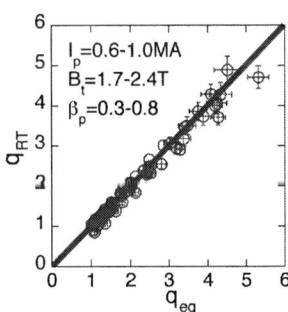

FIGURE 1. Comparison of the safety factor by the real-time calculations (q_{RT}) and by the equilibrium reconstructions (q_{eq}) under wide variety of plasma parameters (I_p, B_t, and β_p).

LHRF system is briefly described here; see ref. [11] in detail. Multi junction (MJ) antenna launches LH waves (2GHz) from outboard mid-plane of JT-60U. Eight MJ modules constitute the launcher (4x2 in toroidal and vertical, respectively). Each module has 12 waveguides. Phase difference Δφ of LH waves between the

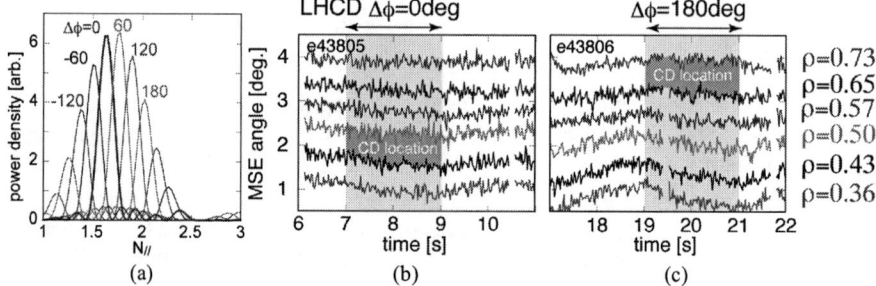

FIGURE 2. (a): $N_{//}$ spectra in front of LH launcher for typical $\Delta\phi$. (b) and (c): temporal evolution of MSE angles at various locations. Normalized minor radius ρ at each MSE channel is indicated at the right of (c). An arrow above each figure indicates application duration of LHCD.

neighboring MJ modules is controlled in order to control $N_{//}$ spectrum in front of the launcher. Plasma control processor sends command value of $\Delta\phi$ every 10ms. The LH system accepts discrete $\Delta\phi$ by 5 degree, and locks $\Delta\phi$ in the command value within 2-3ms. The step of $\Delta\phi$ by 5 degree smoothly changes the $N_{//}$ spectrum, and the temporal response within 3ms was enough for the real-time control. Figure 2(a) shows power density spectrum of $N_{//}$ calculated under the antenna geometry. We can see $\Delta\phi$ controls $N_{//}$ spectrum and that $N_{//}$ at primary peak becomes larger with increase in $\Delta\phi$. Controllability of CD location by $N_{//}$ was investigated. Figure 2(b) shows temporal evolution of MSE signals that represent current enclosed in a magnetic surface at MSE location. LHCD was applied during t=7-9s at $\Delta\phi=0°$. Difference of MSE angles between two MSE channels at $\rho=0.43$ and 0.50 increases most during the LHCD, showing LHCD at $\rho_{CD}=0.43-0.50$. MSE signals for LHCD at $\Delta\phi=180°$ is presented in Fig. 2(c); LHCD location is $\rho_{CD}=0.65-0.73$ in this case. We found that CD location moved outward by increasing $\Delta\phi$ or $N_{//}$ in a range $0°<\Delta\phi<180°$. The control logic described below uses the result, outward shift of ρ_{CD} with increase of $\Delta\phi$.

Control Logics

Using the above components, we constructed a q(r) control system. The control system controls ρ_{CD} through $N_{//}$ (or directly $\Delta\phi$) in such a way to minimize the difference between the real-time q(r) and the given reference $q_{ref}(r)$. Figure 3 illustrates samples of control logic for q(r) control. The system compares the real-time q(r) and the reference $q_{ref}(r)$ with a weight w(r), and detects a location where a residual (q(r)-$q_{ref}(r)$)w(r) is the largest; the weight is constant in time and the weight profile is given considering accuracy of q measurement and importance on the control. For example, the weight at the plasma edge is small, since edge safety factor is not controllable; instead plasma current dominates the edge q. In order to reduce the

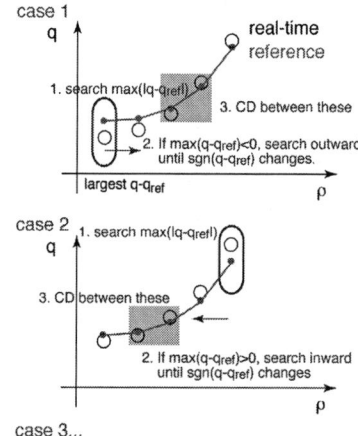

FIGURE 3. Schematics of real-time q profile control logic.

largest residual, control system determines a reference CD location ρ_{CDref} where a current driver should drive current. Thus, $\Delta\phi$ is feedback-controlled in order that CD location ρ_{CD} detected in real-time agrees with the reference location ρ_{CDref};

$$\frac{d}{dt}(\Delta\phi) = -\alpha(\rho_{CD} - \rho_{CDref}) \ . \qquad (3)$$

The proportional gain α [degree/s] is an adjustable positive constant. When ρ_{CD} is smaller than ρ_{CDref}, the system should increase ρ_{CD} (or $\Delta\phi$) so that the negative sign at right-hand-side of eq. (3) appears. We must note that the above control logic is not specific to actuator (here, LHCD). If any current driver has a controllable parameter to change its CD location, the current driver can be easily built into the system.

Finally, in case of LHCD, there is another controllable parameter, injection power P_{LH}. In order to keep plasma loop voltage constant, P_{LH} was controlled so as not to change LH driven current I_{LH}, even if $\Delta\phi$ or $N_{//}$ changed. Since LHCD efficiency $\eta_{CD}=I_{LH}n_eR_p/P_{LH}$ is proportional to $<1/N_{//}^2>$ [12], P_{LH} is proportionally controlled to $<1/N_{//}^2>^{-1}$. Here, $<1/N_{//}^2>$ is an average of $1/N_{//}^2$ weighted by power spectrum in Fig. 2(a).

Experimental Result

The real-time q(r) control system was applied to a plasma (E044082), having I_p=0.6MA, B_t=2.3T. Line averaged electron density was controlled to n_e=0.5x10^{19}m^{-3} by gas-puff. Gap between the launcher and the plasma surface was controlled fixed in order to obtain a good coupling of LH waves to the plasma. Figure 4 shows temporal evolution of the discharge with contour plot of q(r) in its bottom. The number of MSE channels used for control was 9 in this discharge, covering ρ~0.1-0.7. Reference q profile was set to a monotonic positive magnetic shear having q(0)=1.3; see filled rectangles in Fig. 5. Co-LHCD (maximum injection power P_{LH}=1MW) was applied after t=4s, and real-time control of q(r) has started at t=4.5s. At the initial phase of q(r) control (t<10s), LH power was not fully injected into the plasma due to interlocks against arc in LH antenna, although the control system requests 1MW injection. After t=10s, LH power was stably injected, decreasing loop voltage down to 0V. During the good-coupling phase, we see vanish in q=1 surface and shrink of q=1.25 surface in the contour plot of q in Fig. 4. Figure 5 shows q profile at t=10 and 14s. Safety factor near the center increased, and approached to the reference. The control system itself determined $N_{//}$ (or directly controllable $\Delta\phi$), as shown in Fig. 4. The largest residual, max((q(r)-q$_{ref}$(r))w(r)),

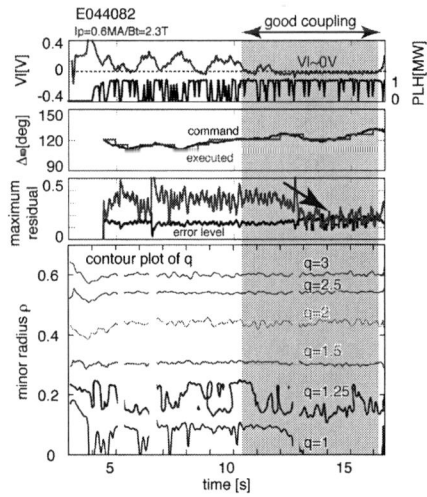

FIGURE 4. Waveforms of loop voltage (Vl), LH power (P_{LH}), phase difference between launcher modules ($\Delta\phi$), maximum of residuals between q by real-time calculation and reference q, and contour plot of safety factor profile q(ρ, t).

decreased down to the error level of q(r) measurement for 3s (t~13-16s). Although the maximum LH power (1MW) was limited due to conditioning of LH system, we are going to have an experiment to raise q(0) (or minimum in q profile) above 1.5 or 2 in future, in order to vanish rational surfaces of m/n=3/2 or 2/1 which are resonant to NTMs to achieve higher performance.

The current profile change in the experiment is discussed here. Figure 6(a) shows total current profile $j(\rho)$ determined by reconstruction of equilibrium with MSE, before and during q(r) control in Fig. 4. We see decrease in $j(\rho<0.2)$ and increase in $j(\rho=0.2-0.6)$ by the q(r) control. We calculated theoretical LH driven current profile at

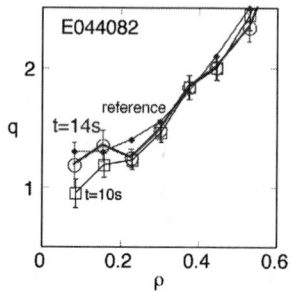

FIGURE 5. q profiles in Fig. 4 by real-time calculation at beginning of the control (t=10s; open rectangles), and during the control (t=14s; circles). Closed rectangles show the reference q profile.

t=16s under the experimental T_e, n_e and q profiles shown in Fig. 6(c). We used a combined ray-tracing and Fokker-Planck code developed by Bonoli and Englade [13]; it had been reported that the theoretical calculation, using the above code, agreed with experimentally determined LH driven current profile in JT-60U [5]. In calculation here, 21 rays having different $N_{//}$ are injected from 4 vertical antenna locations (total 84 rays). Power of each ray is given to satisfy $N_{//}$ spectrum in Fig. 2. Figure 6 (b) shows ray trajectories for $\Delta\phi=120°$ until 99% of power deposition. Absorption of LH waves was strong enough. Calculated LH driven current density profile $j_{LH}(r)$ at $\Delta\phi=120°$ is indicated in Fig. 6(a). Comparing j(r) and $j_{LH}(r)$, j(r) decreased in a location where j_{LH} is smaller than j, and vice versa. The temporal change in j profiles was consistent with the calculated j_{LH} profile, since bootstrap and beam driven current densities were quite small compared to j or j_{LH}; j_{LH} determined j evolution.

NTM CONTROL BY ECCD

The NTM limits attainable β_N in JT-60U high β_p-mode discharges. The NTM degrades confinement so that heating power becomes insufficient to sustain the β_N. Thus, ECCD was applied to stabilize the NTM. Figure 7(a) shows waveforms of a discharge where ECCD stabilized NTM at β_N~3 after mode saturation (late ECCD). Injection angle (46°) of EC waves was set in order that ECCD location is at NTM island [14]. In the discharge, β_N was feedback-controlled by NB power. The NB power gradually increased after appearance of m/n=3/2 NTM, showing degradation of confinement by the NTM. After application of ECCD at t=6s, intensity of magnetic fluctuation decreased and the NTM was completely stabilized. The plasma recovered its initial β_N (~3) due to the stabilization. We see recovery of confinement after t=7s, since the NB heating power decreased at constant β_N. On the contrary, misalignment of ECCD location and island location cannot stabilize NTM; see Fig. 7(b) for injection angle of 44°. The ECCD locations for the two cases are ρ=0.40 and 0.47, respectively. We found that precise ECCD at the island was essential for NTM stabilization.

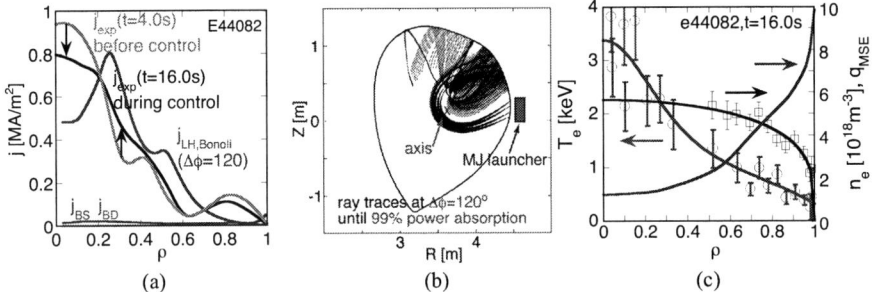

FIGURE 6. (a) current density profiles in a discharge in Fig. 4 (E44082) before and during q(r) control. LH driven current calculated using Bonoli code is presented. (b) ray trajectories of LH waves ($\Delta\phi=120°$). (c) T_e, n_e, q profiles for the calculation (t=16.0s).

We tried application of ECCD before NTM onset (early ECCD) in order to suppress appearance of NTM not to stabilize NTM after its appearance [15]. The result is shown in Fig. 8(a). ECCD at rational surface (q=3/2) was applied after t=3.0s when the β_N was much lower than 3. In the same scenario except the ECCD off the rational surface, m/n=3/2 NTM appeared when β_N reached 3. Thus, the m/n=3/2 NTM was suppressed by the aligned ECCD even if β_N reached 3.

Intensity of magnetic fluctuation ($|\dot{B}|/f$) is a measure of magnetic island size for a given mode-number and plasma configuration. Dependence of the fluctuation intensity on EC power was investigated as seen in Fig. 8(b) [15]. Higher EC power tends to decrease the fluctuation intensity for both the early and late ECCD. A prominent and important difference is that early ECCD requires much smaller power to decrease fluctuation to a given level than the late ECCD does. Full suppression is possible at 2.4MW for early ECCD, while 3MW for late ECCD. It was found that the early ECCD scenario is effective to save EC power than the late one. Real-time evaluation of q profile can (and will) be used to detect rational surface location where early ECCD is applied.

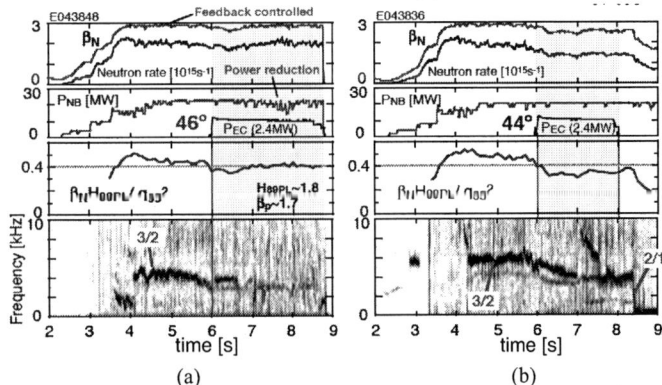

FIGURE 7. Temporal evolution of β_N, neutron production rate S_n [s^{-1}], NB and EC injection power [MW], $\beta_N H_{89P}/q_{95}^2$, and intensity of magnetic fluctuations. (a) for ECCD at island, and (b) for misalignment between ECCD location and the island.

SUMMARY

Real-time control system of safety factor profile has been developed. The system consists of 1) calculation of the safety factor profile and current drive location using MSE diagnostic, and 2) control of CD location by the control of the parallel refractive index of LH waves. Safety factor at the center q(0) was raised from 1 to 1.3, and sustained for 3s near the reference. Real-time q(r) control experiment continues for the demonstration of control at high β plasma. Complete stabilization of

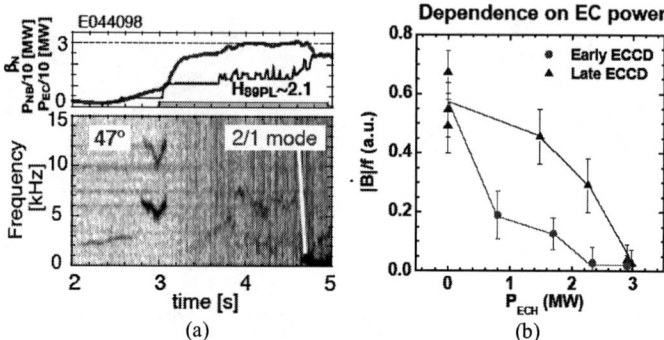

FIGURE 8. (a) Time evolution of β_N, NB and EC power, and intensity of magnetic fluctuations. EC waves are injected before β_N reaches 3, at which m/n=3/2 NTM appears for misaligned ECCD. (b) Intensity of magnetic fluctuation vs. EC power for early and late ECCD cases.

NTM was demonstrated at $\beta_N \sim 3$ using ECCD injection. The stabilization recovered thermal confinement and β_N. Precise ECCD at NTM island was essential to stabilize the NTM. Injection of EC waves at NTM-resonant rational surface before the mode onset suppressed appearance of NTM. EC power required for depress magnetic fluctuation intensity to a given level was lower for the early ECCD than for the late ECCD.

ACKNOWLEDGMENTS

The authors would like to thank the members who have contributed to the JT-60 project for their fruitful discussion, continuous support and encouragement.

REFERENCES

1. Ide, S. et al., *Nucl. Fusion* **44**, 87 (2004), Ide, S. et al., *Plasma Phys. Control. Fusion* **44**, L63 (2002).
2. Suzuki, T. et al., *Proceeding of 20th IAEA Fusion Energy Conference (Vilamoura)* IAEA-CN-116/EX/1-3 (2004); submitted to *Nucl. Fusion*.
3. Ide, S. et al., *Proceeding of 20th IAEA Fusion Energy Conference (Vilamoura)* IAEA-CN-116/OV/1-1 (2004); submitted to *Nucl. Fusion*.
4. Suzuki, T. et al., *Plasma Phys. Control. Fusion* **44**, 1 (2002).
5. Naito, O. et al., *Phys. Rev. Lett.* **89**, 065001 (2002).
6. Takase, Y. et al., 'Plasma current start-up by outboard PF coils in JT-60U and TST-2', this conference (B-16).
7. Fujita, T. et al., *Phys. Rev. Lett.* **87**, 245001 (2001), Ishida, S. et al., *Phys. Plasmas* **5**, 2532 (2004).
8. Fujita, T. et al., *Fusion Eng. Des.* **34-35**, 289 (1997).
9. Kurihara, K. et al., *Fusion Tech.* **34**, 548 (1998).
10. Suzuki, T. et al., *Nucl. Fusion* **44**, 699 (2004), Suzuki, T. et al., *J. Plasma Fusion Res.* **80**, 362 (2004).
11. Seki, M. et al., *Fusion Sci. Tech.* **42**, 452 (2002).
12. Ushigusa, K. et al., *Fusion Sci. Tech.* **42**, 255 (2002).
13. Bonoli, P. T. and Englade, R. C., *Phys. Fluids* **29**, 2937 (1986).
14. Isayama, A. and the JT-60 Team, *Phys. Plasmas* **12**, 056117 (2005).
15. Nagasaki, K. et al., *Nucl. Fusion* **43**, L7 (2003).

Full-wave Electromagnetic Field Simulations of Lower Hybrid Waves in Tokamaks

J. C. Wright*, P. T. Bonoli*, M. Brambilla[†], E. D'Azevedo**, L. A. Berry[‡],
D. B. Batchelor[‡], E. F. Jaeger[‡], M. D. Carter[‡], C. K. Phillips[§], H. Okuda[§],
R. W. Harvey[¶], J. R. Myra[‖], D. A. D'Ippolito[‖] and D. N. Smithe[††]

*MIT - Plasma Science and Fusion Center Cambridge, MA 02139
[†]Institute für Plasma Physik Garching, Germany
**Computer Science and Mathematics Division, Oak Ridge National Lab Oak Ridge, TN
[‡]Oak Ridge National Laboratory - Oak Ridge, TN, USA
[§]Princeton Plasma Physics Laboratory - Princeton, New Jersey, USA
[¶]CompX - Del Mar, CA, USA
[‖]Lodestar Research Corporation - Boulder, CO, USA
[††]ATK-Mission Research Corp. - Newington, VA, USA

Abstract. The most common method for treating wave propagation in tokamaks in the lower hybrid range of frequencies (LHRF) has been toroidal ray tracing, owing to the short wavelengths (relative to the system size) found in this regime. Although this technique provides an accurate description of 2D and 3D plasma inhomogeneity effects on wave propagation, the approach neglects important effects related to focusing, diffraction, and finite extent of the RF launcher. Also, the method breaks down at plasma cutoffs and caustics. Recent adaptation of full-wave electromagnetic field solvers to massively parallel computers [1] has made it possible to accurately resolve wave phenomena in the LHRF. One such solver, the TORIC code, has been modified to simulate LH waves by implementing boundary conditions appropriate for coupling the fast electromagnetic and the slow electrostatic waves in the LHRF. In this frequency regime the plasma conductivity operator can be formulated in the limits of unmagnetized ions and strongly magnetized electrons, resulting in a relatively simple and explicit form. Simulations have been done for parameters typical of the planned LHRF experiments on Alcator C-Mod, demonstrating fully resolved fast and slow LH wave fields using a Maxwellian non-relativistic plasma dielectric. Significant spectral broadening of the injected wave spectrum and focusing of the wave fields have been found, especially at caustic surfaces. Comparisons with toroidal ray tracing have also been done and differences between the approaches have been found, especially for cases where wave caustics form. The possible role of this diffraction-induced spectral broadening [2] in filling the spectral gap in LH heating and current drive will be discussed.

1. INTRODUCTION

Electromagnetic waves in the the lower hybrid range of frequencies (LHRF) are characterized by a frequency approximately at the geometric mean of the ion and electron fundamental cyclotron frequencies, $\Omega_{ci} \ll \omega \ll \Omega_{ce}$. At these frequencies, the wavelengths are on the order of a few millimeters at values of the dielectric constant, $\left(\omega_{pe}/\Omega_{ce}\right)^2$, typical parameters of fusion research devices such as Alcator C-Mod and the proposed International Tokamak Experimental Reactor (ITER), where they have been proposed as a method of edge current profile control [3]. Three waves are supported by the plasma at this frequency range: the slow lower hybrid wave (usually the one referred to as the lower hybrid wave, the fast lower hybrid wave (or Whistler wave), and the ion plasma wave which is supported by finite ion Larmor radius effects. As explained in Section 2, at plasma parameters of experimental interest that are used in the simulations presented in this paper, the ion plasma wave is strongly evanescent, and so only two waves are supported.

The short wavelength relative to machine sizes of the order of a meter have necessitated the use of ray tracing as the primary method calculating the power and current drive deposition in the plasma. While this can explain some features such as broadening of the launched spectrum due to toroidicity [4], it breaks down in cases where the rays undergo multiple reflections from cutoffs and caustics and form a stochastic field. Extended raytracing techniques such as the Maslov method popular in seismology [5] and the wave-kinetic method [6], are valid at the caustic surfaces; but because the LH cutoffs in tokamak plasmas occur in the plasma edge where the gradients are very large, they violate the Wentzel, Kramers and Brillouin (WKB) approximation where the plasma is changing on the same scale as the wavelength [7]. There is the additional difficulty of treating linear mode conversion between the fast and slow branches, but interesting work on this is being pursued by Tracy, Kaufman, and Jaun [8].

Full wave simulations do not have these issues and incorporate the effects of diffraction that are not included in the ray tracing model. As available computer power has increased, several efforts have been made at including full wave effects in lower hybrid simulations. Pereverzev developed a hybrid method in which the ray equations were used in the direction of the group velocity and a local wave equation was used to evolve the wave front perpendicular to it [2]. He found enhanced spectral broadening due to diffraction especially at caustic surfaces. Peysson has performed full wave lower hybrid simulations in cylindrical geometry and coupled those calculations to a two dimensional Fokker-Planck code [9]. The lack of toroidicity reduced the wave equation to one dimension with toroidal and poloidal mode numbers as parameters. Also, the simulations were applicable to large aspect ratio devices.

We present in this paper, the first full wave calculations of lower hybrid waves in toroidal geometry. The plasma dielectric is Maxwellian and non-relativistic. The electrons are treated as being strongly magnetized and the ions are unmagnetized. Thermal effects are retained. This paper is organized as follows. In Section 2 we will describe in more detail, the wave equation and dielectric model used in these simulations.

2. THE WAVE EQUATION AND PLASMA MODEL

The wave equation that results from the linearized Maxwell-Boltzmann system [10, 11] is given in Eq. (1). The plasma response in embodied in the term $\mathbf{J}^P$ in Eq. (2). We have formulated the plasma conductivity, σ, for the LHRF in which the lower hybrid frequency is given as, $\omega_{\text{LH}} = \omega_{\text{pi}}/\sqrt{(1 + (\omega_{\text{pe}}/\Omega_{\text{ce}})^2)}$, for which the relation, $\Omega_{\text{ci}} \ll \omega \sim \omega_{\text{LH}} \ll \Omega_{\text{ce}}$ holds.

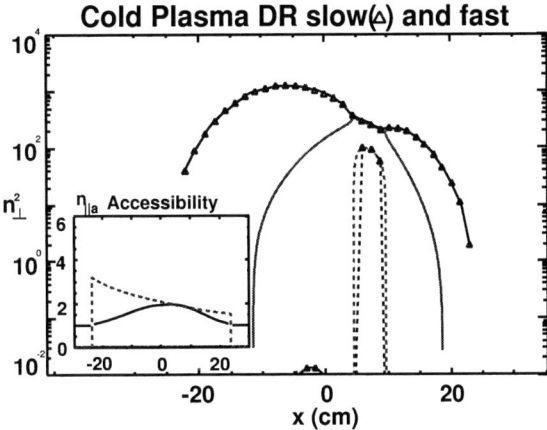

FIGURE 1. Cold plasma electromagnetic dispersion relation for slow and fast LH waves using Alcator C-Mod parameters: [deuterium gas, $n_\parallel$=1.5, f_0 = 4.6 GHz, B_0 = 5.3 T, T_e=3.5 keV, T_i=2.0 keV, I=1 MA, $n_e(0) = 1.5 \times 10^{20} \text{m}^{-3}$.] The accessibility plot in the inset shows that the waves are excluded from the region between 3 and 10 cm. In the full size graph, the corresponding dispersion relation for this case shows that the two propagating branches are heavily damped in this excluded region. The lines without symbols are the fast branch and the lines with symbols are the slow branch. Solid lines are the real part of $n_\perp^2$ and the corresponding dashed lines are the imaginary part with the sign reversed.

$$\nabla \times \nabla \times \mathbf{E} = \frac{\omega^2}{c^2}\left\{\mathbf{E} + \frac{4\pi i}{\omega}\left(\mathbf{J}^P + \mathbf{J}^A\right)\right\} \quad (1)$$

$$\mathbf{J}^P = \overleftrightarrow{\sigma}\left[f_0(\mathbf{x}, v_\perp, v_\parallel)\right] \cdot \mathbf{E} \quad (2)$$

In this range of frequencies, the ions are unmagnetized and the electrons are strongly magnetized $\left[(k_\perp \rho_e)^2 \ll 1\right]$. Thus, we no longer need the ion finite Larmor radius (FLR) effects and the LH plasma dielectric, $\overleftrightarrow{\varepsilon} \equiv \overleftrightarrow{1} + \frac{4\pi i}{\omega}\overleftrightarrow{\sigma}$, simplifies to the zero FLR contributions:

$$\overleftrightarrow{\sigma} \cdot \mathbf{E} = S\,\mathbf{E}_\perp + iD\,(\mathbf{b}\times\mathbf{E}_\perp) + P\,E_\zeta\mathbf{b} \quad (3)$$

$$S \approx 1 + \frac{\omega_{pe}^2}{\Omega_{ce}^2} - \frac{\omega_{pi}^2}{\omega^2} \approx 1 \quad (4)$$

$$D \approx -\frac{\omega_{pe}^2}{\omega^2}\frac{\omega}{\Omega_{ce}} + \frac{\omega_{pi}^2\,\Omega_{ci}}{\omega^2\,\omega} \approx -\frac{\omega_{pe}^2}{\omega\Omega_{ce}}$$

$$P = 1 - \frac{\omega_{pe}^2}{\omega^2} - \frac{\omega_{pi}^2}{\omega^2} \approx \frac{\omega_{pe}^2}{\omega^2}$$

where S, D, and P, are the Stix cold plasma dielectric elements in the LHRF for the normal, co-normal, and parallel directions [12]. The neglected FLR corrections to Eq. (3)

support the mode converted ion plasma wave. Since this wave does not propagate for plasmas of experimental interest in which $\omega/\omega_{LH} > 2$ [13], we may neglect these corrections in the following analysis, and thus there are only two propagating modes, the fast electromagnetic LH branch that damps via electron Landau damping (ELD) and transit time magnetic pumping (TTMP) and the slow electrostatic LH branch that damps via ELD. Note, that although the FLR terms play no role, thermal effects are retained through the plasma dispersion function which is retained in the the ELD and TTMP damping in the plasma model.

The dispersion relation associated with the dielectric in Eq. (3) is given below in Eq. (5).

$$P_4 n_\perp^4 + P_2 n_\perp^2 + P_0 = 0 \tag{5}$$
$$P_4 = S$$
$$P_2 = (S+P)(n_\parallel^2 - S) + D^2$$
$$P_0 = P\left[(n_\parallel^2 - S)^2 - D^2\right]$$

The spectral representation employed in TORIC in Eq. (6) uses a truncated Fourier decomposition of the flux angles and cubic Hermite polynomial finite elements (FE) for the flux dimension, ψ.

$$\mathbf{E}(\mathbf{x}) = \sum_m \mathbf{E}_m(\psi) \exp(im\theta + in\phi) \tag{6}$$

We may obtain an estimate of the necessary resolution by referring again to the dispersion relation in Eq. (5). In the electrostatic limit, it reduces to Eq. (7) [13]. For the parameters given in Figure 1, we can estimate that from Eq. (4) and Eq. (7) that $k_\perp \approx 66 \text{cm}^{-1}$ or $\lambda_\perp \approx 1\text{mm}$.

$$k_\perp^2 \approx -\frac{P}{S} k_\parallel^2 \Rightarrow k_\perp \approx \frac{\omega_{pe}}{\omega} k_\parallel \tag{7}$$

For comparison, in mode conversion cases in which the ion Bernstein wave (IBW) is present, the IBW wavelength in a typical C-Mod discharge [1] is 4 to 5 mm requiring 255 poloidal modes to resolve. This implies a need for 4 times more resolution, which results in 64 times more computational power to invert the resulting stiffness matrix.

The neglected sixth order coefficient is proportional to $\beta \sqrt{\frac{m_e}{m_i}}$ and has no effect on the slow and fast branches. Analysis of Eq. (5) yields an accessibility criterion for the slow LH wave to propagate that is given by $n_\parallel \geq n_{\parallel a}$ where $n_{\parallel a}$ is given by Eq. (8). We will use this condition to construct a case in which a fast polarized lower hybrid wave is launched from the antenna and mode converts to a slow polarized lower hybrid wave at a surface where the radial wave number vanishes (a caustic) and is trapped between this caustic and the edge cutoff. This full wave result will be contrasted with a ray tracing simulation of the same case.

$$n_{\parallel a} \equiv \frac{\omega_{pe}}{\Omega_{ce}} + \sqrt{S} \tag{8}$$

3. LOWER HYBRID CUTOFFS AND CAUSTICS

In Figure 1, we show the cold plasma dispersion relation with $m = 0$ for a case in Alcator C-Mod with accessibility to the slow wave limited to a narrow region that on the low field side is between x of 5 and 10 cm. The parameters used for this simulation are similar to those that are used in Alcator C-Mod LH discharges [deuterium gas, $n_\parallel$=1.5, f_0 = 4.6 GHz, B_0 = 5.3 T, T_e=3.5 keV, T_i=2.0 keV, I=1 MA, $n_e(0) = 1.5 \times 10^{20} \text{m}^{-3}$.] A fast LH wave launched from the low field side will start to propagate at its cut-off at $r = 19$cm (the solid line with symbols in the outer plot of Figure 1), propagate inwards to the caustic at $r \approx 10$cm, mode convert to the slow LH wave (the solid line without symbols in Figure 1), propagate out to the cut-off at $r = 23.8$cm, and reflect inward to the caustic again. Consequently, the waves will be trapped in an annulus as they travel poloidally. This behavior is apparent in the TORIC full wave simulation simulation results shown in Figure 2 on the right.

We now simulate the same wave-plasma conditions with a ray tracing code, ACCOME [14]. For purposes of this comparison, we disable ACCOME's equilibrium evolution and Fokker-Plank modules. We use the same EFIT equilibria in both codes. Figure 2 shows the result in the right figure. Rays launched from different locations in the antenna region converge to form a caustic and radially reflect towards the plasma edge where they reflect from the low density cutoff and receive an upshift in there parallel wavenumber. When the parallel index exceeds 2, the rays are detrapped and propagate towards the plasma center where they eventually damp on the hotter core electrons. The resulting power deposition is compared with the ray tracing result in Figure 3.

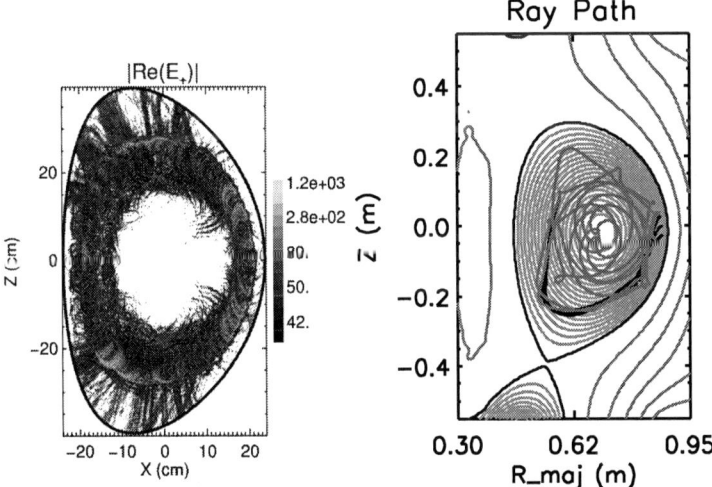

FIGURE 2. Left figure: Full wave TORIC results showing wave structure confined between caustic and cutoff. Same parameters as given in Figure 1. Right figure: ACCOME ray tracing simulation with same parameters as in Figure 1. Rays initially follow caustic and are trapped between caustic and cutoff. Eventually rays are detrapped by wavenumber upshifts and cutoff reflections and damp in the center of the plasma on hotter electrons.

ACCOME's power deposition profile is located within $r/a \sim 0.6$ whereas TORIC's is between $r/a \sim [0.65, 0.85]$. The primary cause of this difference is the diffraction that takes place at the caustic surface in the full wave code. The resulting slow down in the phase velocity bridges the spectral gap and causes all the power in the waves to be absorbed. In contrast, in the raytracing, no such large shift occurs and the rays slowly dissipate their energy in the core. An analysis of the evolution of $n_{\parallel}$ in the two simulation results shows significantly more upshift in the full wave case. For raytracing, the local $n_{\parallel}$ evolves to 2.5 on the high field side from 1.5 at the antenna due primarily to geometric effects of major radius position. The distribution of $n_{\parallel}$ on flux surfaces in the full wave results have a significant upshift from an averaged launched $n_{\parallel}$ of 2 to greater than 4 in the middle of the annulus at $r/a = 0.75$. The resulting enhanced damping at a parallel index of 4 damps all the wave energy in that region.

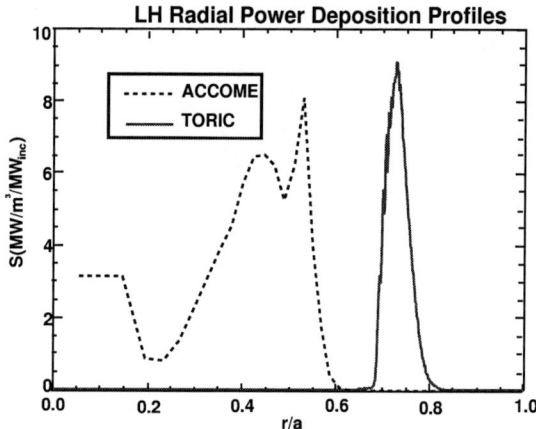

FIGURE 3. A comparison of the power deposition predicted by raytracing and full-wave codes ACCOME and TORIC. The full wave power given by the solid curve is localized completely between the caustic and the cutoff. The raytracing result shows all power deposited inside the caustic radius in the core of the plasma.

4. REDUCED LOWER HYBRID WAVE EQUATION

The evanescent ion plasma wave still restricts the resolution in these simulations and can cause numerical pollution [15] without sufficient resolution. This mode may be removed from the system by algebraically eliminating Epsi from the system. We may do this because the coefficient of $dE_\psi/d\psi$ is essentially zero, even when finite larmor radius effects are retained. This decouples E_ψ, leaving only two modes for which to solve. Back substitution is used to evaluate E_ψ. In addition, the boundary conditions have been generalized in Equations 9 and 10 from the previous current strap model to include arbitrary waveguide polarization. In these equations, Θ_w, Δ_g and θ_0 represent the static magnetic field pitch angle, the angular half height of the waveguide and the angular location of the waveguide, respectively. These new boundary conditions at the plasma wall permit the code to impress $E_\parallel$ as a boundary conditions for slow LH wave launch, as well as $E_\perp$ for fast wave launch, and are used in the next section to launch slow waves in Alcator C.

$$\frac{c}{\omega}\left[(\nabla \times E)^{(m)}_{[\eta,\zeta]}\right]_{\psi_A} = \frac{4\pi i}{c} J^S [-\sin\Theta_w, \cos\Theta_w] \qquad (9)$$

$$\left[E^{(m)}_{[\eta,\zeta]}\right]_{\psi_A} = -\exp(-im\theta_0)\frac{\exp(-im\Delta_0)+1}{4m^2\Delta_g^2 - \pi^2}[-\sin\Theta_w, \cos\Theta_w] \qquad (10)$$

5. APPLICATION TO ALCATOR C

The new algorithm described in the previous section permits converged lower hybrid simulations at lower resolutions. The presence of the evanescent ion plasma wave required a very fine mesh for numerical stability even though the mode in not propagative. By eliminating it, we are left only with the limits imposed by the longer wavelength slow lower hybrid wave (the fast lower hyrid wave has a yet larger wavelength.) In addition, the cpu requirements for a given resolution are reduced by approximately a factor of two because the stiffness matrix has fewer unknowns.

We now apply this new algorithm to lower hybrid experiments on the now defunct Alcator C experiment [16]. The parameters of the simulation are given in Figure 4 and are correspond to a scenario in which the lower hybrid waves are accessible to the center of the device. The range of the contours in the figure have been adjusted to accentuate the filamentary structures present in the solution.

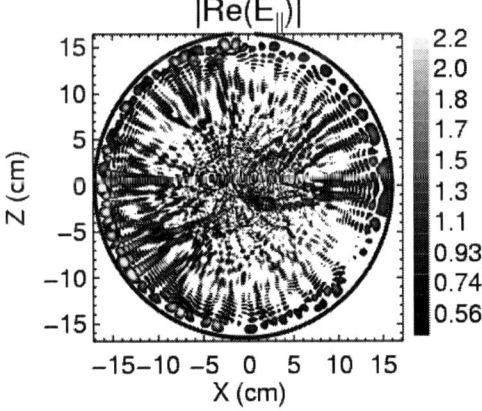

FIGURE 4. Full wave simulation of lower hybrid waves in Alcator C experiment. The wave parameters are chosen such that the wave is accessible to the center. Parameters of run: (240Nr x 127 Nm) H plasma, $n_\parallel$=1.5, f=4.6 GHz B_0 = 10 T, T_e=1.8 keV, Ti=1.0 keV $n_e(0) = 0.510^{20} m^{-3}$, I=170 kA

The full wave electric field structures are also seen in toroidal ray tracing calculations for this case, where rays undergo multiple radial transits between edge cutoffs and caustics near the core.

6. CONCLUSIONS

The necessary algorithms for solving full wave dispersion in realistic toroidal geometries have been developed by the RF community over the past few decades. The availability first, of more powerful serial processors and later of parallel architectures has opened up a broader range of physics regimes to these codes. For the first time we can consider lower hybrid frequency regime calculations and the wide range of physics issues involved, such as the role of wave focusing and diffraction in LH spectral broadening in 2D toroidal geometries.

ACKNOWLEDGMENTS

All simulation presented in this paper were performed on the MIT-PSFC Marshall theory cluster. Research sponsored by the Mathematical, Information, and Computational Sciences Division; Office of Advanced Scientific Computing Research; U.S. Department of Energy, under Contract No. DE-AC05-00OR22725 with UT-Battelle, LLC.

REFERENCES

1. Wright, J. C., Bonoli, P. T., Brambilla, M., Meo, F., D´Azevedo, E., Batchelor, D. B., Jaeger, E. F., Berry, L. A., Phillips, C. K., and Pletzer, A., *Phys. Plasmas*, **11**, 2473–2479 (2004).
2. Pereverzev, G. V., *Nucl. Fusion*, **32**, 1091–1107 (1992).
3. ITER Physics Expert Group on Energetic Particles, Heating and Current Drive and ITER Physics Basis Editors, *Nucl. Fusion*, **39**, 2495–2540 (1999).
4. Bizarro, J. P., *Nucl. Fusion*, **33**, 831–834 (1993).
5. Chapman, C. H., and Keer, H., *Stud. Geophys. Geo.*, **46**, 615–649 (2004).
6. Kupfer, K., Moreau, D., and Litaudon, X., *Phys. Fluids B*, **5**, 4391:4407 (1993).
7. Brambilla, M., and Cardinali, A., *Plasma Phys.*, **24**, 1187–1218 (1982).
8. Tracy, E. R., Jaun, A., and Kaufman, A. N., "Modeling of Ray Splitting in a Tokamak," in *Proceedings of the 16th Topical Conference on Radio Frequency Power in Plasmas*, edited by P. Bonoli and S. Wukitch, , American Institute of Physics, New York, 2005, p. in this publication.
9. Peysson, Y., Sébelin, E., Litaudon, X., Miellou, D. M. J.-C., Shoucri, M. M., and Shkarofsky, I., *Nucl. Fusion*, **38**, 939–943 (1998).
10. Stix, T. H., *The Theory of Plasma Waves*, chap. 10, p. 250, in [12] (1992).
11. Brambilla, M., *Plasma Phys. Controlled Fusion*, **31**, 723–757 (1989).
12. Stix, T. H., *The Theory of Plasma Waves*, American Institute of Physics, New York, 1992.
13. Bonoli, P., *IEEE Trans. Plasma Sci.*, **PS-12**, 95–107 (1984).
14. Devoto, R. S., Blackfield, D. T., Fenstermacher, M. E., Bonoli, P. T., Porkolab, M., and Tinios, G., *Nucl. Fusion*, **32**, 773–786 (1992).
15. Llobet, X., Appert, K., Bondeson, A., and Vaclavik, J., *Computer Phys. Comm.*, **59**, 199–216 (1990).
16. Porkolab, M., Schuss*, J. J., Lloyd, B., Takase, Y., Texter, S., Bonoli, P., Fiore, C., Gandy, R., Gwinn, D., Lipschultz, B., Marmar, E., Pappas, D., Parker, R., and Pribyl, P., *Phys. Rev. Lett.*, **53**, 1153 (1984).

Bridging the spectral gap in lower hybrid current drive by parametric instability

R. Cesario[1], C. Castaldo[1], A. Cardinali[1], F. Paoletti[2]

[1]*Associazione EURATOM/ENEA sulla Fusione, c.p. 65, 00044 Frascati, Italy*
[2]*East Windsor Regional School District, Hightstown, NJ 08520 USA*

Abstract. A broadening of the radiofrequency (RF) power spectrum coupled to tokamak plasma is necessary to occur in order to explain the existing experiments of lower hybrid current drive (LHCD). The presented modeling shows that the parametric instability (PI) driven by ion sound quasi-modes produces in the scrape-off-plasma an important contribution to such spectral broadening. As effect of the quasi-linear interaction of the resulting LH spectrum penetrating in the bulk, the LH power fraction deposited in the plasma at the first pass results enhanced. Consequently, well defined LH deposition profiles are obtained when the ray propagation in toroidal geometry is taken into account. Considering the parameters of LHCD experiments of JET (Joint European Torus), and other machines as well, the PI growth rate is high enough for compensating the convective losses and broadening a fraction ($\approx 10\%$) of the launched power spectrum. The LH spectral broadening is intrinsic to coupling RF power in LHCD experiments, and increases operating with higher plasma densities. As principal implication of such spectral broadening, experiments able evidencing the effects of a well-defined LH deposition profile, as those characterized by high electron temperature in the core and broad profile, are successfully interpreted. Useful experiments are the LHCD-sustained internal transport barriers of JET. The design of scenarios relevant to the modern fusion research program, which require the control of the plasma current profile in the outer half of plasma, can be properly achieved by considering the physics of the plasma edge for modeling the LH deposition profile and the q-profile evolution.

Keywords: Lower hybrid current drive, scrape-off plasma, spectral broadening, parametric instability, RF power deposition profile in the plasma.

PACS: PACS: 52.35.Hr

INTRODUCTION

The physical mechanisms that determine the LH deposition profile in realistic operating conditions should be considered for making the lower hybrid current drive (LHCD) [1-2] a robust tool for controlling the current profile in tokamaks. Since the LHCD effect [1-5] is based on the wave interaction with a tail of the electron distribution function of plasma, the assessment of the LH power $n_{//}$ spectra that effectively propagate in the experiments is crucial for determining the deposition profile ($n_{//}$ is the refractive index component in direction parallel to the confinement magnetic field). At this regard, a long-lasting debate is still open on the so-called spectral gap in LHCD, i.e., about the causes of broadening of the launched $n_{//}$

spectrum, which is necessary to occur for explaining the available experimental data by means of the quasi-linear theory [3-6].

The approach of multi-radial pass produced by ray-tracing in toroidal geometry [6] was widely accepted as cause of spectral broadening, and utilized for modeling the LH deposition profile in experiments. A difficulty with such approach consists in the fact that the WKB approximation fails at the cut-off layers [3]. At these layers the LH waves are considered as optic waves, despite of their much longer wavelengths. The LH deposition profile might thus result arbitrary when undetermined reflections from the edge occur. Therefore, assuming that the multi-radial pass is the only cause of spectral broadening, the LH deposition profile might be not a well-defined feature of LHCD, which, thus, cannot be utilized as a robust tool for controlling the current profile. Conversely, there are indications from experiments that well-defined LH-deposition profile are produced, as occurred, e.g., in the LHCD-sustained internal transport barriers (ITBs) of JET [7-10]. These barriers are characterized by high electron temperatures in the core with broad profiles, and most of the LH-driven current density is localized at two thirds of the minor radius. The observed ITB features are consistent with the hypothesis of an LHCD-sustained low/negative magnetic shear occurring in a layer close to the ITB radial foot [10]. Considering the multi-reflections as the only cause of the spectral broadening in LHCD, the precision of the LH-deposition profile results insufficient for finding consistency with the current profile supported by measurements, and with the observed features of the ITBs as well. Therefore, it seems that the multi-radial pass alone cannot bridge the spectral gap in LHCD. The role of the physics of the edge, which is supported by experimental observations in LHCD experiments of spectral broadening obtained by RF probes and by microwave reflectometry [10,11], should be considered for properly modeling the LH-deposition profile.

BROADENING OF THE ANTENNA SPECTRUM IN LHCD EXPERIMENTS BY THE PHYSICS OF THE EDGE

The physics of the edge appears evident when considering the whole scenario of the experiments that coupled LH waves to tokamak plasmas, including those aimed at heating the bulk ions [12]. Such experiments differ from the LHCD experiments essentially for the higher operating plasma densities, which is necessary for locating the cold lower hybrid resonant layer, $\omega_{LH} \approx \omega_0$, in the core, and meeting the mode-conversion of the launched electron plasma wave (EPW) into an ion wave, which is collisionless absorbed by ion-cyclotron harmonic damping. The LHCD regime, which does not require mode-conversion, works instead at relatively low densities, $\omega_{LH} \ll \omega_0$, but higher than the EPW cut-off value: $\omega_{pe} \gtrsim \omega_0$. The test of the ion heating scheme resulted impossible, since at the required high plasma densities the LH power does not penetrate in the bulk. The occurrence of parametric instability (PI) in the scrape-off plasma resulted the only effect on the plasma of the coupled RF power. An example of the PI spectra detected by RF probes is shown in Figure 1.

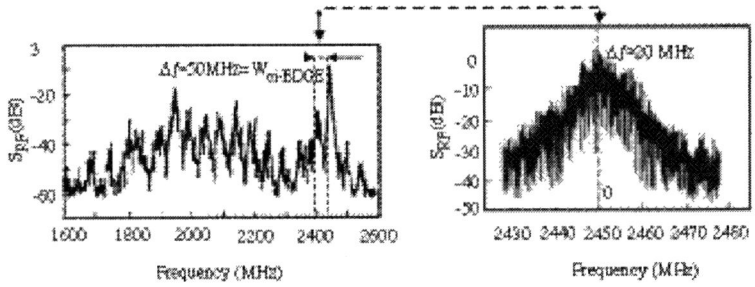

Figure 1. RF probe spectra in the LH experiment aimed at ion heating in FT

The phenomenology consists in a broadening around the operating line frequency and in several sidebands with a typical non-monotonic envelope [13-15]. Such spectra were interpreted in terms of a cascade of parametric instabilities [15]. PIs driven at very low frequencies by ion-sound quasi-modes deplete the pump and produce, in turns, a secondary LH pump with $n_{//}$ spectrum broader than that launched by the antenna. Such pump produces sidebands with maximum growth rates around the 10th Ω_{ci} harmonic, consistently with the observed typical non-monotonic envelop of sidebands. The full deposition of the coupled LH power at the edge is also explained in terms of strong quasi-linear absorption on the plasma particles due to the enormous LH spectral broadening, consistently with the observed ion tails from the plasma edge [5]. At the lower operating plasma densities of the LHCD experiments, the level of the ion-cyclotron frequency-shifted sidebands resulted generally much lower, with only a few sidebands monotonically decreasing at higher frequency shifts from the pump. The pump broadening is reduced too, but remains order of magnitudes bigger than the frequency line width of the RF power sources utilized in the experiments. Therefore, the (non linear) physics of the plasma edge, which was not considered in the LH heating and CD schemes, produces a spectral broadening that depends on the operating plasma density. At relatively low densities the broadening is less pronounced but still persists, and, however, the RF penetration in the bulk is permitted. The spectral broadening must be however considered in every LH experiment, also considering that the LHCD experiments need that a spectral broadening occurs for working.

No behavior similar than the pump broadening was observed in the ICRH and ECRH experiments, which utilize electromagnetic waves. The LH waves are electrostatic, and thus the coherent motion of the particle in the wave field mainly carries the energetic flux. Therefore, there is not any *a priori* reason for neglecting the mode coupling of the LH waves with the thermal background particle motion at low frequencies ($\omega<<\omega_0$). The parametric instability of a lower hybrid pump wave $\Phi_0[-i(\omega_0 t - \mathbf{k}_0\cdot\mathbf{r})]$ is driven by a low frequency mode $\Phi[-i(\omega t - \mathbf{k}\cdot\mathbf{r})]$ and growths by two sideband waves $\Phi_{1,2}[-i(\omega_{1,2}t - \mathbf{k}_{1,2}\cdot\mathbf{r})]$, where $\mathbf{k}_{2,1}=\mathbf{k}\pm\mathbf{k}_0$, $\omega_{2,1}=\omega\pm\omega_0$ (selection rules). We assume $\mathbf{k}_0=k_{0x}\mathbf{x}+k_{0z}\mathbf{z}$, $\mathbf{k}_{1,2}=k_{1,2x}\mathbf{x}+k_{1,2y}\mathbf{y}+k_{1,2z}\mathbf{z}$, and utilize the relation $\mathbf{n}=\mathbf{k}c/\omega_0$ between refractive indexes and wavevectors. The plasma is modeled as a slab including the region of the edge close to the antenna mouth. The x direction coincides with the (radial) direction of the plasma gradients, and y, z correspond to the poloidal and the toroidal directions, respectively. The Vlasov-Poisson system is solved in slab plasma for LH coupled modes up to the 2nd order (referring to the perturbation of the low frequency mode). The relevant parametric dispersion relation for LHCD experiments is [16-18]:

$$\varepsilon(\omega,\mathbf{k}) - \frac{\mu_1(\omega_1,\mathbf{k}_1,\mathbf{k}_0,E_0)}{\varepsilon(\omega_1,\mathbf{k}_1)} - \frac{\mu_2(\omega_2,\mathbf{k}_2,\mathbf{k}_0,E_0)}{\varepsilon(\omega_2,\mathbf{k}_2)} = 0 \quad (1)$$

The solutions of Eq. 1, ω is the complex frequency: $\omega = \omega_{Re} + i\gamma$, where γ is the growth rate, and $\omega_{Re2,1} = \omega_{Re} \pm \omega_0$. In Eq. 1, ε is the dielectric function, $\mu_{1,2}$ are the coupling coefficients referring to the lower and the upper sidebands respectively, and are calculated considering the ion magnetized. The expression of the coupling coefficients is [18]:

$$\mu_{1,2} = \frac{\chi_e(\omega) - \varepsilon(\omega)}{\chi_e(\omega)} \frac{\omega_{pi}^2}{\omega_0^2} \frac{\omega_{pi}^2}{4k^2 c_s^2} (1 + \frac{\omega}{k_z v_{the}} Z)^2 \sin^2 \delta_{1,2} \frac{u^2}{c_s^2} \quad (2)$$

The coupling coefficients are derived in the limit for $\omega \leq k_{//} v_{the}$, which is satisfied by all the solutions of Eq. 1 obtained considering typical parameters of the plasma edge of LHCD experiments performed in tokamak plasmas. In Eq. 2, χ_e is the electron suscecitivity, c_s is the sound speed, Z is the plasma function. $\delta_{1,2}$ are the angles which the perpendicular wavevectors of the lower and upper sidebands make with the perpendicular wavevector of the pump, i.e., $\delta_{1,2} = \angle$ ($\mathbf{k}_{1,2\perp}$, $\mathbf{k}_{0\perp}$), and $u = ek_0 \Phi_0/m_e \omega_{ce}$. The angle $\delta_{1,2}$ is an important parameter for determining the strongest PI channels, since it affects the convective loss. The solution of Eq. 1 is the complex frequency, $\omega + i\gamma$ for a given wavevector component $k_\perp$ of the low frequency mode that drives the instability, which is assumed as free variable. The plasma parameters of realistic LHCD experiments, supported by the measurements of plasma edge (spectroscopy of the edge and Langmuir probes) have been considered. For any run of the numerical calculation, one of the following parameters has been kept fixed: $k_{//}$, $k_{0//}$, $\delta_{1,2}$, Φ_0, B_0, n_e, T_e. In this way all the channels of PI are characterized by frequency and growth rates obtained by the analysis in the homogeneous plasma. The PI channel with highest growth rate results that driven by ion-sound quasi-modes. The effect of the gradient of the electron temperature reduces the growth rate for more internal radial position values.

The conditions for developing a parametric instability are produced by the convective loss due to the finite extent of the pump wave region and to the plasma inhomogeneity. The amplification factor should be A>1 at the threshold condition, and A≈10 for producing a significant depletion of the coupled RF power into LH sideband waves that arise from the thermal noise. The details of this analysis are contained in Ref [10,11]. The depletion of the pump power has been calculated considering the classical reference of L. Chen and L. Berger [19].

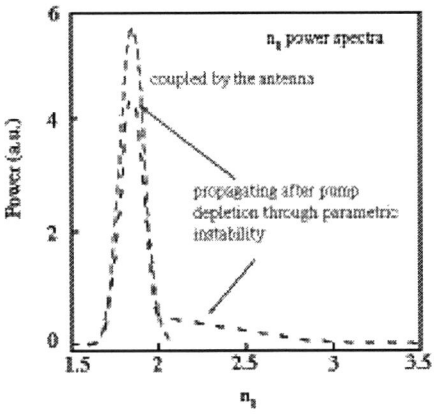
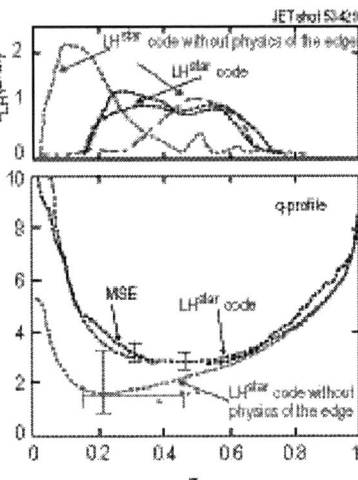

Fig. 2. Parametric instability-induced spectral broadening

Fig. 3. LH current density profiles and respective modeled q-profiles compared with that supported by MSE measurements.

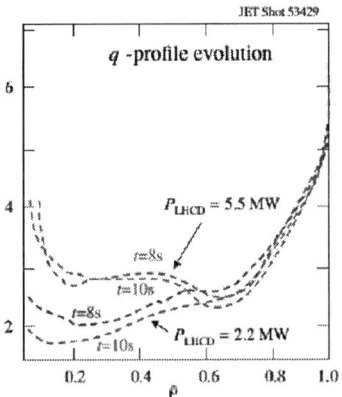

Fig. 4. q-profile evolution for ITB experiments of JET. Interpretive modeling with 2.2 MW of LHCD power. Predictive modeling with 5.5 MW of LHCD power. ρ is the normalized flux co-ordinate.

The important result is that, for a typical LHCD experiment of JET [7,9,10], a fraction of about 10% of the pump power is deposited in the scrape-off layer on sidebands with $n_{//} \approx 2.1 \div 2.3$, as shown in Fig. 2. This fraction decreases at increasing $n_{//}$ (0.1% for $n_{//} \approx 3$), due to increase of convective losses. For $n_{//} > 3.5$, no depletion occurs as the pump power density goes below the threshold. Thus, a cut-off in $n_{//}$ of the LH power spectrum penetrating in the plasma bulk is determined. The LH-deposition profile has been calculated considering or not the contribution of the PIs to the spectral broadening, by utilizing a ray-tracing+2D-Fokker-Planck code [20], which retains the $n_{//}$ upshift due to ray propagation in toroidal geometry. The experiments of JET considered in the present analysis were performed with a small fraction of current drive, $I_{LHCD}/I_P \approx 30\%$, utilizing LHCD both during the current ramp-up, and at the end of the ramp-up (I_P=2.3MA, B_T=3.3T), in combination with the main heating power (16 MW of neutral beam and 5 MW of ion-cyclotron resonant heating), [7,9,10]. The significant effects of the LHCD on the ITBs, resulting radially broad and long-lasting (about 40 t_E), can be reasonably explained only by considering the effect of the physics of the edge. Indeed, the LH-deposition profiles obtained considering or not the physics of the edge have been inputted, with experimental kinetic profiles and magnetic measurement data, in the JETTO code [21] for modeling the evolution of the q-profile. Considering the phase of the current ramp-up, the LH deposition profiles and the respective q-profiles are shown in Fig. 3. The effects in the LH-depositions when retaining the minor changes due to measurement uncertainty of the inputted kinetic profiles are also shown in the figure. It is evident that only including the effect of the physics of the edge a more precise deposition profile is obtained, with most of the current driven at about half radius. In a further simulation performed considering LHCD during the main heating phase, in which broader T_e profiles occur, the peak of LH deposition moves to about two thirds of the minor radius. The corresponding q-profiles are shown in the Fig. 4 (case of the experiment with P_{LHCD}= 2.2 MW). In the experiment, the radial foot of the LHCD-sustained ITBs is produced close to the radial position of the layer with low/negative shear. The low shear condition is lost at t≈10 s, consistently with the ITB collapse. Steady state ITBs are expected to occur, instead, operating with double LHCD power. These results support the hypothesis that a low shear layer stabilizes the turbulence, thus improving the local confinement and producing the observed ITB behavior [22-26].

The PI produces a similar spectral broadening in LHCD experiments performed in different machines [10,11]. The typical parameters relevant to the PI-driven quasi-mode are: $k_\perp$ ≈ 10 - 15 cm^{-1}, $\omega/2\pi$ ≈ 0.2 – 1 MHz, consistently with the frequency broadening measured in the LH range by RF probes, and in the density fluctuation range by microwave reflectometry [27,11]. Such common behavior is determined by the circumstance that all the LHCD experiments meet similar operating conditions, in the operating frequencies and scrape-off parameters, which determine mainly the growth rates and the launcher coupling performance as well. These conditions are: $n_{0//}$ in the range 1.5 – 3, and layers with ω_{pe}/ω_0 ≈1 are located near the effective antenna-plasma interface (as necessary for launching the slow electron plasma wave); ω_{pe}/ω_0 of the order of ten, or more, in the layers close to the last closed magnetic surface, as typically obtained in JET and in other tokamaks as well [28,29]; the electron temperatures are in the range from a few eV to 100 eV; the scrape-off radial dimensions lye typically in the range of 3 - 6 cm.

Figure 5 shows the LH deposition profile modeled considering the physics of the edge for an experiment utilizing LHCD on FTU [28,29]. The peak of the deposition profile is close to the maximum of the emission of the FEB camera (fast electron Bremsstrahlung). However, the relatively narrow electron temperature profile does not allow performing a satisfactory test of the precision of the LH deposition profile. The electron temperature profile of the FTU experiment is indeed much narrower than that of the aforementioned experiment of JET. The layer with 2 keV is located at about one thirds of the minor radius. Such circumstance produces a deposition reasonably located inside the inner half of plasma utilizing whatever LHCD model.

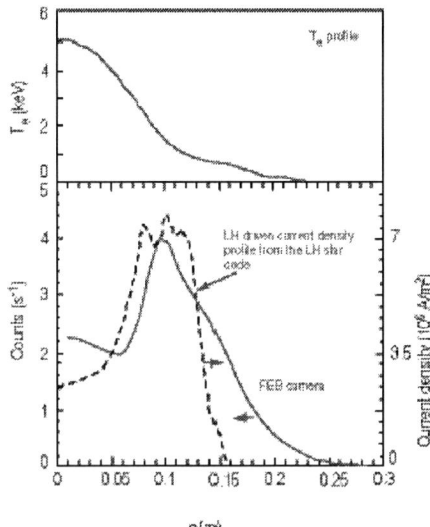

Fig. 5. Modeling of the LH deposition profile compared with the FEB camera profile for an electron ITB experiment produced utilizing 1.5 MW of LHCD power on FTU. ρ is the flux co-ordinate.

CONCLUSIONS

The conditions necessary for the occurrence of the spectral broadening induced by parametric instabilities coincide with those necessary for coupling LH power to tokamak plasmas. The parametric dispersion relation available by classical works of the literature and the convective losses due to finite extent of the pump and plasma inhomogeneity have been considered for carrying the numerical computation of frequency, growth rates, amplification factor and pump power depletion of the PIs. The ion-sound-driven PI represents the most important channel for LHCD experiments. The present model explains the existing results on long-lasting ITBs sustained by LHCD. A layer with low magnetic shear is produced in the outer half of plasma, consistently with both the radial foot location and with the time duration of ITB. Conversely, by considering only the multi-radial reflections to bridge the $n_{//}$ gap, the experimental results are not interpreted with sufficient accuracy.

Due to the dependence of the PI growth rate on the electron temperature, a strong reduction of the PI growth is expected operating with low recycling/higher electron temperature at the edge (e.g., by vessel Lithium-coated [30,31]). In these conditions, LHCD operations at operating plasma densities higher than in the present experiments would be possible. In addition, a proper tailor of the launched $n_{//}$ antenna spectrum (electronically achievable) would possibly allow a successful control of the LH deposition and of the magnetic shear profiles in the plasma.

The use of the proposed LHCD model as a predictive tool will allow the design of experimental scenarios requiring the control of the q-profile for improved stability and confinement in steady-state and advanced-tokamak regimes.

ACKNOWLEDGMENTS

The authors acknowledge Dr. S. Bernabei, Dr. M. Brambilla, Prof. N. Fish, Dr F. Santini and Dr. R. Wilson for the helpful discussions. This work has been done under the EFDA JET work programme [31].

REFERENCES

1. N. J. Fisch, Phys. Rev. Lett. **41** (1978) 873
2. S. Bernabei, et al., Phys. Rev. Lett. **49**,2 (1982) 1255
3. M. Brambilla, Kinetic Theory of Plasma Waves, Clarendon Press, Oxford (1998) 551-557
4. S. Succi, et al., Plasma Phys and Contr. Fusion 27 (1985) 863-871
5. F. Santini in Course and Workshop on Applications of RF waves to Tokamak Plasmas, Varenna (Italy) 1985, Editors: S Bernabei, U. Gasparino and E. Sindoni, International School of Plasma Physics 1985, 251; and F. De Marco, same Course at p. 316
6. Bonoli P.T., Englade R.C., Phys. Fluids, **29**, 2937 (1986)
7. Mailloux, et al., Phys. of Plasmas, 9,5, (2002) 2156
8. F. Crisanti, et al., Phys. Rev. Lett., **88** (2002) 145004
9. C. Castaldo, et al., Phys. of Plasmas, 9 8 (2002) 3205
10. R Cesario, et al., PRL **92** 17 (2004) 175002
11. R. Cesario, et al. submitted to Nuclear Fusion
12. T. H. Stix 1965, Phys. Rev. Letters **15**, 878 (1965)
13. Y. Takase, et al. Phys. Fluids **26**, 2992 (1985)
14. R. Cesario and V. Pericoli, Nucl. Fusion (1987)
15. R. Cesario and A. Cardinali, Nucl. Fusion **29**, 10 , 1709 (1989)
16. M. Porkolab, Phys Fluids **17** ,1432 (1974)
17. C.S. Liu and V. K. Tripathi Phys. Fluids Report **24**,1709 (1984)
18. V. K. Tripathi, C.S. Liu and C. Grebogi, Phys. Fluids **22**, 1104 (1979)
19. L. Chen, R. and L., Berger, Nucl. Fusion **17**, 779 (1977)
20. A. Cardinali, in Recent Results in Phys Plasmas, **1**, 185 (2000)
21. G. Cenacchi, A. Taroni, in Proc. 8th Computational Physics, Computing in Plasma Physics, Eibsee 1986, (EPS 1986), Vol. 10D, 57
22. Beklemishev, Horton, Phys Fluids B **4**, 2176 (1992)
23. F Romanelli, F. Zonca, Phys Fluids B **5**, 4041 (1993)
24. E.J. Strait, et al., Phys. Rev. Letters **75**, 4421 (1995)
25. F. Zonca, et al., Physics of Plasmas **11**, 2488-96, (2004)
26. L. Chen, et al. Physical Review Letters **92**, 075004-07, (2004)
27. R. Cesario, et al., Nucl. Fusion **32**, 2127 (1992)
28. V. Pericoli-Ridolfini, et al., Journal of Nuclear Materials 20 (1995)
29. M. Leigheb et al., .Journal of Nuclear Materials 914 (1995)
30. M. Abdou, et al., Fus. Eng. Design,. **54**, 181 (2001)
31. S.I. Krasheninnkov, L.E. Zakharov, G.V Pereverzev Phis. of Plasmas, **10** 1678 (2003)
32. J. Pamela et al., *in Proceedings of the 19th International Conference on Fusion Energy*, Lyon, 2002 [International Atomic Energy Agency (IAEA), Vienna, 2002]. (All the members of the EFDA JET Collaboration appear in the appendix of the paper).

Long pulse, multi-MW operation in Tore Supra

G.T. Hoang
on behalf of the Tore Supra Team
Association Euratom-CEA, CEA Cadarache, CEA-DSM-DRFC,
F-13108 St Paul lez Durance, France

Abstract. Long pulse operation on Tore Supra has now entered a new phase, characterised by the use of heating power level in excess of 10 MW, during pulses lasting several tens of resistive times. This has been made possible by the combined use of 3 radiofrequency heating and current drive systems, at the ion cyclotron frequency (9 MW coupled to the plasma at 57 MHz), the lower hybrid frequency (3 MW at 3.7 GHz) and the electron cyclotron frequency (0.7 MW at 118 GHz). Key technological and physics issues related to long pulse operation, required for a reactor, are addressed.

This paper gives an overview of Tore Supra results obtained in most recent experiments, in which specific real time (RT) algorithms dedicated to the device safety and plasma control have been routinely used. In terms of plasma duration and injected energy, significant progress has been recorded, namely, a LH energy exceeding 1GJ has been injected into zero loop voltage plasmas lasting more than 6 minutes. Pellet fuelled LH driven discharges lasting up to 2 minutes have been also demonstrated. High power experiments combining ICRH and LHCD have been carried out at plasma density close to the Greenwald limit.

Real time control. Controlling simultaneously plasma parameters, PFC temperature and heating systems is necessary for steady-state operation. Multiple feedback has been used to obtain steady-state discharges at zero loop voltage.

In particular, RT plasma equilibrium reconstruction[1] allowed to control both plasma position and shape within a few millimeters, a key issue for RF wave coupling. A new comprehensive RT control is now available in Tore Supra. RT treatment of various diagnostics (hard X-ray tomography, interfero-polarimetry, impurities..) is dedicated to the device safety, and to prevent MHD instabilities. In addition, a set of seven actively cooled infrared (IR) endoscopes survey five RF antennae continuously during the discharge. Specific RT algorithms are used to prevent overheating of these antennae (Fig. 1).

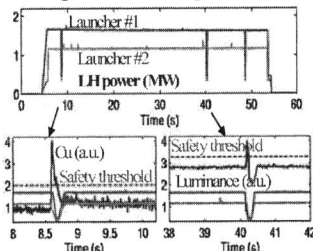

Fig. 1: Discharge lasting 55 s with about 3 MW of LH power (top). The LH power is lowered momentarily at t=8.6s and t=40.2s (at t=8.6s, because of the high level of Cu detected; at t = 40.2s, due to the arcings in the waveguides detected by IR camera).

Fully LHCD experiments. Long pulse LHCD experiments have been carried out at: B_T = 3.4 T, I_p = 0.5-0.7 MA, and density $n_e(0) \le 3 \times 10^{19}$ m^{-3}. The accessible range of I_p and density is presently constrained by the available LH power and the value of B_T which was optimized to avoid MHD instabilities in these conditions[2]. Figure 2 illustrates a discharge performed at 0.5 MA lasting six minutes[3], with an LH energy of 1.07GJ injected by two launchers. One of two LH launchers devoted to long pulse operation[4] has injected more than 2 MW over six minutes without any arcing and with a reflection coefficient < 5 %. All the plasma parameters were perfectly controlled in stationary conditions. As can be seen in Fig. 2d, the neutron flux was constant, confirming very stable density, temperature and impurity content over 6 min. This discharge exhibited the features of the so-called Lower

Hybrid Enhanced Performance (LHEP) regime[5]: i) a high value of inductance, l_i=1.45; ii) and a negative magnetic shear within r/a <0.2 (Fig. 2e). A peaked electron temperature profile ($T_e(0)$ = 4.65 keV) was observed, together with an enhancement of energy confinement with respect to the ITER L-mode scaling (H~1.35, Fig. 2b), mainly by the electron channel. At t = 252s, a burst of metallic impurities, coming from the LH waveguides, was detected. As a consequence, the LH power was lowered by the RT control to avoid damaging the LH launcher and disruption. The discharge was immediately recovered with an acceptable metallic impurity level. However, after this event, the 3/2 MHD activity localized near the plasma centre developed, which did not affect the global energy confinement. This example shows clearly that integrated RT control of device safety and plasma equilibrium is crucial for continuous operation.

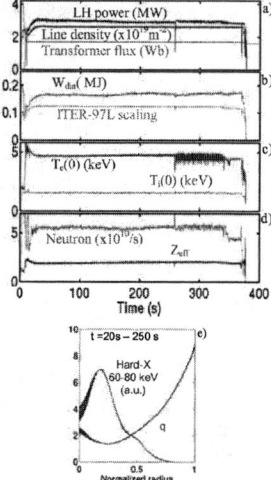

Fig. 2: Six-minute LHCD discharge with injected energy of 1.07 GJ

Pellet fuelled LH driven discharge. In Tore Supra, pellets can be injected continuously (≤10 Hz) from four poloidal locations regularly distributed from the low to the high field side, at a speed between 100-600 m/s.

So far, most of the experiments have been carried out in the low field side configurations. They have been performed under RT control for maintaining the loop voltage at a constant value close to zero (~ 70mV) together with a plasma density near the target (volume average density of $1.5 \times 10^{19} m^{-3}$). One hundred and fifty-five pellets have been injected into the plasmas lasting up

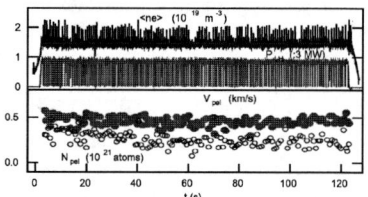

Fig. 3a: Time traces of a two-minute LHCD discharge fuelled by 155 pellets.

to 2 minutes[6] (Fig. 3a). In order to avoid rapid pellet ablation by the fast electrons, the LH power (3MW) was notched 30 ms (~ slowing down time of fast electrons) before the arrival of each pellet (speed of 500m/s). Average instantaneous fuelling efficiency was estimated about 65% (injected atoms of 4×10^{22}). This demonstration is very encouraging for ITER continuous operation in the presence of energetic electrons required for efficient current drive.

Several tens of pellets have also been injected in experiments with up to 8MW of ICRH and 2 MW of LHCD. During the pellet injection, in spite of perturbed edge plasma the ICRH coupling was pretty well controlled (Fig. 3b), except the ICRF antenna located near the pellet injector.

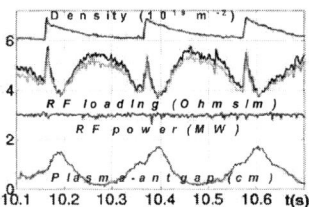

Fig. 3b: Combined LHCD (2MW) and ICRH (8MW) discharge fuelled by pellets: density and ICRH coupling parameters of one of 3 used antennae.

Combined ICRH & LHCD experiments. High power combining ICRH and LHCD has been injtected in plasmas performed at a density close to the Greenwald limit (f_G=0.7 – 0.93), with a loop voltage less than 0.1 V: B_T = 3.7 T, I_p = 0.6-1.1 MA, line density nl = 4-8x10^{19} m^{-2}. The bootstrap fraction (f_B) was about 20%. So far, the highest injected energy was 470 MJ: 4 MW/65 s ICRH pulse during a plasma sustained by 3 MW of LHCD.

The fraction of non-inductive current (f_{NI}) was close to 80% (Fig. 4). A higher ICRH power of 8 MW has been also applied in LHCD (2-3MW) plasmas (Fig. 5). In terms of confinement property, the energy confinement time of these L-mode discharges was found to exceed the ITER L-mode prediction, with T_i close to T_e (except in the core r/a<0.2 where the LH deposition is dominant), which corresponds to a confinement factor H_{Hy2} ~ 0.70 with respect to the ELMy H-mode scaling. Moreover, in some discharges at 0.6 MA, no sawteeth have been observed (q(0) >1 from polarimetry measurement), when preforming the current profile by early application of LHCD. This feature is analogous to the so-called Hybrid regime. As can be seen in Fig. 6, sawteeth during the LH phase (before t=10s) are stabilized when the ICRH power of 6.3MW is applied.

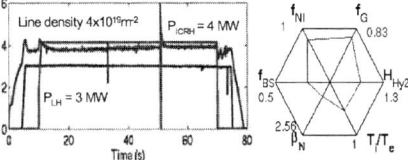

Fig. 4: Combined LHCD/ICRH discharge (0.6MA/4x10^{19}m^{-2}) lasting 80s and its performance compared to the ITER parameters expected in steady-state scenario.

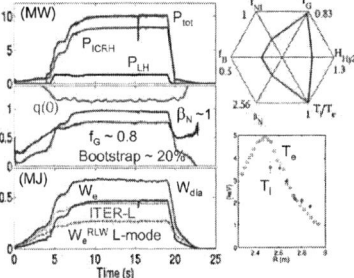

Fig. 5: Combined LHCD/ICRH discharge performed at 0.9MA/6x10^{19}m^{-2}, with T_i ~ T_e.

The improved confinement is maintained together with q(0) ~1.5 for 20 s.

A favourable characteristic occurs in ICRH discharges with spontaneous plasma toroidal rotation: V_Φ up to 70 km/s in the core region. The radial profiles of V_Φ, measured by CXRS diagnostic, of four consecutive discharges with 8MW of ICRH alone, are shown in Fig. 7a. They seem to indicate the existence of a shear rotation at R=2.8m where the error bars are large, correlated to the increase of T_i inside this radius (Fig. 7b). These results should be confirmed in future experiments. Stability analysis[7] indicated that both the Ion Temperature Gradient (ITG) and Trapped Electron (TE) modes inside R=2.8m are expected to be stabilized by the E×B shear, through V_Φ. Stabilizing ITG and TE instabilities could explain the observation of the confinement enhancement.

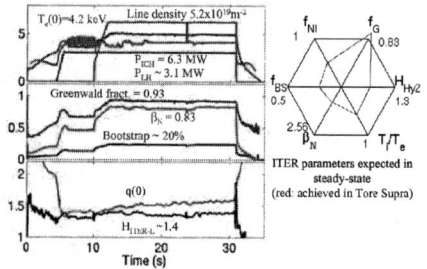
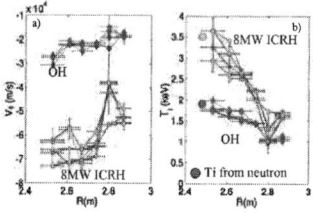

Fig. 6: L-mode sawtooth free discharge exhibiting the feature of the Hybrid regime (Ip= 0.6MA, q_{95} ~6.9).

Fig. 7: Radial profiles for 4 discharges: (a) V_Φ (counter current direction); (b) ion temperature..

Previouly, ICRH experiments combined with a fast current ramp have been developed for triggering the ITBs. Recently, the exploration of this potentially high performance scenario has been resumed, with the aim of expanding the previous operational parameter range ($\leq$ 4-5MW, nl <5 x 10^{19} m^{-2}) towards higher injected power. LHCD has been used for extending the plasma duration. Up to 11 MW of combined ICRH + LHCD power were injected, for about 1s during the fast I_p ramp up (from 0.4MA to 1.4 MA), at nl ~ 8 x 10^{19} m^{-2}, corresponding to f_G~85%. Current profile analysis showed that the q-profile is well reversed, mainly thanks to the fast ramp-up. During the high power phase the q-profile is slowly evolving and reaches q(0)~1 after ~1.5s, probably because the LH driven current is not sufficient. In this phase, the plasma pressure increased by up to 40 % within r/a~0.5, mainly due to the increase of n_e (Fig. 7).

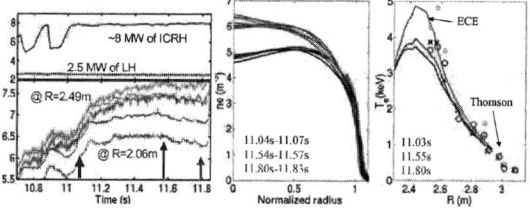

Fig. 7: Combined LHCD and ICRH discharges with fast current ramp-up: (left) RF powers and line density by interferometry; (right) n_e and T_e profiles during the flat top Ip=1.4MA.

Prospects. Tore Supra now operates routinely in steady-state, addressing novel issues both in physics and technology related to very long duration discharges under actively cooled PFC interaction environment. The presently on-going upgrade[8] of the 3 RF systems (9MW ICRH, 6MW LHCD, 2 MW ECRH) to 600-1000s capability will allow to increase the operating domain contributing again to addressing ITER key steady state issues.

---

[1] F. Saint-Laurent, 9th Int. Conf. on Accelerator and Large Exp.l Phys. Control Sys., Korea (2003)
[2] P. Maget, Nucl. Fusion **45** (2005) 69
[3] D. van Houtte et al, Nucl. Fusion **44** (2004) L11–L15
[4] Ph. Bibet et al, Proc. 20th Symposium on Fusion Technology (1998) 339
[5] G.T. Hoang , Nucl. Fusion **34** (1994) 75
[6] A.Géraud et al, 16th PSI (2004, Portland, USA), to be published in J. Nucl. Mater.
[7] C. Fenzi-Bonizec, EPS Conference, London, 2004
[8] B. Beaumont et al, Fusion Engineering and Design, vol.56-57, 667-672 (2001)

Lower hybrid current drive efficiency on Tore Supra and JET

M.Goniche, J.F.Artaud, V.Basiuk, Y.Peysson, T.Aniel, A.Ekedahl,
G.Giruzzi, F.Imbeaux, J.Mailloux[1], D.Mazon, W.Zwingman and
JET EFDA contributors*

*Association Euratom-CEA, CEA/DSM/DRFC, CEA/Cadarache,
F-13108 St. Paul-lez-Durance, France*
[1]*Euratom/UKAEA Fusion Association, Culham Science Centre, Abingdon, Oxfordshire, UK*

The lower hybrid current drive efficiency of 66 Tore Supra pulses has been investigated. The ohmic part of the plasma current (0.6-0.9 MA) is very small (V_{loop} <50mV) for most of the pulses. Different scaling laws were tested with three input parameters: the wave directivity, the plasma current (I_p) or the volume average temperature (<T_e>) and the effective charge (Z_{eff}). When applying these scaling laws to four JET pulses, no discrepancy is found except for the high plasma current (I_p=2.0MA) pulse. Finally the best fit was found by replacing T_e (or I_p) by the thermal electron confinement time. This result is supported by the hard X-ray (HXR) diagnostic indicating a fairly good correlation between the plasma edge HXR emission, normalized to the central emission, and the thermal electron confinement time.

Keywords: **Lower hybrid, current drive, efficiency.**
PACS: 52.50.Sw

INTRODUCTION

Lower hybrid current drive (LHCD) efficiency η as high as $3\text{-}3.5\times10^{19}$ A.W^{-1}.m^{-2} has been reported on JET [1] and JT-60U [2]. The scaling of the efficiency with plasma parameters has been long debated and a simple scaling, the so-called JT-60 scaling, η=12 <T_e>/(5+Z_{eff}), has been proposed. However the database provided by the different tokamaks indicates a large scattering of the experimental points. Surprisingly this scaling does not take into account the $N_{//}^{-2}$-weighted directivity of the wave as predicted by theory. Moreover most of the pulses were not fully LH driven and the calculation of the efficiency with a residual DC electric field is a difficult task, sensitive to different plasma parameters: with the 1-D CRONOS code [3], we verified for example that, for a JET discharge with a loop voltage as low as 29mV, neglecting the electric field leads to an over-estimation of the efficiency by 40%. A previous study [4] indicates that the plasma current could play a major role for the current drive efficiency, following similar findings on JT-60 [1]. In order to investigate more in details the scaling parameters of the LHCD efficiency, a data base of 66 Tore Supra pulses was constituted. In addition four JET pulses were analyzed.

THE DATA BASE

66 Tore Supra pulses with LHCD only (no other additional heating) were selected. For these pulses the loop voltage, averaged on a time interval larger than 1s (2s for 90% of pulses), is ranging between −28mV and +95mV. The time slice was chosen in order to have stationary conditions. Only four shots have a loop voltage above 50mV. For these shots the current drive efficiency was carefully computed by means of CRONOS. For the others, a 0-D correction of the plasma conductivity was applied. The exactitude of this correction was verified by means of CRONOS. The normalized directivity D_n defined as

$$D_n = \int_{Nacc}^{\infty} 4\frac{P(N_{//})dN_{//}}{N_{//}^2} - \int_{-\infty}^{-Nacc} 4\frac{P(N_{//})dN_{//}}{N_{//}^2}$$

where N_{acc} is the minimum parallel index for which the wave is accessible to the plasma ($N_{acc}=1.25$), was computed for each pulse. This directivity varies between 0.49 and 0.94 when $N_{//}$, for which $P(N_{//})$ is maximum, varies between $N_{//0}=1.7$ and $N_{//0}=2.3$. It should be noted that the unequivocal relationship between D_n and $N_{//0}$ holds only for regular power feeding in amplitude and phase.

Other parameters vary in the following range: 0.3-0.9MA for the plasma current I_p, $1.3-3.1 \times 10^{19}$ m^{-3} for the line-averaged density with a peaking factor in the 0.2-0.6 range, 2.1-4.6MW for the LHCD power, 0.65-1.75 keV for the volume-averaged temperature $<T_e>$, 1.6-5.3 for the line-averaged effective charge Z_{eff}. For these rather low density pulses with quasi central LH deposition, the DELPHINE ray-tracing code indicates the deposition to be centered on $r/a \approx 0.2$, in agreement with hard X-ray measurements. The electron temperature profile is peaked and $T_e(0)/<T_e>$ varies between 3.0 and 5.5. Toroidal field is between 3.5 and 3.9T at the center. When the bootstrap current I_{bs}, whose contribution is small ($0.05<I_{bs}/I_p<0.20$) is taken into account, the measured LHCD efficiency varies between 0.4 and 0.9×10^{19} A.W^{-1}.m^{-2}.

The four JET pulses were achieved at higher plasma current (1.1-2.0MA) with a volume-averaged temperature in the range of 1.4-2.5keV. Two pulses were performed at rather low loop voltage (29 and 64mV), the two others at high loop voltage during the current ramp-up phase for the first one and in counter current drive (CCD) configuration for the second one. Power deposition is centered on r/a=0.4, except for the CCD case where the broad deposition is centered on r/a=0 [5]. Efficiency varies between 0.9 and 1.55×10^{19} A.W^{-1}.m^{-2}

SCALING LAWS

From the Tore Supra pulses, scaling laws are established with different input parameters by a least-square method. The scaling with three input parameters D_n, Z_{eff} and $<T_e>$ (or I_p) indicates a dependence on directivity which is only with a power law close to 0.5 whereas theory would predicts a linear dependence. The power law for the temperature dependence is also close to 0.5, i.e. weaker than the JT-60 scaling. The $Z_{eff}^{-0.27}$ dependence is indeed the closest power law to the $6/(5+Z_{eff})$ scaling and for the range of experimental Z_{eff}, they differ by less than 5%. The $<T_e>$ and I_p scalings (figure 1) have almost the same power law for this third parameter and the standard deviations ($\sigma=0.07/0.08$) are very close. This result is due to the very high correlation between $<T_e>$ and I_p data for this database

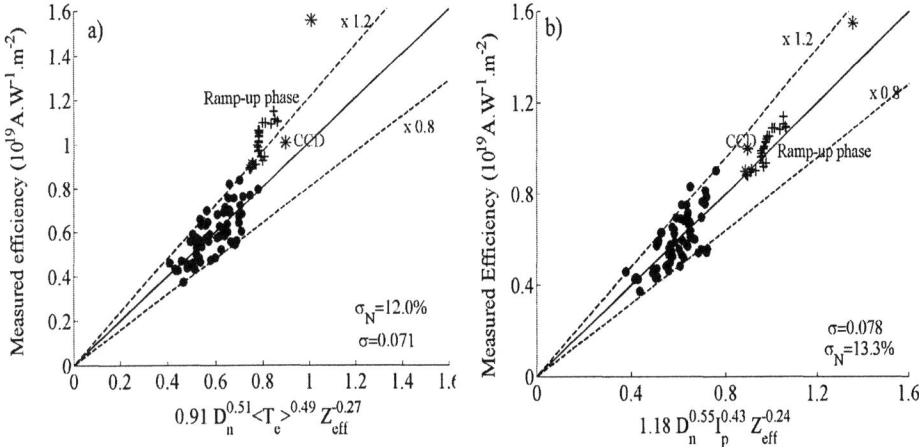

FIGURE 1. Measured current drive efficiency as a function of the scaling law established with D_n, Z_{eff} and a) $<T_e>$, b) I_p. from the Tore Supra database (circles). The JET pulses(*) are plotted with these scalings. For the JET pulse in the ramp-up phase (+), different times are shown.

When these two scalings are applied to the JET pulses, a rather good agreement is found with a significant discrepancy only for the 2MA pulse with the highest efficiency: the computed efficiency is lower (by 30%) from the experimental efficiency for the T_e scaling. For the I_p scaling, the discrepancy is only 15%.

The current drive efficiency was computed, considering an upper bound $N_{//L}$ ($=c/2.5v_{th}$) of the up-shifted $N_{//}$ spectrum, with the experimental profiles of density and temperature. Profiles of Z_{eff} are calculated assuming a peaking factor of 0.4 and power deposition is assumed to be centered on $r/a=0.2$ with a width of 0.3. The calculated directivity D_n was also taken into account. A good agreement is found with the same standard deviation than the scaling law with $<T_e>$ ($\sigma=0.07$).

HARD X-RAY MEASUREMANTS

From the 59 chord hard X-ray (HXR) diagnostic installed on Tore Supra, the width of the LH deposition profile was characterized by computing the ratio of a central chord (HXR40, $r/a\sim0$) to an outer chord (HXR29, $r/a\sim0.5$). When the efficiency is plotted as a function of this ratio, a close correlation is found for most pulses (figure 3). For the low $N_{//}$ cases ($N_{//}<1.75$) with rather peaked HXR profiles (ratio >4), the efficiency is clearly larger. This diagnostic is also sensitive to direct losses of fast electrons at the plasma edge: the signal of the two chords viewing the toroidal limiter is very large during LHCD, typically from 2 to 10 times that of the central chord. We found a good correlation between this normalized signal and the confinement time of the thermal electrons estimated from the n_e, T_e profiles. When the confinement time increases by a factor 2, this ratio decreases by a factor ~10. In order to further document the close relationship between the fast electron and thermal electron confinement, a new scaling of the efficiency is performed where $<T_e>/I_p$ is replaced by the electron confinement time $\tau_{E,th}$. For the Tore Supra pulses, the best fit ($\sigma=0.056$) is found with this input parameter (figure 4). This scaling fits the JET pulses although the 2MA pulse is above (+30%) the scaling.

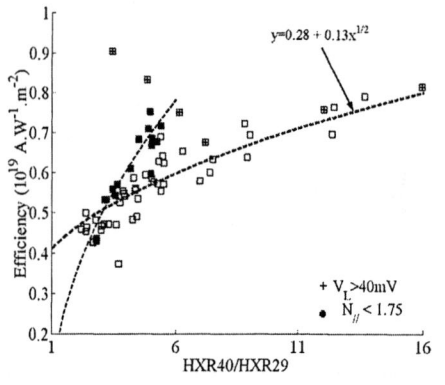 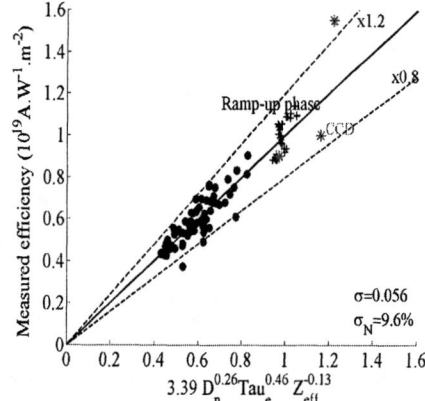

FIGURE 3. Measured efficiency as a function of the width of the HXR emission.

FIGURE 4. Measured efficiency as a function of the scaling law established with D_n, Z_{eff} and $\tau_{E,th}$. Same symbols as figures 1 and 2.

CONCLUSION

For the range of plasmas parameters accessible on Tore Supra, the scaling of the current drive with $<T_e>$ is clearly weaker than the linear scaling. The Z_{eff} dependence ($1/(5+Z_{eff})$), suggested by strong physics arguments, is actually found. Because of the strong correlation between I_p and the $<T_e>$ of this data base, the most relevant parameter cannot be clearly identified, although the 2MA JET pulse suggests that the current is the scaling factor. Peaked LH deposition profiles lead, in most cases, to higher current drive efficiency. The confinement of fast electrons, closely related to that of thermal electrons, play a key role for the efficiency.

REFERENCES

1. ITER physics expert group on energ. particles, heating and current drive, Nucl.Fus., **39** (1999) 2495
2. Y.Ikeda et al., IAEA-CN-60/A3-1, Proc. of the 15th IAEA conference, Séville (1994) 415
3. V.Basiuk, J.F.Artaud et al., Nucl.Fus., **43** (2003) 822
4. Y.Peysson et al., Plasma Phys .Control. Fusion **42** (2000) B87
5. V.Basiuk et al., e-proc. 12th Int..Cong. on Plasma Phys., 2004, http://hal.ccsd.cnrs.fr/ICPP2004/en/

Statistical Analysis of Lower Hybrid Current Drive Efficiency on FTU

Giuseppe Calabrò[1], V. Pericoli Ridolfini[1], FTU Team[1], ECRH Team[2]

[1]*Associazione EURATOM-ENEA sulla Fusione, Via E. Fermi 45 FRASCATI-Roma, Italy*
[2]*Associazione EURATOM-ENEA sulla Fusione, IFP-CNR, Via R. Cozzi,53 Milano, Italy*

Abstract. A multi parameter statistical analysis of Lower Hybrid Current Drive (LHCD) efficiency has been carried out on all the data available for the Frascati Tokamak Upgrade (FTU). The volume of the parameters space scanned is quite large and of direct interest for ITER. The positive dependence of the CD efficiency with $<T_e>$ has been confirmed. Comparing the results with the expectations for the CD process, a reasonable agreement is found for the ion charge, Z_{eff}, dependence (CD efficiency $\propto 1/(Z_{eff}+5)$), while the influence of other parameters (like q_a, $\bar{n}_e$, B_T, $N_\parallel$) is discussed in detail. Clear evidence of the synergy between the lower hybrid and electron cyclotron (EC) waves is found, provided the EC power is directly absorbed on the fast e⁻ tails. The effect of different $N_\parallel$ (LH parallel refraction index) spectra is also investigated.

Keywords: Lower hybrid waves, electron cyclotron waves, current drive.
PACS: 52.55Fa, 52.55Wq.

1. INTRODUCTION

The efficiency of the lower hybrid waves in driving toroidal currents is a key issue for dimensioning a LH system for ITER. The purpose of a multi parameters analysis for the LHCD efficiency was not only to derive a reliable scaling law applicable in the next FTU experiments, but also to single out those factors common to all tokamaks which could deserve more investigations. The paper is organised as follows. The analysis results are presented in Sec. 2. The effect of EC heating absorption on electron tails is discussed in Sec. 3. In Sec. 4 conclusions are drawn.

2. LHCD – DATA HANDLING

The parameter space spanned presently by LHCD data on FTU is bounded within the following limits: line averaged plasma density $0.29 \cdot 10^{20} \leq \bar{n}_e \leq 1.29 \cdot 10^{20}$ m⁻³, central electron temperatures $1.1 \leq T_{e0} \leq 7.4$ keV, corresponding to volume averaged temperatures between $0.27 \leq <T_e> \leq 1.2$ keV, plasma current $0.3 \leq I_p \leq 0.7$ MA, toroidal magnetic field $4 \leq B_T \leq 7.2$ T, edge safety factor between $4.7 \leq q_a \leq 10.7$, effective ion charge state $1.1 \leq Z_{eff} \leq 9$. The range of radio frequency (RF) parameters are instead the following: LH power $0.4 \leq P_{LH} \leq 2.1$ MW, peak value of the power spectrum of the parallel index of refraction $1.32 \leq N_{\parallel,pk} \leq 2.42$, EC heating power $0 \leq P_{ECH} \leq 0.8$ MW. The present analysis has been limited to discharges with $Z_{eff} < 6$. Two subsets of discharges have been analysed. The first with more than 70% of plasma current driven by LH up to full CD, the second with $50\% < I_{LH}/I_p \leq 70\%$. All the selected discharges satisfy the good accessibility criterion fixed in Ref. [1]: $N_{\|,pk}-N_{\|,acc} \geq 0$, being $N_{\|,acc}$ the minimum $N_\|$ value requested to a LH wave to access a plasma with $n_e = \bar{n}_e$ and $B = B_{T0}$ (B_{T0} value of B_T at the vessel centre). The regression analysis was started directly on the magnitude of I_{LH} (as calculated in Ref. [2]) rather than on η_{CD}, which is a derived quantity, $\eta_{CD} = \bar{n}_e \cdot I_{LH} \cdot R_M / P_{LH}$, with R_M=plasma major radius. The analysis result is shown in Fig. 1 where the experimental values of I_{LH} are plotted versus the regression variable. The correlation coefficient R_{lin} is very close to 0.9 and the standard deviation is around 5% only. The loop voltage does not play any role: if included in the regression variable it produces $\Delta R_{lin} \approx 0.001$ only, and it is assigned with an exponent <0.001. The safety factor q_a of the discharge can be easily included as free parameter instead of either B_T or I_p to which is linked in FTU by the relation $q_a = 0.53 \cdot B_T / I_p$. With the same points considered in Fig. 1 the regression analysis has been performed on the CD efficiency calculated for $Z_{eff}=1$ [1]. The regression scaling laws are the following:

$$I_{LH} = 0.87 \cdot <T_e>^{0.497} \cdot \bar{n}_e^{-0.883} \cdot q_a^{-0.388} \cdot B_T^{0.774} \cdot (Z_{eff}+5)^{-0.928} \cdot P_{LH}^{0.625} \cdot N_{\|,pk}^{-0.291} \quad (1)$$

$$\eta^*_{CD} = 0.1543 \cdot <T_e>^{0.517} \cdot \bar{n}_e^{0.0796} \cdot q_a^{-0.392} \cdot B_T^{0.774} \cdot P_{LH}^{-0.364} \cdot N_{\|,pk}^{-0.296} \quad (2)$$

The measuring units are: I_{LH}[MA], T_e[keV], n_e[10^{20} m^{-3}], B_T[T], P_{LH}[MA] and η^*_{CD}[10^{20} m^{-2}·A/W]. Figure 2 reports all the η^*_{CD} (=η_{CD} @Z_{eff}=1) values plotted versus the regression variable. The two continuous curves, shown in Fig.2, give the average value of the expected CD efficiency for $N_{\|pk}$=1.52 and $N_{\|pk}$=1.82 according to Eq. (3) for the spanned range of the regression variable.

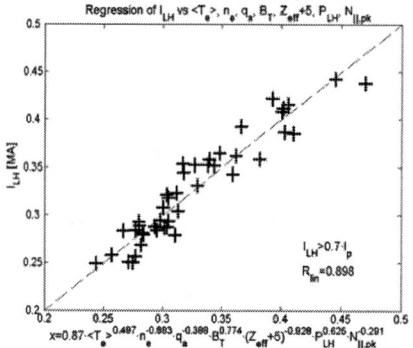
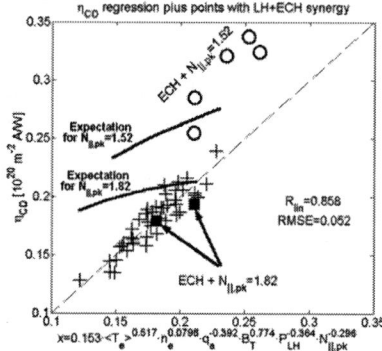

FIGURE 1. Multiple linear regression analysis for the amount of LH driven current. Only discharges with Z_{eff}<6 and I_{LH}/I_p>70% are considered.

FIGURE 2. Multiple linear regression analysis for the CD efficiency (done only on the same points of Fig. 1 represented by +). Values calculated when also EC power is launched (B_T=7.1 T) are also added (O symbols).

Discharges with the ECH power absorbed by the thermal bulk electrons are included in the data set used for regression, whereas are excluded those where the cold resonance is absent and direct absorption can take place only on the LH fast e⁻ tails, which are shown with void circles on Fig. 2. This choice aims to highlight possible synergetic effects, as detailed in Sect. 3. To discuss the effect of the various quantities on the CD efficiency and discrepancies with the expected values, we should start from the classical definition of CD efficiency [3]. We suppose that the fast tail is bounded at high velocity by $v_1 = c/N_{\|,pk}$, and extends towards the low velocity to the value

$v \approx 3.5 v_{th,e}$ in order to have significant absorption. Following Ref. [4] we also assume that $<T_e>$ is a good estimate for the average value of $v_{th,e}$. Consequently the dependence of η_{CD} on T_e can be rewritten:

$$\eta_{CD} = \frac{7.92}{5+Z_{eff}} \cdot \frac{1/N_{\|,pk}^2 - 2.37 \cdot 10^{-2} \cdot T_{e,keV}}{\ln\left[6.49/\left(N_{\|,pk} \cdot \sqrt{T_{e,keV}}\right)\right]} \quad \left[10^{20} m^{-2} A/W\right] \quad (3)$$

The amount of the driven current is then predicted to increase with $<T_e>$, and to decrease with $N_{\|,pk}$ and Z_{eff}, besides of being directly proportional to P_{LH} and inversely to $\bar{n}_e$. With respect to the predictions of Eq. (3), the experimental data show that:

a) I_{LH} increases when decreasing q_a (no dependence from Eq. (3)), dependence on I_p?
b) I_{LH} decreases slightly less than linearly when increasing $\bar{n}_e$
c) a positive dependence on $<T_e>$, $<T_e>^{0.5}$, is found in agreement with [1]
d) I_{LH} grows less than linearly with P_{LH}
e) I_{LH} increases remarkably with B_T (no dependence from Eq. (3))
f) I_{LH} increases with $1/N_{\|,pk}$ much slower than expected by Eq. (3)
g) the dependence on $Z_{eff}+5$ is almost inverse linear, consistently with Eq. (3).

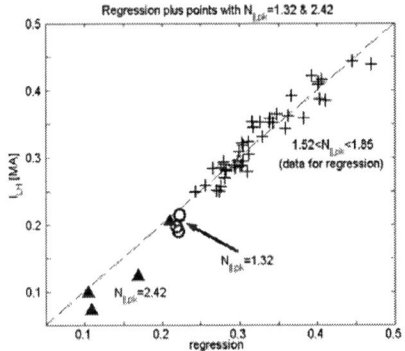

FIGURE 3. Discharges with $N_{\|pk}$=1.32 (O symbols) and =2.42 (▲ symbols) compared with the regression laws and the related points used in Fig. 2. Despite the less driven current fraction the same scaling with $N_{\|pk}$ holds.

The first three features may be accounted for by the toroidal upshift of $N_{\|}$ that the LH waves suffer when propagating in a toroidal geometry, as described in Ref. [2]. The not direct proportionality of I_{LH} to P_{LH} is outside the experimental errors, suggesting a reduced absorbed power [2]. The benefit of increasing B_T can be explained by the effect of the fluctuations onto the radially traveling rays [2]. The large discrepancy between the expected effect of the launched $N_{\|,pk}$ on η_{CD} and that found has prompted us to check the validity of this scaling law extending as much as possible the interval of $N_{\|,pk}$. Within the error uncertainties, the weak dependence of the overall driven current on the $N_{\|}$ spectrum, close to $I_{LH} \propto 1/N_{\|,pk}^{0.3}$, is confirmed by these new points (as shown in Fig. 3). Finally, the impurities behaviour concerning the CD efficiency is close to the expectations. A more careful analysis is needed to account for the above discussed discrepancies and, in particular, for the differences between expected and experimental values shown in Fig. 2.

3. EFFECT OF ECH ABSORPTION ON ELECTRON TAILS

The EC power source in FTU has a frequency f_{EC}=140 GHz. Taking into account the relativistic mass increase and the Doppler shift of the ECH waves when they are launched with an angle $\theta \neq 0$ respect to the normal to B_T, the electrons in the tail generated by LHCD come into resonance at the plasma minor radius r when [5]:

$$\frac{B_T(r)}{B_{res}} = \frac{N_{\|LH} - n_{\|EC}}{\sqrt{N_{\|LH}^2 - 1}} \tag{4}$$

They have velocities $c/N_{\|LH}$, where c is the light speed, $n_{\|EC}=v_{\|EC}/c=\sin\theta$ and $N_{\|LH}$ are the parallel refractive indexes of the LH and EC waves respectively. Amongst the possible several schemes for direct EC absorption by fast e⁻ tails, clear results are attained only if the cold EC resonance is outside the plasma cross-section in order that bulk electrons play no role, and no direct current drive can be ascribed to the EC waves. The two requests imply respectively $B_{T0} \geq 6.8$ T and $n_{\|EC} \leq 0$. We fixed our investigation to the case when B_{T0} ranges between 6.9-7.1 T and $n_{\|EC}$ is =0, i.e. the EC waves are injected perpendicularly to B_T. This is the so-called down-shifted regime, where the electron relativistic mass increment is balanced by an increase of B. Even though synergetic effects have been already reported in FTU in previous works [5, 6], the multi-shot regression analysis has revealed particularly suitable to highlight them. The results are shown in Fig. 2. In the down-shifted regime the amount of the driven current is considerably larger in the $N_{\|,pk}=1.52$ case, whereas for $N_{\|,pk}=1.82$ no increment is found. From Eq. (4), for $B_{T0}=7$ T, the velocities pertaining to the higher $N_\|$ would be resonant outer half the minor radius approximately where little or no LH power deposition occurs and hence no tails develop. Conversely, for $N_\|=1.52$ resonance is possible up to $r/a \leq 0.2$. These statements are supported by the measured radial profiles of the hard X-rays emitted perpendicularly to the magnetic field by bremsstrahlung of the fast electrons. Discharges with significant synergy EC-LH waves appear to follow the same scaling law obtained for the rest of the data but with a larger numerical factor. The values of CD in synergy with ECH are the highest so far reported in the tokamak literature.

4. CONCLUSIONS

The data set of the LHCD discharges is rich enough in FTU to allow a statistical analysis that has the aim to highlight the factors determining the CD efficiency. The regression analysis has pointed out the laws governing the LHCD efficiency. The dependence of η_{CD} on the different quantities involved can be accounted for by considering the various physical processes involved. The effect of increasing the electron temperature that enhances the CD efficiency, previously found [1] has been confirmed on a statistically more solid base. Clear evidence of the synergy between the LH and EC waves is found, provided the EC power is directly absorbed on the fast e⁻ tails, that must be located inside the resonance region.

REFERENCES

1. Pericoli Ridolfini V., et al., *Phys. Rev. Lett.* **V. 82, No. 1**, p. 93-96 (1999).
2. Pericoli Ridolfini V., et al., "Study of Lower Hybrid Current Drive efficiency over a wide range of FTU plasma parameters", submitted to *Nuclear Fusion* (2004).
3. Fisch, N.J., *Phys. Rev. Lett.,* **V. 41** (No. 13), p. 873-876 (1978).
4. Barbato, E., *Plasma Phys. Controll. Fus.,* **V. 40**, A181, (1998).
5. Pericoli Ridolfini V., et al., *in Proc. of the 14th Topical Conference on Radiofrequency Power in Plasmas*, Oxnard, CA, (USA), **V. 595** (2001), p. 225-232.
6. Pericoli Ridolfini V., et al., *in Proc. of the 18th IAEA Fusion Energy Conf.*, Sorrento, Italy, 2000.

Impurity Radiation From The LHCD Launcher During Operation In JET And Investigation Of Launcher Damage

K.K. Kirov[1], J. Mailloux[2], A. Ekedahl[3], LHCD team and JET EFDA Contributors[*]

[1]*Max-Planck-Institute for Plasmaphysics Euratom Association, D-85748 Garching, Germany.*
[2]*Euratom/UKAEA Fusion Association, Culham Science Centre, Abingdon, OX14 3DB, UK.*
[3]*Association Euratom-CEA, CEA/DSM/DRFC, CEA/Cadarache, F-13108 St. Paul-lez-Durance, France*
[*]*See J. Pamela, Fusion Energy 2004 (Proc. 20th Int. Conf. Vilamoura 2004) IAEA Vienna, OV1/2*

Abstract. In this study, the most likely causes of the enhanced radiation in front of the LHCD launcher are investigated: fast ions from the warm plasma, fast electrons parasitically accelerated in front of the grill and arcs. Evidence for the presence of each of these mechanisms is discussed. The experimental conditions favouring the appearance of these phenomena and their impact on the launcher have also been highlighted.

Keywords: LHCD, LHCD launcher, impurity radiation, antenna damage.
PACS: 53.55.Fa, 52.50.Sw, 52.40.Db

INTRODUCTION

The operation of the LHCD system in JET requires avoidance of harmful plasma-launcher interactions, characterised by increased radiation in front of the LHCD antenna. This radiation is due to impurities from the LHCD launcher, and its enhancement indicates possible damage to the grill. Since first operation in 1994, the present JET LHCD launcher has accumulated damage like blobs and protuberances on its top left corner and melted material in the middle of the rows, Fig. 1, and determining the causes of this is relevant to the future applications of LHCD in fusion devices.

THE LHCD LAUNCHER AND THE UTILISED DIAGNOSTICS

The LHCD system on JET consists of 24 3.7GHz klystrons, each of which produces up to 650kW for pulses of up to 20s, arranged in 6 modules (A-F). Each klystron feeds two hybrid junctions, ordered in 6 rows and 8 columns at the grill mouth as shown in Fig. 1. The grill is surrounded by a protective guard frame made of Carbon (C). A poloidal limiter on the right hand side of the launcher separates it from the nearby ICRH antenna B and the launcher is kept radially behind this limiter at all time.

The radiated power in JET is measured by a bolometric diagnostic. Two lines of sight were used in the study: one, which looks at the plasma in front of the launcher and another one which measures the radiation from the main plasma. Their difference, referred to as *d35*, is used as an estimate for the enhanced radiation in front of the grill. In addition, UV spectrometry is used to measure the line radiation of the impurities in the plasma. The launcher grill contains Iron (*Fe*) and Nickel (*Ni*) and increased UV radiation of these elements indicates their release from the grill. Enhanced signals from both, the bolometer and the UV spectrometer, point out that the radiation comes from the LHCD grill. These two diagnostics are also included in the LHCD radiation protection system.

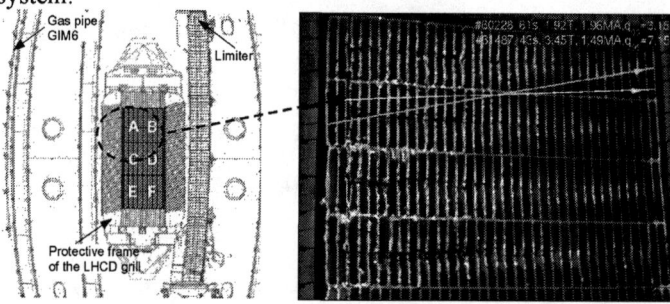

FIGURE 1. A drawing of the LHCD launcher and the surrounding in-vessel components and a picture of the most damaged upper part, modules A and B, of the grill with the reconstructed field lines in front of the launcher for two typical JET pulses.

In the study, data from JET campaigns C7-C14 have been investigated. Data are taken at the time of the maximum of the *d35* signal and *FeXXIII* UV radiation during LHCD operation.

ASSESSMENT OF THE MOST LIKELY CAUSES.

Analysis of the experimental data shows that the enhanced impurity radiation from the LHCD launcher depends in a complex manner on a number of parameters. This is attributed to the fact that many different processes are involved in the impurity release from the launcher. This conclusion is also consistent with the assorted damage to the grill. The most likely causes of the impurity radiation from the LHCD launcher, not arranged in a particular order, are summarised here.

Impurity radiation related to the LHCD power.

The main causes of enhanced radiation associated with the LHCD operation are the parasitic fast electrons in the SOL and the arcs. Both of these are related to the electric field at the grill mouth, which is proportional to the square root of the applied LHCD power, $P_{LH}^{1/2}$, and also depends on the reflection coefficients. An example of a radiation event when LHCD alone was applied is given in Fig. 2a. The dependence of the radiation on the applied LHCD power is shown in Fig. 2b, according to which, it is less probable to observe enhanced radiation at small LHCD power, $P_{LH} < 1.5MW$.

Fast electrons in SOL can be produced by parasitical absorption of the portion of the coupled LH power which is launched at very large parallel refractive index, e.g. $N_\parallel > 30$ [1], [2]. The fast electrons can act on the protruding parts of the LHCD launcher, e.g. the blobs and the protuberances of the grill and the inner walls of the protective frame and the surrounding limiters, as they increase the heat load and ease the melting and evaporation of the targeted components.

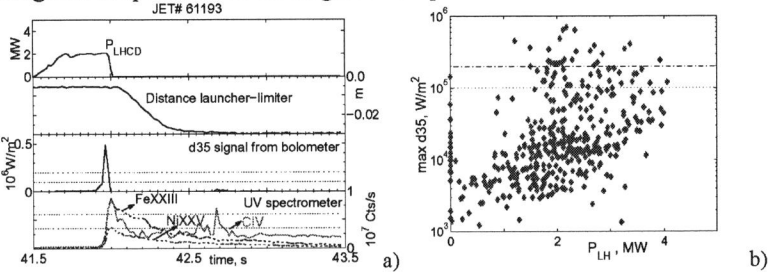

FIGURE 2. Radiation from the LHCD launcher when LHCD was applied alone a) and max. $d35$ vs. LHCD power, P_{LH}, from the LHCD-only experimental database b). The present launcher protection limits of $d35$ and $FeXXIII$ are given by dotted and dash-dotted lines.

The grooves-like damages to the grill, observed on all of the rows, are in the same direction as the magnetic field lines in front of the launcher for the most typical JET experiments, Fig. 1, and this is consistent with interaction with fast electrons. Hotspots [3] on the divertor components, magnetically connected to the LHCD grill, are another evidence for the presence of fast electrons in the SOL.

An arc can deposit enough energy in the waveguides to melt the grill. Impurity ions of Fe and Ni could be expected to enter the plasma, thus increasing the radiation. Arcs are more likely to occur between structures with sharp edges and at high rate of gas puffing near the launcher, which results in higher neutral pressure in the vicinity of the grill and inside the multijunctions. For that reason, the gas injection rate in LHCD experiments is optimised to provide good coupling but not to increase the pressure in front of the launcher too much. A statistical analysis of radiation events shows that arcs occur less often during operation with a well conditioned launcher.

Interaction of the launcher with the fast ions created by ICRF and/or NBI.

The enhanced radiation in the front of the LHCD launcher is partially a result of the fast ion bombardment of the grill. Fast ions are either fusion products or can be created by the additional heating of ICRH and NBI. Factors, which enhance the radial excursion of their banana orbits, are larger energies, heavier ions such as T or He^3 and smaller plasma current I_p. The non-confined fast ions leave the plasma and sputter the in-vessel components thus producing impurities in the SOL.

An illustrative example of the impurity production by fast ions is shown in Fig. 3a for a case where LHCD power was switched off but enhanced radiation from the LHCD antenna was detected and the bolometer and UV spectrometer signals were correlated with the launcher movement, i.e. changing distance launcher-limiter.

The impurity radiation tends to increase with the additional heating power in both cases with and without gas injection, Fig 3b. Investigation of the JET pulses at low LHCD power, $P_{LHCD} < 1.5MW$, which excludes fast electrons and arcs as a possible cause of radiation, shows that the fraction of the shots with enhanced impurity radiation from the LHCD launcher increases above 5% when $P_{NBI} + P_{ICRF} > 8$ MW, Fig. 3c. The reduction of the radiation events for $P_{NBI} + P_{ICRF} > 12$ MW can be attributed to the fact that in this case LHCD operated with launcher retracted further from the plasma, which can be seen by the increase of the averaged distance launcher-limiter, Fig. 3c. Orbit calculations of the fast ions for typical experimental configurations in JET show that the most liable part of the launcher to suffer fast particle bombardment is the upper left corner, where the most serious harm has been observed.

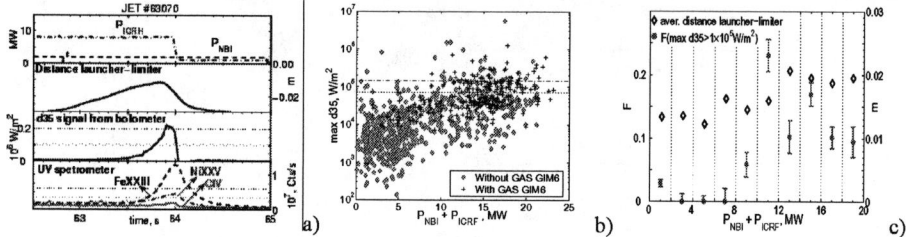

FIGURE 3. Radiation from the LHCD launcher caused by fast ions with NBI and IRCH applied a) and max $d35$ vs. the additional heating power $P_{NBI} + P_{ICRH}$ with and without gas injection b). The fraction F (points) of the low LHCD power, $P_{LHCD} < 1.5$MW, pulses with enhanced radiation, max. $d35 > 1\times 10^5$ W/m^2, and the averaged distance launcher-limiter (diamonds) for selected intervals of $P_{NBI} + P_{ICRH}$, given by vertical grid lines c).

CONCLUSIONS

The impurity generating processes and their impact on the LHCD launcher are summarised in Table.1.

TABLE 1. Possible causes of enhanced radiation in front of LHCD launcher and their impact on it.

Sources	Processes and impurities	Enhancing factors	Impact on launcher
Fast electrons parasitically created in SOL by LHCD.	Increased heat load on the in-vessel components leading to evaporation and/or sublimation: *Fe, Cu, Ni, C*	1. High $N_\parallel$ spectrum of the launched LHCD power; 2. High electric field at the grill mouth; 3. High n_e in front of launcher.	Melted material and grooves-like formations parallel to the magnetic field lines.
Arcs	Evaporation: *Fe, Cu, Ni*	1. High neutral pressure in front of launcher; 2. High electric field at the grill mouth and in the multijunctions; 3. Non-conditioned launcher.	Melted material and grooves-like formations in the middle of the rows.
Fast ions: 1. fusion products; 2. ICRH/NBI created.	Increased heat load on the in-vessel components, physical and chemical sputtering: *Fe, Cu, Ni, C*	1. High ICRH+NBI power; 2. Small plasma current; 3. Launcher positioned near plasma.	Severe damage to the upper-left corner of the grill.

ACKNOWLEDGMENTS

This work has been conducted under the European Fusion Development Agreement

REFERENCES

1. Mailloux J et al, *J. Nucl. Mater* 241-243, 745 (1997)
2. Goniche M et al, *Nuclear Fusion* 38, 919 (1998)
3. Rantamäki K et al, 2003, *30th EPS Conf. on Contr. Fusion and Plasma Physics*, St Petersburg, Vol. 27A

Lower Hybrid Experiments on MST

M.C. Kaufman, J.A. Goetz, M.A. Thomas, D.R. Burke and D.J. Clayton

Department of Physics, University of Wisconsin, Madison, WI 53706

Abstract. Current drive using RF waves has been proposed as a means to reduce the tearing fluctuations responsible for anomalous energy transport in the RFP. A traveling wave antenna operating at 800 MHz is being used to launch lower hybrid waves into MST to assess the feasibility of this approach. Parameter studies show that edge density is a major factor in antenna/plasma coupling. Gas puffing near the antenna is shown to alter coupling without changing plasma conditions. Hard x-ray emission has been correlated to RF power and is seen to vary strongly with direction of power flow through the antenna.

INTRODUCTION

Experimental and theoretical work indicate that anomalous energy and particle transport observed in the reversed field pinch (RFP) is due to tearing fluctuations. These fluctuations can be reduced when parallel current is added to the edge of standard RFP plasmas [1]. Inductive parallel current drive (PPCD) experiments on MST [2] have demonstrated nine-fold improvement in energy confinement and a tripling of electron temperature. This technique, however, is inherently transient and non-local. RF current drive is an obvious candidate for steady control, and the lower hybrid slow wave has been proposed as a possible choice for the RFP [3].

An antenna with a 20 cm aperture and 4.8 cm wavelength has been designed and built to launch lower hybrid slow waves into MST in order to address the goal of improving transport in an RFP [4]. With the stringent restraints of the MST vacuum vessel, a traveling wave antenna based on an interdigital line [5] was chosen. RF power enters the structure at one end and then propagates to the other end; along the way some power is radiated as a lower hybrid slow wave. Input power can be fed from either end (port direction) with the output end connected to a dummy load.

EXPERIMENTAL OBSERVATIONS

The present antenna operates at 80 kW for 10ms. Anticipated RF power needed for fluctuation stabilization is ~1-2 MW. At present levels of power, coupling and loading issues are the main focus. Power is measured at either feed end and at loops placed along the backplane of the antenna. Figure 1(a) shows RF during a typical discharge. The coupling can be gauged by the power damping length relative to the length of the antenna. Plasma conditions have a strong effect on coupling. The power damping length decreases for high densities and high currents as shown in Figure 1(b). Previous physical models [6] of the coupling in response to plasma conditions do not fully explain the behavior on MST, and this is now an active area of research.

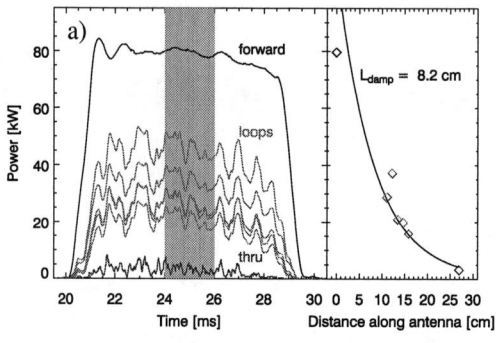 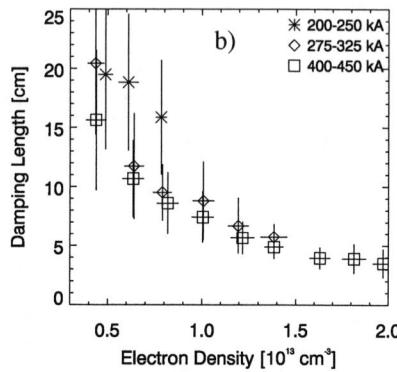

FIGURE 1. (a) Calculation of damping length of RF power flowing down the antenna. (b) Damping Length vs. Density for several different plasma currents.

In high confinement RFP plasmas, the coupling between the antenna and plasma is low. Density profiles show a steep gradient in the high confinement regime, giving a low edge density near the antenna face. Transitions from high to low confinement plasmas with RF power applied have shown a distinct change in the coupling behavior at the transition time. The density profile also flattens at the same time giving an indication that local edge density is a major driver of antenna performance.

To explore the possibility of good RF operation in high confinement plasmas, we attempted to increase the edge density near the antenna with local gas puffing. Similar experiments were done in H-mode plasmas on JET with promising results [7]. Figure 2(a) shows a toroidal cross section of MST with a conduction pipe attached to a puff valve threaded through the pumping duct and terminating approximately 15 cm from the port 2 feed on the antenna. Helium was used and the relative size of the gas input was monitored by an edge-chord spectrometer.

At the time of the experiment, high confinement PPCD plasmas were not reliably obtained, so lower current, low density plasmas were used instead. Although not in a high confinement regime, the antenna also couples very poorly in such plasmas. Figure 2(b) shows the response of the antenna to local puffing for ports 1 and 2 into low density plasmas. For a given line-averaged density, the damping length of the power flow decreases with edge puffing in both port directions. For puffs approximately 50% larger than those displayed in the Figure, the change in damping length becomes more pronounced. This gives confidence that not only can good antenna performance in a high confinement regime be achieved, but also that coupling can at least be increased in other operating conditions.

Hard x-rays are the primary means to detect interaction of the LH wave with the plasma. A fixed array of CdZnTe detectors is positioned opposite and view the face of the antenna. Additionally, several movable detectors allow for crossviews and views toroidally away from the antenna.

Figure 3 shows contour plots of x-ray flux and energy on RF power ramp-up for a

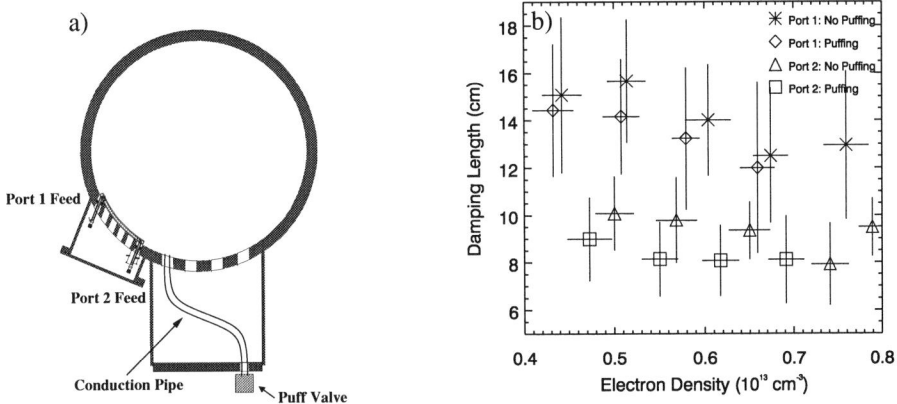

FIGURE 2. (a) Diagram of localized puffing apparatus near the lower hybrid antenna. (b) Damping length vs. line-averaged density for local puffing on ports 1 and 2.

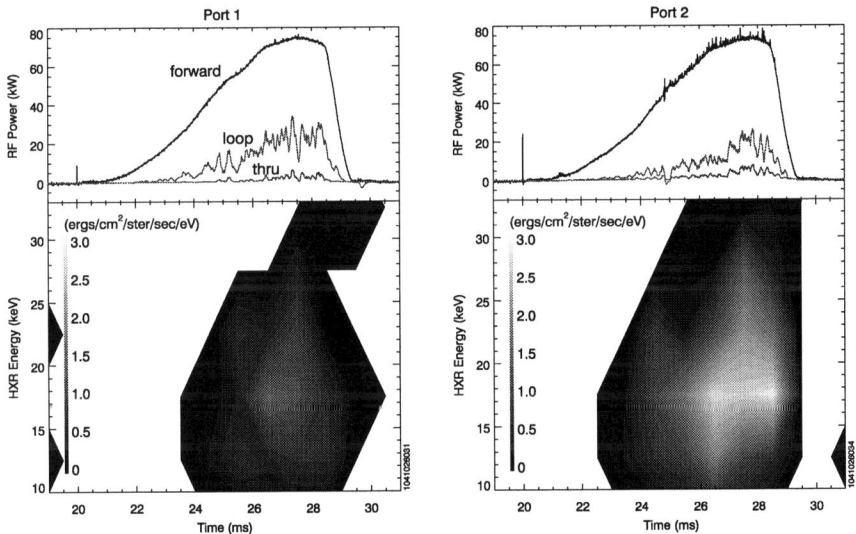

FIGURE 3. Plot of hard x-ray flux vs RF power for both feed directions.

detector viewing the middle of the antenna. There is a threshold for x-ray production that occurs when power is ∼5 kW at the center of the antenna. There is also a notable increase in x-ray flux and energy as RF power increases.

While the overall x-ray energies produced by RF are similar for both power feed directions, power fed from outboard end of the antenna produces a much higher flux than from the inboard feed. The explanation for the flux asymmetry is somewhat of a mystery. In the normal operating regime, plasma electron flow is in the same direction as power flowing from port 1. To test if flow direction is the reason for the difference in flux, the toroidal magnetic field was reversed, causing the electrons to flow in the opposite direction across the face of the antenna. No change in flux was observed with the field reversed. Since the antenna is mounted on the inboard side of the machine, the Shafranov shift brings the port 1 feed farther from the plasma than port 2 and could account for the flux difference in ports. The next generation antenna will test this by constructing the antenna face to be more parallel with the shifted flux surfaces.

CONCLUSION

The LH antenna on MST now runs routinely at 80 kW for 10 ms in a variety of plasma conditions. High confinement plasmas are a natural regime where we would want to operate with RF power. Unfortunately, in those regimes the antenna does not perform well. Recent experiments with local fueling to raise the local plasma density have sucessfully increased the amount of power transmitted to the plasma.

Hard x-ray data indicate plasma interaction even where RF power is well below that needed to reduce fluctuations and transport. Asymmetries in x-ray flux between power feed direction cannot be explained by plasma electron flow direction. Explaining hard x-ray fluxes at this level of RF power and beyond are a major area of current and future research.

The next generation antenna is expected to handle about 300 kW and we hope to see a greater range of plasma effects. More instrumentation on the antenna will enable direct local density measurements and will resolve coupling and power flow to a greater degree.

ACKNOWLEDGEMENTS

This work is supported by US DOE Contract DE-FG02-96ER54345.

REFERENCES

1. Ding, W.X., Brower, D.L, Terry, S.D., et al., Phys. Rev. Lett. **90**, 0355002-1 (2003)
2. Dexter, R.N., Kerst, D.W., Lovell, T.W., Prager, S.C., and Sprott, J.C., *Fus. Tech.* **19**, 131 (1991)
3. Uchimoto, E., Cekic, M., Harbey, R.W., et al., *Phys. Plasmas* **1**, 3517 (1994)
4. Goetz, J.A., et al. "Design of a Lower Hybrid Antenna for Current Drive Experiments on MST" *Radio Frequency Power in Plasmas, 14th Topical Conference* (2001)
5. Matthaei, G.L., *IRE Transactions on Microwave Theory and Techniques*, **Vol MTT-10**, 479 (1962)
6. Golant, V.E., Sov. Phys.-Tech. Phys. **16**, 1980 (1972)
7. V. Pericoli Ridolfini, et al. "LHCD coupling during H-mode and ITB in JET plasmas" *Radio Frequency Power in Plasmas, 14th Topical Conference* (2001)

Lower Hybrid Antenna Design for MST

J.A. Goetz, M.A. Thomas, M.C. Kaufman, and S.P. Oliva

Department of Physics, University of Wisconsin, Madison, WI 53706

Abstract. Inter-digital line antennas are being used to test the feasibility of lower hybrid current drive in MST. The antennas use λ/4 resonators and launch slow waves at 800 MHz with $n_\parallel \sim 7.5$. Routine operation has been achieved with a good impedance match between antenna and plasma. High power antenna design improvements include larger vacuum feed-throughs, better impedance matching, and rf instrumentation on all resonators. The antenna and feed-through modeling was performed with CST Microwave Studio^{TM}. The pulse-forming network that powers the klystron is being upgraded to a 50 kV – 30 ms pulse. The goal for the LHCD system on MST is a modular design that can handle 300 kW per antenna.

Keywords: lower hybrid waves, antennas, current drive, reversed field pinch
PACS: 52.35.Hr, 52.40.Fd, 52.50.Sw, 52.55.Hc, 52.55.Wq

INTRODUCTION

Inductive current profile control has been highly successful in reducing transport in the Madison Symmetric Torus (MST) RFP [1], but it is transient and non-localized [2]. Current profile control with rf waves offers the possibility of steady and more precise control. Theoretical feasibility and optimization studies have identified the lower hybrid slow wave as a good candidate for current density profile control in the RFP [3]. Other studies have focused on fast wave current drive, which could also drive poloidal current in the outer region as well as toroidal current in the central region [4]. Ray tracing and Fokker-Planck calculations have identified suitable propagating waves that can provide localized current drive with relatively high current drive efficiency. To meet the goals of improving transport in an RFP, antennas have been designed and built to launch lower hybrid slow waves into MST.

MST LH ANTENNA DESIGN

The traveling wave antenna concept [5] is well suited for MST. It overcomes many of the inherent technical difficulties for the RFP, requiring only two coaxial feeds through the vacuum vessel, having a small radial height (<3 cm), and can be mounted on the inner wall of MST. The inter-digital line is an electrostatic variant of the comb-line antenna used successfully for launching fast waves on JFT-2M [6]. It behaves electrically like a microwave band-pass filter [7], and has been designed to produce the desired $k_\parallel$ spectrum at 800 MHz. The antenna is a slow wave structure in which a resonant array of conducting rods is alternately grounded to opposite sides of a rectangular cavity. The rods are coupled to each other both inductively and capacitively, and are matched to 50 Ω coaxial feeds at each end of the array. Coupling to the lower hybrid wave is accomplished by the electric field between elements fringing though an aperture in the cavity. The fields are evanescent in vacuum but couple to the slow wave at $n_\parallel = 7.5$ in the presence of plasma.

The first LHCD antenna for MST was designed using the SPICE circuit simulation code [8]. The antenna was modeled as an array of coupled transmission lines using analytic impedance estimates. Initial experiments investigated coupling and loading issues from 10 W to 10 kW. These experiments showed that one feed had a power limit of 2-3 kW, above which the antenna was fully reflecting. The other feed was fully reflecting at all power levels. This antenna had 1 cm diameter vacuum feed-throughs and the impedance match between the feeds and the traveling wave structure was not very good, resulting in a VSWR ~ 6. Bench measurements of the wave spectrum showed that the desired $n_\parallel = 7.5$ was being produced but also indicated that external circuit tuning is an important parameter in the antenna behavior. With these limitations on the power flow in the antenna identified, a second antenna was designed, built, and installed in MST.

In designing the second MST LH antenna, a finite-element method was used to refine the impedance values that are input to the SPICE model [9]. This led to better impedance matching, a lower VSWR ~2, and lower Ohmic losses inside the structure. In addition, a larger diameter (2 cm) vacuum feed-through was designed in order to achieve a better impedance match at the coaxial feed and to obtain better power handling. Instrumentation was added to the antenna backplane to measure power density near the center of the antenna. Experiments designed to investigate coupling and loading issues were carried out with the second antenna [10]. This antenna can handle up to 80 kW, which is the limit of the klystron given the capabilities of the pulse-forming network power supply. The impedance match to the antenna is good over a wide range of plasma parameters and input power when fed from both ports.

The behavior of the antenna phasing and $n_\parallel$ spectrum during plasma operation has been investigated. Vector network analyzers are used to measure the power amplitude and phase on the central five elements of the antenna. As shown in Figure 1, the phase between elements remains near the $\pi/2$ design value. The $n_\parallel$ can be calculated from the measured phase data, ϕ, by $<n_\parallel> = \frac{<k>}{|k_0|} = \frac{<\phi>/d}{2\pi/\lambda_0}$, where d = 1.2 cm, the element spacing, and $\lambda_0 = 37.5$ cm, the vacuum wavelength. The antenna phasing and hence $n_\parallel$ are quite robust to plasma loading and show only slight changes over a wide range of parameters. An example of this behavior versus plasma density is shown in Figure 2. The measured value of $n_\parallel$ is near the design value of 7.5. The variation of $n_\parallel$ with respect to wave launch direction is under further investigation.

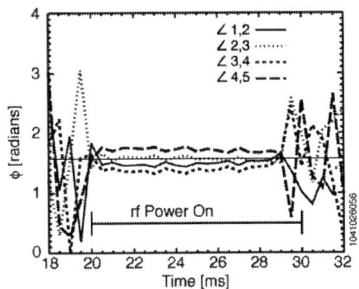
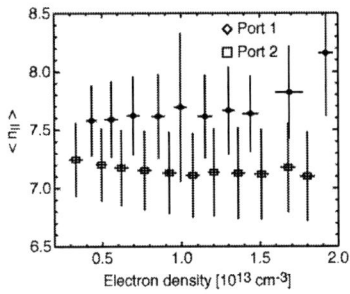

FIGURE 1. Phase between the central five elements of the antenna measured with a vector network analyzer phase-locked to the transmitter. Each line represents the phase difference between adjacent antenna elements. The design value for the phase is $\pi/2$ radians.

FIGURE 2. The parallel index of refraction is robust to plasma loading. The index is calculated from the phase data of Figure 1. The discrepancy between the different launch directions persists over a wide range of plasma parameters.

In order to carry the MST lower hybrid program forward there will be several upgrades performed to the transmitter system and antenna. The pulse-forming network that powers the klystron will be upgraded to a 50 kV, 30 ms long pulse. This should allow for 300 kW tube operation. To handle this increased power, the antenna is being redesigned. The CST MicroWave Studio™ package for full 3D electromagnetic computer modeling is being used to design the new antenna. The model of the antenna used as input to the code is shown in Figure 3.

To handle the increased power, larger diameter feed-throughs (4 cm) have been fabricated in-house. This will keep the power density at the vacuum interface similar to that of the present antenna ($\sim$250 MW/m^2). In addition, a longer impedance matching section is needed to accommodate the larger feeds (see Fig. 3). It is desirable to lower the VSWR to less than 1.4. This will remove the need for external tuning and achieve better directivity. Modeling with Microwave Studio™ has resulted in a design with good impedance matching of the antenna to the feeds, a VSWR < 1.1, and a reflection coefficient of less than –30 dB over the frequency range of interest (see Figure 4). The third generation antenna will have improved instrumentation. Vector power measurements of each antenna element will characterize the power flow along the structure. Phase and amplitude electronics are under construction and will be used to look for changes in the antenna dispersion with plasma loading. In addition, Langmuir probes will be installed on the antenna to characterize the edge plasma.

EXPERIMENTAL PLAN FOR MST LHCD PROGRAM

A 300 kW lower hybrid system will be operational in MST by the summer of 2005. A new pulse-forming network to power the klystron is being constructed and the third generation antenna is being built. Testing and experiments will then proceed in stages. First, coupling and loading measurements will be performed at powers up to

300 kW. Wave propagation and rf-plasma interaction will be studied in detail to confirm that the antenna is coupling to the desired wave. The power flow and loading along the structure will be measured and power handling limits investigated in order to produce an optimized design for higher power experiments. Second, the program will be expanded to 600 kW by adding another 300 kW klystron-antenna module. It is expected that this power should be sufficient to estimate the current drive efficiency. Ultimately, a high power (1–2 MW) LHCD experiment will be designed and implemented with fluctuation and transport reduction as the goals.

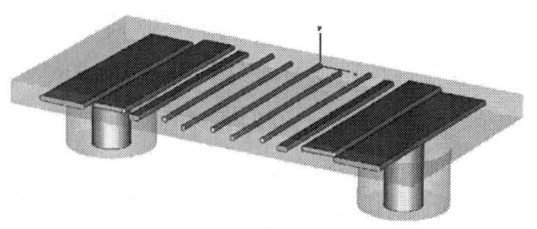

FIGURE 3. Model of the antenna used for input to the CST Microwave Studio™ 3D electromagnetic simulation code. Note the larger coaxial feed-through and the longer impedance matching section.

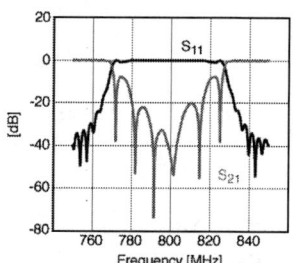

FIGURE 4. The S-parameters resulting from the simulation are shown over the frequency range of interest. S_{11} is the black line and S_{21} is the gray line.

ACKNOWLEDGMENTS

This work was supported by U.S. DoE Contract DE-FG02-96ER54345.

REFERENCES

1. Dexter, R.N., Kerst, D.W., Lovell, T.W., Prager, S.C., and Sprott, J.C., *Fus. Tech* **19**, 131 (1991).
2. Sarff, J.S., Almagri, A.F., Cekic, M., *et al., Phys. Plasmas* **2**, 2440 (1995).
3. Uchimoto, E., Cekic, M., Harvey, R.W., Litwin, C., Prager, S.C., Sarff, J.S., and Sovinec, C.R., *Phys. Plasmas* **1**, 3517 (1994).
4. Shiina, S., Kondo, Y., and Ishii, H., Nucl. Fusion **34**, 1473 (1994).
5. LaHaye, R., Armentrout, C.J., Harvey, R.W., Moeller, C.P., and Stambaugh, R.D., *Nuclear Fusion* **20 (2)**, 218 (1980).
6. Ogawa, T., Hoshino, K., Kanazawa, S., *et al.*, Nucl. Fusion **41**, 1767 (2001).
7. Matthaei, G.L., IRE Transactions on Microwave Theory and Techniques, **Vol MTT-10**, 479 (1962).
8. Goetz, J.A., Thomas, M.A., Forest, C.B., *et al.*, in *Proceedings of the 14th Topical Conference on Radio Frequency Power in Plasmas*, AIP Conf. Proc. **595**, 253 (2001).
9. Goetz, J.A., Thomas, M.A., Chattopadhyay, P.K., *et al.*, in *Proceedings of the 15th Topical Conference on Radio Frequency Power in Plasmas*, AIP Conf. Proc. **694**, 263 (2003).
10. Kaufman, M.C., Goetz, J.A., Thomas, M.A., and Burke, D.R., "Lower Hybrid Experiments on MST", these proceedings (2005).

Implementation of LHCD Experiments on Alcator C-Mod*

R. Parker[1], N. Basse[1], W. Beck[1], S. Bernabei[2], R. Childs[1], R. Ellis[2], E. Fredd[2],
N. Greenough[2], M. Grimes[1], D. Gwinn[1], J. Hosea[2], J. Irby[1], P. Koert[1],
C.C. Kung[2], B. Labombard[1], J. Liptac[1], G.D. Loesser[2], E. Marmar[1],
G. Schilling[2], D. Terry[1], J. Terry[1], R. Vieira[1], G. Wallace[1],
J.R. Wilson[1], J. Zaks[1]

[1]*Plasma Science and Fusion Center, MIT, Cambridge, MA, USA*
[2]*Princeton Plasma Physics Laboratory, Princeton University, Princeton, NJ, USA*

Abstract. An antenna-transmitter system for driving current in the LHRF has been installed in Alcator C-Mod. The antenna is a grill consisting of 4 poloidal rows of waveguides, each with 24 guides in the toroidal direction. Power is supplied by 12 klystrons capable of 250 kW operation at a frequency of 4.6 GHz. Thus the total source power is 3 MW, with about 1.5 MW available to be coupled to the plasma. Power supply and heat throughput limits in C-Mod limit the pulse length to 5 s, which however represents several current redistribution times. With 90° phasing, the $n_\parallel$ spectrum is sharply peaked at 2.3 and the range $1.5 < n_\parallel < 3.5$ can be accessed dynamically by varying the phase of the klystrons. The system is in the commissioning phase with klystron power limited to ~20 kW and pulse length to 10 ms. Early results from plasma operation are discussed.

Keywords: Lower Hybrid Current Drive, Alcator C-Mod
PACS: 28.52.Cx, 52.40.Fd, 52.50.Sw, 52.70.La

SYSTEM DESCRIPTION

An LHCD experiment is being implemented on Alcator C-Mod. The goal is to produce and study steady-state regimes in C-Mod, with $I_{LHCD}/I_p \sim 30\%$, $f_{BS} \sim 70\%$, $\beta_n \sim 3$ and $H_H \sim 1\text{-}2$ for $T_{pulse} \sim 5\text{ s} > L/R > \tau_{res}$ [1].

RF power is coupled by a grill consisting of 4 poloidal rows of waveguides with each row composed of 24 waveguides in the toroidal direction [2]. The grill is powered by 12 klystrons operating at 4.6 GHz and capable of 250 kW steady-state [3]. Thus, 3 MW is available at the source; with transmission losses about 2 MW is available to be injected to the plasma. A second grill and 4 additional klystrons are planned to be added in an upgrade, bringing the total source power to 4 MW. A master oscillator provides inputs to 12 vector modulators whose phase- and amplitude-controlled outputs feed the inputs to the 12 klystrons[4]. The $n_\parallel$-spectrum can be dynamically varied during a plasma pulse over the range $1.5 < n_\parallel < 3$ with a response time of ~ 1 ms.

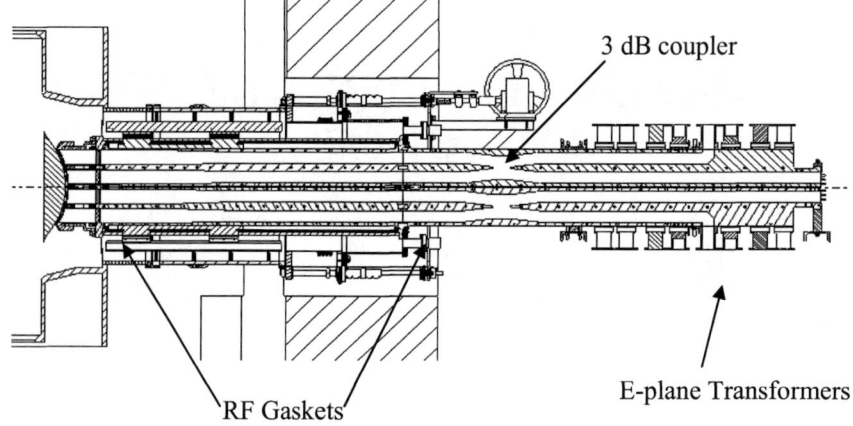

FIGURE 1. Cross-section of launcher as installed on Alcator C-Mod.

The klystron output is divided 4 ways by conventional components before being applied to sub-height (5.5 mm) waveguides in the launcher. See Figure 1. The launcher consists of two sections of 25 stacked plates, 24 of them milled on one side to form four waveguides, and 4 Ti couplers containing the vacuum windows and forming the grill. E-plane transformers couple RF power to the top and bottom waveguides in the rear waveguide assembly, where it divides again by 2 in a 3 dB slot coupler. An H-plane transformer in the forward assembly of stacked-plate waveguides brings the waveguides to their final dimension of 5.5x6 mm^2. Specially designed copper gaskets provide an RF seal between rear and forward waveguide assemblies and the forward waveguide assembly and the couplers [5].

The 4 Ti couplers are EDM-machined out of solid blocks of titanium. Al_2O_3 windows are brazed into the Ti waveguides and the ends of the couplers are machined to conform to standard-shaped C-Mod plasmas. Molybdenum limiters are installed on each side of the couplers and the whole launcher can be moved from a position flush with the limiter to several cm behind it. Figure 2 shows the waveguide grill as it appears to the plasma and a rear view of one of the couplers showing the vacuum

FIGURE 2. The grill as installed in the C-Mod torus (left) and a rear view of one of the couplers.

windows.

Each of the 12 klystrons powers two adjacent columns of 4 waveguides in the grill. The phases of the RF in one vertical column, e.g., column 1, are set by the phase of the input to the klystron powering that column, while the phase in the adjacent column of waveguides, e.g., column 2, is varied relative to the first column by means of a high power phase shifter. The klystron phase can be varied electronically with a time response of ~ 1ms, while the settings of the high power shifters can only be varied between plasma shots. Nevertheless, varying only the phase of the klystrons with the high power phase shifter fixed, e.g., at 90°, produces a close approximation to the ideal $n_{||}$ spectrum obtained with uniform phase progression[2].

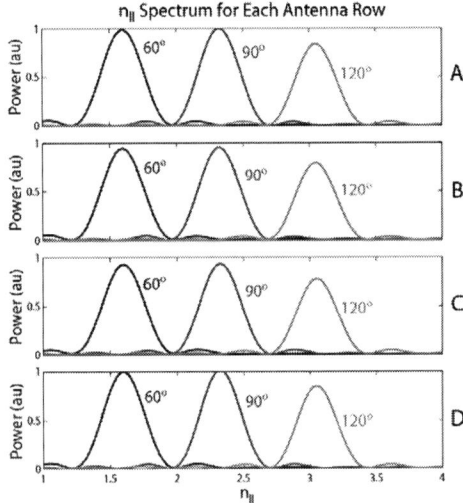

FIGURE 3. The $n_{||}$ spectrum as inferred from measurements made at the output of the rear waveguide structure.

The amplitude and phase of the RF in all 96 waveguides with respect to the amplitude and phase of the 12 klystrons have been measured at the junction of rear and forward waveguide assemblies. From these measurements, and knowledge of the attenuation and phase shift of the forward waveguides and grills, the $n_{||}$ spectrum can be calculated as a function of incremental phase shift. The results for the four rows of couplers are shown in Figure 3.

INITIAL COUPLING RESULTS

Low power (100-200 kW source power) measurements of the reflection coefficient as a function of phase and plasma density at the grill have been made. The forward and reflected power to the waveguides feeding the grill were monitored by means of 48 directional couplers located at the input to the E-plane transformers feeding the rear waveguide assembly. A global reflection coefficient was obtained by averaging the

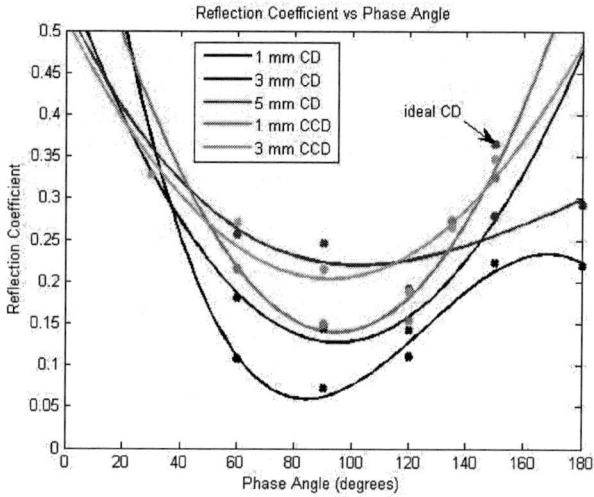

FIGURE 4. The global reflection coefficient as a function of phase between adjacent pairs of waveguides and grill position. The phase between the two waveguides forming each pair is ± 90° depending on whether the pair-wise corresponds to co (CD) or counter (CCD) current drive.

reflection coefficients measured at the couplers. The density was measured at 6 positions on the grill by Langmuir probes flush-mounted in the spacers between couplers (see Figure 2.) The density at the grill can be varied by an order of magnitude, from 3e17 cm–3 to 3e18 cm-3, by moving the grill from a position 5 mm behind the protection limiter to one that is 1 mm behind it. In the preliminary results shown in Figure 4 a minimum global reflection coefficient of 7% was obtained at the highest density and 90° co-current drive phasing.

ACKNOWLEDGMENT

This work was supported by DOE Cooperative Grant No. DE-FC02-99ER54512 and Contract no. DE-AC02-76CH03073.

REFERENCES

1. Bonoli, P.T., Parker, R. R., Porkolab, M., et al., Nucl Fusion **40**, 1251 (2000).
2. Bernabei, S., Hosea, J. C., Kung, C. C., et. al., Fusion Sci and Technol **43**, 145 (2003).
3. Grimes, M., Gwinn, D., Parker, R., et.al., Proc 19th IEEE/NPSS Symposium on Fusion Engineering (SOFE), Atlantic City, NJ, 16-19 (2002).
4. Terry, D., Burke, W., Grimes. M., et.al.,Proc 19th IEEE/NPSS Symposium on Fusion Engineering (SOFE), Atlantic City, NJ, 23-6 (2002).
5. Hosea, J., Beals, D., Beck, W., et.al., Proc 20th Symposium on Fusion Technology (SOFT), Venice, (2004)

Microstrip Directional Coupler Design For A Reduced Height Waveguide

G. Wallace, P. Koert, R. Parker, D. Terry, and S.J. Wukitch

MIT Plasma Science and Fusion Center

Abstract. We have developed a compact, mass-producible probe inserted into two holes ¼ wavelength apart in the narrow side of the reduced height waveguide. The probe consists of two current loops and a directional coupler mounted directly on a single microstrip circuit board thereby greatly reducing the physical size of the probe. Calibrations have shown directivity greater than 20 dB at 4.6 GHz for each of the over 100 probes built.

Keywords: lower hybrid technology, Alcator C-Mod tokamak
PACS: 28.52.Lf, 52.50.Sw, 52.70.Gw

INTRODUCTION

The lower hybrid current drive (LHCD) system recently installed on Alcator C-Mod is a key component of operation in advanced tokamak regimes. The system launches 4.6 GHz lower hybrid waves into the plasma from a phased array of 96 reduced height waveguides, and it is necessary to accurately measure the phase and amplitude of the forward and reflected waves in each of these waveguides for proper operation [1]. A set of plates stacked together forms the LHCD launcher such that the narrow side of the waveguide is exposed on the top and bottom of the stack. Since the wide side of the waveguide is inaccessible for 22 of the 24 waveguides in the stack, conventional E-probes cannot be used. Sets of holes were drilled a distance $\lambda_g/4$ apart in the exposed, narrow side of the waveguide to accommodate current loops sampling the H_z field of the TE_{10} waveguide mode [2]. These holes served as the starting point for our design of a compact directional coupler.

DESIGN

One method for creating a directional coupler probe is to take sample signals, measured at two (or more) points separated by $(2n+1)\lambda_g/4$, where n=0,1,2,3... and λ_g is the waveguide wavelength at the desired frequency. The two out-of-phase signals are then combined through a coupling structure such that the sampled portions of the wave propagating one direction in the waveguide cancel, as shown in Fig. 1 [3]. The initial design constraint for the directional coupler probes was given by two holes drilled $\lambda_g/4$ (0.882") apart in the narrow side of the waveguide before our involvement with the project. These holes were intended to be used with a directional coupler probe

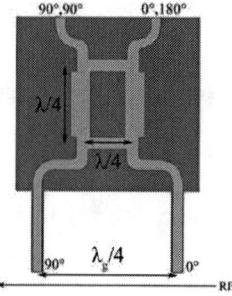 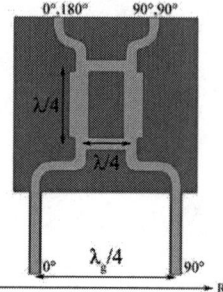

FIGURE 1. Front view of microstrip board. Plated thru holes form coupling loops located at the bottom of the two legs. Phases of the signals from both legs are shown at the coupling loops and the output ports.

designed by the PPPL [2], however it was discovered that the original design would not fit due to obstructions caused by other diagnostics mounted to the forward end of the LHCD launcher assembly.

We decided to incorporate the probe legs and the directional coupler onto a single microstrip circuit board to reduce the space required for connecting the probe legs and the directional coupler. An additional benefit of mounting the entire coupling structure on one microstrip board is that circuit board manufacturing techniques provide very accurate machining and therefore highly repeatable results.

Current Loop

Because the wide side is inaccessible for a majority of the waveguides in the stack, a conventional electric field probe cannot be used to sample the propagating wave, but the oscillating H_z field of the TE_{10} waveguide mode, in which the waveguides will operate, can easily be coupled via a perpendicular current loop. The current loop consists of a plated thru hole from the conductor to the ground plane on the microstrip board. This creates a loop of the same thickness as the dielectric substrate (0.032"). Computer simulations using CST Microwave Studio 5 showed that the proper insertion depth of the loop for the desired coupling level, -55 dB, was 0.560". The current loops were padded with 39 Ω surface mount resistors at the end of each probe leg to reduce the amount of circulating power in the coupler and lessen the effect of the impedance mismatch at the end of the probe leg. A Teflon "boot" is inserted into the waveguide hole to maintain a pressure seal in the waveguide.

Branch-Line Hybrid

The branch-line hybrid is a simple type of directional coupling structure consisting of a rectangular microstrip "box" with inputs at two corners and outputs at two corners. The coupler is designed to split power from each input evenly while introducing a $\pi/4$ phase difference between the two output signals. The properties of the ideal branch-line coupler can be described by the following scattering matrix:

$$[S] = \begin{bmatrix} 0 & \frac{\sqrt{2}j}{2} & \frac{\sqrt{2}}{2} & 0 \\ \frac{\sqrt{2}j}{2} & 0 & 0 & \frac{\sqrt{2}}{2} \\ \frac{\sqrt{2}}{2} & 0 & 0 & \frac{\sqrt{2}j}{2} \\ 0 & \frac{\sqrt{2}}{2} & \frac{\sqrt{2}j}{2} & 0 \end{bmatrix} \quad (1)$$

When combined with the $\pi/4$ phase difference between the incoming signals from the two current loops, a sum or difference is produced at the two output ports.

The leg widths of the box are such that the line impedance of the transverse legs is $Z_0 = 50\ \Omega$, while the widths of the longitudinal legs is adjusted such that $Z_1 = Z_0*2^{-1/2} = 35.4\ \Omega$. The dimensions of the box are such that it is $\lambda_{microstrip}/4$ on each side. The box is rectangular because the wavelength of a microstrip TEM mode varies with the width of the conductor for a given dielectric thickness [4]. The dimensions of the coupling structure were tuned for a 4.6 GHz signal using Agilent ADS.

Mounting Case and Connectors

A case was required for both mounting the microstrip board to the waveguide assembly and to shield the coupler from interfering signals in the experimental cell. The case was designed to be low profile such that it would fit under the vertical obstruction caused by the thermocouple flange and other diagnostics. Additionally, the cases must nest together horizontally. Two designs were made for the mounting case due to the different space constraints on the forward and rear ends of the waveguide. The probes on the forward end of the waveguide were designed to have the connectors mounted horizontally, while the connectors on the rear probes are mounted vertically. The board is held with set screws such that the ground plane of the microstrip board sits flat against the back of the case.

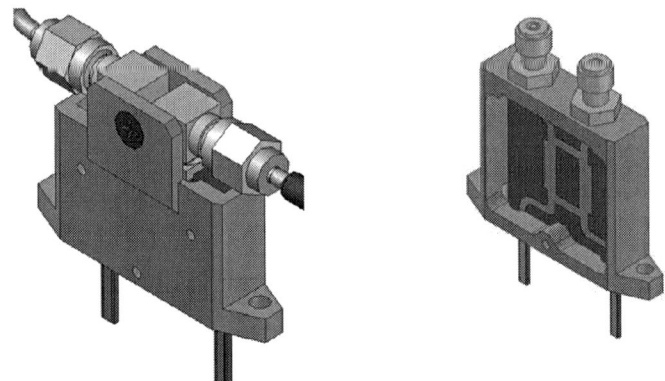

FIGURE 2. Forward probe (left) and rear probe (right) complete with boards and connectors. The shielding cover on rear probe is not shown.

RESULTS AND CONCLUSIONS

Test Results

The first batch of probes assembled yielded about 75% with directivity greater than -20 dB. By simply replacing the microstrip boards and connectors in the bad probes, all 104 probes achieved directivity of greater than -20 dB at 4.6 GHz, with many of the probes at -35 dB or better. Average coupling was -51.76 dB with a standard deviation of 0.69. The sum and difference traces from a typical probe are shown in Fig. 3. Additionally, phase measurements show a standard deviation of less than four degrees for the 104 probes. The probes also did not add any measurable mismatch to the waveguide.

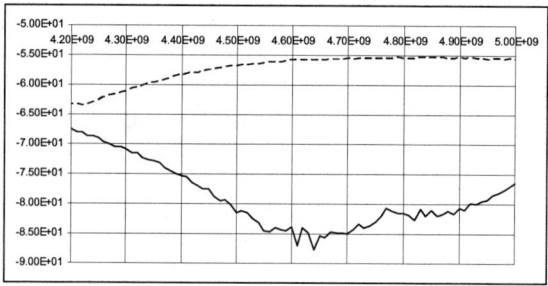

FIGURE 3. A-11 probe trace. Sum (dashed) and difference (solid) coupling is plotted. The vertical axis is coupling is dB and the horizontal axis is frequency in Hz.

Conclusions

A compact directional coupler probe for the reduced height waveguide on Alcator C-Mod LHCD system has been designed, manufactured, and tested. The probes were designed with tight space constraints but have exhibited excellent repeatability and directivity better than -20 dB at 4.6 GHz. 96 probes have been installed on the rear waveguide of the LHCD launcher providing amplitude data on the reflected wave. In the future we hope to also use the probes to measure the phase and amplitude of the forward and reflected waves.

ACKNOWLEDGEMENT

This work was supported by US DOE Contract No. DE-FC02-99ER54512.

REFERENCES

1. S. Bernabei et al., *Fusion Science and Tech.*, **43** 145-152 (2002).
2. C. Kung et al., *Fusion Engineering*, 268-271 (2002).
3. P. Jacquet et al., *Rev. Sci. Inst.*, **68** 1176-1182 (1997).
4. D. Pozar, *Microwave Engineering*, Hoboken: John Wiley & Sons, Inc, 1998, pp. 161-163.

ELECTRON BERNSTEIN RANGE
OF FREQUENCIES

Electron Bernstein Wave Research on the National Spherical Torus Experiment

G. Taylor[a], A. Bers[b], T.S. Bigelow[c], M.D. Carter[c], J.B. Caughman[c], J. Decker[b], S. Diem[a], P.C. Efthimion[a], N.M. Ershov[d], E. Fredd[a], R.W. Harvey[e], J. Hosea[a], F. Jaeger[c], J. Preinhaelter[f], A.K. Ram[b], D.A. Rasmussen[c], A.P. Smirnov[d], J.B. Wilgen[c], J.R. Wilson[a]

[a] *Princeton Plasma Physics Laboratory, Princeton University, Princeton, NJ 08543, USA*
[b] *Plasma Science and Fusion Center, MIT, Cambridge, MA 02139, USA*
[c] *Oak Ridge National Laboratory, Oak Ridge, TN 37831, USA*
[d] *Moscow State University, Moscow, Russia*
[e] *CompX, Del Mar, CA 92014, USA*
[f] *Institute of Plasma Physics, Prague, Czech Republic*

Abstract. Off-axis electron Bernstein wave current drive (EBWCD) may be critical for sustaining non-inductive high β NSTX plasmas. Modeling results predict that the ~ 100 kA of off-axis current needed to stabilize a solenoid-free high β NSTX plasma could be generated by by 3 MW of 28 GHz EBW power. Synergy with the bootstrap current may enhance CD efficiency by ~ 10%. EBW radiometry measurements on NSTX support coupling to EBWs by launching elliptically polarized electromagnetic waves oblique to the confining magnetic field. Plans are being developed to implement a 1 MW, 28 GHz *proof-of-principle* EBWCD system to test the EBW coupling, heating and CD physics at high rf power densities on NSTX.

Keywords: Spherical tokamaks, electron Bernstein waves
PACS: 52.55.Fa, 52.35.Hr

INTRODUCTION

Off-axis rf-driven current may be critical for non-inductively sustaining high β plasmas in the National Spherical Torus Experiment (NSTX) [1,2]. However local rf current drive techniques established on conventional high aspect ratio tokamaks and stellarators are not readily applicable to spherical torus plasmas. Electron cyclotron current drive (ECCD) cannot be employed on high β NSTX plasmas, since these plasmas are overdense ($\omega_{pe} \gg \omega_{ce}$) and therefore not accessible to low harmonic electron cyclotron waves. Furthermore, high harmonic fast wave current drive has proved incompatible with neutral-beam-injection due to fast ion absorption [3] and limits on wave accessibility limit the efficiency of a lower hybrid current drive scheme [4]. The electrostatic electron Bernstein wave (EBW) offers the potential for local current drive (EBWCD) in NSTX since the wave propagates in overdense plasma and is strongly absorbed at electron cyclotron resonances [5,6].

One goal of NSTX EBW research is to model the propagation, damping and CD efficiency of EBWs for NSTX plasma scenarios. Numerical modeling results for a β ~ 41% NSTX plasma predict efficient off-axis EBWCD can be achieved [7] via the

Ohkawa [8] CD process and that positive synergies may exist with the bootstrap current at multi-megawatt rf power levels [9]. Another goal of NSTX EBW research is to establish a viable and resilient technique for efficiently coupling rf power to EBWs in the plasma. Recent radiometric thermal EBW emission data acquired on NSTX [10] support efficient EBW coupling via O-mode launch oblique to the confining magnetic field [11]. These thermal EBW coupling experiments and the associated numerical EBW modeling research support the design and implementation of a *proof-of-principle*, 1 MW, 28 GHz EBW heating and CD system planned for NSTX. This *proof-of-principle* system will test the EBW coupling, heating and CD physics at high rf power densities and the rf technologies needed for a future 4 MW EBWCD system.

EBWCD MODELING RESULTS

The GENRAY ray tracing computer code [6,12] and the CQL3D Fokker-Planck code [13] have been used to model the propagation and damping of EBWs and EBWCD for high β NSTX plasma equilibria. Figure 1 summarizes CQL3D modeling results for 3 MW of 28 GHz EBW power coupled into a β = 41% NSTX plasma. The EBW launcher couples rf power below the plasma midplane to generate 135 kA of EBWCD near a normalized radius, ρ = 0.7 (Fig. 1(a)). The driven-current density (33 kA/cm^2) exceeds the local bootstrap current density and therefore may be sufficient to suppress neoclassical tearing modes (NTMs) that can strongly degrade plasma confinement in spherical tokamaks [14,15]. Figure 1(b) shows that the rf quasilinear velocity space diffusion coefficient peaks near the trapped–passing boundary, efficiently driving current via the Ohkawa current drive process [7,8].

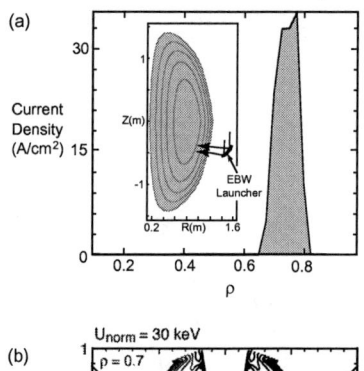

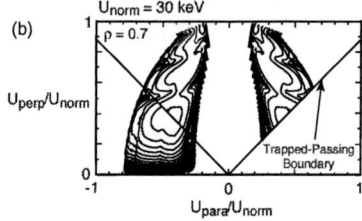

FIGURE 1: CQL3D modeling results for 3 MW of 28 GHz EBW power coupled into a β = 41% NSTX plasma with I_p = 1 MA and $B_t(0)$ = 0.35 T; (a) EBW-driven current density versus normalized minor radius (ρ) and (b) contours of the quasilinear rf velocity space diffusion operator plotted at ρ = 0.7.

The synergy between EBWCD and the current driven by the electron bootstrap effect have also been calculated with CQL3D [9]. A simple bootstrap model in CQL3D is used in these studies: the transiting electron distributions are connected in velocity-space at the trapped-passing boundary to trapped-electron distributions that are displaced radially outwards/inwards by a half-banana width for the co-/counter-passing regions. This model agrees well with standard bootstrap current calculations, over the outer 60% of the plasma radius [16]. At 4 MW of applied EBW power, there is a ~ 10% synergistic increase in the bootstrap current due to enhanced rf pitch angle scattering of the electrons. Locally, bootstrap current density increases in proportion to increased plasma pressure, and this effect can significantly affect the radial profile of the rf-driven current density.

EBW COUPLING CALCULATIONS AND MEASUREMENTS

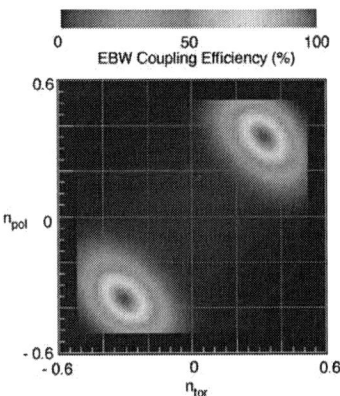

FIGURE 2: EBW coupling efficiency calculated by AORSA-1D for 28 GHz rf power coupled into a β = 41% NSTX plasma. Coupling efficiency is plotted versus the n_{pol} and n_{tor} of the launched electromagnetic wave.

A critical issue for implementing EBWCD is to establish a resilient technique to efficiently couple rf power to EBWs in the plasma. Recently, the AORSA-1D full wave code [17] has been modified to solve the EBW mode coupling in a 1-D slab geometry. The AORSA-1D code results for a launch frequency of 28 GHz into a β = 41 % NSTX plasma with I_p = 1 MA and $B_t(0)$ = 0.35 T are summarized in Fig. 2. The launch polarization was adjusted to obtain the maximum EBW coupling efficiency as a function of the poloidal and toroidal wavenumber of the incident microwave power [n_{pol} and n_{tor}, respectively]. Efficient coupling of near-circularly polarized microwave radiation was found for a launched electromagnetic wave with n_{pol} = ± 0.35 and n_{tor} = ± 0.3.

Since the mode conversion process is reciprocal [18], studying mode-converted thermal EBW emission with absolutely calibrated radiometers allows experimental benchmarking of the numerical modeling predictions of EBW coupling efficiency for specific values of n_{pol} and n_{tor}. Recently, obliquely-viewing, dual polarization radiometry has been employed on NSTX to evaluate the coupling efficiency of 16-18 GHz thermal EBW emission [10]. An EBW coupling efficiency 80±20% was achieved in good agreement with ~ 65% coupling efficiency predicted by a model that included a 1-D full wave calculation of the EBW mode conversion layer, radiometer antenna pattern modeling and 3-D EBW ray tracing and deposition modeling [19]. Thermal EBW emission at 16.5 GHz was consistent with the near-circular polarization predicted by modeling. Thermal EBW radiometric studies on NSTX will be extended to 20-40 GHz in 2005 [20] and will test EBW coupling efficiency predictions from AORSA-1D at 28 GHz.

PLANS FOR EBWCD ON NSTX

EBWCD numerical modeling predicts current drive efficiencies that are typically ~ 40-50 kA/MW at 28 GHz in β = 20 - 40% NSTX plasma equilibria [7]. Assuming that the EBWCD system needs to generate ~ 100 kA of plasma current [2], that the microwave coupling efficiency to EBWs is ~ 90% [10] and that the waveguide transmission loss is 10-15%, 4 MW of microwave source power would be needed in order to deliver the necessary 3 MW

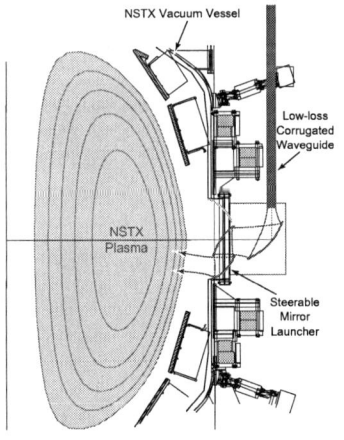

FIGURE 3: Conceptual design for a steerable EBW mirror launcher on NSTX. Second mirror may be switched to launch power above the midplane allowing EBWCD in the opposite toroidal direction.

of EBW power into the plasma. Four EBW steerable mirror launchers will couple microwave power into the plasma. A sketch showing one possible conceptual design for an NSTX EBW launcher is shown in Fig. 3. This design would allow the second mirror to switch between steerable launchers above and below the midplane, to drive current in either toroidal direction [6]. A simpler single steerable mirror launcher, requiring less midplane port access, could be used if only one direction of current drive were needed.

The radiometric thermal EBW coupling experiments cannot test the importance of parametric decay to lower hybrid waves [21] and ponderomotive effects that can become important at high rf power densities. Furthermore, there is a negative particle pinch [22] that has not been included in our EBWCD modeling so far that can reduce the bootstrap current offsetting the Ohkawa current, thus reducing CD efficiency. In order to test these effects we propose to intall a ~ 1 MW, 28 GHz *proof-of-principle* EBWCD system on NSTX using one EBW launcher and transmission line.

ACKNOWLEDGMENTS

This work is supported by US Department of Energy (USDOE) contract nos. DE-AC02-76CH03073, DE-FG02-91ER-54109, DE-FG03-02ER54684, DE-FG02-99ER-54521 and a USDOE grant to encourage innovations in fusion diagnostic systems.

REFERENCES

[1] Ono, M., *et al.*, Proc. 17th Nucl. Fusion **40**, 557 (2000).
[2] Kessel, C. et al., *"Advanced ST Plasma Scenario Simulations for NSTX"* Proc. 20th IAEA Fusion Energy Conference, Vilamoura, Portugal (IAEA, Vienna, 2004), paper IAEA-CN-116/TH/P2-4.
[3] Rosenberg, A., *et al.*, Phys. Plasmas **11**, 2441 (2004).
[4] Bers, A., *"Theory of plasma heating in the lower hybrid range of frequencies (LHRF)"*, *Proc.3rd Topical Conference on Radio Frequency Power in Plasmas*, Pasedena, CA, (January 1978)
[5] Cairns, R.A., *et al.*, Phys. Plasmas, **7**, 4126 (2000).
[6] Forest, C.B., *et al.*, Phys. Plasmas, **7**, 1352 (2000).
[7] Taylor, G., *et al.*, Phys. Plasmas **11**, 4733 (2004).
[8] Ohkawa, T., "*Steady state operation of tokamaks by RF heating*", GA Report GA-A13847 (1976). See National Technical Information Document No.PB2000-108008.
[9] Harvey, R.W. and Taylor, G., *"Electron Bernstein wave-bootstrap current synergy in the National Spherical Torus Experiment (NSTX)"* to be published in Phys. Plasmas **12** (May 2005)
[10] Taylor, G., *et al.*, *"Efficient coupling of thermal electron Bernstein waves to the ordinary electromagnetic mode on the National Spherical Torus Experiment (NSTX)"* to be published in Phys. Plasmas **12** (May 2005)
[11] Preinhaelter, J., and Kopécky, V., J. Plasma Phys. **10**, 1-12 (1973).
[12] Smirnov, A.P., and Harvey, R.W., Bull. Am. Phys. Soc. **40**, 1837 (1995).
[13] Harvey, R.W., and McCoy, M.G., *Proceedings of the IAEA Technical Committee on Advances in Simulation and Modeling of Thermonuclear Plasmas*, Montreal, Quebec (International Atomic Energy Agency, Vienna, 1993), p. 489; USDOC NTIS Doc. No. DE93002962
[14] Sauter, O., *et al.*, Phys. of Plasmas **4**, 1654 (1997).
[15] R.Buttery, R.J., *et al*., Phys. Rev. Lett. **88**,125005 (2002).
[16] Sauter, O., Angioni, C., and Lin-Liu, Y.R., Phys. of Plasmas **6**, 2834 (1999).
[17] Jaeger, E.F., Berry, L.A., and Batchelor, D.B., Phys. of Plasmas **7**, 3319 (2000).
[18] Bers, A., and Ram, A.K., Physics Letters **301**, 442 (2002).
[19] Preinhaelter, J., *et al.*, *"EBW Simulation for MAST and NSTX Experiments"*, this conference (2005).
[20] Caughman, J., *et al*, *"Design and Testing of an Electron Bernstein Wave Emission Radiometer for the National Spherical Torus Experiment"*, this conference (2005).
[21] Stevan, V., and Bers, A., Phys. Fluids **27**, 175 (1984).
[22] Fisch, N.J., Rev. Modern Phys. **59**, 175 (1987).

EBW Experiments in the Madison Symmetric Torus

J.K. Anderson*, M. Cengher*, W. A. Cox*, C. B. Forest*, S. M. McMahon*, R.I. Pinsker[†], and V. Svidzinski*

*Department of Physics, University of Wisconsin, Madison WI 53706, USA
[†]General Atomics, San Diego, California 92014, USA

Abstract. A phased pair of waveguides is being used to experimentally investigate coupling to the electron Bernstein wave (EBW) in the Madison Symmetric Torus (MST) reversed field pinch (RFP). Coupling experiments have been performed from 3.3 to 3.9 GHz at ~1 Watt. These show that the coupling efficiency depends sensitively upon several factors, including the edge n_e profile, the polarization (X or O mode), and $N_\perp$ (controlled by phase between the waveguides). The amplitude and phase of reflected power (in each arm of the antenna) are compared to simulations and show general agreement. The predicted (a)symmetry in reflection for (X)O-mode launch versus interguide phasing is observed. A boron nitride cover has been installed on the end of the antenna, acting as a limiter and preventing plasma from contacting the waveguide vacuum windows; this has a measurable effect on coupling both into vacuum and into the target plasma. A heating experiment at 3.6 GHz and 150 kW has been installed, and measures have been taken to prevent breakdown at the window. Tube performance has been improved over past tests at this power level, and the capability of a 10 millisecond pulse has been achieved.

INTRODUCTION

The electron Bernstein wave (EBW) may facilitate localized heating and current drive in over-dense plasmas, found for example in the reversed field pinch (RFP) or spherical torus [1]. Current profile control may be necessary for stabilization of MHD activity in these configurations and is difficult by other means. A staged high power EBW experiment is under development on the Madison Symmetric Torus[2] RFP based on a simple waveguide grill antenna. The first stage consists of coupling studies at low power (~ 1 Watt) which have been performed to optimize several launch parameters. A moderate power (stage 2: ~ 100 kW) experiment is currently being brought online for initial heating studies, and a 300 kW experiment is under construction.

Electromagnetic launch of either the X- or O- mode at the plasma boundary can deliver power to the Bernstein mode[3]. The EM wave is cutoff where the

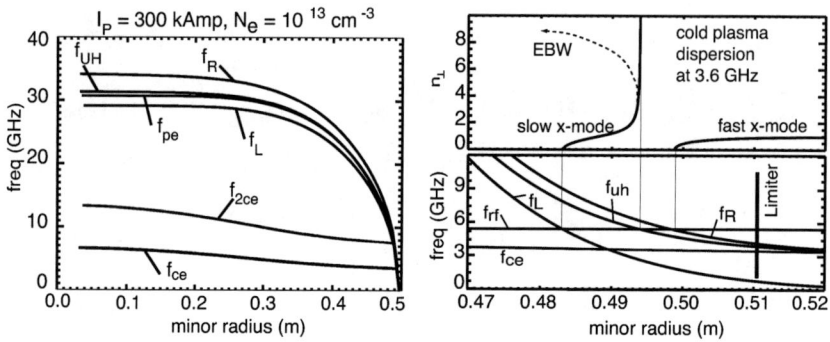

Figure 1: Cutoff and resonance frequencies in ECRF for a typical MST discharge. The upper hybrid resonance is typically within 3 cm of the antenna; mode conversion to the EBW takes place in the near field of the antenna.

launched frequency equals the local plasma frequency; in the MST the overdense plasma places the upper hybrid layer (where mode conversion to the EBW occurs) in the near field of the antenna. The EBW then propagates inward to the electron cyclotron resonance where the power is strongly absorbed. The important frequencies versus minor radius in the MST are shown in figure 1 along with an expanded view in the mode conversion layer of the edge region.

EBW COUPLING STUDIES

A simple antenna is being used for EBW heating and current drive experiments on the MST: a pair of rectangular S-band waveguides are adjacently connected and positioned flush with the inner surface of the close-fitting conducting shell (entering through a 4.5" port hole). Detailed studies (both experimental and numerical simulation) of coupling from this structure have been performed[4]. Key factors for coupling to the EBW are: 1. polarization (waveguides can be oriented to launch X- or O-mode), 2. edge density profile (determines the precise location of the upper hybrid resonance and distance between it and other relevant layers), and 3. the phase between the wavefronts in each arm of the antenna (allows for finite $N_\perp$ at launch).

Simulations for this antenna in X- mode, O- mode, and simple coupling into vacuum make several observable predictions. Vacuum coupling studies have been useful for benchmarking the phase and reflected power diagnostic. Initial experiments showed an unexpectedly high reflected fraction from plasma, likely due to plasma entering the waveguide and placing the upper hybrid layer at an undesired location. One solution for this vexation is a thin dielectric (BN) cover on the end of the antenna. Figure 2 shows the affect on coupling: (left) measured coupling into vacuum versus interguide phasing for three cases: no cover– which quantitatively agrees with simulations at 15% maximum reflection; the second case is a smooth BN cover $\frac{1}{16}$" thick with slightly higher reflection; and the highest reflection is

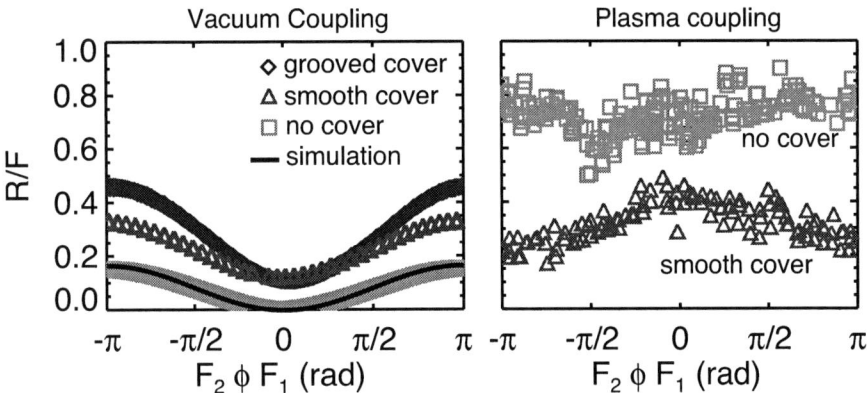

Figure 2: Left: Coupling into vacuum for antenna with no, smooth, and grooved BN cover. The reflection into vacuum is slightly worse with a cover. Right: Coupling into a quiescent plasma: the presence of the cover substantially improves coupling.

observed for a grooved cover (average thicknesses of $\frac{3}{32}$" BN). The higher vacuum reflection with dielectric present is not surprising. The right hand plot shows coupling into similar quiescent plasmas with and without the BN cover (these examples are X-mode launch). There is a very clear improvement by adding the cover, and the experimentally observed optimum launch phase has shifted by nearly 90 degrees (the minimum in R/F moves from $\sim -\pi/2$ without the cover to $\sim -\pi$ with it).

Simulations predict symmetry properties for this launch structure, which have been experimentally observed. O-mode reflection is expected to be optimized at (and to be symmetric about) zero phasing, while the minimum reflection for X-mode launch is nonzero and the symmetry is expected to be broken. Figure 3a) shows the reflected fraction versus interguide phasing for both X- and O-mode polarization; both agree qualitatively with the simulations. Figures 3b) and 3c) show that the phase of the reflected power contains reflectometer-like information. The top trace is the phase between the reflected and forward power in one arm of the antenna versus time during a quiescent period of an MST discharge, and the lower plot is the edge electron density measured by a triple Langmuir probe. The correlation between the two signals is apparent, particularly in the fluctuating part of the signals. This is currently not being used as a diagnostic of any sort, but with full wave modeling this type of data may help pinpoint the location of the upper hybrid resonance.

EBW HEATING STUDIES

Stage 2 of the EBW heating experiment is now operational. Two 75 kW traveling wave tube (TWT) amplifiers (3.2 - 3.8 GHz) feed power into the two-waveguide antenna. Several technical improvements have been made since previous experiments at this power level. The dielectric antenna cover has reduced plasma and

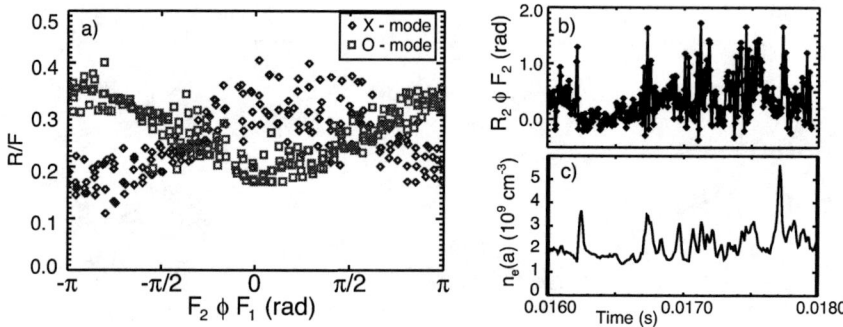

Figure 3: a) Coupling into plasma for X- and O- mode launch: The minimum in reflected power is zero for O- mode and non-zero for X-mode, as expected from simulations. b) Phase between forward and reflected power in one arm of the antenna, when compared with the edge electron density (c), there is a notable correlation.

neutral gas in the proximity of the quartz vacuum interface window (a serious problem leading to window failure in the last attempt at high power). Tube performance has been improved by altering the pulse of magnetic field for focusing the electron beam: body current in the tube is now at the specified level. An improved high voltage power supply has enabled a 10 millisecond pulse, more than doubling the previous 4 millisecond pulse capability. A four-waveguide antenna, to be driven by four 75 kW TWTs has been designed for the third stage, where heating effects are expected to be observable.

SUMMARY

Studies of coupling to the EBW in the overdense RFP plasma are maturing; simulations agree qualitatively with measurements at low power. Coupling into plasma and vacuum window longevity have been improved by adding a dielectric cover to the twin-waveguide antenna, and several other technical challenges have been solved to improve the moderate power ($\sim$ 100 kW) experiment. A 300 kW EBW heating system is under construction for direct tests of EBW heating feasibility in the RFP.

ACKNOWLEDGEMENTS

This work is supported by USDOE.

REFERENCES

1. A. Ram and S. D. Schultz, *Phys. Plasmas* **7**, 4024 (2000).
2. R. Dexter, D. Kerst, T. Lovell, S. Prager, and J. Sprott, *Fusion Technol.* **131** (1991).
3. H. P. Laqua, *Radio Frequency Power in Plasmas, 15th Topical Conference* (AIP Press, Melville, NY, 2003).
4. M. Cengher, J. K. Anderson, C. B. Forest, and V. Svidzinski, In preparation for submission to *Phys. Plasmas* (2005).

Survey of EBW Mode-Conversion Characteristics for Various Boundary Conditions

H. Tanaka*, H. Igami[†] and T. Maekawa*

Graduate School of Energy Science, Kyoto University, Kyoto, Japan
[†]*National Institute of Fusion Science, Toki, Japan*

Abstract.
A survey of linear mode-conversion characteristics between external transverse electromagnetic (TEM) waves and electron Bernstein waves (EBW) for various plasma and wave parameters has been presented. It is shown that if the wave propagation angle and polarization are adjusted appropriately for each individual case of the plasma parameters, efficient mode conversion occur for wide range of plasma parameters where the conventional 'XB' and 'OXB' scheme cannot cover. It is confirmed that the plasma parameters just at the upper hybrid resonance (UHR) layer strongly affect the mode conversion process and the influence of the plasma profiles distant from the UHR layer is not so much. The results of this survey is useful enough to examine wave injection/detection condition for efficient ECH/ECCD or measurement of emissive TEM waves for each individual experimental condition of overdense plasmas.

Keywords: Electron Bernstein Wave, Mode-Conversion
PACS: 52.25.Os, 52.35.Hr

INTRODUCTION

Electron cyclotron heating (ECH) is widely used for plasma heating and current drive (ECCD) because the incident wave power propagates through plasmas and is absorbed locally and it is very useful to control the local plasma pressure and/or current density. For overdense plasmas, mode-conversion from incident transverse electromagnetic (TEM) waves to electron Bernstein waves (EBW) is an important subject. In the reverse case where TEM waves are emitted via mode-conversion from EBW, measurement of such electron Bernstein wave emission (EBE) brings information on the electron temperature profile in overdense plasmas. Then, it is necessary to know the mode-conversion characteristics for both wave injection and detection.

So far, many reports has been published based on so called 'XB' or 'OXB' scheme [1, 2, 3, 4]. In the framework of the cold plasma resonance absorption model in a slab geometry, the electric fields of incident TEM waves can be expressed by linear combination of a orthogonal set, $\vec{g}_1$ and $\vec{g}_2$ modes [5]. Here, $\vec{g}_1$ mode excites EBW most efficiently but $\vec{g}_2$ mode excites no EBW at all. The principle of time-reversal invariance shows that the mode-conversion rate to the TEM waves is equal to that to EBW for the $\vec{g}_1$ mode. Then, the mode-conversion characteristics are described by using $\vec{g}_1$ mode.

In this report, we present a survey of the mode-conversion characteristics for various plasmas and wave parameters based on $\vec{g}_1$ mode which gives optimal mode-conversion. The results described here is useful to examine the wave injection/detection condition for efficient ECH/ECCD or measurement of emissive TEM waves in overdense plasmas.

PLASMA MODELING AND METHOD OF ANALYSIS

The coordinate systems and the plasma slab are described in Fig. 1 schematically. The x-axis is the direction of density gradient and the z-axis is the direction of the magnetic field. The normalized parameters, ω/Ω, L_n/λ_0 and $\vec{N} = \vec{k}/k_0$ are used, where ω is wave frequency, Ω is electron cyclotron frequency at UHR layer, $L_n = n_c/(dn/dx)$ is density scale length, n_c is plasma cutoff density, dn/dx is density gradient at UHR layer, λ_0 is wavelength in vacuum, $\vec{N}$ is refractive index vector, $\vec{k}$ is wavenumber vector and k_0 is wavenumber in vacuum.

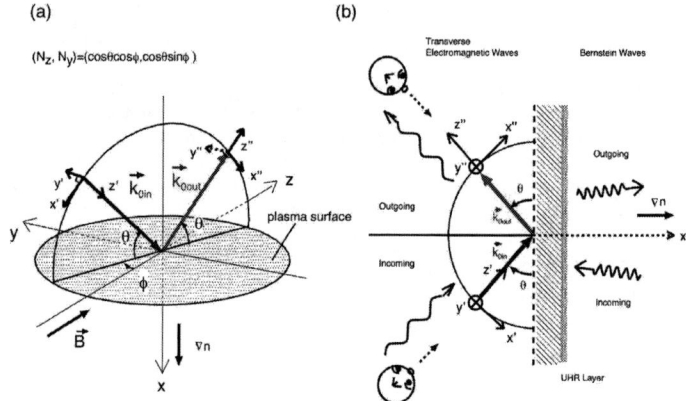

FIGURE 1. Coordinate systems and plasma slab used here

The linear relationship among electric field components of incident TEM wave and EBW are written with the scattering matrix and $\vec{g}_1$ mode and the optimal mode-conversion rate are expressed by the elements of the scattering matrix. In our calculation, the elements of the scattering matrix are obtained by solving the wave equation under appropriate boundary conditions numerically. It is assumed that ω/Ω and L_n/λ_0 are constant from the plasma surface to the region far beyond the UHR layer.

A SURVEY OF THE MODE-CONVERSION CHARACTERISTICS

In Figure 2, the optimal mode-conversion rate in N_z - N_y space is plotted in contour map for parameters indicated by the points numbered from (1) to (9) in Fig. 3. The left column ((1) to (3)) is at $\omega/\Omega = 1.67$, the middle column ((4) to (6)) is at $\omega/\Omega = 1.25$ and the right column ((7) to (9)) is at $\omega/\Omega = 1.1$. The upper row ((1), (4), (7)) is for the steep density gradient, the middle row ((2), (5), (8)) is for the moderate density gradient and the lower row ((3), (6), (9)) is for the gentle density gradient. The area where mode-conversion rate is more than 95 % is shaded and the area where it is less than 10 % is meshed. These contour plots represent 'efficient mode-conversion windows' for various angles of the $\vec{g}_1$ mode injection. The mode-conversion rate from the $\vec{g}_1$ mode to EBW and that from EBW to the $\vec{g}_1$ mode are the same if waves propagate at the same direction and have the same $\vec{N}$. Thus, these contour plots also represent the mode-conversion rate of EBE for various emission angles. The pattern of the contour map has the symmetry about $N_z = 0$ due to the principle of mirror symmetry for physical systems. This symmetry holds for polarization too.

For the case of the steep density gradient, the efficient mode-conversion window occupies one continuous area and changes the shape to like a bow in the positive N_y region as ω/Ω decreases. For the case of the moderate density gradient, the efficient mode-conversion window separates into two symmetrical area. Beside, the third area appears around $N_z = 0, N_y \simeq 0$ when ω/Ω is low as 1.1. For the case of the gentle density gradient, the efficient mode-conversion window is rather narrow areas near the points where $N_z = \pm N_{//opt}$ and $N_y = 0$. Here $N_{//opt} = \sqrt{\Omega/(\omega+\Omega)}$ is the parallel refractive index at he plasma cutoff layer and corresponds to the optimal injection angle of 'OXB' scheme.

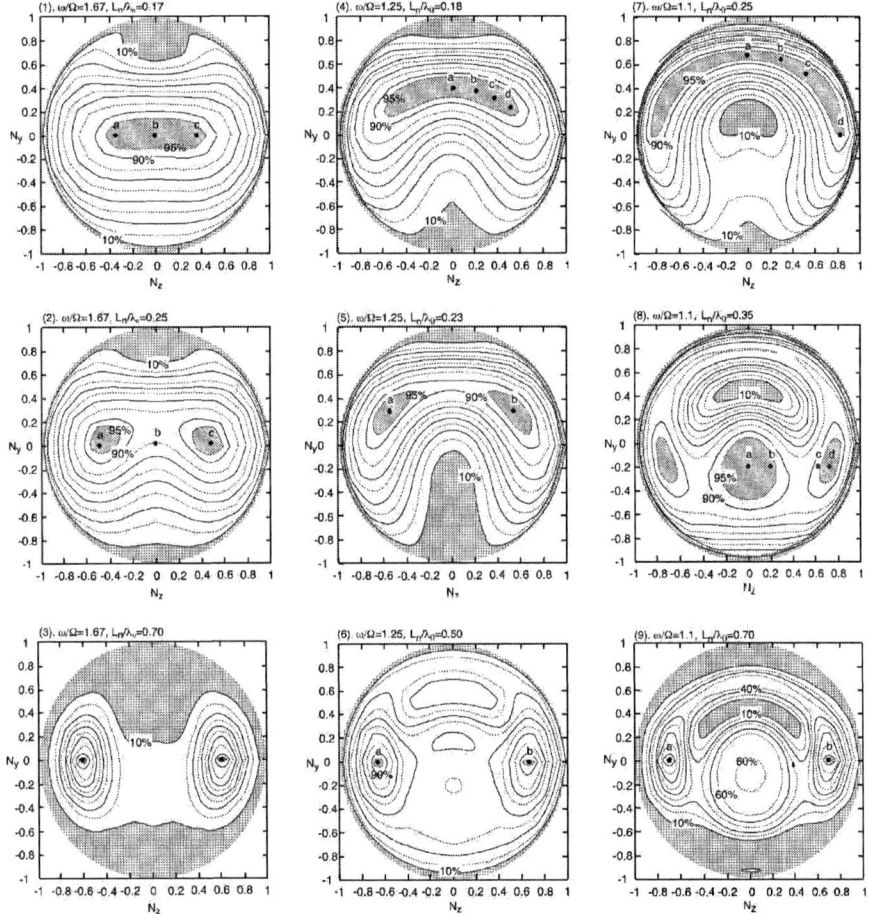

FIGURE 2. The optimal mode-conversion rate in N_z - N_y space

In Figure 3, optimal mode-conversion rate is plotted in a contour map in L_n/λ_0 - ω/Ω

space for different optimization of wave propagation angles. As shown in Fig. 3 (a), optimizing polarization but fixing the direction of the wave propagation in perpendicular, efficient mode-conversion occurs only at a narrow area. This area corresponds to the conventional 'XB' scheme. If optimization for N_y or N_z is included, the efficient mode-conversion region becomes wider (Fig. 3 (b) or (c)). And if we take $\vec{g}_1$ mode and both N_y and N_z are optimized, mode-conversion rate more than 95 % can be obtained in the broader area where the conventional 'XB' and 'OXB' scheme cannot cover as shown in Fig. 3 (d).

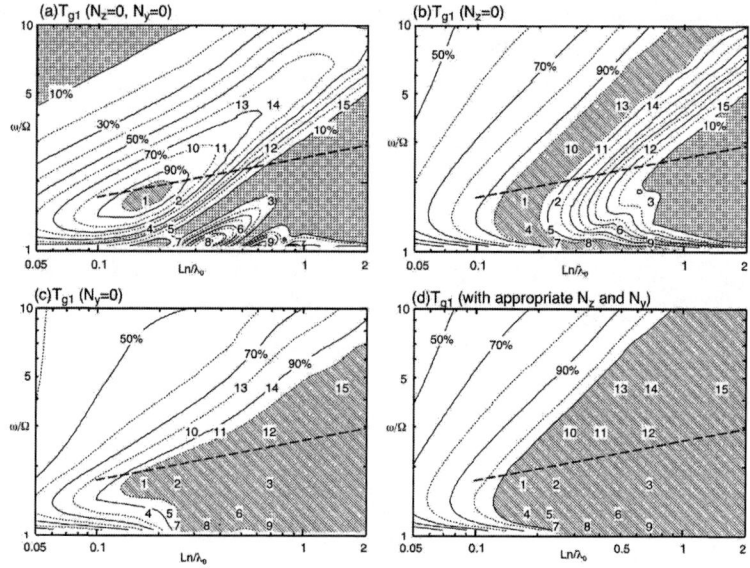

FIGURE 3. The optimal mode-conversion rate in L_n/λ_0 - ω/Ω space

In real toroidal plasmas, density profiles are different from the linear profile used here. Then, the optimal mode-conversion rate calculated for simulated model profiles of an NSTX plasma is compared with that for the linear profile used here. The patterns of the contour maps of the mode-conversion rate are similar to each other for the same L_n/λ_0 and ω/Ω at the UHR layer but different profiles. Therefore, the mode-conversion process is governed by the wave dynamics around the narrow mode-conversion region where the UHR layer and the cutoffs exist close to each other, and the results obtained by the simple profile used here is useful and helpful in determining TEM wave parameters for ECH/ECCD or EBE measurement.

REFERENCES

1. H. P. Laqua, et al., *Phys. Rev. Lett.*, **78**, 3467–3470 (1997).
2. A. K. Ram and S. D. Schults, *Phys. Plasmas*, **7**, 4084–4094 (2000).
3. R. A. Cairns and C. N. Lashmore-Davies, *Phys. Plasmas*, **7**, 4126–4134 (2000).
4. J. Preinhaelter, M. A. Irzak, L. Vahara and G. Vahara, *Rev. Sci. Instrum.*, **72**, 391-393 (2001).
5. H. Igami, et al., *Plamsa Phys. Contr. Fusion*, **46**, 261–275 (2004).

EBW simulation for MAST and NSTX experiments

J. Preinhaelter[1], G. Taylor[2], V. Shevchenko[3], J. Urban[1], M. Valovic[3], P. Pavlo[1], L. Vahala[4], G. Vahala[5]

1) EURATOM/IPP.CR Association, Institute of Plasma Physics, 182 21 Prague, Czech Republic
2) Princeton Plasma Physics Laboratory, Princeton, New Jersey 08543, USA
3) EURATOM/UKAEA Fusion Association, Culham Science Centre, Abingdon, OX14 3DB, UK
4) Old Dominion University, Norfolk, VA 23529, USA
5) College of William & Mary, Williamsburg, VA 23185, USA

Abstract. The interpretation of EBW emission from spherical tokamaks is nontrivial. We report on a 3D simulation model of this process that incorporates Gaussian beams for the antenna, a full wave solution of EBW-X and EBW-X-O conversions using adaptive finite elements, and EBW ray tracing to determine the radiative temperature. This model is then used to interpret the experimental results from MAST and NSTX. EBW for ELM free H-modes in MAST suggests that the magnetic equilibrium determined by the EFIT code does not adequately represent the B-field within the transport barrier [1]. Using the EBW signal for the reconstruction of the radial profile of the magnetic field, we determine a new equilibrium and see that the EBW simulation now yields better agreement with experimental results. EBW simulations yield excellent results for the time development of the plasma temperature as measured by the EBW radiometer on NSTX [2].

Keywords: Spherical tokamaks. Electron Bernstein waves.
PACS: 52.55.Fa, 52.35.Hr

Introduction

The characteristic low magnetic field and high plasma density of a spherical tokamak do not permit the typical radiation of O and X modes from the first five electron cyclotron harmonics. Thus only electron Bernstein waves (EBW), (modes not subject to a density limit), which mode convert into electromagnetic waves in the upper hybrid resonance region, can be responsible for the measured radiation [3]. The 3D plasma and antenna model was described in [1, 4, 5].

Magnetic Field Reconstruction from EBW Signal in MAST

We found good agreement between the simulation of EBW emission and the detected signals for L-modes and ELMy H-modes. On the other hand the emission from ELM free H-modes in MAST suggests that the magnetic field in the transport barrier as determined by EFIT is too low. Typically the detected signal in the 16-60GHz band has five peaks, each corresponding to the emission from subsequent electron cyclotron harmonics (see Fig. 1). The gaps between the peaks correspond to the frequencies at which the upper hybrid frequency coincides with some of the electron cyclotron harmonics. From the position of the gaps in the spectrum of the detected signal we can determine the magnitude of the magnetic field. The corresponding R position occurs where the upper hybrid frequency coincides with the frequency of the electron cyclotron harmonics (we investigate the shot where the detected signal originates in the equatorial plane). The profiles of the EFIT magnetic field and the reconstructed profile from EBW signal are given in Fig. 2. We assume that only the poloidal magnetic field is influenced at the transport barrier. The magnetic configuration in the rest of the plasma is assumed to be that determined by EFIT.

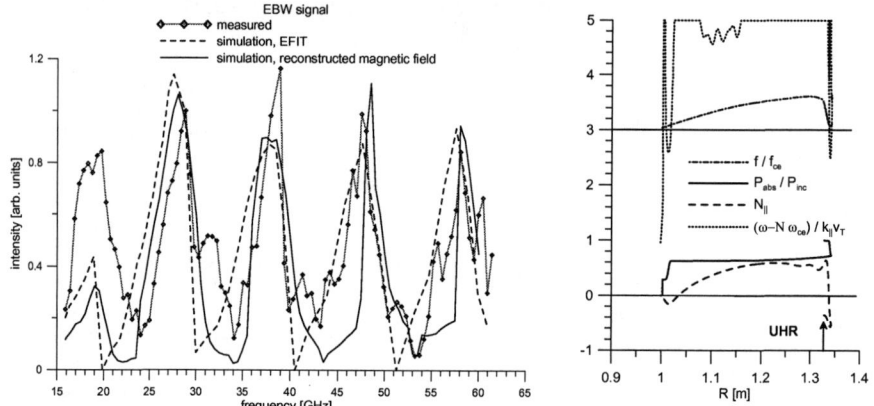

FIGURE 1. Comparison of the detected ECE signal and two simulations (EFIT magnetic equilibrium-blue line, and the equilibrium with a bump in the magnetic field profile at the transport barrier. #8694, t=280ms, the time corresponds to the final stage of the prolonged ELM free period of the H-mode shot. **(left)**. Radial dependence of the wave and the plasma parameters along the EBW ray **(right)**.

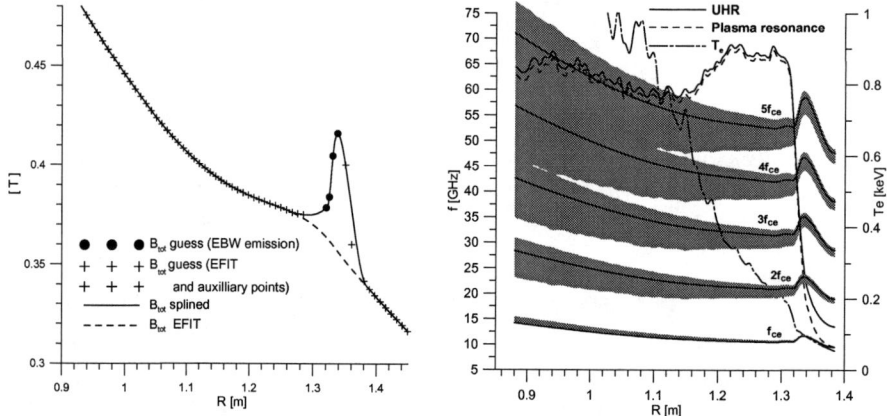

FIGURE 2: Comparison of the EFIT profile of the magnetic field and the field reconstructed from the EBW signal for #8694 **(left)**. Radial profiles of the characteristic resonances in the equatorial plane **(right)**.

To gain some insight into the significance of the proposed bump on the magnetic field we consider the radial profiles of the characteristic resonances (Fig. 2). The broadening of nf_{ce} is given by the factor $1/(1 \pm 3N_\parallel (v_T/c))$. Rays with frequency below nf_{ce} are usually emitted with $N_\parallel = 1$, while $N_\parallel$ is oscillating for rays with frequency above nf_{ce}. Thus we consider an initial value of $N_\parallel = 0.36$. In the shaded areas EBW is strongly damped and the detected wave is emitted from the edges of these areas. The broadening of the gaps in the emitted spectrum (Fig. 1) is due to the bump on the magnetic field in the transport barrier. For the first two harmonics the magnetic field decreases in the transport barrier from the UHR region in the direction towards the plasma center. EBW with frequency slightly below nf_{ce} are then emitted from the rather cold plasma. On the other hand, EBW emitted slightly above the third and higher harmonics is strongly reabsorbed by the rarefied plasma in front of the UHR (see Fig. 1 right).

Here the ray is launched from the UHR region and starts to propagate out of the plasma. Its frequency is approaching the 3rd electron cyclotron harmonic (magnetic field increases in this direction due to the bump) and it is partially absorbed here. The ray is then reverted back to the dense plasma and is fully absorbed at the 3rd harmonics at the plasma center. Emission is the reverse process and the ray emitted from the plasma center where plasma temperature is 1keV is partly reabsorbed at the plasma boundary so finally $T_{rad} = 0.7$keV. Waves with slightly lower frequency (e.g., 37GHz) are fully absorbed at the plasma boundary with $T_{rad} = 0.1$keV only. The broadening of gaps due to the development of the bump in the magnetic field in the transport barrier can thus explain the observation of the missing EBW emission for some well-developed ELM free H-modes.

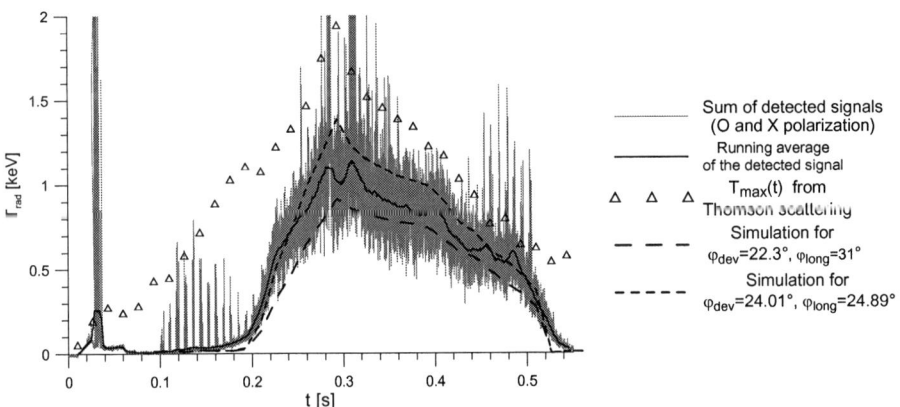

FIGURE 3: The time development of EBW signal (sum of O and X channels) detected by the NSTX antenna operating at 16.5GHz. #113544.

One of the most important results deduced from EBW emission measurements on NSTX is the time development of the radiation plasma temperature [2]. The obliquely viewing EBW radiometer on NSTX detects signals polarized parallel and perpendicular to the magnetic field at the EBW mode conversion layer. Since the outgoing O-mode at oblique incidence is approximately circularly polarized, the sum of these signals can be interpreted as a radiation temperature. We have performed simulations of the time development of ECE in the interval $0.09s < t < 0.6s$. In Fig. 3, the time evolution of the sum of the EBW radiometer signals is plotted along with the maximum temperature detected by Thomson scattering (this independent measurement for $0.1s < t < 0.5s$ corresponds well to the temperature at $R = 1.007\ m$) and the theoretical simulation of EBW radiometer signals of Gaussian beams corresponding to actual antenna orientation (angles $\phi_{dev} = 23.3°$, $\phi_{long} = 31°$) as well as to orientations that are optimal for $EBW - X - O$ mode conversion (angles $\phi_{dev} = 24.01°$, $\phi_{long} = 24.89°$). The signal is proportional to the central temperature only for $t > 0.3s$ when the plasma current and the poloidal magnetic field reach their stationary values. For $t < 0.3s$, EBW is emitted from the 2nd electron cyclotron harmonic at the plasma periphery, where the plasma temperature is low. The simulation shows that the conversion efficiency for EBW-X-O is optimal and practically constant for $t \in (0.2s, 05s)$.

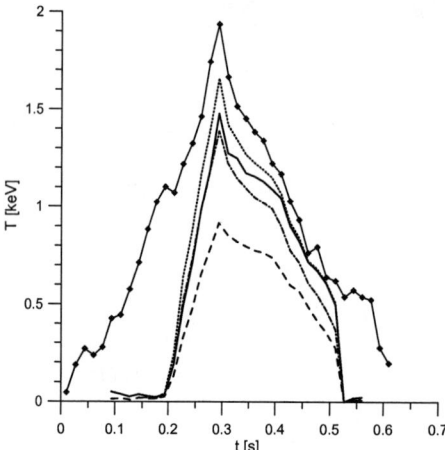

FIGURE 4: Time development of the central temperature (diamonds), a simulation for EBW effective temperature for optimum antenna angles (dotted), and for actual antenna angles (solid) as well as their reduced values (only these can be detected in the experiment) by the corresponding conversion efficiencies (dot-and-dash and dashed).

Simulations also clarify the relationship between the central temperature derived from the EBW signal and the temperature determined from Thomson scattering. We can state that the EBW temperature is always less than the actual temperature. In Fig 4 we depict the ideal effective temperature of EBW radiation for conversion efficiency equal 1. Even these values are smaller than $T_{Thomson}$ because of reabsorption of EBW and the parasitic radiation from the second harmonic [2]. The intensity of EBW radiation is further reduced by losses due to imperfect conversion. This last effect is more pronounced if the antenna does not have optimum orientation.

Acknowledgments

Supported by U.S. Dept. of Energy, by UK Engineering and Physical Sciences Research Council, by EURATOM and by AS CR project #AV0Z-20430508.

References

1. J. Preinhaelter, V. Shevchenko, M. Valovic, et al., ECA **Vol. 28G**, P-4.184 (2004).
2. G. Taylor, P.C. Efthimion, B.P. LeBlanc, et al., Phys. Plasmas (2005) to be published
3. H.P. Laqua, et al., review, 15th RF Power in Plasma, Moran, 2003, edit. C. Forest, AIP **694**, 5.
4. J. Preinhaelter, V. Shevchenko, M. Valovic, et al., edit. C. Forest, AIP **694**, 388.
5. J. Preinhaelter, J. Urban, P. Pavlo, et al., Review of Scientific. Instruments., Vol. 75, (2004), p 3804.

Self-consistent Formulation of EBW Excitation by Mode Conversion

Abraham Bers and Joan Decker

Plasma Science and Fusion Center, MIT, Cambridge MA 02139

Abstract. Based upon a FLR-hydrodynamic formulation for high frequency waves in a collisionless plasma, we formulate the self-consistent, coupled set of ordinary differential equations whose solution gives the mode conversion of O- and/or X-waves at an angle to $\mathbf{B}_0$ to electron Bernstein waves (EBW) at the upper-hybrid resonance UHR layer occurring at the edge of an ST plasma.

1. FLR Hydrodynamics for High-Frequency Waves

Our starting point are the first three moments of the non-relativistic Vlasov equation for the particle distribution function $f(\mathbf{w}, \mathbf{r}, t)$, i.e the moments of $\mathbf{w}^0$, $m\mathbf{w}$, and $m\mathbf{u}\mathbf{u}$, where $\mathbf{w} = \mathbf{v} + \mathbf{u}$ and $\langle \mathbf{u} \rangle = 0$. We retain the full anisotropic pressure tensor $nm\langle \mathbf{u}\mathbf{u} \rangle = \overline{\overline{\mathbf{P}}}$, and close the equations for the high frequencies of interest by setting $\nabla \cdot \overline{\overline{\overline{\mathbf{Q}}}} = 0$, where $\overline{\overline{\overline{\mathbf{Q}}}} = nm\langle \mathbf{u}\mathbf{u}\mathbf{u} \rangle$ is the heat flow tensor. The equations are then:

$$\frac{\partial n_s}{\partial t} + \nabla \cdot (n_s \mathbf{v}_s) = 0 \tag{1}$$

$$n_s m_s \left(\frac{\partial}{\partial t} + \mathbf{v}_s \cdot \nabla\right) \mathbf{v}_s + \nabla \cdot \overline{\overline{\mathbf{P}}}_s = q_s n_s (\mathbf{E} + \mathbf{v}_s \times \mathbf{B}) \tag{2}$$

$$\left(\frac{\partial}{\partial t} + \mathbf{v}_s \cdot \nabla\right) \overline{\overline{\mathbf{P}}}_s + (\nabla \cdot \mathbf{v}_s) \overline{\overline{\mathbf{P}}}_s + \left[\overline{\overline{\mathbf{P}}}_s \cdot \nabla \mathbf{v}_s + \left(\overline{\overline{\mathbf{P}}}_s \cdot \nabla \mathbf{v}_s\right)^t\right] \tag{3}$$

$$= -\frac{q_s}{m_s}\left[\mathbf{B} \times \overline{\overline{\mathbf{P}}}_s + \left(\mathbf{B} \times \overline{\overline{\mathbf{P}}}_s\right)^t\right]$$

where the superscript t on an expression indicates its transpose. These have to be solved self-consistently with Maxwell's equations.

Consider first the linear susceptibility that (1)-(3) give for a homogeneous plasma in a magnetic field. Linearizing (1)-(3) about a neutral, uniform, thermal equilibrium given by $\overline{\overline{\mathbf{P}}}_0 = p_0 \overline{\overline{\mathbf{I}}}$, $p_0 = n_0 \kappa T_0$, $\mathbf{v}_0 = 0 = \mathbf{E}_0$, $\mathbf{B}_0 = \hat{z} B_0$; then, assuming a wave dependence $e^{i\mathbf{k}\cdot\mathbf{r} - i\omega t}$ for all linear fields, solving the linearized (3) for $\overline{\overline{\mathbf{P}}}_1 (\mathbf{J}_1)$, where $\mathbf{J}_1 = q n_0 \mathbf{v}_1$, and substituting this into linearized (2), one obtains:

$$\mathbf{J}_1 - i\boldsymbol{\beta} \times \mathbf{J}_1 - \delta_T^2 \overline{\overline{\mathbf{T}}} \cdot \mathbf{J}_1 = i\omega\epsilon_0 \alpha^2 \mathbf{E}_1 \tag{4}$$

where $\beta = \omega_c/\omega$, $\alpha = \omega_p/\omega$, $\delta_T = v_T/c$, and the thermal correction tensor $\overline{\overline{\mathbf{T}}}$ is a function of $\mathbf{n} = c\mathbf{k}/\omega$ and β; $\overline{\overline{\mathbf{T}}}$ is found proportional to $(k_\perp v_T/\omega_c)^2$, $(k_\parallel v_T/\omega)^2$, and

$(k_\perp v_T/\omega_c)(k_\parallel v_T/\omega)$ – each of which must be small compared to unity in a hydrodynamic description. Defining the linear susceptibility tensor $\overline{\overline{\chi}}$ from $\mathbf{J}_1(\mathbf{E}_1) \equiv -i\omega\epsilon_0 \overline{\overline{\chi}} \cdot \mathbf{E}_1$, and solving (4) to order δ_T^2, one obtains $\overline{\overline{\chi}} = \overline{\overline{\chi}}_C + \delta_T^2 \overline{\overline{\chi}}_T$, where $\overline{\overline{\chi}}_C = -\alpha^2 \overline{\overline{\beta}}^{-1}$ is the cold plasma susceptibility tensor, $\overline{\overline{\beta}} = (1 + i\beta\times)\overline{\overline{\mathbf{I}}}$, and $\overline{\overline{\chi}}_T = \overline{\overline{\beta}}^{-1} \cdot \overline{\overline{\mathbf{T}}} \cdot \overline{\overline{\chi}}_C$.

These are identical to what is found from a direct solution of the linearized Vlasov equation, linearized around an equilibrium Maxwellian, and assuming ab-initio the same small parameters [1], or from the complete Vlasov linear susceptibility, for an equilibrium Maxwellian, expanded in the same small parameters; in both cases ignoring Landau damping and Doppler-shifted cyclotron resonance damping, i.e. for $|k_\parallel v_T| \ll |\omega - n\omega_c|$ with $n = 0, \pm 1, \pm 2$, and so valid for $\mathcal{O}(k_\perp^2 \rho^2)$.

The self-consistent wave description is obtained by using the resulting $\overline{\overline{\chi}}$ in Maxwell's equations, and the dispersion for the natural waves is found from the determinant of $[\mathbf{nn} + (1 - n^2)\overline{\overline{\mathbf{I}}} + \overline{\overline{\chi}}] \equiv \overline{\overline{\mathbf{D}}}$. This FLR hydrodynamic formulation is found to give the correct dispersion behavior for the transition from an SX-wave to an EBW, near the upper-hybrid resonance (UHR) when the resonance is either below ω_c or above ω_c and below $2\omega_c$. This shows the same behavior as is obtained from a numerical solution of the fully electromagnetic Vlasov equation solution for $\overline{\overline{\chi}}$ in $\overline{\overline{\mathbf{D}}}$ [4].

2. MODE CONVERSION DIFFERENTIAL EQUATIONS

Mode conversion to EBW occurs at the UHR layer, which is at the outboard edge in high-beta ST plasmas, such as NSTX; these are over-dense, $\omega_p \gg \omega$, and the poloidal B-fields are appreciable there. We model the mode conversion region as a slab, around the UHR location, with the following inhomogeneous equilibrium: $n_0 = n_0(x)$, $\overline{\overline{\mathbf{P}}}_0 = p_0(x)\overline{\overline{\mathbf{I}}}$; $p_0(x) = n_0(x)\kappa T_0(x)$, $\mathbf{B}_0(x) = B_p(x)\hat{y} + B_t\hat{z}$, $\mathbf{v}_0 = v_{0t}\hat{z}$, and $\mathbf{E}_0 = 0$. The equilibrium pressure balance and Ampere's equations are:

$$p_0' = -qn_0 v_{0t} B_p \tag{5}$$
$$B_p' = \mu_0 q n_0 v_{0t} \tag{6}$$

and $(\ )' = (d/dx)(\ )$. The linearized fields are now taken as $\sim F(x)\exp(ik_y y + ik_z z - i\omega t)$, and the linearized (1)-(3) become:

$$\left(1 - \delta_T^2 v_{0t} n_z\right) n + i(n_0 v_x)' - n_0(n_y v_y + n_z v_z) = 0 \tag{7}$$

$$n_0\left[\mathbf{v} - i(\mathbf{E} + \mathbf{v} \times \mathbf{b}_0)\right] = \delta_T^2\left[n_0 v_{0t} n_z \mathbf{v} - i\nabla \cdot \overline{\overline{\mathbf{P}}} - i v_{0t} b_p n\hat{x} + i n_0 v_{0t}\hat{z} \times \mathbf{B}\right] \tag{8}$$

$$\left(1 - \delta_T^2 v_{0t} n_z\right)\overline{\overline{\mathbf{P}}} + i v_x p_0' \overline{\overline{\mathbf{I}}} + i p_0(\nabla \cdot \mathbf{v})\overline{\overline{\mathbf{I}}} + i p_0\left[\nabla\mathbf{v} + (\nabla\mathbf{v})^t\right] + \tag{9}$$
$$i\left[\mathbf{b}_0 \times \overline{\overline{\mathbf{P}}} + \left(\mathbf{b}_0 \times \overline{\overline{\mathbf{P}}}\right)^t\right] = 0$$

where the following normalizations have been used: $\mathbf{v} = \tilde{\mathbf{v}}c$, $n = \tilde{n}n_{00}$, $\overline{\overline{\mathbf{P}}} = \tilde{\overline{\overline{\mathbf{P}}}} n_{00}\kappa T_{00}$, $\mathbf{B} = \tilde{\mathbf{B}}m\omega/q$, $\mathbf{E} = \tilde{\mathbf{E}}mc\omega/q$, $\mathbf{J} = qn_{00}c\tilde{\mathbf{J}}$, $k_y = \omega n_y/c$, $k_z = \omega n_z/c$, $\nabla = (\omega/c)\nabla$,

$n_{00} = n_0(x = 0)$, $T_{00} = T_0(x = 0)$, and the subscript 1 and the tilde have been omitted. The current density is now $\mathbf{J} = n_0\mathbf{v} + \delta_T^2 v_{0t}\hat{z}n$. To find the operator relation $\mathbf{J}_{op}(\mathbf{E})$ to $\mathcal{O}(\delta_T^2)$, one needs to find $\mathbf{v}$ to $\mathcal{O}(\delta_T^2)$ and n to $\mathcal{O}(\delta_T^0)$. Formally this will give $\overline{\overline{\mathcal{X}}} = \overline{\overline{\mathcal{X}}}_{pt}^C + \overline{\overline{\mathcal{X}}}_{op}^T$, where $\overline{\overline{\mathcal{X}}}_{pt}^C$ is the cold plasma tensor in the inhomogeneous poloidal and toroidal $\mathbf{B}_0$-fields, and $\overline{\overline{\mathcal{X}}}_{op}^T$ is the thermal correction operator (in x) tensor. The self-consistent field Maxwell equations are then,

$$B_x = n_y E_z - n_z E_y \tag{10}$$

$$B_y = n_z E_x + i E_z' \tag{11}$$

$$B_z = -i E_y' - n_y E_x \tag{12}$$

$$n_z B_y - n_y B_z = \left(\overline{\overline{\mathbf{K}}}_C \cdot \mathbf{E}\right)_x + \delta_T^2 \left(\overline{\overline{\mathcal{X}}}_{op}^T \cdot \mathbf{E}\right)_x \tag{13}$$

$$-n_z B_x - i B_z' = \left(\overline{\overline{\mathbf{K}}}_C \cdot \mathbf{E}\right)_y + \delta_T^2 \left(\overline{\overline{\mathcal{X}}}_{op}^T \cdot \mathbf{E}\right)_y \tag{14}$$

$$i B_y' + n_y B_x = \left(\overline{\overline{\mathbf{K}}}_C \cdot \mathbf{E}\right)_z + \delta_T^2 \left(\overline{\overline{\mathcal{X}}}_{op}^T \cdot \mathbf{E}\right)_z \tag{15}$$

where $\overline{\overline{\mathbf{K}}}_C$ is the cold plasma permitivity tensor, $\overline{\overline{\mathbf{K}}}_C = \overline{\overline{\mathbf{I}}} + \overline{\overline{\mathcal{X}}}_{pt}^C$ with $\overline{\overline{\mathcal{X}}}_{pt}^C = -\alpha^2(x)\overline{\overline{\beta}}_{pt}^{-1}(x)$, and $\overline{\overline{\beta}}_{pt} = (1 + i\mathbf{b}_0\times)\overline{\overline{\mathbf{I}}}$, where $\mathbf{b}_0 = \hat{y}b_p(x) + \hat{z}b_t(x)$ with $b_{p,t} = qB_{0p,t}(x)/m\omega$. From the above indicated solution procedure for $\overline{\overline{\mathcal{X}}}_{op}^T$, one finds in detail:

$$\overline{\overline{\mathcal{X}}}_{op}^T \cdot \mathbf{E} = \overline{\overline{\mathcal{X}}}_{pt}^C \cdot \left[\overline{\overline{\mathbf{T}}}_0 \cdot \mathbf{E} + \overline{\overline{\mathbf{T}}}_1 \cdot \mathbf{E}' + \overline{\overline{\mathbf{T}}}_2 \cdot \mathbf{E}''\right] \tag{16}$$

where

$$\overline{\overline{\mathbf{T}}}_0 = \overline{\overline{\mathbf{A}}} + \frac{v_{0t}}{c}\left(n_z \overline{\overline{\beta}}_{pt}^{-1} + \overline{\overline{\mathbf{n}}} + \hat{z}\mathbf{N}\right) \tag{17}$$

$$\overline{\overline{\mathbf{T}}}_1 = \overline{\overline{\mathbf{B}}} + \frac{v_{0t}}{c}\left(\hat{z}\mathbf{M} - i\hat{x}\hat{z}\right) \tag{18}$$

$$\overline{\overline{\mathbf{T}}}_2 = \overline{\overline{\mathbf{Y}}}_1 \tag{19}$$

and where

$$\overline{\overline{\mathbf{A}}} = \overline{\overline{\mathbf{X}}}_1' + in_y \overline{\overline{\mathbf{X}}}_2 + in_z \overline{\overline{\mathbf{X}}}_3 \tag{20}$$

$$\overline{\overline{\mathbf{B}}} = \overline{\overline{\mathbf{X}}}_1 + \overline{\overline{\mathbf{Y}}}_1' + in_y \overline{\overline{\mathbf{Y}}}_2 + in_z \overline{\overline{\mathbf{Y}}}_3. \tag{21}$$

With $c = 1, 2$, or 3,

$$\overline{\overline{\mathbf{X}}}_c = T_0 \overline{\overline{\mathbf{M}}}^{-1} \cdot \overline{\overline{\mathbf{W}}}_c \text{ and } \overline{\overline{\mathbf{Y}}}_c = T_0 \overline{\overline{\mathbf{M}}}^{-1} \cdot \overline{\overline{\mathbf{U}}}_c, \tag{22}$$

where $T_0 = T_0(x)/T_0(x=0)$, $\overline{\overline{\mathbf{M}}}$ is obtained from $\overline{\overline{\beta}}_{pt}$ by replacing each unity on the diagonal by a $\overline{\overline{\beta}}_{pt}$ and the off-diagonal terms $ib_{p,t}$ by $ib_{p,t}\overline{\overline{\mathbf{I}}}$; and where

$$\overline{\overline{\mathbf{W}}}_c = \overline{\overline{\mathbf{V}}}_{c1} \cdot \overline{\overline{\beta}}_{pt}^{-1} + \overline{\overline{\mathbf{V}}}_{c2} \cdot \left(\overline{\overline{\beta}}_{pt}^{-1}\right)'; \quad \overline{\overline{\mathbf{U}}}_c = \overline{\overline{\mathbf{V}}}_{c2} \cdot \overline{\overline{\beta}}_{pt}^{-1} \tag{23}$$

with

$$\overline{\overline{V}}_{11} = \begin{pmatrix} p'_0/p_0 & in_y & in_z \\ in_y & 0 & 0 \\ in_z & 0 & 0 \end{pmatrix}, \quad \overline{\overline{V}}_{12} = \begin{pmatrix} 3 & 0 & 0 \\ 0 & 1 & 0 \\ 0 & 0 & 1 \end{pmatrix},$$

$$\overline{\overline{V}}_{21} = \begin{pmatrix} in_y & 0 & 0 \\ p'_0/p_0 & i3n_y & in_z \\ 0 & in_z & in_y \end{pmatrix}, \quad \overline{\overline{V}}_{22} = \begin{pmatrix} 0 & 1 & 0 \\ 1 & 0 & 0 \\ 0 & 0 & 0 \end{pmatrix}, \quad (24)$$

$$\overline{\overline{V}}_{31} = \begin{pmatrix} in_z & 0 & 0 \\ 0 & in_z & in_y \\ p'_0/p_0 & in_y & i3n_z \end{pmatrix}, \quad \overline{\overline{V}}_{32} = \begin{pmatrix} 0 & 0 & 1 \\ 0 & 0 & 0 \\ 1 & 0 & 0 \end{pmatrix}.$$

Finally,

$$\overline{\overline{n}} = \begin{pmatrix} -n_z & 0 & 0 \\ 0 & -n_z & n_y \\ 0 & 0 & 0 \end{pmatrix} \quad (25)$$

where $n_z = ck_z/\omega$ and $n_y = ck_y/\omega$; and

$$\mathbf{N} = \mathbf{N}_1 \cdot \overline{\overline{\beta}}_{pt}^{-1} + \mathbf{N}_2 \cdot \left(\overline{\overline{\beta}}_{pt}^{-1}\right)'; \quad \mathbf{M} = \mathbf{N}_2 \cdot \overline{\overline{\beta}}_{pt}^{-1} \quad (26)$$

where $\mathbf{N}_1 = -i\hat{x}(n'_0/n_0) + \hat{y}n_y + \hat{z}n_z$ and $\mathbf{N}_2 = -i\hat{x}$.

The solution of the O.D.E.'s (10) - (15), with appropriate boundary conditions specifying the WKB waves asymptotic to the mode conversion layer, gives the transmission, reflection, and mode conversion of these waves incident on the UHR layer. This will be implemented in a numerical code and compared to the much simpler but not self-consistent formulation of the same problem, in which the equilibrium gradients were left arbitrary, with zero flow, and the EBW was introduced in an ad-hoc manner to the cold plasma equations [5].

References

[1] W.P. Allis, S.J. Buchsbaum, and A. Bers, "Waves in Anisotropic Plasmas", M.I.T. Press, Cambridge, 1963, p. 90 corr. (Note: corrections to this book can be obtained from A. Bers, upon request.)

[2] A. G. Sitenko and K.N. Stepanov, Soviet Physics JETP **4**, 512 (1957)

[3] T.H. Stix, "Plasma Waves", A.I.P., New York 1992, p. 258

[4] S. Puri, F. Leuterrer, and M. Tutter, J. Plasma Phys. **9**,89 (1973)

[5] A. Bers, A.K. Ram, and S.D. Schultz, 2-nd Europhys. Topical Conf. on RF, Brussels, Belgium, EPS Contrib. Papers **22**, 237 (1998)

Fully Relativistic Ray-tracing and Dispersion Relations of Electron Bernstein Waves

E. Nelson-Melby*, R.W. Harvey*, A.P. Smirnov*, A.K. Ram[†] and S. Coda**

*CompX, P.O. Box 2672, Del Mar, CA 92014-5672, USA
[†]Plasma Science and Fusion Center, MIT, Cambridge, MA, USA
**CRPP-EPFL, Lausanne, Switzerland

Abstract. Electron Bernstein waves (EBW) are important in current and future research in high density, high temperature plasmas heated by microwaves. Usually, EBWs are described using an electromagnetic, hot-plasma but non-relativistic dispersion relation. However, EBWs in magnetically confined fusion devices usually damp near electron cyclotron resonances, where relativistic effects are important in order to describe the damping correctly. Recently, a fully-relativistic, high-frequency (ion motion neglected) dispersion relation has been added to the ray-tracing code GENRAY [1]. Comparisons of electron Bernstein wave solutions from both the fully and non-relativistic dispersion relations and ray-tracing will be presented. Significant differences are found for temperatures of order 10 keV and above which could be relevant for future experiments.

Keywords: Electron Bernstein Waves; Relativistic Ray-Tracing
PACS: 52.35.Hr; 52.55.Fa

FULLY RELATIVISTIC EBW DISPERSION RELATION

Previous work to examine relativistic effects on electron Bernstein waves has usually considered only relativistic approximations [2, 3], or other restrictions ($n_\parallel = 0$) [4]. The fully relativistic (no approximations or restrictions) dispersion relation used in the ray-tracing code GENRAY [1] was written by A.K. Ram (the R2D2 code) and E. Nelson-Melby, using the methods of B.A. Trubnikov [5] and I. Weiss [6], respectively. It can be used for examining fully relativistic effects in any of the high-frequency modes (i.e. O-mode and X-mode), but in this paper we will concentrate on EBWs. A.K. Ram et. al. [7, 8] have shown that there can be differences in the non-relativistic and fully-relativistic EBW dispersion relations. In the following sections, solutions from the fully-relativistic dispersion relation and its use in GENRAY will be presented, along with comparisons with the non-relativistic results.

DISPERSION RELATION ROOTS

Figure 1 shows an electron Bernstein wave near the fundamental cyclotron frequency ($\omega/\omega_{ce} = 1$). These solutions were found using Muller's algorithm for finding complex roots of the determinant of the dispersion tensor $|D|$. At each point, to find the root, all plasma and wave parameters are fixed except for the perpendicular wave number $n_\perp$, for which an initial guess is given. Then a (possibly complex) $n_\perp$ is found such

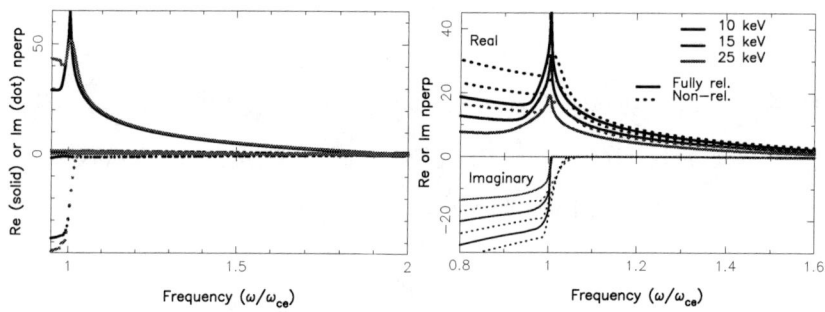

FIGURE 1. EBW near fundamental frequency, $n_\parallel = 0.1$, $\omega_{pe}^2/\omega_{ce}^2 = 4.0$. Left: $T_e = 5$ keV, Black: fully relativistic, Magenta: non-relativistic. Solid: real part, Dotted: imag. part. Right: Higher temperatures, $T_e = 10, 15,$ and 25 keV.

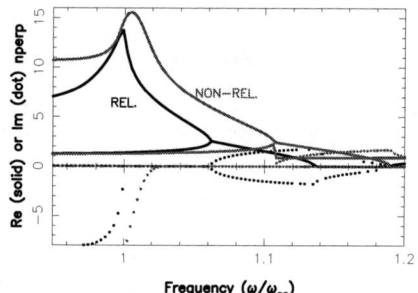

FIGURE 2. Fully relativistic vs. Non-relativistic EBW (ITER-like case). $n_\parallel = 0.05$, $\omega_{pe}^2/\omega_{ce}^2 = 0.3$, and $T_e = 12$ keV.

that $|D|(n_\perp) = 0$. This is the solution that is plotted in Fig. 1 (fully rel. in black, non-rel. in magenta). In this case, $n_\parallel = 0.1$ and $\omega_{pe}^2/\omega_{ce}^2 = 4$ were kept fixed, and the frequency (ω/ω_{ce}) was scanned from about 1.8 down to 1. It can be seen that even for a moderate electron temperature of 5 keV, there is not much difference between the fully and non-relativistic roots. However, for temperatures of 10 keV or higher, there are clear differences. The non-relativistic solutions (dashed lines) are consistently larger in magnitude that the fully relativistic (both real and imaginary parts) and the damping (represented by the imaginary part) begins sooner for the non-relativistic.

Another case with a large difference between the non-relativistic and fully relativistic EBW is shown in Fig. 2. Here the parameters were chosen to be closer to the ITER-like model used for ray-tracing (see the next section): $n_\parallel = 0.05$, $\omega_{pe}^2/\omega_{ce}^2 = 0.3$, and $T_e = 12$ keV. Here it can be seen that the non-relativistic EBW extends higher above the fundamental cyclotron frequency than the fully relativistic, and the damping (cyan dotted line) begins above the $\omega/\omega_{ce} = 1$ line. However, because of the relativistic shift, the fully relativistic EBW only begins damping below $\omega/\omega_{ce} = 1$. The glitch near $\omega/\omega_{ce} = 1.11$ is the root solver transitioning from one nearby solution to another.

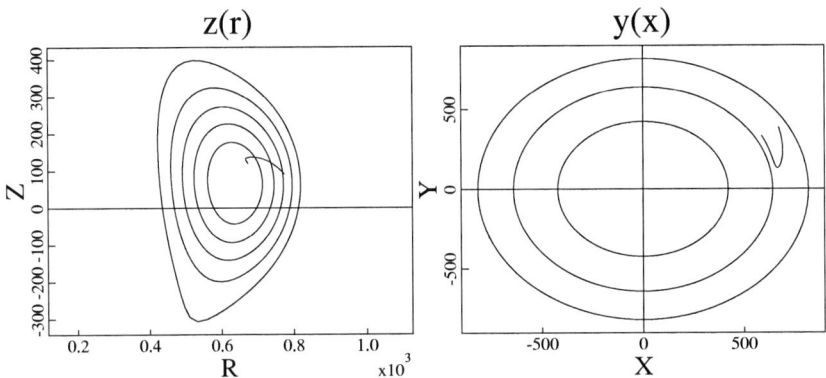

FIGURE 3. Left: Cross-sectional view of the tokamak model, Right: Top view. The fully-relativistic ray-traced EBW is shown by the solid line. Distances are in cm.

RAY-TRACING WITH GENRAY

The ray-tracing code GENRAY, primarily developed by Smirnov and Harvey [1], has several different options for the choice of dispersion tensor to use in the ray-tracing equations. Included are several variations of cold plasma, hot (non-relativistic) plasma waves, a $k_\perp \rho_{Larmor}$ expansion of the fully relativistic dispersion, and a recently added all-$k_\perp \rho_{Larmor}$ fully relativistic dispersion. Figure 3 shows a single ray for a fully relativistic EBW launched within the plasma for an ITER-like scenario. The model equilibrium used had a nearly flat density profile of 1.02×10^{20} m^{-3}, a fairly peaked temperature profile ranging from 5 keV at $\rho = 0.8$ to 25 keV in the center, and a toroidal magnetic field on axis of 5.3 Tesla. The temperature along the ray path shown in Fig. 3 ranges from 7.2 keV to 20.5 keV. The wave frequency is 145 GHz. Where the ray ends in the center (near $\omega/\omega_{ce} = 1.02$), it has completely damped.

Comparisons using a hot, non-relativistic dispersion tensor for the ray-tracing are shown in Figs. 4 and 5. In all the figures, the fully relativistic results are shown in solid black, and the non-relativistic in dashed red. Figure 4 shows the ray paths in two different planes. The paths immediately begin to diverge, and by the time the fully relativistic ray has completely damped, the non-relativistic ray is tens of centimeters away. The non-relativistic ray does not yet exhibit any significant damping along its path, and indeed is stopped after reaching an arbitrary maximum number of steps. The non-relativistic ray at the stopping point has only reached down to $\omega/\omega_{ce} = 1.04$, apparently not yet close enough for damping. In any case, even if the non-relativistic damping were to occur at the same value of ω/ω_{ce} as for the fully relativistic, it would occur in a different physical location because of the ray paths diverging. The reason for the diverging paths can be seen in Fig. 5, which shows the local values of $n_\parallel$ (left) and $n_\perp$ (right) along the rays. At the launch point the values of $n_\perp$ differ by a factor of 1.5 (fully rel. $n_\perp = 3.384$, non-rel. $n_\perp = 5.245$). The ray paths follow the group velocity direction, which depends on $n_\perp$ and $n_\parallel$. In order to accurately model EBW propagation and damping in plasmas with electron temperatures greater than 7 keV, fully relativistic ray-tracing will be necessary.

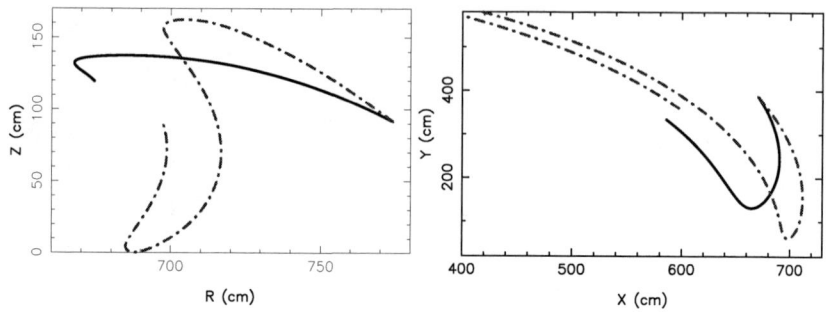

FIGURE 4. Left: Closeup of ray path, side view (R–Z plane). Right: Closeup of ray path, top view (X=R cos ϕ, Y=R sin ϕ plane). Solid black: fully relativistic, Dashed red: non-relativistic.

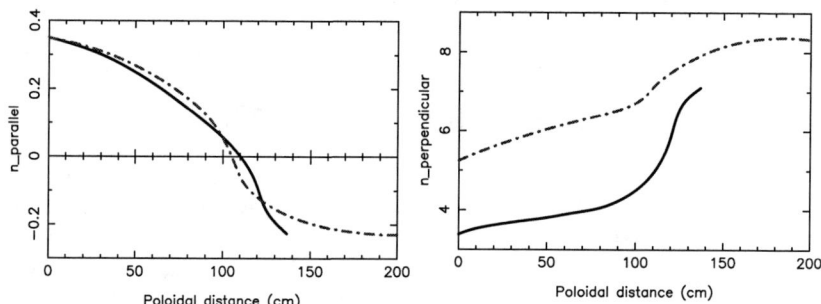

FIGURE 5. Left: Closeup of $n_\parallel$ along ray path. Right: Closeup of $n_\perp$ along ray path. Solid black: fully relativistic, Dashed red: non-relativistic.

ACKNOWLEDGMENTS

Supported by DOE grant DE-FG03-02ER54684 with additional support from the CRPP-EPFL.

REFERENCES

1. A. P. Smirnov, and R. W. Harvey, "Calculations of the Current Drive in DIII-D with the GENRAY Ray Tracing Code," in *Bull. APS*, 1995, vol. 40, ab. 8P.35.
2. P. A. Robinson, *Phys. Fluids*, **31**, 107–114 (1988).
3. E. Lazzaro, and A. Orefice, *Phys. Fluids*, **23**, 2330 (1980).
4. A. Georgiou, *Plasma Phys. Control. Fusion*, **38**, 347–353 (1996).
5. B. A. Trubnikov, *Plasma Physics and the Problem of Controlled Thermonuclear Reactions*, vol. III, Pergamon, London, 1959.
6. I. Weiss, *J. Comput. Phys.*, **61**, 403–416 (1985).
7. A. K. Ram, et al., "Relativistic Effects in Heating and Current Drive by Electron Bernstein Waves," in *30th EPS Conf. on Controlled Fusion and Plasma Physics*, St. Petersburg, 2003, p-3.204.
8. A. K. Ram, and J. Decker, "Relativistic Modifications to Electron Bernstein Waves," in *16th Topical Conference on Radio Frequency Power in Plasmas*, 2005, poster B-09.

Relativistic Modifications to Electron Bernstein Waves

A. K. Ram* and J. Decker*

*Plasma Science & Fusion Center, MIT, Cambridge MA 02139, U.S.A.

Abstract. In the electron cyclotron range of frequencies (ECRF), electron Bernstein waves (EBW) offer an attractive means for driving plasma currents in spherical tori (ST) such as NSTX. Previously it has been shown that, for ECRF waves, relativistic effects significantly modify the propagation and damping of the extraordinary and ordinary modes. A fully relativistic dispersion code R2D2 has been developed for studying the relativistic characteristics of all ECRF waves. In this paper we report on initial results obtained from this code which illustrate the relativistic modifications to the propagation and damping of EBWs. We find that even at temperatures relevant to present spherical tori, the relativistic dispersion properties of EBWs are significantly different from their non-relativistic counterpart.

INTRODUCTION

The overdense nature of NSTX-type ST plasmas makes them unsuitable for heating and/or current drive by the conventional ordinary O and extraordinary X modes in the EC range of frequencies. However, EBWs have no density cutoffs and damp strongly on electrons at the fundamental or any harmonic of the Doppler-shifted electron cyclotron resonance [1]. Since EBWs cannot propagate in vacuum they are excited, indirectly, by mode conversion of externally launched O mode or X mode [1].

Away from the mode conversion region, as EBWs propagate into the core of the plasma, the perpendicular (to the magnetic field) wavelength $\lambda_\perp$ shortens and becomes comparable to, or less than, the electron Larmor radius ρ_e. Consequently, in contrast to the case of X and O modes, we cannot cannot carry out small $\rho_e/\lambda_\perp$ expansions of the dielectric tensor elements for EBWs [2].

RELATIVISTIC DIELECTRIC TENSOR

There are two numerically useful representations for obtaining the relativistic conductivity tensor $\overline{\overline{\sigma}}$ for a relativistic Maxwellian distribution function [3]. The first form of the conductivity tensor is:

$$\overline{\overline{\sigma}} = \frac{1}{4\pi} \frac{\omega_p^2}{\omega_c} \frac{c^4}{v_t^4} \frac{1}{K_2\left(\frac{c^2}{v_t^2}\right)} \int_0^\infty d\xi \left\{ \frac{K_2\left(R^{1/2}\right)}{R} \overline{\overline{T}}_1 - \frac{K_3\left(R^{1/2}\right)}{R^{3/2}} \overline{\overline{T}}_2 \right\} \quad (1)$$

where ω_p, ω_c, v_t are the rest mass electron plasma frequency, cyclotron frequency, and the thermal velocity, respectively, K_v is the modified Bessel function of the second kind of order v, and

$$\overline{\overline{T}}_1 = \begin{pmatrix} \cos\xi & -\sin\xi & 0 \\ \sin\xi & \cos\xi & 0 \\ 0 & 0 & 1 \end{pmatrix} \quad (2)$$

$$\overline{\overline{T}}_2 = \frac{c^2}{\omega_c^2} \begin{pmatrix} k_\perp^2 \sin^2\xi & -k_\perp^2 \sin\xi(1-\cos\xi) & k_\perp k_\parallel \xi \sin\xi \\ k_\perp^2 \sin\xi(1-\cos\xi) & -k_\perp^2(1-\cos\xi)^2 & k_\perp k_\parallel \xi(1-\cos\xi) \\ k_\perp k_\parallel \xi \sin\xi & -k_\perp k_\parallel \xi(1-\cos\xi) & k_\parallel^2 \xi^2 \end{pmatrix} \quad (3)$$

$$R = \left(\frac{c^2}{v_t^2} - i\xi\frac{\omega}{\omega_c}\right)^2 + 2\left(\frac{k_\perp c}{\omega_c}\right)^2(1-\cos\xi) + \frac{k_\parallel^2 c^2 \xi^2}{\omega_c^2} \quad (4)$$

For any equilibrium distribution function $f_0(p_\perp, p_\parallel)$, which generates no plasma current, the second form of the conductivity tensor is:

$$\overline{\overline{\sigma}} = -\frac{i}{2}\frac{\omega_p^2}{\omega_c}\left\langle \sum_{n=-\infty}^{\infty} \frac{1}{n-\overline{\omega}}\left(\frac{1}{\kappa T}\frac{p_\perp}{m\gamma}\right) \overline{\overline{\sigma}}_N f_0(p_\perp, p_\parallel) \right\rangle \quad (5)$$

where

$$\overline{\overline{\sigma}}_N = \begin{pmatrix} \frac{n^2}{\zeta^2} p_\perp J_n^2 & -i\frac{n}{\zeta} p_\perp J_n J_n' & \frac{n}{\zeta} p_\parallel J_n^2 \\ i\frac{n}{\zeta} p_\perp J_n J_n' & p_\perp J_n'^2 & i p_\parallel J_n J_n' \\ \frac{n}{\zeta} p_\parallel J_n^2 & -i p_\parallel J_n J_n' & \frac{p_\parallel^2}{p_\perp} J_n^2 \end{pmatrix} \quad (6)$$

$$\zeta = \frac{k_\perp p_\perp}{m\omega_c}, \overline{\omega} = \frac{1}{\omega_c}\left(\omega\gamma - k_\parallel\frac{p_\parallel}{m}\right), \omega_c = \frac{eB_0}{m}, \langle \ldots \rangle = \int_0^\infty dp_\perp p_\perp \int_{-\infty}^\infty dp_\parallel \quad (7)$$

The use of the perturbed conductivity tensor (1) or (5) in Maxwell's equations leads to the relativistic dispersion tensor

$$\overline{\overline{D}}(\vec{k},\omega) = \frac{c^2}{\omega^2}\vec{k}\vec{k} + \left(1 - \frac{c^2 k^2}{\omega^2}\right)\overline{\overline{I}} + \frac{4\pi i}{\omega}\overline{\overline{\sigma}} \quad (8)$$

where $\overline{\overline{I}}$ is the unit tensor and $\vec{k}\vec{k}$ is a dyadic. The dispersion relation for waves in the EC range of frequencies is obtained by setting the determinant $\det(\overline{\overline{D}})$ to zero.

We have developed a code R2D2 that solves for the dispersion characteristics of EC waves using two separate and distinct numerical routines [4]. One routine uses (1) for $\overline{\overline{\sigma}}$ in (8) while the other routine uses (5). Since we are not aware of any similar code in existence, the independent routines allow us to benchmark our own code. For a variety of cases we find that the two routines provide numerically identical results leading us to have confidence in our code.

In the mode conversion region near the edge of NSTX-type plasmas, we find that the results obtained from the relativistic description are the same as those from the non-relativistic description. Thus, the mode conversion formalism developed in [1] is not modified by relativistic effects.

Figures 1a and 1b show the differences between the non-relativistic and relativistic dispersion characteristics of EBWs as a function of temperature. As the figures illustrate, differences start to occur at electron temperatures below 1 keV. From different numerical results we find that the differences persist but decrease as $n_\parallel$ increases. These results signify the relevance of including relativistic effects in the propagation and damping of EBWs. A relativistic ray tracing code based on R2D2 is being developed to further elucidate the differences along ray paths in toroidal plasma equilibria.

Figures 2a and 2b show the right hand circularly polarized and the parallel components of the wave electric field. The differences between the relativistic and non-relativistic results implies that the field polarizations in the quasilinear diffusion operator have to be properly treated. This is important for studying current drive by EBWs [5].

This work is supported by DoE Grant Numbers DE-FG02-91ER-54109 and DE-FG02-99ER-54521.

REFERENCES

1. A. K. Ram and S. D. Schultz, *Phys. Plasmas* **7**, 4084 (2000).
2. M. Bornatici, R. Cano, O. De Barbieri, and F. Engelmann, *Nuclear Fusion* **23**, 1153 (1983), and references therein.
3. B. A. Trubnikov, in *Plasma Physics and the Problem of Controlled Thermonuclear Reactions*, edited by M. A. Leontovich (Pergamon Press Inc., New York, 1959) Vol. III, p. 122.
4. A. K. Ram, J. Decker, and Y. Peysson, accepted for publication in *J. Plasma Physics* (2005)
5. J. Decker et al., presented in this conference.

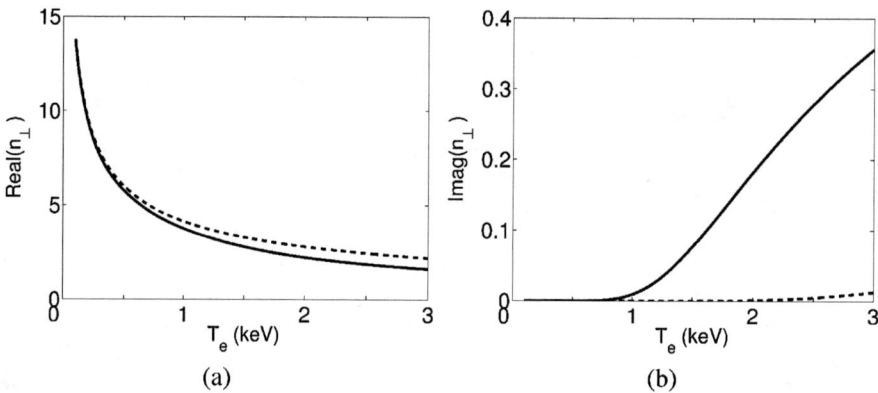

FIGURE 1. (a) Real and (b) imaginary part of $n_\perp$ versus the electron temperature T_e. Plotted are the relativistic (solid) and the non-relativistic (dashed) characteristics of EBWs for $\omega_p/\omega_c = 6$, $\omega/\omega_c = 1.9$, and $n_\parallel = 0.2$.

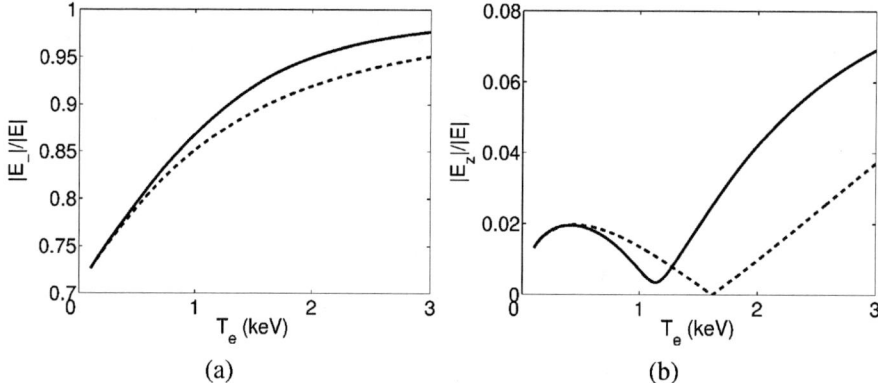

FIGURE 2. (a) The normalized magnitude of the right hand circularly polarized wave electric field (in a plane perpendicular to the magnetic field) and (b) the normalized wave electric field along the magnetic field versus the electron temperature T_e. Plotted are the relativistic (solid) and the non-relativistic (dashed) characteristics for the same parameters as Fig. 1

Plasma Current Start-up by Outboard PF Coils in JT-60U and TST-2

Y. Takase, A. Ejiri, K. Hanada,[1] S. Ide,[2] O. Mitarai,[3] S. Shiraiwa,[4] M. Ushigome, JT-60 Team and TST-2@K Team

University of Tokyo, Kashiwa 277-8561 Japan
[1] *Kyushu University, Kasuga 816-8580 Japan*
[2] *Japan Atomic Energy Research Institute, Naka 311-0193 Japan*
[3] *Kyushu Tokai University, Kumamoto 862-8652 Japan*
[4] *1-36-2-308 Hakusan,Tokyo 113-00101 Japan*

Abstract. Plasma current (I_p) initiation and ramp-up utilizing induction by poloidal field (PF) coils located on the outboard side of the torus were demonstrated in both conventional aspect ratio tokamak (100 kA in JT-60U) and spherical tokamak (10 kA in TST-2). In the presence of sufficient source of plasma by RF ionization (1 MW in JT-60U, 100 kW in TST-2), I_p formation is possible without the existence of a PF null, with the initial vertical field (B_v) in the direction opposite to that required for equilibrium. Utilization of the extra flux swing provided by the PF coils before B_v switches sign is important in arriving at a high enough I_p that matches B_v at the end of PF coil current ramp. In JT-60U, recharging of the OH coil was observed with only perpendicular and counter NB injection, suggesting overdrive by bootstrap current. In TST-2, a quasi-steady-state I_p of 4 kA was formed and sustained by RF power (P_{RF}) alone, without the use of induction. The plasma current centroid of this type of plasma is located on the outboard side. The dependences of I_p on P_{RF} and B_v were found to be weak. The results presented here offer encouragement to CS-less operation of a tokamak fusion reactor, which should have a large impact on the economic competitiveness.

Keywords: tokamak, spherical tokamak, plasma start-up, ECH, current drive, bootstrap current, plasma current ramp-up
PACS: 52.55.Fa, 52.50.Nr, 52.50.Sw, 52.55.Wq

INTRODUCTION

Elimination of the center solenoid (CS), otherwise known as the OH coil, results in substantial improvement in the mass power density of a tokamak-based fusion reactor.[1] Successful I_p start-up utilizing induction by outboard poloidal field (PF) coils has been demonstrated in both conventional aspect ratio tokamak JT-60U[2] and spherical tokamak TST-2 ($R \leq 0.38$ m, $a \leq 0.25$ m, $B_t \leq 0.3$ T)[3]. Sustainment and further ramp-up of I_p by noninductive means have also been achieved in these devices.

PLASMA CURRENT START-UP IN JT-60U

Plasma current start-up to 100 kA was achieved successfully in JT-60U without the use of CS, as shown in Fig. 1. Only PF coils located on the outboard side of the torus were used, in this case VR and VT_{out} coils shown in Fig. 1. In previous experiments,[4] the inner turns of the VT coil (VT_{in}) were also used. In addition to acting as a small OH coil, VT_{in} in combination with VT_{out} produces a field null in the inboard region inside the vacuum vessel, creating a favorable condition for plasma start-up. The use of outboard PF coils alone has demonstrated clearly that I_p start-up without a field null is possible (even with B_v as high as 0.08 T), provided there is strong enough ionization source (over 1 MW of ECH power in JT-60U). In the absence of EC power, I_p does not start rising until a field null is formed inside the vacuum vessel. The flux conversion efficiency (i.e., ΔI_p per flux swing), however, is substantially lower while B_v is in the direction opposite to that required for equilibrium. Controlled surveys indicate that this type of I_p start-up favors lower neutral pressures than normal start-up with the CS, and the EC fundamental resonance to be placed slightly to the high-field side of the center of the vacuum vessel. The plasma current channel starts forming on the low field side of the EC resonance layer.

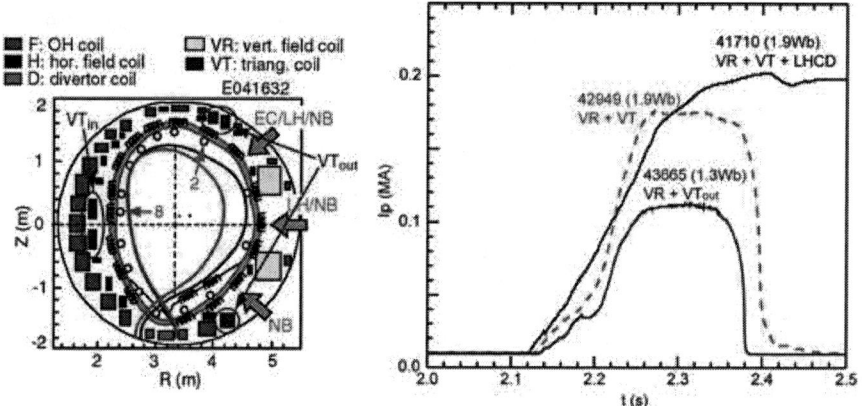

FIGURE 1. Coil configuration (left) and plasma current waveforms (right) for JT-60U. In 42949 the inboard turns of VT coil were used, but in 43665 only outboard turns of VT coil were used.

Even in the absence of LHCD, I_p of 260 kA was maintained for 1 sec by NB only, and I_p ramp-up from 215 to 310 kA was achieved by EC and NB only. However, I_p did not increase beyond 310 kA with over 10 MW of total (EC + NB) injected power (the absorbed NB power was substantially lower). This can be contrasted with ramp-up by LHCD from 200 kA to 400 kA with only 1.6 MW of injected power.[4] In this discharge, I_p was still rising at the end of the LH pulse. Such a high ramp-up efficiency is made possible partly by the higher current drive efficiency of LHCD and partly by its ability to operate at much lower densities (by roughly an order of magnitude).

An indication of bootstrap overdrive was obtained by predominantly perpendicular neutral beam (NB) heating.[2] The negative loop voltage and the positive slope of the CS current (i.e., recharging) indicate clearly that I_p is overdriven non-inductively. β_p of up to 4 and β_N of up to 2.5 have been achieved in such plasmas, with a very large normalized ITB radius of $\rho_{foot} = 0.8$. A more convincing indication of bootstrap overdrive was observed when only counter and perpendicular NB were used, in which case the contribution of the beam driven current is negligible.

PLASMA CURRENT START-UP IN TST-2

Plasma current start-up experiments were performed on the TST-2 spherical tokamak while it was temporarily located to Kyushu University (May 2003 – March 2004), utilizing the 8.2 GHz LHCD system for the TRIAM-1M tokamak (TST-2@K Project). RF power was injected through eight waveguide horn antennas located on the low field side of the torus below the midplane. The electric field polarization is perpendicular to the total magnetic field (*i.e.*, X-mode) at full I_p, but is a mixture of X- and O-modes for the present experiment.

I_p start-up to 10 kA has been demonstrated without the use of CS.[3] Only the outboard PF coils (PF2, PF3 and PF5, shown in Fig. 2) were used. Figure 2 shows evolutions of coil currents and I_p. The initially positive PF3 coil current (creating an opposite B_v) is reduced at 0.138 s and increases in the negative direction to provide the flux and the equilibrium field. The PF2 and PF5 (connected in series) coil current starts at 0.142 s when the PF3 current has reached a plateau level. I_p starts at t = 0.139 s around the time I_{PF3} crosses zero, reaches 10 kA at 0.143 s, and then decays. The current ramp-up rate is 5 MA/s. The plasma position was deduced using a variable current filament model, in which the plasma current is represented by a single filament whose location was adjusted to fit the magnetic measurements. At 0.142 s, approximately half way during the current rise when it first becomes possible to determine the plasma position, the plasma current centroid is located 5-10 cm to the low field side of the EC resonance layer. Near the I_p peak at 0.143 s, the plasma current centroid is located near the center of the vacuum vessel at $R = 0.35$ m and z = +0.08 m, subsequently moving inward and upward. In some cases I_p starts when I_{PF3} crosses zero as shown here, but in other cases starts before I$_{PF3}$ reversal. As in JT-60U, I_p start-up can take place without a PF null.

Plasma start-up and sustainment without an inductive field have also been studied in TST-2. Quasi-steady-state discharges with I_p of 4 kA were obtained at $B_t = 0.16$ T. The line integrated density measured along a vertical chord at $R = 0.39$ m was about 3×10^{17} m^{-2}, and the electron temperature determined from soft X-ray pulse height analysis was 160 eV. The plasma current centroid was located on the outboard side of the torus. The dependences of the plasma current on the RF power and the vertical field strength were weak, suggesting a possible existence of a self-regulating mechanism. In some cases, spontaneous sawtooth-like oscillations of the plasma current with a period of about 10 ms, or a transition to a higher density state were observed. The mechanisms of plasma current formation and sustainment were investigated, but not resolved. It is inferred that the bootstrap current contributes significantly. The contribution of EC driven current is likely to be negligible because of very weak single-pass absorption and lack of directionality. The condition was not favorable for generation of EBW by mode conversion, but this possibility cannot be excluded.

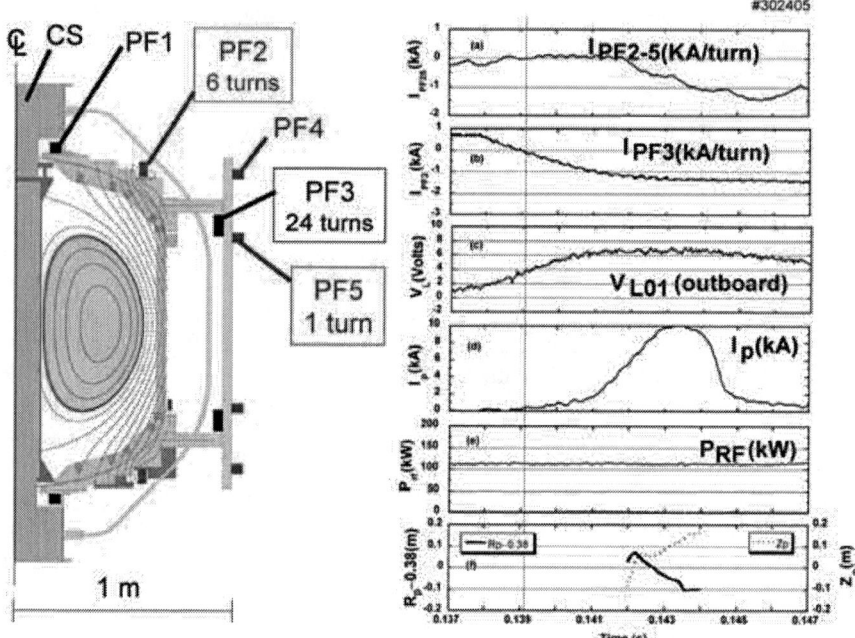

FIGURE 2. Coil configuration (left) and plasma current waveforms (right) for TST-2 at $B_t = 0.28$ T. (a) PF2 and PF5 coil current (connected in series), (b) PF3 coil current, (c) loop voltage on the outboard wall of the vacuum vessel, (d) plasma current, (e) RF (8.2 GHz) power, (f) plasma radial (solid line) and vertical (dotted line) positions. The schematic plasma on the left figure does not correspond to the discharge shown on the right figure.

ACKNOWLEDGMENTS

JT-60U experiments were carried out as joint work under the Facility Utilization Program of JAERI. TST-2@K experiments were carried out under the framework of joint-use research of RIAM Kyushu University and the NIFS Collaborative Research Program.

REFERENCES

1. Nishio, S., et al., "Technological and Environmental Prospects of Low Aspect Ratio Tokamak Reactor VECTOR," 20th IAEA Fusion Energy Conf. (Vilamoura, Portugal, 1-6 November 2004) IAEA-CN-116/ FT/P7-35.
2. Takase, Y., et al., "Development of a Completely CS-less Tokamak Operation in JT-60U," 20th IAEA Fusion Energy Conf. (Vilamoura, Portugal, 1-6 November 2004) IAEA-CN-116/EX/P4-34.
3. Ushigome, M., et al., "Development of completely solenoidless tokamak operation in JT-60U," submitted to Nucl. Fusion (2005).
4. Mitarai, O., et al, J. Plasma Fusion Res. **80**, 549 (2004).
5. Shiraiwa, S. et al., Phys. Rev. Lett. **92**, 035001-1 (2004).

ELECTRON CYCLOTRON RANGE
OF FREQUENCIES

The 10 MW, CW ECRH System For W7-X: Status And First Integrated Tests

V. Erckmann[1], P. Brand[3], H. Braune[1], G. Dammertz[2], G. Gantenbein[3]
W. Kasparek[3], H. P. Laqua[1], G. Michel[1], M. Thumm[2],
and the W7-X ECRH teams at IPP, FZK and IPF

[1] *Max-Planck-Institut für Plasmaphysik, Teilinstitut Greifswald, Association EURATOM,
Wendelsteinstr. 1, D-17491 Greifswald, Germany*
[2] *Forschungszentrum Karlsruhe, Association EURATOM-FZK,
Institut für Hochleistungsimpuls- und Mikrowellentechnik,
Postfach 3640, D-76021 Karlsruhe, Germany,*
[3] *Institut für Plasmaforschung, Universität Stuttgart, Pfaffenwaldring 31, D-70569 Stuttgart, Germany*

Abstract. Electron Cyclotron Resonance Heating (ECRH) is the main heating system for the W7-X stellarator and the only one for CW-operation in the first stage. The mission of W7-X, which is presently under construction at IPP-Greifswald, is to demonstrate the inherent steady state capability of stellarators at reactor relevant plasma parameters. W7-X is therefore equipped with a superconducting coil system operating at 3 T and a divertor for 10 MW steady state heat removal. A 10 MW ECRH plant with CW-capability at 140 GHz is under construction to meet the scientific objectives. The commissioning of the ECRH plant is well under way, the status of the project and first full power, cw test results are reported.

Keywords: ECRH, Stellarator, Plasma heating, Microwaves, Gyrotron
PACS: 52.50.-b

INTRODUCTION

The physics goals for W7-X (major radius 5.5 m, minor radius 0.55 m) determine the main machine parameters as well as a consistent set of heating systems, diagnostics, data acquisition and machine control. The W7-X mission can be formulated as follows:

1. Demonstration of quasi steady state operation at reactor relevant parameter, with temperatures $T_e = 2$-10 keV, $T_i = 2$-5 keV and densities $n_e \sim 0.1$ - $3 \cdot 10^{20}$ m^{-3}
2. Demonstration of good plasma confinement
3. Demonstration of stable plasma equilibrium at a reactor relevant plasma ß about 5 %
4. Investigation and development of a divertor to control plasma density, energy and impurities

W7-X does not aim at DT-operation and provisions for remote handling in radioactive environment are not foreseen. The physics objectives lead to the formulation of a set of optimization principles [1], which are met by the 3-D architecture of the W7-X magnetic configuration. The quasi steady state operation (< 30 min) asks for a superconducting coil system (for economic reasons), a continuous operating heating system and stationary particle and energy control. ECRH was chosen as the basic heating system for steady state operation in the long mean free path regime. The development of a divertor for steady state particle and energy control is probably the most challenging tasks for W7-X and constitutes a major part of the physics and technology program. An island divertor, which was investigated successfully at W7-AS, will be available at W7-X from the very beginning and is designed for 10 MW heat removal and active pumping. A balanced NBI system with 5 MW and ICRH with 4 MW in 10 s pulsed operation are foreseen for experimental flexibility. The high-ß criterion will be addressed at reduced magnetic field in a later state of the machine operation, where NBI will be available with 20 MW, 10 s pulsed.

An ECR-heating power of 10 MW is required to meet the envisaged plasma parameters [1] at the nominal magnetic field of 2.5 T. The standard heating and current drive scenario is X2-mode with low field side launch. High-density operation above the X2 cut-off density at $1.2 \; 10^{20}$ m^{-3} is accessible with O2-mode ($< 2.5 \; 10^{20}$ m^{-3}) and at even higher densities with O-X-B mode conversion heating [2,3]. As W7-X has no OH-transformer for inductive current drive, EC-current drive is a valuable tool to modify the internal current density distribution and to counteract residual bootstrap currents. High flexibility is therefore requested:

- arbitrary toroidal launch   (current drive, O2-mode, O-X-B conversion)
- arbitrary poloidal launch   (on- and off-axis heating)
- arbitrary wave polarization   (e.g. elliptical for oblique launch)
- high- and low field side launch   (phase space physics)
- AM capability   (heat waves and switching experiments)

An optical transmission system was developed for W7-X, which is the most simple, reliable and cost effective solution. The overall-design of the ECRH-system is briefly reviewed in Sec.2. The gyrotrons and the transmission line are described in Sec. 3 and 4, respectively. Recent results on integrated full power, cw-tests are reported in Sec. 5.

GENERAL DESIGN OF THE ECRH SYSTEM

ECRH must operate with a maximum reliability and availability. We have thus chosen a strict modular design, which allows to operate each gyrotron separately and independent from all others to optimize the availability during installation and commissioning, in case of maintenance and breakdowns. This design also minimizes the costs because series production of identical modules is possible, and at the same time maximizes the experimental flexibility. The total ECRH power is generated by 10 gyrotrons operating at 140 GHz with 1 MW output power in CW operation each. Two subgroups of 5 gyrotrons are arranged symmetrically to a central beam duct in the ECRH hall.

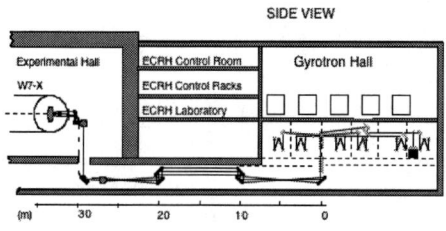 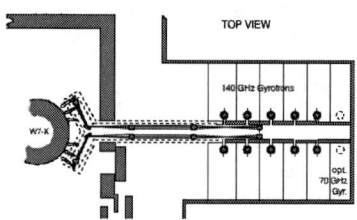

FIGURE 1: The ECRH plant for W7-X

The ECRH building is prepared to house two additional 70 GHz gyrotrons in a later state of the experiment to ensure reliable start-up for NBI-heated high-□ plasmas at reduced magnetic field of 1.25 T. The arrangement of the gyrotrons and the transmission system in the ECRH building is seen from Fig. 1. Each gyrotron is equipped with its own water/oil cooling module and is fed by one main-power supply module and one body-modulator/crowbar unit. The individual modules of the main power supply are located in a separate building and are connected with the ECRH hall via triaxial HV lines having an approximate length of 30 m. Snubber circuits assist in handling of the stored energy in the long HV cable in case of a breakdown.

The power supply is designed as a multi user supply and meets also the requirements of the other heating systems NBI and ICRH (inverse polarity). The main power supply consists of 10 solid state, pulse step modulated (PSM) units, 8 of them are capable of 65 kV/50 A (bipolar) in continuous operation, two are designed for use with ICRH with half the voltage and twice the current. Series connection of both supplies is possible for use with ECRH. In combination with a low power body supply of +30 kV/0.5 A an acceleration voltage in the range of 80-90 kV as needed for the gyrotrons is available with some margin. The body supply is located in close vicinity of each gyrotron in the gyrotron hall and is an air insulated floating deck device to ease maintenance. Power modulation of each gyrotron up to 10 kHz is foreseen for transient physics investigations. The heater supply, which is on cathode potential and the crowbar circuit (thyratrons) is incorporated, a schematic circuit is seen from Fig.2.

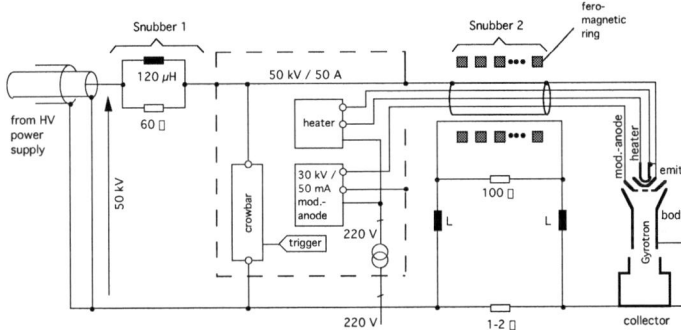

FIGURE 2: Schematic HV-circuit of the W7-X gyrotron supply

The transmission of the RF-power to the torus is performed by two open multi-beam mirror lines, each of them combining and handling 5+2 individual RF-beams. Four front steering Plug-In Launchers (PILA's) provide the necessary scanning range in toroidal and poloidal angles and are mounted within the large A and E-ports of the W7-X vessel. Two beams can be switched to smaller ports, which give access to high field side launch.

THE W7-X GYROTRON

As gyrotrons with the required performance did not exist at the time of the W7-X approval, a European R&D program was launched in 1998 as a combined effort of several research laboratories and Thales Electron Devices (TED) as industrial partner to develop the W7-X gyrotron [4]. Also in 1998 a contract was placed with CPI (USA) to develop in parallel a W7-X gyrotron with the same specifications.

Both development lines came to a successful end in 2002. The design parameters of the W7-X gyrotron are summarized in Table 1 and incorporate technologies, which were not state of the art at the time of the definition:

The gyrotron uses a diode-type magnetron injection gun, which operates at V_{acc} = 81 kV with

RF output power	1 MW
Frequency	140.3 GHz
Accelerating voltage	81 kV
Beam current	40 A
Retarding collector voltage	> 25 kV
Power modulation depth	0.3-1 MW
Modulation frequency	up to 10 kHz

TABLE 1: Gyrotron design parameters

respect to the body and at a current of I_b = 40 A. The cylindrical cavity operates in the $TE_{28,8}$-mode (TED) or $TE_{28,7}$-mode (CPI). The (cold-cavity) frequency is designed to operate at 140.3 GHz to compensate for the frequency downshift during power loading of the cavity. The magnetic field of the SC-magnet at the cavity is 5.56 T. The RF-beam is separated from the electron beam through a highly efficient quasi-optical mode converter consisting of a rippled-wall, helically-cut waveguide launcher with tapered diameter [5] followed by a 3 mirror imaging system for beam shaping. The output window unit uses a single, edge cooled CVD-diamond disk with an outer diameter of 106 mm, a window aperture of 88 mm and a resonant thickness of 1.8 mm. The collector, the output window and the third mirror are on ground potential. At the nominal depression voltage, the cathode is at –50 kV. The beam tunnel, cavity, the launcher and the first two mirrors have the depression potential of +30kV.

The European R&D went through a 2-step program with the 'Maquette' as the first and the 'Prototype' as the second R&D gyrotron. All TED-gyrotrons undergo first tests at the FZK-test stand, which can operate for 180 s at 40 A and full cw at a reduced current of < 25 A. The results obtained with both tubes are summarized in Table 2.

	TED-'Maquette'			TED-'Prototype'		
Power (MW)	0.98	0.86	0.74	0.92	0.89	0.54
Pulse length (s)	10	45	100	55	180	939
Current (A)	39.5	41.5	40.7	40	41	26
Efficiency (%)	47	39	32	41	41	39

TABLE 2: TED Prototype gyrotron performance measured at FZK (teststand limit is 180 s at 40 A).

Both tubes are well qualified for operation in the W7-X system, although the ultimate specifications were not fully met (in particular full power cw-operation). The 'Maquette' was transferred to IPP after the test period at FZK and first operation at IPP was achieved by fall of 2003. Since then the gyrotron is routinely operated for high power transmission line tests on site at IPP. The 'Prototype' is still at FZK and will serve as a test vehicle for peripheral improvements, in particular tests of an improved collector sweeping system.

An order for seven series gyrotrons was placed with TED in 2003, the first gyrotron was delivered in February 2005 and is presently under test at FZK. The next gyrotrons will become available with an average delivery sequence of one gyrotron every 4 months.

Factory tests of the CPI gyrotron were restricted by the CPI test-stand limitations, which allows only short pulse operation (few ms) at full current and cw-operation at a reduced beam current < 25 A. An output power of 920 kW was achieved at CPI in 3 ms pulses with 80 kV accelerating voltage (20 kV depression voltage) and 38 - 45 A beam current, the efficiency is about 37 %. An output power of 500 kW was achieved with very good reliability in repetitive 10 minutes pulses with reduced beam current of 25.6 A. The final tests of the tube were performed at IPP, where the power supply has full current cw capability (see Sec. 5).

4. TRANSMISSION LINE

Based on the excellent performance of the 800 kW / 140 GHz beam waveguide [6] on the stellarator W7-AS, a quasi-optical transmission system was chosen for W7-X as a low-cost solution with high efficiency. The millimeter waves are transmitted as Gaussian beam by iterative transformation with focusing metallic mirrors [7]. The design of such mirrors is relatively simple and straightforward. To minimise surface deformations by thermal loading a 60 - 70 mm thick honeycomb stainless steel structure is coated with a thin (2 mm) layer of electroformed copper as low-loss mirror surface. Cooling channels are located directly below the copper and spiral from the centre (inlet) to the edge of the mirror (outlet). The width and depth of the spirals are adapted to the size of the mirrors and the gaussian distribution of the heat load. A very low thermal deformation below 10^{-3} m^{-1} was confirmed experimentally.

The main advantages of a quasi-optical transmission are low ohmic and diffractive losses, high power capability due to relatively low field strength, and inherent mode filtering as high-order modes are diffracted out of the system. A beam waveguide with many channels becomes, however, complex and bulky. This is avoided by the Multi-Beam-Waveguide (MBWG) concept, where several quasi-optical beams are transmitted by a common mirror system. At the input plane of the MBWG, the Gaussian beams are closely packed and parallel. Four focusing mirrors at distances of two focal lengths (and additional plane mirrors to straighten the beam path) provide a low-loss propagation of all on-axis and off-axis beams and a correct imaging from the input to the output plane. After four mirrors, the spurious modes have cancelled and the beams cross the output plane exactly perpendicularly in the nominal position with a mode purity of 99.8%. Further calculations [8] show, that even a much higher number of beams could be transmitted via a common mirror system without remarkable diffraction loss. The MBWG's for W7-X are designed to transmit 5+2 beams (7 MW), the power handling capability is, however, by a factor of 2-3 higher and would allow to replace the 1 MW gyrotrons by more powerful ones in a later state. The overall transmission efficiency, which is determined by several loss channels was calculated to 88 % under conservative assumptions. To benchmark the calculations and to test performance and stability, a full-scale, uncooled prototype was built and the total transmission efficiency measured in low-power tests yielded 90±2 %. This is in good agreement with the theoretical value for the prototype system, which is 92 %. From amplitude and phase measurements of the various beams at the output of the MBWG, a mode purity between 97 % and 99 % was deduced.

The gyrotrons are located behind the wall of the beam duct and radiate their power laterally through small holes in the shielding concrete wall. For each gyrotron, a beam conditioning optics consisting of five mirrors is mounted on a common base frame. The base frames for 10 gyrotrons are completely installed in the beam duct as shown in Fig. 3 (left). Two mirrors (Beam Matching Optics, BMO) match the gyrotron output to a Gaussian beam with the correct beam parameters. The following two mirrors have sinusoidally corrugated surfaces to set the polarization needed for optimum absorption of the radiation at different heating and current drive scenarios. Finally, a fifth mirror focuses the beam to a plane mirror array (Beam Combining Optics, BCO) as seen from

Fig.3 (middle), which is located at the input plane of the MBWG. The MBWG is designed to transmit up to seven beams (five 140 GHz beams, one 70 GHz beam plus

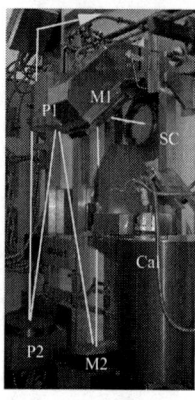
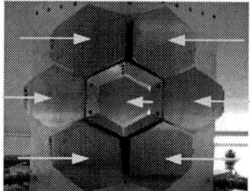
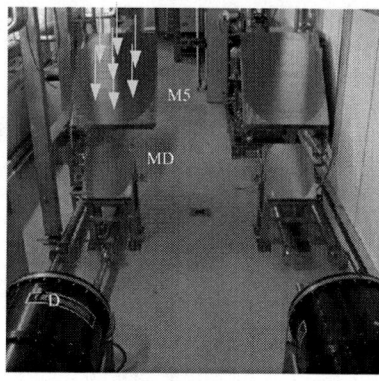

FIGURE 3. View into the transmission duct. Left: Base frame with matching optics M1, M2, and polarizers P1, P2, as well as M3, switch mirror SC and calorimeter Cal. Middle: beam combining optics (BCO) as seen from the first MBWG mirror M5. Individual beams coming from M3 are impinging on the plane mirrors from left and right and are directed onto M5 of the MBWG system. Right: Beam duct with two multi-beam mirrors M5 and the mirrors MD which direct one selected beam into one of the CW dummy loads D seen in the foreground.

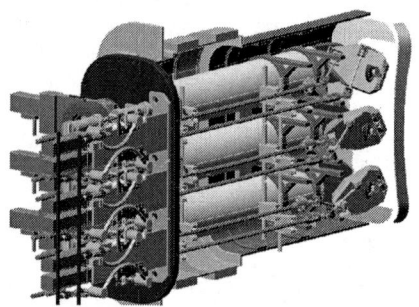
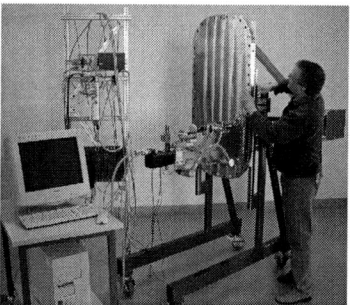

FIGURE 4: Design of the front steering plug-in launcher (left) and a simplified full size mock-up (right)

one spare) as seen from Fig.3 (right). Two symmetrically arranged MBWGs transmit the power of all gyrotrons over a distance of 45 m, the total lengths of the transmission lines are 57 to 65 m depending on the locations of the corresponding gyrotron and torus port. At the output plane of the MBWG, a beam distribution mirror array (identical to the BCO-unit) separates the individual beams and directs each of them via two mirrors and through a vacuum barrier window towards the front steering launcher. Four large ports of W7-X will be equipped with the plug in launchers, which cover a wide poloidal and toroidal steering range to meet the requirements for optimum current drive (typically at 15 ° toroidal angle), O-X-B launch (at 35 ° toroidal angle) and off-axis heating (± 30 ° poloidal angle). The design is seen from Fig. 4 (left). A simplified full size mock-up launcher was manufactured and is ready for cyclic tests under vacuum and stray radiation loading (see Fig. 4 (right).

5. INTEGRATED CW-TESTS

The CPI gyrotron was successfully tested at IPP-Greifswald towards full power operation at 30 minutes pulse duration, which is the target value for W7-X. The microwave beam was transmitted through 7 single beam mirrors as described in Fig. 3 into a commercial cw-load from CCR [9]. As the beam parameters were not known with the required accuracy at the beginning of tests, a provisional BMO-unit (M1 and M2) was used resulting in transmission losses, which exceed the calculated values. The overall losses are estimated in the range 50 – 70 kW. The maximum power in 30 min pulses measured in the CCR load was 820 kW and represents the Gaussian mode content of the gyrotron beam because of the inherent filtering capability of the open transmission line. A typical time trace for an experimental sequence of two shorter (2 min) and one longer pulse (27 min) is shown in Fig. 5 (left).

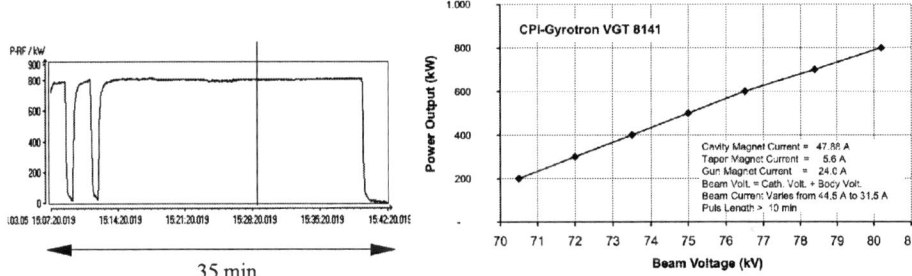

FIGURE 5: left: RF-power as measured by the calorimeter vs. time, right: RF-power vs. beam voltage (beam current varies also). The measured power is the Gaussian fraction of the gyrotron beam.

All measured parameters, in particular the gas pressure in the tube, the temperature in the collector wall and the temperature of the most loaded mirrors became stationary with the longest measured time scale of about 5 minutes. Problems arose from the imperfect beam matching to a Gaussian beam and side lobes hitting the beam duct concrete wall or uncooled elements like the M1-mirror support. Additional water cooled absorbing targets where installed at the measured hot spots. After some 30 min pulses a very reliable operation was achieved with all components behaving as expected. The beam pattern was measured at the end of the test period with high accuracy and the design of a matched phase correcting BMO is under way. It is expected, that some fraction of the lost power will be recovered, which would increase the useful power in the Gaussian mode. Even more important, however, is the reduction of the power in hot spots from stray radiation, were even a few tens of kW may cause significant heating of uncooled surfaces in cw-operation. An example for a power scan is shown in Fig. 5 (right), which was obtained in 10 min pulses to save experimental time.

It is worth noting, that none of the peripheral systems at IPP like main PS, central cooling system, body-modulator, transmission line components, rf-diagnostics, central control and data-acquisition system has ever seen cw-operation before and consequently all systems had to go through this qualification process together with the gyrotron.

5. SUMMARY AND CONCLUSIONS

The 10 MW, cw ECRH heating and current-drive system for W7-X is presently under construction. The system is designed for high flexibility with respect to wave coupling and fast power control. All major heating and current-drive scenarios such as X2-mode, O2-mode and O-X-B mode-conversion are accessible with the optical in vessel laucher. Prototype gyrotrons were successfully developed in Europe and USA. The CPI gyrotron was operated at 820 kW in the Gaussian mode for 30 min. The project has now entered the phase of series installation and commissioning. The quasi-optical multi-beam waveguide system offers favourable transmission characteristics and the most loaded components showed an excellent performance under full power, cw conditions.

REFERENCES

[1] V. Erckmann, H.J. Hartfuß, M. Kick, H. Renner, J. Sapper, F. Schauer, E. Speth, F. Wesner, F. Wagner, M. Wanner, A. Weller, and H. Wobig,: The W7-X project: Scientific Basis and Technical Realization. Proc. 17th IEEE/NPSS Symposium on Fusion Engineering, San Diego, USA (1997). Ed. IEEE, Piscataway, NJ 1998, 40 - 48

[2] H. P. Laqua, V. Erckmann, H.J. Hartfuß, H. Laqua, W7-AS Team, and ECRH Group:
Resonant and nonresonant Electron Cyclotron Heating at Densities above the Plasma Cut off by O-X-B Mode Conversion at the W7-AS Stellarator,
Phys. Rev. Lett. **78**, 3467 (1997)

[3] M. Rome', V. Erckmann, U. Gasparino, N. Karulin:
Electron Cyclotron resonance heating and current drive in the W7-X stellarator
Plasma Phys. Control. Fusion 40 (1998) 511-530

[4] G. Dammertz, S. Alberti, A. Arnold, E. Borie, et al.:
Development of a 140 GHz, Continuous Wave Gyrotron for the W7-X Stellarator,
IEEE Trans. Plasma Science 30 (2002) 808-818

[5] G.G.Denisov et al.: 110 GHz gyrotron with a built-in high-efficiency converter, Int. J. Electronics 72, pp. 1071-1091,1992

[6] Thumm, M., W. Kasparek: Passive high-power microwave components, IEEE Trans. on Plasma Science, PS-30, 755-786 (2002).

[7] Goldsmith, P.F.: Quasi-optical Systems.
New York: IEEE Press, Chapman and Hall, Publishers, ISBN 0-7803-3439-6, 1998

[8] Empacher, L. and Kasparek, W.: Analysis of a multiple-beam waveguide for free-space transmission of microwaves. IEEE Trans. Antennas Propagat., Vol. AP-49 (2001) 483 - 493

[9] R.L. Ives et al., Development of a 1 MW CW Waterload für Gaussian Mode Gyrotrons, Conf. Digest 23rd Int. Conf. Infrared Millimeter Waves, Colchester (UK), pp. 226-227, 1998.

Performance Optimisation in ASDEX Upgrade with ECRH and ECCD

Leuterer F., Angioni C., Dux R., Gantenbein G.[1], Günter S., Manini A., Maraschek M., Mück A., Münich M., Neu R., Peeters A.G., Ryter F., Wagner D., Zohm H. and ASDEX Upgrade Team

Max Planck Institut für Plasmaphysik,EURATOM Association 85748 Garching, Germany
1 Institut für Plasmaforschung, Universität Stuttgart, 70569 Stuttgart, Germany

Abstract. In the ASDEX Upgrade tokamak we use localised electron cyclotron heating and current drive at 140 GHz. Both confinement and stability properties of the plasmas can be favourably altered. Particle transport can be modified with central ECRH. Linear gyrokinetic studies indicate that the transport is determined by collisionality and by the kind of turbulence which is dominant. A similar result is obtained for the electron heat transport. At higher collisionality the turbulent contribution to the diffusivity decreases. With central ECRH accumulation of high Z impurities can be prevented. Trace impurity injection experiments show that this is due to strongly increased impurity diffusion.
Stability improvement can be achieved by controlling the sawtooth instability and thus the generation of seed islands for neoclassical tearing modes, by triggering their frequent interruption with ECCD, or by their complete suppression with ECCD on the resonant flux surface.

Introduction

The ASDEX Upgrade ECRH system consists of 4 gyrotrons, each 140 GHz / 0.5 MW / 2 sec, with 4 separate transmission lines and launchers. The launched beams are focused with the beam waist $w_0 \approx 22$ mm located in the plasma. The narrow power deposition profile has been confirmed experimentally [1,2]. This allows simultaneous electron heating at various locations in the plasma and was used extensively in transport studies presented in the first chapter. The second chapter describes experiments aimed at suppression or avoidance of neoclassical tearing modes by localised ECCD.

Investigation of Transport with ECRH

Particle diffusion

Central electron heating by ECRH can have quite different effects on the density profile. The profile may get flatter, may stay unaffected, or may even peak slightly. A summary is shown in Fig.1, taken from reference [3]. Here density peaking (defined by the ratio of the central chord H1 to the peripheral chord H2 of the interferometer) during the phase with ECRH, normalised to the peaking during the non-EC heated phase, is plotted on the x-axis, while the central line average density is on the y-axis. The non-EC heated phase is represented by an open symbol on the line x=1, while the

EC phase is represented by a full symbol. A displacement towards lower values on the x-axis from the open to the full symbols describes density flattening due to central ECRH. Low density L-mode plasmas generally show a strong flattening, and at intermediate density the profile does not change. In H-mode plasmas the profiles flatten at intermediate density, remain unchanged at higher density.

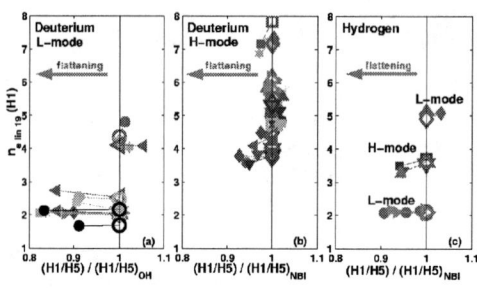

FIGURE 1. Line averaged density as a function of the relative density peaking in L-mode and H-mode plasmas (from ref [3])

An explanation of this behaviour is indicated by the inspection of the expression for the thermodiffusive contribution to the total anomalous particle flux. This thermodiffusive coefficient depends on the type of the unstable mode and its frequency. Hence the thermodiffusive part, which is usually directed inward for ion temperature gradient driven instabilities (ITG), can change sign in the instability domain of the trapped electron modes (TEM), becoming outward directed when the mode propagates in the electron drift direction. In this way a mechanism can be identified which leads to flatter profiles when the plasma is in an instability domain with dominant TEMs and when the electron to ion temperature ratio is increased by the application of central ECRH. Indeed the mode frequency and the thermodiffusive coefficient increase with T_e/T_i. Thus central electron heating can produce a clear effect of density flattening. This result is confirmed by quasi-linear gyrokinetic calculations performed with the gyrokinetic code GS2, [3,4]. Instead, when the dominant plasma instability is an ITG, the thermodiffusive term is predicted to be small and directed inwards. In this case the effect of central electron heating on the density profile is predicted to be small, leading to a slight peaking.

These theoretical predictions agree with the experimental diagrams presented in Fig.1. Detailed calculations with the gyrokinetic code GS2 using experimentally determined parameters and identifying the mode with the largest linear growth rate, have shown that plasmas exhibiting density flattening in response to central ECRH are usually dominated by TEM instabilities. At intermediate densities in L-mode, where no flattening with central ECRH is observed, an ITG is found as dominant mode [3]. For the same reason, H-mode plasmas at intermediate densities in Fig.1, which show small density flattening with central ECRH, are found in the TEM instability domain, while at higher density in the ITG domain no flattening is observed. This is explained by the increasing stabilising effect of collisions on TEMs, which are dominant at low density and low collisionality, while at higher density ITGs are dominant.

Electron heat transport

Turbulence is also the reason for enhanced electron heat transport. This was shown quite clearly in low density plasmas where the heat flux through the confinement region around half the minor radius was changed by a factor of 10 while keeping the heat flux at the edge of the plasma constant [5]. The results suggested the existence of a threshold in the inverse electron temperature gradient length above which the electron heat flux increases. An empirical model for this heat diffusivity proved to be very successfull in describing both the experimental steady state and heat pulse diffusivities [5]. In agreement with this model the latter ones were substantially higher than the steady state diffusivities in these low density plasmas. This model also supports the hypothesis of a threshold at $R/L_{Tc} \approx 4$.

These experiments were interpreted with linear gyrokinetic stability calculations using the GS2 code [6]. Again TEM turbulence was shown to be the dominant mechanism for the enhanced diffusivity. The derived normalized heat flux was compared to the experimental one, matched only by one scaling factor, as shown in Fig.2. The theory fits very well to the experimental data, confirming the existence of a threshold in R/L_{Te}, its correct value, and the importance of TEM turbulence together with collisions even in this low density plasma.

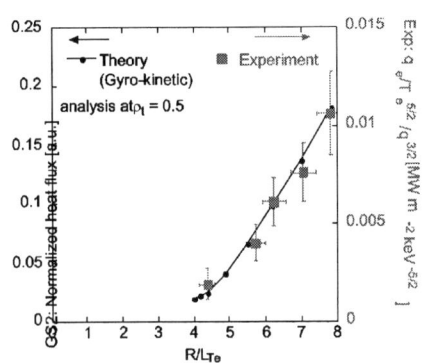

FIGURE 2. Comparison of normalized heat flux estimate from GS2 calculations with the heat flux from the experiments (from ref [6])

Like in the case of particle diffusion described above, an increasing collisionality should stabilise the TEM modes and ITG turbulence should become more dominant. This was studied in a series of plasmas with

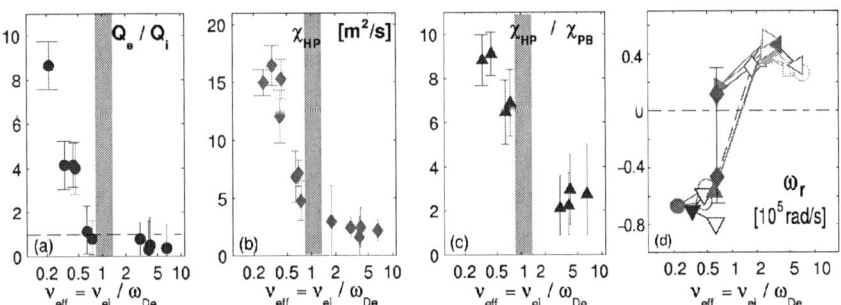

FIGURE 3. Electron to ion heat flux ratio (a), electron heat pulse diffusivity (b), heat pulse to power balance diffusivity ratio (c), and rotation frequency from GS2 analysis (d) versus v_{eff} (from ref [7])

increasing density [7]. The results are shown in Fig.3. The ratio of electron to ion heat flux, the heat pulse diffusivity, and the ratio of heat pulse to steady state heat diffusivity are plotted versus the normalised collisionality. All these quantities decrease substantially with collisionality. Also shown is a plot of the frequency of the most unstable mode obtained in a study of these plasmas with the gyrokinetic code GS2. This indicates a transition from dominant TEM modes (negative frequency) to dominant ITG modes (positive frequency) around a normalised collisionality of 1.

Impurity transport

A major task of ASDEX Upgrade is to study the behaviour of the plasma with tungsten as a first wall material. Presently about 65% of the wall is covered with tungsten. Here we observe a strong influence of central ECRH on the impurity accumulation in the centre. In H-mode plasmas the W-concentration is found to decrease substantially when relatively low power central ECRH is applied. For example, in an improved H-mode plasma at $8 \times 10^{19} m^{-3}$ with 5 MW neutral beam heating (NBI), additional central ECRH with 0.8 MW reduced the tungsten concentration by more than one order of magnitude to a hardly detectable level [8].

To study the impurity transport we did some pulses where Silicon was injected by a laser blow off technique. With central counter ECCD (to reduce the effect of sawteeth) the Silicon concentration in the plasma centre did rise (and fall) much faster than without ECCD [9]. The time evolution was fitted to the solution of a transport equation with a diffusion and a convection term. An example is shown in Fig.4. The plasmas were type-I ELMy H-mode discharges with $I_P = 0.8$ MA, $B_t = 2.4$ T, $q_{95} = 5.3$, and line averaged density of $n_e = 5.5 \times 10^{19} m^{-3}$, heated with 5 MW NBI and 0.8 MW ECCD. Without ECCD the Silicon diffusion coefficient D in the central region is close to neoclassical. With ECCD however, D can increase in the central region by up to a factor of 10 with respect to the case without ECCD, while it remains practically unchanged near the edge.

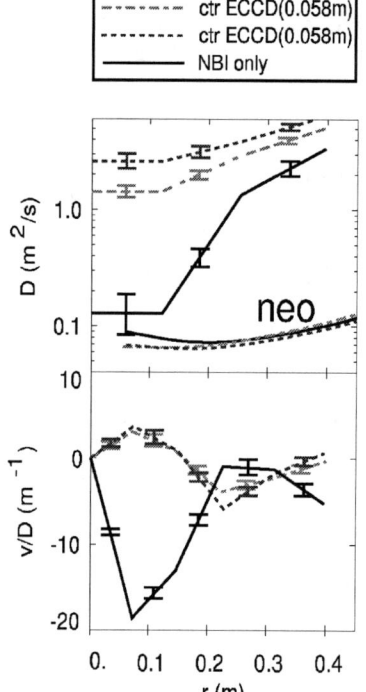

FIGURE 4. Transport parameters for Si-impurities in H-mode plasmas with only NBI heating and with combined NBI and ECCD (from ref [9])

In addition the ratio of drift velocity and diffusion coefficient v/D drops to near zero in the central region.

NTM stabilisation by ECCD

A main application of ECCD in ASDEX Upgrade is the control of neoclassical tearing modes (NTMs). These are β-limiting resistive MHD instabilities driven unstable by the loss of bootstrap current within the associated magnetic island, which can be triggered by an externally created seed island, e.g. due to a sawtooth crash at sufficiently high local β_p [10]. Three methods of NTM control by ECCD have been studied, namely suppression of an existing NTM by replacing the missing bootstrap current within the island (local ECCD at the resonant q-surface), avoiding the seed island creation due to sawteeth (local ECCD at the q=1 surface), and inducing the transition to the frequently interrupted regime (FIR) of NTMs (local ECCD at the (m+1)/(n+1) surface).

NTM suppression by replacing the missing bootstrap current

This method is based on local current drive at the resonant surface of interest, q_{res}= m/n. The suppression of (3,2) NTMs was demonstrated up to a β_N of 2.6, obtained by 12.5 MW of neutral beam injection (NBI), with a ECCD power of 1.2 MW lauched at a toroidal angle of $\Phi = 15°$, [11]. Defining an efficiency for NTM suppression as the maximum β_N at which the mode can be suppressed, divided by the applied ECCD power [12], $\varepsilon = \beta_N/P_{ECCD}$, we get for this (3,2) suppression case a value of $\varepsilon = 2.17$ MW^{-1}. Similarly, a (2,1) NTM could be suppressed in a plasma with $\beta_N = 1.9$, obtained with 6.25 MW of NBI, with a ECCD power of 1.8 MW launched at $\Phi = 12°$ [13], yielding an efficiency of $\varepsilon = 1.06$ MW^{-1}.

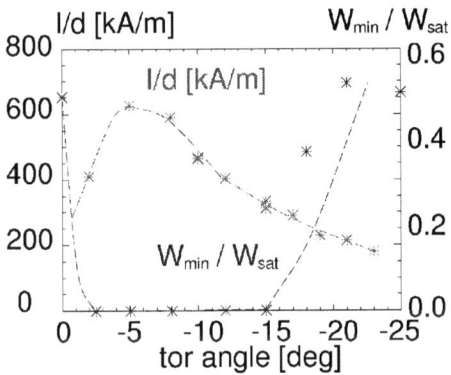

FIGURE 5. ECCD current density I/d as calculated from TORBEAM, and experimental reduction of island size. Full NTM suppression is obtained for toroidal launching angles of 2.5° to 15° (from [12])

The need to localise the driven current inside the island as long as possible has been shown by a series of experiments where the toroidal injection angle was varied in otherwise identical discharges. A larger toroidal angle Φ increases the driven current I_{ECCD}, however, with a widening current profile width d. Fig.5 shows the results in terms of W/Wsat, i.e. the final island width normalised by the saturated island width before ECCD is switched on [12]. It can be seen that in this series of experiments, (3,2) NTMs could be completely suppressed for a range of toroidal angles $5° \leq \Phi \leq 15°$. This is in qualitative agreement with the angle dependence of the current density, $j_{ECCD} \sim I_{ECCD}/d$, also shown in the figure. In particular, the current density j_{ECCD} has a maximum at 5°, although the driven current I_{ECCD} is a factor of 2 lower than at 15°. For mode suppression a narrow deposition with d < W is of advantage since the driven current can remain fully inside the shrinking island for a longer time. This should increase the efficiency of mode suppression. Indeed, with 5° toroidal launching angle the (3,2) NTM could be suppressed at $\beta_N = 2.6$ with only 1 MW of ECCD power and an efficiency of $\varepsilon = 2.6$ MW^{-1}. And the (2,1) NTM could be suppressed up to $\beta_N = 2.3$ with 1.4 MW of ECCD at 5° with an efficiency of $\varepsilon = 1.64$ MW^{-1}. Thus narrow deposition with high current density increases the suppression efficiency. However, this also requires more effort to place and to keep the driven current centered in the shrinking island.

Avoidance of NTMs by sawtooth control

NTMs need a seed island as an initial perturbation. This is usually supplied by a sawtooth crash or a fishbone located at the q=1 surface which produces a perturbation reaching the resonant q-surface of the NTM. Thus a method to avoid NTMs is the control of sawteeth.

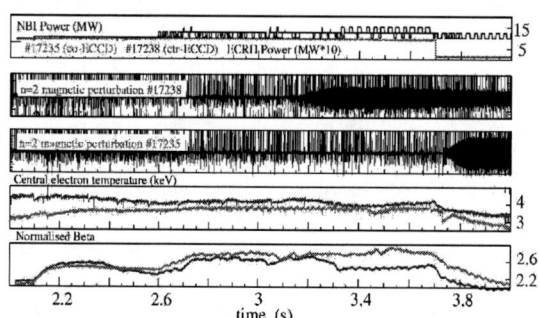

FIGURE 6. Avoidance of NTMs by sawtooth suppression with early co-ECCD at the q=1 surface, or counter ECCD near the centre (from ref [14])

The sawtooth amplitude and frequency can be controlled with ECCD, depending on power and location of the driven current. Co-ECCD in the centre leads to higher frequency low amplitude sawteeth. Complete suppression of sawteeth was achieved with co-ECCD close to the q=1 surface, and with counter ECCD in the centre. This is in qualitative agreement with a reduction of the magnetic shear at the q=1 surface.

Based on these observations we applied 1.5 MW ECCD in a discharge prone to NTMs [14]. The time traces are shown in Fig.6. ECCD was applied early while the NBI power was still ramped up and NTMs were not yet destabilised. With co-ECCD at q=1 the NTM could be avoided throughout the 2 sec ECCD phase even at a NBI power level of 12.5 MW and $\beta_N = 2.9$, although the ECCD power was not sufficient to suppress sawteeth completely. As soon as ECCD was switched off a large sawtooth triggered the NTM. With counter ECCD at the same location the sawtooth period became long, but sawteeth were still present and finally triggered a (3,2) NTM already during the ECCD phase. Obviously the ECCD power was not sufficient at this high β_N, and we do not yet know the scaling of the required power with β_N.

Inducing transitions to the FIR-NTM regime with ECCD

With increasing β_N the confinement reduction due to an NTM becomes stronger. But this trend is stopped at $\beta_N = 2.3$. At higher β_N a new regime occurs where the growth of the NTM is repeatedly interrupted and thus never reaches a saturated value. This is the so-called frequently interrupted regime of NTMs (FIR-NTM). Correspondingly the confinement degradation is less strong [15]. This interruption of the growth is due to frequently occuring (m+1,n+1) ideal mode bursts which are destabilized at high pressure gradients, and which couple with the (m,n) NTM reducing its amplitude.

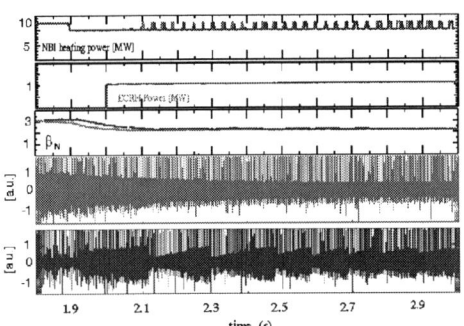

FIGURE 7. ECCD on the q=4/3 surface. Co-ECCD leads to frequently interrupted (3/2)-NTM (from ref [16])

Experiments in ASDEX Upgrade have shown that we can trigger such a destabilization of the ideal mode already at $\beta_N < 2.3$ with ECCD on the q=(m+1,n+1) resonant surface. Our present understanding is that this decreases the local shear which in combination with a strong pressure gradient leads to ideal instability. Fig.7 shows an example in which, at constant β_N, co-ECCD triggers the transition to FIR-NTM, while ctr-ECCD does not [15]. The applicability of this scheme to prevent strong confinement degradation has still to be studied.

Future plans with ECRH in ASDEX Upgrade

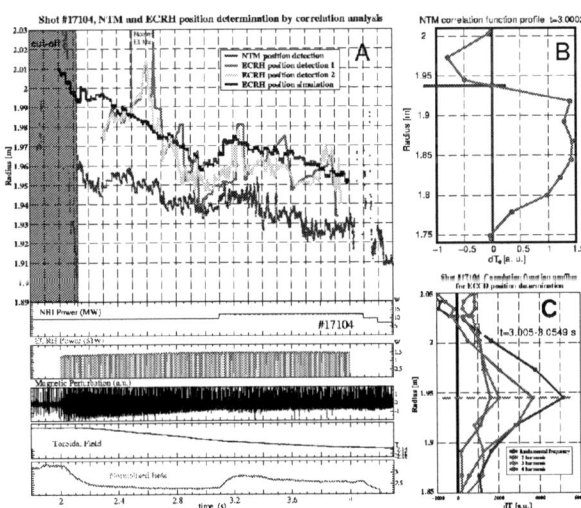

FIGURE 8. Evaluation of ECCD deposition and of NTM location from modulated ECE channels. From ref. [16])

As suppression of NTMs with high efficiency requires narrow current drive profiles, a deposition control is absolutely necessary to keep the driven current inside the shrinking island. This is particularly true when a further increase of the heating power and thus β_N leads to an increased Shafranov shift and thus a displacement of the island location. This requires an online evaluation of the mode location and of the ECCD deposition. An example for such an analysis, not yet made online, is shown in Fig.8. Shown are the temporal evolution of r_{dep} from ECRH modulation (2 different methods), the modelled evolution of r_{dep} (TORBEAM code) and the temporal evolution of r_{res} from the phase jump of ECE channels that see the modulation of T_e

due to the rotating island [16]. Here, the NTM is not fully suppressed, but its amplitude first decreases as r_{dep} is shifted into the island and approaches r_{res}. Then, at 3.1 sec, β is increased by applying more heating power and it can be seen that r_{res} moves away from r_{dep} due to the increased Shafranov shift. Consequently, the mode amplitude grows again. Such an analysis will provide the basis for a feedback controlled deposition. The actuator will be a fast steerable mirror such that the poloidal launching angle can be changed by 10° within 100 msec.

Furthermore 4 new gyrotrons will be installed which can be operated at several frequencies in the range 105 GHz to 140 GHz giving more flexibility with respect to the magnetic field. The available power is planned as 4 MW at 140 GHz and 3.2 MW at 105 GHz[17].

References

1. Kirov K. et al., *Plasma Phys. Contr. Fusion* **44**, 2583 (2002)
2. Leuterer F. et al., *Nucl. Fusion* **43**, 744 (2003)
3. Angioni C. et al., *Nucl. Fusion* **44**, 827 (2004)
4. Kotschenreuther M. et al., *Comput. Phys. Commun.* **88**, 128 (1995)
5. Ryter F. et al., *Nucl. Fusion* **43**, 1396 (2003)
6. Peeters A.G. et al., *Phys Plasmas* **12**, 022505 (2005)
7. Angioni C. et al., *Phys. Plasmas* **12**, 040701 (2005)
8. Neu R. et al., *Plasma Phys. Contr. Fusion* **44**, 811 (2002)
9. Dux R. et al., *Plasma Phys. Contr. Fusion* **45**, 1815 (2003)
10. Zohm H. et al., Phys. Plasmas **8**, 2009 (2001)
11. Gantenbein G. et al., *Phys. Rev. Lett* **85**, 1242 (2000)
12. Maraschek M. et al., 20th IAEA Conf. Fusion Energy, Vilamoura, Portugal 2004, paper IAEA _CN_116/EX/7-2, subm. *Plasma Phys. Contr. Fusion*
13. Gantenbein G. et al., 15th Conf. RF Power in Plasmas, AIP Conf. Proceedings 694, 317 (2003)
14. Mück A. et al., 30th EPS Conference St. Petersburg, *ECA* **27A**, p. 1.131 (2003)
15. Günter S. et al., *Nucl. Fusion* **43**, 161 (2003)
16. Keller A. et al., 30th EPS Conference St. Petersburg, *ECA* **27A**, p. 1.130 (2003)
17. Leuterer F. et al., Fus. Eng. Des. **66-68**, 537 (2003)

Third Harmonic X-mode ECH top-launch on the TCV tokamak

G. Arnoux*, S. Alberti*, L. Porte*, S. Nowak[†], B. Marlétaz*, Ph. Marmillod*, Y. Martin* and TCV Team*

*Centre de Recherches en Physique des Plasmas, Association EURATOM-Confédération Suisse, Ecole Polytechnique Fédérale de Lausanne, CRPP-EPFL, CH-1015 Lausanne, Switzerland.
[†]Istituto di Fisica del Plasma, Associazione EURATOM-ENEA-CNR, Via Bassini 15, 20133 Milano, Italy.

Keywords: ECH, X-mode, Third harmonic, Absorption, Feedback control, TCV
PACS: 52.35.Hr

1. INTRODUCTION

The electron cyclotron heating (ECH) system on the "Tokamak à Configuration Variable" (TCV) is composed of six 0.45MW, 82.7GHz, 2s. pulse length gyrotrons for second harmonic X-mode heating (X2) [1] and three 0.45MW, 118 GHz, 2 s. pulse length gyrotrons for third harmonic X-mode heating (X3) [2]. With the moderate magnetic

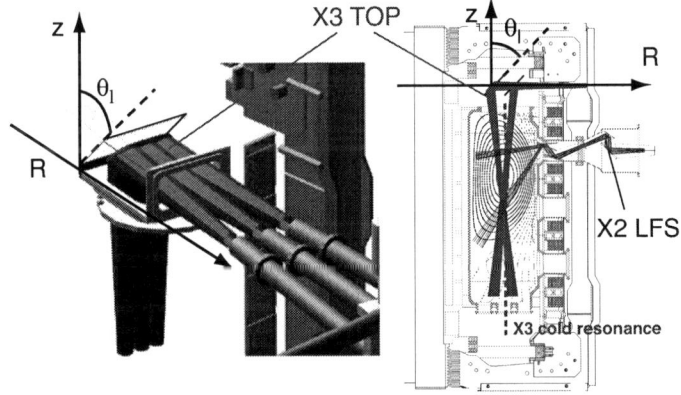

FIGURE 1. The X3 launcher placed at the top of the vessel can be steered radially (R) from shot to shot and the mirror angle θ_l can be varied during a shot. $\theta_l = 45^o$ corresponds to a pure vertical launch and the top launch maximizes the RF beam path within the resonance layer.

field of TCV (1.5 T) the third harmonic allows to extend the ECH operational domain up to the X3 cutoff density $n^{X3}_{e,cutoff} = 11.5 \cdot 10^{19}$ m^{-3}, which is significantly higher than the X2 cutoff density $n^{X2}_{e,cutoff} = 4.2 \times 10^{19}$ m^{-3}, and allows in particular the heating of H-mode plasmas. The X3 absorption coefficient $\alpha_3^{(X)}$ is lower than the X2 absorption coefficient by a factor $\alpha_3^{(X)}/\alpha_2^{(X)} \propto (v_{th}/c)^2 = T_e[keV]/511$ where v_{th} is the thermal velocity, c is the speed of light and T_e is the electron temperature. Due to the lower absorption of X3 a top launch configuration is chosen such to maximize the beam path within the resonance layer (Figure 1). With such a launching schema the absorption is maximized by maximizing the optical depth $\tau_{X3} = \int \alpha_3^{(X)} ds$.

The X3 launcher is composed by a single elliptical mirror made of copper with a focal length of 700 mm. The waist of the RF gaussian beam is approximately located at the vacuum vessel center and has a value of $w_0 = 30$ mm in electric field. The radial position of the mirror R_l can be steered from shot to shot and the launcher poloidal angle θ_l can be varied during a shot with a maximum rate $d\theta_l/dt = 20^o/s$. At $\theta_l = 45^o$ (Figure 1), the beam is injected in a pure vertical direction and $\theta_l > 45^o$ means that the beam goes from the low field side (LFS) to the high field side (HFS) deviating slightly from the vertical path. With a top-launch injection, the absorption is strongly dependent on the launcher poloidal angle θ_l. This sensitivity has been experimentally studied in previous experiments [2, 3] and is also modeled using the ray-tracing code TORAY-GA [4, 5, 6]. For a typical target plasma of TCV and for a fixed radial position of the launcher R_l, this sensitivity is shown in Figure 2 where the calculated X3 absorption is plotted against θ_l for different central electron temperatures $T_{e,0}$ and densities $n_{e,0}$. Note that the typical FWHM of these curves gives angular tolerance for the mirror which is typically 1.4^o [3]. The maximum absorption increases monotonically with the

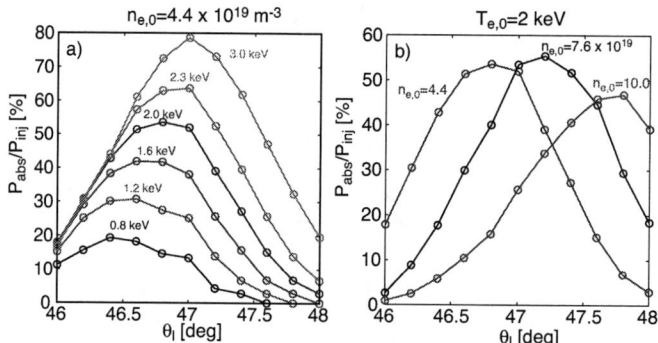

FIGURE 2. (a) The X3 absorption calculated by the ray-tracing code TORAY-GA as a function of θ_l for different electron temperatures T_e. The global X3 absorption P_{abs}/P_{inj} is significantly increases by T_e and $\theta_{l,opt}$ is increased when T_e increases because of the relativistic shift which shifts the resonance towards the HFS. (b) The X3 P_{abs}/P_{inj} as a function of θ_l for different electron densities n_e. $\theta_{l,opt}$ increases when n_e increases in order to compensate the refraction.

temperature and is weakly dependent on the density. The T_e and n_e dependence of the global X3 absorption will be further discussed in section 3. Figure 2 (a) shows that the optimal launcher angle $\theta_{l,opt}$ increases linearly with the temperature because of the shift of the resonance towards the HFS (relativistic shift). The $\theta_{l,opt}$ variation is such that $d\theta_{l,opt}/dT_e = 0.2$ deg/keV. As shown on Figure 2 (b), $\theta_{l,opt}$ increases with the density (refraction) with a corresponding variation of $d\theta_{l,opt}/dn_{e,0} = 0.2$ [deg/(10^{19} m^{-3})]. In conclusion, both T_e and n_e variations significantly modify the optimal launcher angle.

To maximize the absorption during a discharge where both T_e and n_e may change, a real time feedback control on the X3 launcher has been developed and successfully used on TCV.

Section 2 describes this system and shows its dynamic on L-mode plasmas. In section 3, the X3 absorption properties on L-mode plasmas are discussed and preliminary results on H-mode heating are presented. In section 4, a comparison of the modelling performed with the ray-tracing code TORAY-GA and the beam-tracing code ECWGB is presented.

2. A REAL-TIME FEEDBACK CONTROL

To maximize the X3 absorption during a plasma discharge taking into account the temperature (relativistic shift) and density (refraction) dynamics, a real-time feedback control on the launcher poloidal angle has been developed and installed on TCV. The main idea of this system is to generate an error signal which is proportional to the slope $dI_x/d\theta_l$ of the plasma response function $I_x(\theta_l)$. The reference $R(s)$ is set such to maintain the mirror angle on the zero value of the error signal $(dI_x/d\theta = 0)$ via a PID controller. According to this idea, an analog system has been realized (Figure 3) which

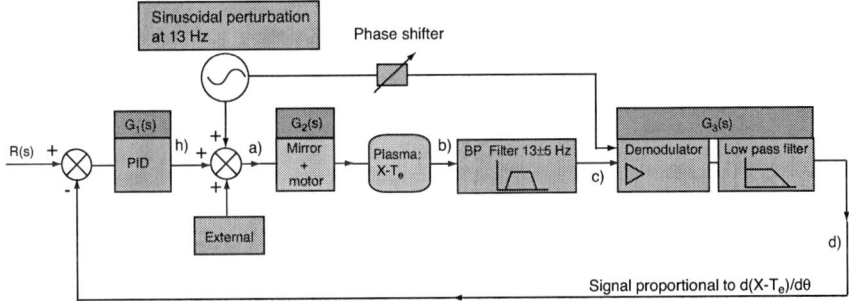

FIGURE 3. Schematic view of the feedback system. (a) The sinusoidal perturbation is added to the external constraint on θ_l like the signal (a) in Figure 4. (b) the T_e X signal contains both the global plasma response and the mirror-motor dynamic response. (c) is obtained by processing the global plasma response through a 13 ± 5 Hz band-pass filter to get only the response to the 0.2^o perturbation. (d) the synchronous demodulation of the plasma response with the sinusoidal perturbation build an error signal as shown in Figure 4 (d) and (g). (h) A regulation signal is built from the error signal via a PID controller and allows to regulate the launcher poloidal angle.

is based on a synchronous demodulation of the plasma response $I_x(\theta_l)$ to a sinusoidal reference signal which is applied on the mirror angle. The line integrated soft X-ray emissivity $I_x(\theta_l)$ is proportional to $n_e^2 T_e$ and can be used in the feedback loop as long as the temperature variations dominate the density variations which is the case for L-mode plasmas on which the feedback has been used. According to the mirror-motor transfer function $G_2(s)$ which has been measured and can be approximated by a 2nd order pole [7], the mirror angle modulation frequency has been chosen to be 13 Hz. The open loop characteristics of the system are shown in Figure 4 (a)-(d) where the time traces of the relevant quantities are represented in a case on which a linear sweep of the mirror angle has been performed. (The subplots (a)-(d) correspond to the signals indicated in Figure 3). The sweep is such to cross the optimum angle corresponding to the maximum absorption. The closed loop dynamics of the feedback system is shown in Figures 4 (e, f, g and h) where the feedback is switched on at the same time as the ECH at 0.6 s. The

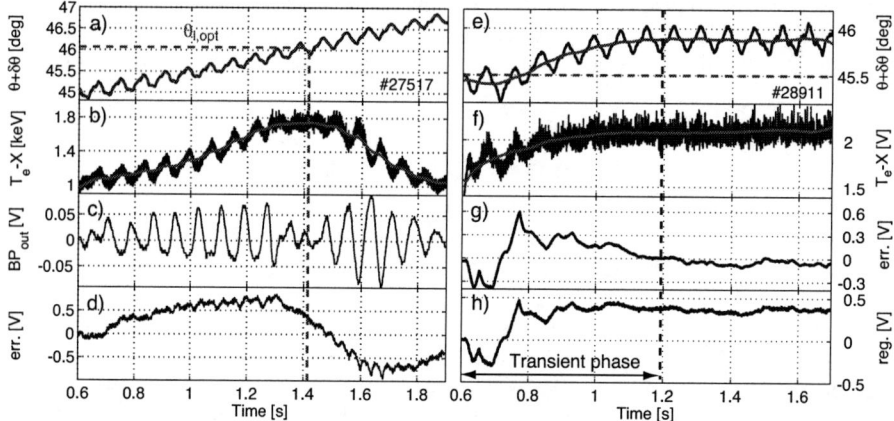

FIGURE 4. In open loop: (a) the 13 Hz sinusoidal oscillation added to the launcher angle, swept from 45° to 47°, (b) the global plasma response via the temperature from soft X-ray measurement integrated on a vertical line (two foil method), (c) the plasma response to the only perturbation, (d) the regulation signal obtained from the synchronous demodulation. In closed loop: (e), the launcher angle goes to $\theta_{l,opt} = 45.9^o$ which is 0.4^o higher than the imposed angle (dashed line), (f) T_e-X is maintained at 2.1 keV, (g) the error signal remains around zero (h), the regulation signal is built using the PI terms of the controller.

preprogramed mirror angle is 45.5^o and remains constant during the shot (dashed line on Figure 4 (e)). After a 600 ms transient phase the mirror angle reaches the optimum value. From 1.2 s, the system is locked on $\theta_{l,opt} = 45.9 \pm 0.1^o$ and maintains a temperature T_e-X = 2.1 keV. The system is able to control the launcher accurately enough to follow a difference of 0.4^o which increases the temperature of 33 %. As shown on Figure 4 (g)-(h) obtained by using a PI controller, after the transient phase, the error signal remains around zero corresponding to a stationary optimum mirror angle.

The real time feedback control on the X3 launcher has been extensively used on L-mode discharges, but needs some improvement for experiments on H-mode plasmas due to important density perturbations at the L-H transition as well as due to the presence of ELMs. To optimize such a feedback system, a detailed analysis of the system is under study and a model is being developed which will allow to perform simulations using the software package Simulink.

3. X3 ABSORPTION PROPERTIES

In this section, first the X3 absorption properties are studied on L-mode plasmas using the feedback control to maximize the absorption. In the second part, preliminary results on heating of H-mode plasma using X3 are presented.

As reminded in the introduction, the absorption depends mainly on the temperature and the density [3, 7]. To compare the ray-tracing predictions shown in Figure 2, a set of experiments has been performed in which the absorption is determined by modulating the power of a full gyrotron (0.45MW) and measuring the diamagnetic flux response via a diamagnetic loop (DML) [8, 9]. The absorption measurement is performed during

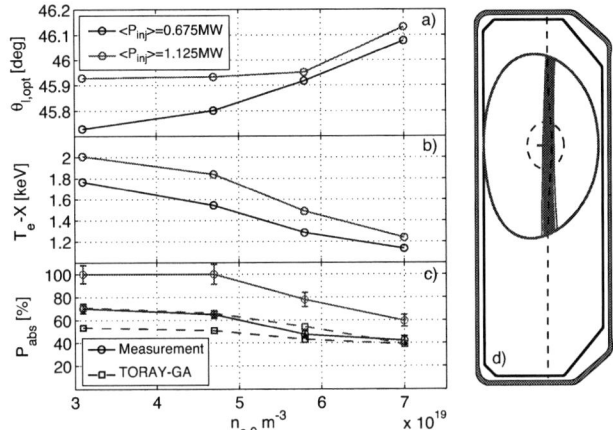

FIGURE 5. As a function of the central density $n_{e,0}$ for two averaged injected power $\langle P_{inj} \rangle = \{0.675; 1.125\}$ MW: (a) The optimal angle $\theta_{l,opt}$ given by the X3 feedback system. (b) The T_e-X signal measured during a 200 ms stable interval. (c) The X3 absorption measurements (o) compared with the TORAY-GA predictions (square).

a phase in which the feedback has stabilized to the optimum mirror angle $\theta_{l,opt}$. This scenario is repeated for 2 injected power (different T_e) and 4 different densities. Figure 5 (d) shows a typical L-mode plasmas used in the experiments where the beam trajectory is calculated with the ray-tracing code TORAY-GA. Figures 5 (a), (b) and (c) show $\theta_{l,opt}$, the T_e-X measurement and the X3 absorption respectively as a function of the central density and for $\langle P_{inj} \rangle = 0.675$ and 1.125 MW. As expected, $\theta_{l,opt}$ increases with the density because of refraction but with a weaker sensitivity as predicted by TORAY-GA 2 (b) since the temperature is not constant. The refraction effect is attenuated by the temperature decrease which shifts the hot resonance layer closer to the cold resonance. At the maximum injected power of 1.125 MW and densities lower than $4.8 \cdot 10^{19}$ m^{-3}, full single pass absorption is measured. The difference between the measured value (DML) and the TORAY-GA predictions is associated with the presence of a suprathermal electron population (SEP) [12] generated by the X3 itself. Experimental evidence of the existence of an SEP is observed on diagnostics such as the high field side ECE [10] or the hard X-ray camera [11]. One notice that at higher densities ($n_{e,0} > 5 \cdot 10^{19}$ m^{-3}) the difference between the measurement and the calculation decreases. This effect might be due to a faster thermalization (stronger coupling to the bulk) of the SEP that occurs at higher densities. An evidence of this effect is that, at these higher densities, no signature of the presence of an SEP is observed.

On TCV, it has been observed already that X3 ECH is able to affect the ELMs frequency [13]. Figure 6 shows the time traces of an H-mode plasma in which 1.35 MW X3 is injected. This scenario can be decomposed in 4 phases. The first phase ($t < 1$ s) with the standard high-frequency low-amplitude ELMs typically observed on TCV [14, 15] is an ohmic H-mode plasma. At the X3 switch on ($t = 1$ s) a short ELM free period (100 ms) is followed by a stationary phase (1.2 − 1.45 s) with low-frequency large-amplitude ELMs. During this phase, the central electron temperature increases from 1 keV to 2.2 keV, with a central density of $7 \cdot 10^{19}$ m^{-3} corresponding to the line averaged density shown in Figure 6 (e). At $t = 1.45$ s, one gyrotron is witched off and

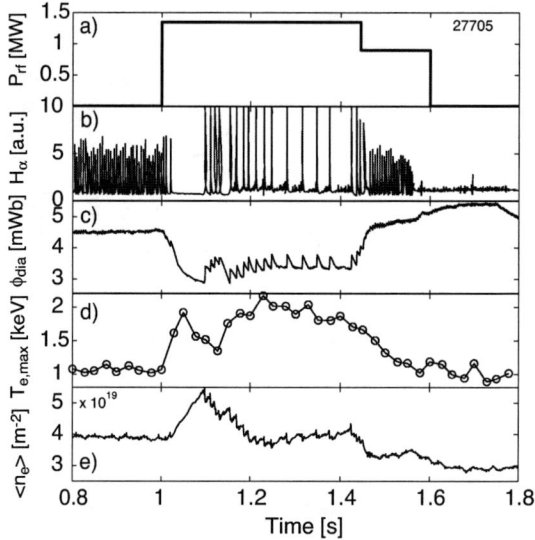

FIGURE 6. H-mode ECH experiments scenario with (a) the X3 injected power, (b) the H$_\alpha$ line emission, (c) the diamagnetic flux, (d) the maximum temperature from Thomson scattering measurement and (e) the averaged density on a central view line.

the H-mode plasma return back to the high-frequency low-amplitude ELMs. This effect indicates that the available X3 power is marginal for maintaining the discharge in the low-frequency large-amplitude regime which is a new ELM regime for TCV. Further analysis and additional experiments on X3 H-mode plasmas are presently outgoing.

4. RAY-TRACING VS BEAM-TRACING

TORAY-GA is a ray-tracing code that does not take into account diffraction effects. To evaluate the importance of these effects with the X3 top-launch, a comparison between TORAY-GA and the beam-tracing code ECWGB [16, 17, 18] is presented. The RF beam in TORAY-GA is simulated by a set of rays with a gaussian power density and a cylindrical envelop which diameter corresponds to the minimum waist [3]. On the other hand, ECWGB properly simulates the phase front evolution which corresponds in free space (fundamental TEM$_{00}$ gaussian mode) to a quadratic surface. No diffraction effects associated to a partial absorption of the beam are considered in ECWGB. The beam-tracing code calculate the phase fronts and determines a set of ray trajectories (normal to the phase surfaces) along which the absorption is calculated with a common ray-tracing procedure. Figures 7 (a) and (b) show the ray trajectories calculated by TORAY-GA and ECWGB respectively. The density $n_{e,0} = 8.2 \cdot 10^{19}$ m^{-3} and the temperature

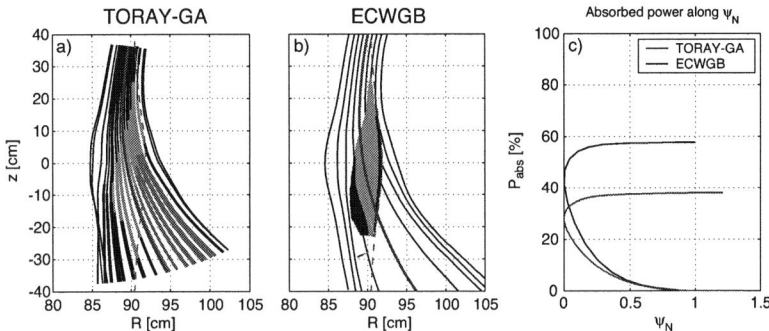

FIGURE 7. The X3 beam trajectory for $n_{e,0} = 8.2 \cdot 10^{19}$ m^{-3} and $T_{e,0} = 2.3$ keV is calculated by (a) the ray-tracing (TORAY-GA) and (b) the beam-tracing (ECWGB). The orange area indicates where $10\% \leq P_{abs} \leq 90\%$ and for TORAY-GA, the rays in light blue indicates that $P_{abs} > 90\%$. (c) The total absorption along the normalize flux coordinate of the beam trajectory (red rays in (a) and b)).

$T_{e,0} = 2.3$ keV have been chosen respectively to maximize the diffraction effect and to have a significant absorption. The region where the main absorption occurs is colored in orange. The two calculations give similar beam trajectories and power deposition as well. Nevertheless Figure 7 (c) shows the global absorption along the normalized flux coordinate of the beam trajectories and reveal a significant discrepancy between TORAY-GA and ECWGB. It has been verified that the absorption calculation along the central ray for each code is the same indicating that the absorption calculation of the two codes are equivalent. Some calculations have also been performed using ECWGB by suppressing the diffraction effect and indicate that the diffraction is not the origin of the discrepancy. The discrepancy is probably to be associated to the geomatrical distribution of the rays in TORAY-GA. Further studies on possible diffraction effects are presently underway and in particular in the case of stronger beam focusing.

5. CONCLUSION

The X3 top-launch absorption depends strongly on the launcher poloidal angle. To maximize the absorption by tracking the optimal launcher angle according to the temperature and density variations during a discharge, a real time feedback control on the X3 launcher has been developed and installed on TCV. The feedback system has been extensively used on L-mode plasmas with an accuracy on the mirror angle of 0.1^o. Improvements of the feedback system for heating H-mode plasmas are under development through a detailed modeling of the system allowing simulations with the software Simulink.

X3 ECH experiments on L-mode plasmas using the X3 feedback confirm the dependance of $\theta_{l,opt}$ on the density (refraction) and temperature (relativistic shift). At low density ($n_{e,0} < 5 \cdot 10^{19}$ m^{-3}) full single-pass absorption is measured. An upgrade of the X3 system ($P_{inj} \simeq 2$ MW) is under investigations. The discrepancy between the absorption measurement and TORAY-GA calculations is attributed to the presence of the suprathermal electrons population generated by X3 itself which couples to a significant fraction of the power.

In H-mode plasmas heated with X3, a new stationary ELMy regime for TCV with low-frequency and high-amplitude ELMs has been found. Further analysis and experiments X3 heated H-mode plasmas are presently underway.

Diffraction effects on the RF beam propagation inside the plasma have been studied using the beam-tracing code ECWGB. Comparisons of the absorbed fraction calculation with the ray-tracing code TORAY-GA show a discrepancy. This discrepancy is attributed to the beam model in the ray-tracing and not to diffraction effects. Further studies investigating stronger focalization of the beam using ECWGB are underway.

ACKNOWLEDGMENTS

This work was partly supported by the Swiss National Fundation for Research

REFERENCES

1. Goodman, T. P., and al., *Proc. in 19th Symposium on Fusion Technology, Lisbon, Portugal*, pp. 565–568 (1996).
2. Hogge, J.-P., Alberti, S., Porte, L., and Arnoux, G., *Nucl. Fusion*, **43**, 1353–1360 (2003).
3. Arnoux, G., Alberti, S., Porte, L., Nelson-Melby, E., Hogge, J.-P., and TCV Team, *Plasma Phys. Control. Fusion*, **47**, 295–314 (2005).
4. Cohen, R. H., *Phys. Fluids*, **30**, 2442 (1987).
5. Matsuda, K., *IEEE Trans. Plasma Sci.*, **17**, 6 (1989).
6. Lin-Liu, Y. R., and et al., *Proc. in 26th EPS conference on Plasma Physics and Controlled Fusion, Maastricht*, **23J**, 1245 (1999).
7. Alberti, S., Arnoux, G., Porte, L., Marletaz, B., Marmillod, P., Martin, Y., Nowak, S., and TCV Team, *Nucl. Fusion*, **Submitted for publication** (2005).
8. Manini, A., Moret, J.-M., Alberti, S., Goodman, T. P., and Henderson, M. A., *Pasma Phys. Control. Fusion*, **44** (2002).
9. Moret, J.-M., Buhlmann, F., and Tonetti, G., *Rev. Scientific Instr.*, **74**, 4634–4643 (2003).
10. Blanchard, P., Alberti, S., Coda, S., Weisen, H., Nikkola, P., and Klimanov, I., *Plasma Phys. Control. Fusion*, **44**, 2231 (2002).
11. Coda, S., Alberti, S., Blanchard, P., Goodman, T. P., Henderson, M. A., Nikkola, P., Peysson, Y., and Sauter, O., *Nucl. Fusion*, **43**, 1361–1370 (2003).
12. Alberti, S., and al., *Nucl. Fusion*, **42**, 42–45 (2002).
13. Porte, L., Alberti, S., Arnoux, G., Martin, Y., Hogge, J.-P., Goodman, T. P., Henderson, M. A., Nelson-Melby, E., A, P., and Tran, M. Q., *Proc. in 19th IAEA Fusion Energy conference, Lyon* (2002).
14. Martin, Y., and TCV Team, *Plasma Phys. Control. Fusion*, **44**, A143–A150 (2002).
15. Martin, Y., Henderson, M. A., Alberti, S., Amorim, P., and et al., *Plasma Phys. Control. Fusion*, **45**, A351–A365 (2003).
16. Nowak, S., and Orefice, A., *Phys. Fluids*, **5**, 1242 (1993).
17. Nowak, S., and Orefice, A., *Phys. Fluids*, **1**, 1945 (1994).
18. Cirant, S., Nowak, S., and Orefice, A., *J. Plasma Physics*, **53**, 354–364 (1995).

Initial Results of Multi-Frequency Electron Cyclotron Heating in the Levitated Dipole Experiment

A.K. Hansen*, S. Mahar[†], A.C. Boxer[†], J.L. Ellsworth[†], D.T. Garnier*, I. Karim[†], J. Kesner[†], M. Mauel* and E.E. Ortiz*

*Department of Applied Physics and Applied Mathematics, Columbia University, New York, New York 10027
[†]Plasma Science and Fusion Center, MIT, Cambridge, Massachusetts 02139

Abstract. The Levitated Dipole Experiment (LDX) has created high-beta, hot-electron plasmas that are confined by a strong dipole electromagnet via multiple-frequency electron cyclotron resonance heating (ECRH). Multiple frequency ECRH is used to investigate how variation of the power deposition profile may be used to adjust the plasma density and pressure profiles. The initial experiments have been performed using up to 3 kW at 2.45 GHz and 3 kW at 6.4 GHz. Variations included switching on and off a single source while injecting constant power with the other source. We have also investigated the role of magnetic shaping, using external coils, on ECRH phenomena and plasma profile control. The preliminary results of these experiments will be presented.

Keywords: dipole confinement, electron cyclotron heating, superconducting magnet
PACS: 52.25.Xz, 52.50.Sw

INTRODUCTION

The Levitated Dipole Experiment (LDX), a joint Columbia/MIT concept exploration experiment sited at MIT, has begun experimental operations (the first plasma was in August 2004), though the dipole coil has heretofore been supported mechanically rather than levitated.

Plasmas confined by a dipole field are stabilized by compressibility. From ideal MHD, it is expected that marginal stability for interchange modes will occur when the pressure, p, satisfies an adiabaticity condition: $\delta(pV^\gamma) = 0$, where $V \equiv \oint \frac{d\ell}{B}$ is the incremental flux tube volume and $\gamma = \frac{5}{3}$ is the ratio of specific heats.[1] In addition, equilibria satisfying this criterion are also stable to ballooning modes.[2] For marginal profiles with $\eta \equiv \frac{d \ln T}{d \ln n} = \frac{2}{3}$, drift waves are stable as well.[3, 4] Other experiments, but not fusion-oriented, that confine plasma in dipole geometry are the CTX supported dipole[5] and the Mini-RT levitated dipole.[6]

A major goal of research on LDX is to study the effect of the pressure profile on the stability and confinement of the plasma. One of our main "knobs" to do this is electron cyclotron resonance heating (ECRH) at multiple frequencies. This has historically proven to be an effective means to create high β hot-electron plasmas in mirror machines,[7] levitrons,[8] and CTX.[5, 9]

ECRH SOURCES ON LDX

There currently are two ECRH sources in use for LDX. One is a Gerling magnetron, which delivers 3 kW of CW power at 2.45 GHz. The other is a Varian klystron, with a CW power output of 3 kW at 6.4 GHz.

The microwaves from each source are transmitted to the experimental vacuum chamber via waveguide runs of about 6 m. The antennae are inside the vacuum chamber, and are segments of waveguide that are cut an appropriate angle as maximize the transmission and minimize the directivity. LDX relies on a "cavity heating" scheme, i.e. small first-pass absorption and multiple reflections of microwaves from the vacuum chamber walls, which has the benefit of making the heating somewhat isotropic.

EXPERIMENTAL RESULTS

We will present initial results from two areas of investigation. One was to hold the nominal input power fixed and vary the amount put out by each source. The other involved chopping the power in one source on and off while the other source was held fixed, with and without an applied vertical field to change the plasma compressibility. We will focus on the gross macroscopic changes that are visible in a poloidal flux loop.

Fixed input power

Three cases were chosen for this experiment: (1) 3 kW on the 2.45 GHz source, (2) 3 kW on the 6.4 GHz source, and (3) 1.5 kW on both sources together. Data from one of the poloidal flux loops is shown in Fig. 1(a) for the three cases. Note that the firing time of the two sources was different for the case where both sources were used–the 2.45 GHz source was fired at 2 s into the discharge. Clearly, the amplitudes are different between the cases, with the 2.45 GHz-only case having the smallest value, and the combined case having the largest. Rise and decay time constants are shown as well; these were calculated via a nonlinear fit to the data assuming exponential behavior. The time constants in both the rise and decay phases satisfy $\tau_{2.45} < \tau_{2.45+6.4} < \tau_{6.4}$.

Modulation experiments

For these experiments, there are again three cases: (1) both sources are turned on together, (2) the 2.45 GHz source is held constant through the whole discharge with the 6.4 GHz source being chopped on and off, and (3) the 6.4 GHz source held constant throughout the discharge with the 2.45 GHz being chopped on and off.

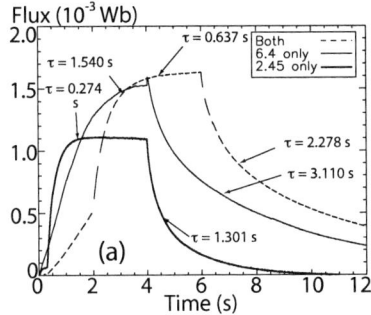 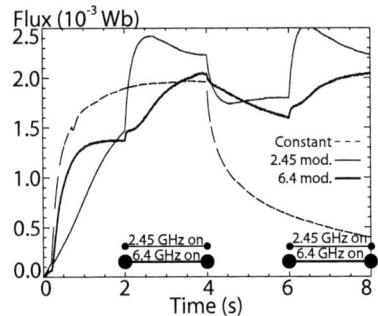

FIGURE 1. (a) Diamagnetic flux for discharges at nominally identical power. Exponential rise and decay time constants are shown (b) Diamagnetic flux for modulations of each source plus a control case. The periods when the modulated source are chopped on and off are shown below in the same linestyle.

No vertical field

Data from the 3 cases are plotted plotted in Fig. 1(b), with indicators in the same linestyle as each trace. Note that the programming for the discharge where both sources were fired concurrently had both switch off at 4 s, while the modulated cases switch off at 8s. Firing the 2.45 GHz source after the 6.4 GHz source results in a higher maximum flux than firing the two sources concurrently, and much higher than firing the 6.4 GHz source after the 2.45 has been on. It appears that the 3 traces are converging toward a similar value at t=4 seconds, at which time whichever source being modulated is shut off, so possibly if the 6.4 GHz were kept on longer than 2 s the flux might reach the same value as in the unmodulated case.

Vertical field applied

By applying an external field via a set of Helmholtz coils, we can dramatically change the volume encompassed by the last closed field line (Fig. 2 (a)), i.e. the confined plasma volume, which should change the plasma compressibility. The general behavior in such discharges is qualitatively similar to the prior results (Fig. 1(b)) until the applied external field is large enough, at which point the behavior shown in Fig. 2(b) occurs. Note that drops in the flux (1.5 s, 4 s) occur during the period in which both RF sources are on. This suggests that the pressure profile has crossed a stability boundary, and the sudden rise at 5 s presumably indicates crossing back to stability.

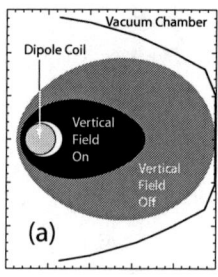 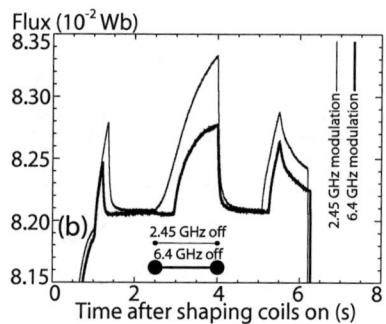

FIGURE 2. (a) Area within last closed field line without (gray) and with (black) applied vertical field. (b) Diamagnetic flux for plasma with a large shaping coil current and modulated RF input power (both sources). The periods when the modulated source are chopped on and off are shown below in the same linestyle.

FUTURE WORK

Our next immediate goal is to get an additional source (10 kW @ 10.5 GHz) online. Once that is done further optimization studies can be performed.

ACKNOWLEDGMENTS

This work is supported by U.S. DOE Grants DE-FG02-98ER54458 and DE-FG02-98ER54459.

REFERENCES

1. M.N. Rosenbluth and C.L. Longmire, *Ann. Phys.* **1**, 120 (1957).
2. D. Garnier, J. Kesner, and M. Mauel, *Phys. Plasmas* **6**, 3451 (1999).
3. J. Kesner, *Phys. Plasmas* **4** 419 (1997).
4. J. Kesner, *Phys. Plasmas* **5** 3675 (1998).
5. M. Mauel, H. Warren, and A. Hasegawa, *IEEE Trans. Plasma Sci.* **20**, 626 (1992).
6. N. Yanagi, T. Mito, J. Morikawa, Y. Ogawa, K. Ohkuni, D. Hori, S. Yamakoshi, M. Iwakuma, T. Uede, I. Itoh, M. Fukagawa, and S. Fukui, *IEEE Trans. Appl. Supercond.* **14** (2), 1539, (2004).,
7. B. Quon et al., *Phys. Fluids* **28**, 1503 (1985).
8. M. Okabayashi et al., *Phys. Fluids* **16** 1337 (1973).
9. H.P. Warren et.al., *Phys. Plasmas* **3** 2143 (1996).

Control of the eITB formation and performance in fully non-inductively sustained ECCD discharges in TCV

M.A. Henderson, R. Behn, A. Bottino, Y. Camenen, S. Coda, E. Fable, T.P. Goodman, An. Martynov, P. Nikkola, A. Pochelon, O. Sauter, C. Zucca and the TCV team

CRPP, Association EURATOM- Confédération Suisse, EPFL, CH-1015 Lausanne, Switzerland

Abstract. The X2 ECH antennas on TCV are used to sustain and tailor the plasma current profile, forming either a centrally peaked or hollow profile. During the transition from peaked to hollow profile, an eITB is observed to form rapidly and in a localized region, which correlates with the appearance of a zero shear flux surface off-axis according to the ASTRA transport code. The barrier position can be controlled via the co-ECCD off-axis deposition location, and the barrier strength with central heating or counter-ECCD (increasing the depth of the hollow current profile), achieving H-factors of up to ≤6. In these discharges, the current in the ohmic transformer coil is held constant to avoid an inductively driven current contribution. After the eITB is created, a small amount of ohmically driven counter (co-) current has been added as a perturbative current source, transfering only a few kW of ohmic power compared to 1.4MW of ECCD. The ohmic current increases (decreases) the eITB performance demonstrating the clear dependence of the eITB on the current profile.

Keywords: ECCD, eITB.
PACS: 52.55.Fi, 52.50.Sw, 52.55.Wq

INTRODUCTION

TCV's flexible ECH antenna system[1] is used to sustain and tailor the current profile to form a non-inductively driven electron Internal Transport Barriers (eITB)[2,3] via inversion of the shear profile under steady-state conditions. Generation of eITB in TCV initially starts with a steady-state Ohmic plasma discharge with the current density profile (j_P) peaked in the center. The current in the ohmic transformer coil is held constant and the plasma current (~80kA) is then sustained using co-Electron Cyclotron Current Drive (ECCD) deposited off-axis (ρ_{CD}~0.4, 1.0MW), see figure 1a, in addition to the self-generated bootstrap current (j_{BS}). The co-ECCD current density profile (j_{CD}) is flat or slightly hollow from the deposition location inward due to finite particle diffusion[4]. The plasma current evolves gradually from the centrally peaked to a slightly hollow profile giving rise to a reversal in the safety factor profile and the formation of a barrier at the appearance of a zero-shear (s=0) flux surface off-axis. This occurs between phases B and C as shown in figure 1a. Once the eITB is formed, central heating or current drive can be added inside the barrier, phase D of figure 1a, increasing the central temperature.

Removing the ohmic current contribution simplifies the modeling process, albeit still complicated, by reducing the source currents to only j_{CD} and j_{BS}. The modeled safety factor profile has provided an eITB performance figure of merit[5] equal to the product of its enclosed volume and strength (associated with the degree to which the safety factor profile is inverted). A simpler figure of merit for characterizing the eITB performance has also been used[6] which is based on the normalized local gradient ρ^* [7]. The strength (ρ^*_{max}) is represented by the maximum of ρ^* and the volume (ρ_{ρ^*}) from the radial location of ρ^*_{max}, see figure 1b.

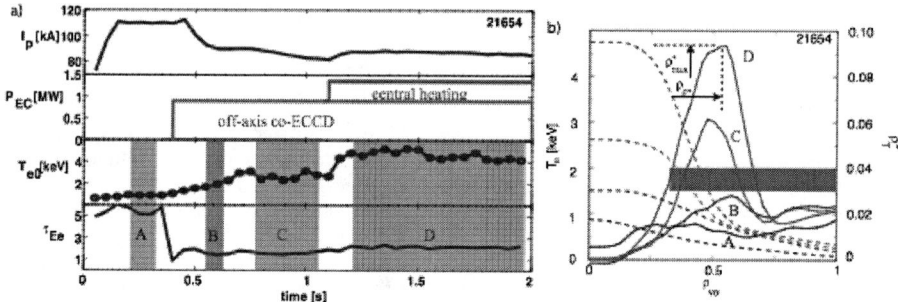

FIGURE 1. (a) Typical evolution of a non-inductively driven eITB on TCV. The discharge is divided into four phases: A ohmic, B pre-ITB, C unheated eITB and D centrally heated. (b) Electron temperature and ρ^* profiles averaged over the four phases of the eITB.

The performance of the eITB is strongly dependent on the volume inside the barrier. Scanning the radial location of the co-ECCD deposition location (ρ_{co}) on a shot-to-shot basis results in the formation of the barrier at ever increasing radial locations. As the barrier radius increases, the eITB performance, measured by H_{RLW} (τ_{eE}/τ_{RLW}[9]) increases proportionally to ρ_{co}^2 [6]. There is also a corresponding increase in the barrier strength attributed to an increase in j_{BS}. As the barrier moves outward there is a net increase in the driven bootstrap current. More bootstrap current results in a stronger shear reversal, thus further increasing the barrier strength. The barrier strength can be controlled independently by replacing the central heating with counter-ECCD, or decreased with co-ECCD[6]. The counter-ECCD (or co-ECCD) will decrease (increase) the current density inside the barrier, resulting in a strengthening (weakening) of the barrier, while only moderately affecting the barrier position.

RAPID AND LOCAL BARRIER FORMATION

Despite the gradual evolution of the current profile and all external actuators being held constant, a sudden and rapid improvement in confinement is observed, as illustrated by the line integrated soft x-ray emission shown in figure 2a. The fast transition suggests the reaching of a threshold during the magnetic shear profile evolution, which triggers the onset of the eITB[8]. Furthermore, the barrier forms in an extremely localized region in the plasma cross section, as shown from the reconstructed local emissivity profiles from a similar eITB discharge, see figure 2b. The barrier initially forms locally near ρ=0.3, later influencing the inner and outer flux surfaces. The localized formation of the barrier points toward a local improvement in confinement that could be related to the appearance of a unique flux surface in the safety factor profile, such as a zero-shear flux surface. This hypothesis has been supported by the current profile evolution simulations by ASTRA, which correlate the appearance of a zero-shear flux surface to the time and location at which the barrier appears to form[10,11].

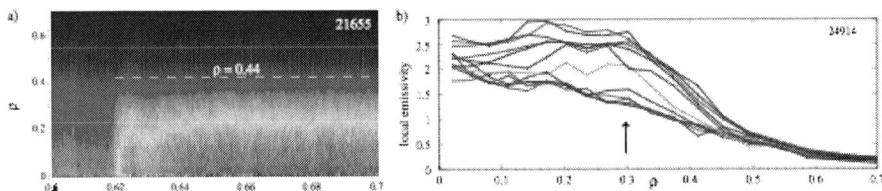

FIGURE 2. (a) Line integrated soft x-ray emission during the rapid barrier formation at t~0.62s. (b) The reconstructed local emissivity profiles at 0.75ms intervals. The barrier forms near ρ=0.3.

OHMIC PERTURBATION EXPERIMENTS

The eITBs described thus far have been sustained with a combination of ECCD and bootstrap current. The current in the ohmic transformer current was held constant thus eliminating the inductive current contribution. Reapplying inductive coil during the phase D of the eITB introduces a novel application of the ohmic current. The ohmic current is a "new" auxiliary source in these discharges, "new" in the sense that the inductive current is used as a pure current source to probe the eITB[12], with negligible power transferred to the plasma, only a few kW relative to 1.4MW for the ECCD and ECH. The steady state ohmic current profile can readily be calculated from the measured temperature profiles and equilibrium profiles[13]. Thus, the plasma current profile can be modified in a controlled manner without significantly modifying other plasma parameters, providing a method for studying the effect of the current profile on the eITB mechanics.

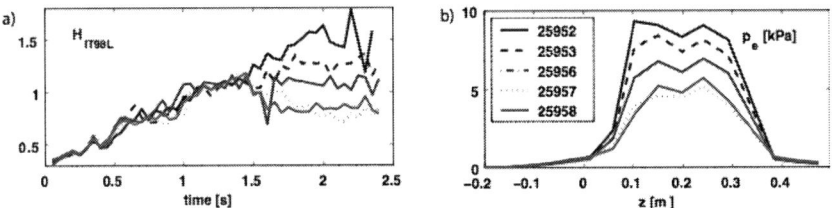

FIGURE 3. (a) The corresponding enhancement factor for the three discharges and (b) the pressure profiles during the co-counter ohmic scans. The ohmic perturbation is applied starting at 1.4s.

Practically, this is accomplished using a preprogrammed, feedback-controlled ramp in the transformer current starting at 1.4s, during the steady state central heating phase of the discharge. The resulting applied loop voltage induces either a co or counter ohmic current. The performance of the eITB is shown to improve (deteriorate) as the counter (co) ohmic current increases (decreases)[12,11], see figures 3a and b. The ohmic current profile depends on the Te profile, which is highest inside the barrier. Thus counter (co) ohmic current will deepen (fill in) the central current hole, improving (degrading) the eITB performance, consistent with the central ECCD scan[6]. Hence, the eITB performance is strengthened (or nearly destroyed) by only tailoring the current profile, without any additional input power.

CONCLUSIONS

The plasma current, completely sustained by EC driven and bootstrap currents (up to 80%), is tailored using the ECCD system. The eITB performance, characterized by the product of the barrier strength and volume, is controlled using a combination of the off-axis co-ECCD and central heating. The barrier volume is controlled via the off-axis co-ECCD deposition location, and the barrier strength by the depth of the central current hole using either counter or co-ECCD centrally deposited.

Even though the plasma current evolves from a peaked to a hollow profile on a relatively slow time scale (~200ms), the eITB forms rapidly in a very localized region of the plasma cross section, while all external actuators are held constant. The temporal and spatial location of the barrier formation correlates with the appearance of a zero-shear flux surface based on modeling of the plasma current evolution.

The ohmic current can be used as a "new" and "pure" current source, since the plasma current is fully sustained via the bootstrap and EC driven currents Preprogrammed ramps in the ohmic current have been used to introduce a perturbative current source with virtually no increase in injected power.

This work was supported in part by the Swiss National Science Foundation.

REFERENCES

1. Goodman, T.P. et al, Proceedings of the 19th Symposium on Fusion Technology, Lisbon, (1996) p. 565.
2. Coda, S. et al, Plasma Phys. Control. Fusion 42 (2000) B311.
3. Goodman, T.P. et al, Nucl. Fusion 43 (2003) 1619.
4. Nikkola, P. et al, Nucl. Fusion 43 (2003) 1343.
5. Henderson, M.A. et al, Phys. of Plasmas 10 (2003) 1796.
6. Henderson, M.A. et al, Plasma Phys. Control. Fusion 46 (2004) A275.
7. Tresset, G. et al, Nucl. Fusion 42 (2002) 520.
8. Henderson, M.A. Phys. Rev. Lett. 215001 (2004).
9. Pochelon, A. et al, Nucl. Fusion 39 (1999) 1807.
10. Henderson, M.A. et al, submitted to Nucl. Fusion.
11. Coda, S. et al, accepted for publication in Phys. of Plasmas.
12. Sauter, O. et al, Phys. Rev. Lett. 94, 105002 (2005).
13. Sauter, O. et al., Phys. Plasmas 6, 2834 (1999); 9, 5140 (2002).

The Physics Performance Of The Front Steering Launcher For The ITER ECRH Upper Port

M. Henderson[1], R. Chavan[1], P. Nikkola[1], G. Ramponi[2], G. Saibene[3], F. Sanchez[1], O. Sauter[1], H. Shidara[1], H. Zohm[4]

(1) CRPP, Association EURATOM- Confédération Suisse, EPFL, CH-1015 Lausanne, Switzerland
(2) Istituto di Fisica del Plasma, EURATOM- ENEA-CNR Association, 20125 Milano, Italy
(3) EFDA Close Support Unit, Boltzmannstrasse 2, D-85748 Garching, Germany
(4) IPP-Garching, Max-Planck-Institut für Plasmaphysik, D-85748 Garching, Germany

Abstract. The capability of any given e.m.-wave plasma heating system to be utilized for physics applications depends strongly on the technical properties of the launching antenna (or launcher). An effective ECH launcher must project a small mm-wave beam spot size far into the plasma and 'steer' the beam across a large fraction of the plasma cross section (along the resonance surface). Thus the choice in the launcher concept and design may either severely limit or enhance the capability of a heating system to be effectively applied for physics applications, such as sawtooth stabilization, control of the Neoclassical Tearing Mode (NTM), Edge Localized Mode (ELM) control, etc. Presently, two antenna concepts are under consideration for the ITER upper port ECH launcher: front steering (FS) and remote steering (RS) launchers. The RS launcher has the technical advantage of easier maintenance access to the steering mirror, which is isolated from the torus vacuum. The FS launcher places the steering mirror near the plasma increasing the technical challenges, but significantly enhancing the focusing and steering capabilities of the launcher, offering a threefold increase in NTM stabilization efficiency over the RS launcher as well as the potential for application to other critical physics issues such as ELM or sawtooth control.

Keywords: ECCD, ECH Antenna, NTM stabilisation
PACS: 52.50.Sw, 52.55.Wq

INTRODUCTION

The purpose of the ITER electron cyclotron resonance heating (ECRH) upper port antenna (or launcher) will be to drive current (ECCD) locally inside the island which forms on the q=3/2 or 2 rational magnetic flux surfaces in order to stabilize the neoclassical tearing mode (NTM). The launcher should be capable of steering the ECCD current deposition profile (j_{CD}) on the resonance surface over the range in which the q=3/2 and 2 surfaces are found, for the various plasma equilibria susceptible to the onset of NTMs (scenarios 2, 3a and 5), as shown in figure 1a. Also, j_{CD} must be narrow relative to the island width and its amplitude greater than that of the bootstrap current (j_{BS}) found outside the island in order to effectively stabilize the NTM, as illustrated in figure 1b. The ratio of these two currents, $\max(j_{CD})/j_{BS}$, provides an NTM stabilisation figure of merit (η_{NTM}). The physics objective for the launcher is to achieve $\eta_{NTM} > 1.2$ for both q=2 and 3/2 of all scenarios. The European Fusion Development Agreement (EFDA) is currently in the process of developing two launcher designs: remote steering (RS)[1] and front steering (FS) launchers. Both launchers are designed to handle 24 beams (eight 1.0MW beams in three ports, but compatible with 2.0MW injection per beam) and are being developed in parallel, with the goal of providing the optimum launcher based on ITER's engineering and physics requirements. The RS launcher[2] offers the advantage of not requiring moving parts within the vessel vacuum boundary. However, it has a limited angular range ($\leq \pm 12°$) and projects a relatively broad beam spot size (>55mm) at the resonance surface. By contrast, the FS launcher[3,4] offers an extended angular range (up to $\pm 24°$) and projects a much narrower spot size on the resonance surface, but requires a rotatable mirror near the plasma. An FS launcher is already planned for the equatorial port[5].

ECCD is the only current source on ITER that is both localized and steerable and should be used to its maximum capabilities. This implies the effective use of each MW injected, but also employing the ECCD in the widest possible range of physics applications. The FS launcher has been shown to be more effective than the RS launcher in stabilizing the NTM[3]. In addition, the steering range of the FS launcher can be increased in order to expand the physics applications beyond NTM stabilization and address additional physics issues such as control of the sawtooth, Edge Localised Mode (ELM), and Frequently Interrupted Regime (FIR).

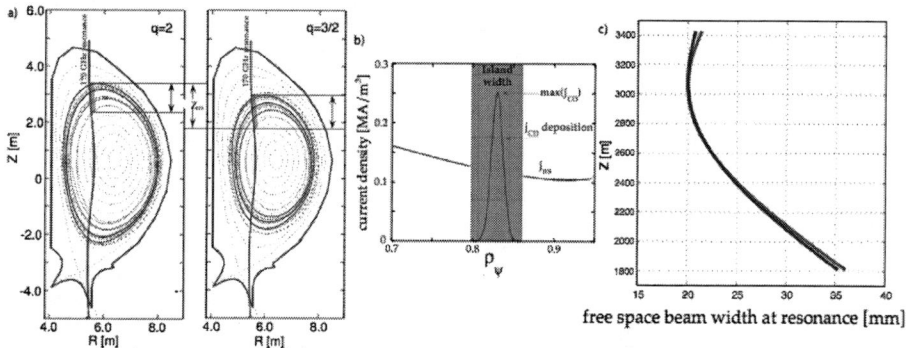

FIGURE 1. (a) The q=2 and 3/2 flux surfaces susceptible to the NTM, with Z_{res} indicating the region that the launcher must access along the resonance surface. (b) Illustration of the current density profile relative to the bootstrap current: the NTM stabilization efficiency, η_{NTM}, is given by $max(j_{CD})/j_{BS}$. (c) The free space beam spot size as a function of the vertical height along the resonance surface.

FS LAUNCHER DESCRIPTION

A simplified view of the FS launcher design is shown in figure 2. Eight circular waveguides (arranged in two rows of four) enter at the port entrance on the right. A miter bend 'dog-leg' assembly is used to direct the 8 beams (in the toroidal and poloidal directions) to one single focusing mirror, with the incident beams partially overlapping. The reflected beams are then sent downward to two separate flat steering mirrors, which redirect the beams into the plasma with a toroidal injection angle of $\beta \sim 20°$. The beams expand from the waveguide aperture (~65mm on focusing mirror), so that they can be refocused to a narrow waist (~20mm) far into the plasma (>1.6m after steering mirror). The steering mirror provides access along the resonance layer from a height of 1.8 to 3.4m (±12.6° for the beam), see figure 1a.

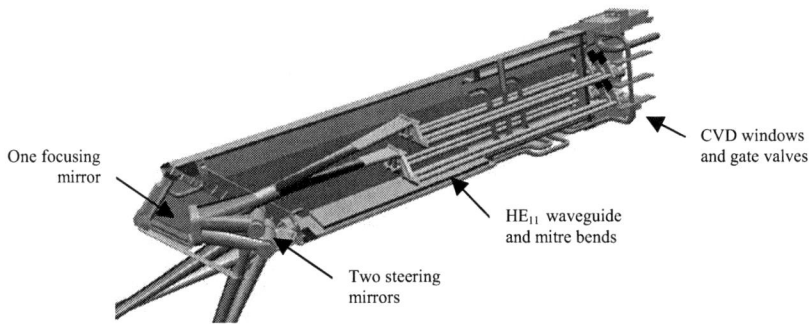

FIGURE 2. The FS launcher installed in the ITER upper port plug.

The steering mechanism is the critical component of the FS design. The steering mechanism on existing ECH launchers have been crippled when the steering mechanism grips, which typically occurs between two moving surfaces. The system proposed for the FS launcher[4] avoids all frictional surfaces. Traditional bearings are replaced with flexure pivots and the movement is controlled via a He pneumatic system using bellows pushing against springs. A coiled cooling tube with either a single or double wall is envisioned to provide a flexible coolant feed to the mirror, following a similar design to that proposed for the equatorial launcher[5] but with reduced stresses.

It is important to note that a failed steering mechanism can only be replaced during a normal tokamak opening. However, four of the six steering mechanisms could fail on the FS launcher and still provide an averaged performance equivalent to that of a fully operational 24-beam RS launcher (8 beam/3 port all purpose launcher).

NTM STABILIZATION AND OTHER PHYSICS APPLICATIONS

The achieved free space beam spot size varies from 20mm to ~44mm along the resonance surface. The resulting deposition location and driven current profile has been calculated using the ECWGB code[6] for the three ITER scenarios (l_i/β_p): 2 (0.7/0.65), 3a(0.7,1.0) and 5(1.0,0.8). Of particular interest is a narrow current density profile (j_{CD}) needed for stabilizing the Neoclassical tearing Mode (NTM). The narrow beam width of the FS launcher results in a more peaked current density profile achieving above marginal performance ($\eta_{NTM} > 1.2$) on all the q=2 and 3/2 surfaces of the calculated equilibria, see table 1, with $1.55 \leq j_{CD}/j_{BS} \leq 3.47$ and a total injected power of 20MW (assumes a transmission efficiency of 83% from gyrotron to plasma). The FS launcher provides a 3 fold increase in η_{NTM} relative to the RS launcher (averaged over all q=2 and 3/2 surfaces), implying that 6.7MW in a FS launcher system is equivalent to that of a 20MW RS launcher system.

TABLE 1. Comparison of the NTM stabilization efficiency (j_{CD}/j_{BS}) for the relevant rational surfaces and equilibria.

	Scenario EoB2		Scenario EoB3a		Scenario EoB5	
	q = 3/2	q = 2	q = 3/2	q = 2	q = 3/2	q = 2
FS Launcher	2.17	3.47	1.55	2.43	1.75	2.15
RS Launcher	0.81	1.07	0.60	0.81	0.59	0.62

At present the equatorial port launcher cannot effectively (j_{CD}>0.2MA/m^2) access the region beyond ρ_{dep}~0.4, while the upper port launcher does not extend inside ρ_{dep}~0.54. Increasing the FS scanning range downward bridges the gap allowing application of ECCD to all flux surfaces inside ρ_{dep}<0.9. This could offer control of the sawtooth instability[7] (j_{CD} deposition near the q=1 surface) or the Frequently Interrupted Regime (FIR)[8] for all potential equilibria. Alternatively, the deposition of the FS upper launcher can be extended outward for potential ELM de-stabilisation by driving current locally in the plasma edge[9]. Preliminary ray tracing calculations using TORAY-GA[10,11] have demonstrated that current can be driven over the range of 0.27 ≤ ρ_{dep} ≤ 0.98. Even though the launch position is relatively high and the path length through the plasma is long, the absorption profile (near q=1) remains narrow since the resonance surface is nearly tangential to the flux surfaces. Dedicated design and ray tracing calculations are in progress for further optimization of the launcher design.

CONCLUSIONS

The upper port launcher is being designed with the goals of optimum flexibility and reliability. Thus two possible launcher designs (RS and FS) are being investigated in parallel for the ITER upper port. The FS launcher provides a factor of 3 increase in the NTM stabilization efficiency over that offered by the RS launcher. The steering range of the FS launcher can be increased to provide overlap with the Equatorial port launcher and permit extended physics applications beyond NTM stabilization, such as sawtooth, FIR and ELM control.

This work is supported in part by the Swiss National Science Foundation and EFDA.

REFERENCES

1. Moeller, C. P., "A method of remotely steering a microwave beam launched from a highly overmoded corrugated waveguide", Proc. 23rd Int. Conf. on IRMMW, 1998.
2. Verhoeven A. G. A.,. "Design of the mm-wave system of the ITER upper port launcher", Proc. 13th ECE & ECH, Nizhny Novograd (2004).
3. Henderson, M. A., Chavan, R., and Sanchez, F., "FS Launcher Study", Lausanne Research Report, LRP 791/04
4. Chavan, R., Henderson, M. A., and Sanchez, F., "An Alternative ECRH Front Steering Launcher for the ITER Upper Port", submitted to Fusion Engineering and Design.
5. Takahashi, K. *et al*, FED, 66 (2003) 473.
6. Farina, D. *et al*, "ECWGB: a beam tracing code for EC heating and current drive", Report FP 03/06 (October 2003), http://www.ifp.cnr.it/
7. Angioni, C. *et al*, Nucl. Fusion **43** (2003) 455-468.
8. Gunter, S. *et al*, Nucl. Fusion **44** (2004) 524-532.
9. Nave, M.F. *et al*, Nucl. Fusion **39** (1999) 1567.
10. Matsuda, K., IEEE Trans. Plasma Sci., **PS-17**, 6 (1989).
11. Cohen, R.H., Phys. Fluids **44**, 139 (2002).

Performance of the ECH Transmission Lines and Launchers in DIII–D

K. Kajiwara,[1] A. Mui,[2] C.B. Baxi, J. Lohr, I.A. Gorelov, M.T. Green, D. Ponce, and R.W. Callis

General Atomics, San Diego, California USA
[1]*Oak Ridge Institute for Science Education, Oak Ridge, Tennessee USA*
[2]*Rensselaer Polytechnic Institute, New Your, New York USA*

Abstract. The efficiency of the transmission line for the 110 GHz ECH system was measured in DIII–D using a low power rf source. The measured efficiency was about 10% lower than expected from theoretical analysis of the components. The launcher temperature increase during rf pulses was measured and the peak mirror surface temperature was inferred from a simulation.

INTRODUCTION

The ECH/ECCD transmission system on the DIII–D tokamak is comprised of six corrugated waveguide lines and three launchers. Each waveguide is designed for 1.0 MW, 10 s. pulses at 110 GHz and each launcher delivers power from two of the waveguides. Losses in the 31.75 mm diameter evacuated transmission system are expected theoretically to occur mainly at miter bends, approximately 1% per miter bend [1]. About 90% of the miter loss is due to mode conversion and 10% is resistive, which depends on the polarization orientation with respect to the plane of the miter bend. Measurements described in [1] on six mite bends assembled in series gave 0.4 dB (0.92%) loss per miter, in good agreement with the theoretical analysis. But low power measurements of the performance of complete transmission lines about 90 m in length and containing up to 12 miter bends indicated lower efficiency, prompting the present series of efficiency measurements.

The launchers are designed without water cooling for simplicity and avoidance of water leakage inside the tokamak vacuum vessel. With only radiative and limited conductive cooling, the launcher temperature must be monitored to prevent the melting of the rf reflecting surfaces of the mirrors. Although in normal long pulse operation over the course of a day, the surface temperatures are predicted to remain within limits, the mirror surface temperatures could be increased abnormally by surface arcing or plasma disruptions.

Poloidal and toroidal steering of the rf beams is provided using movable mirrors. Eddy current induced forces arising during disruptions are particularly problematic for the actuator assemblies on the movable mirrors, which have limited ability to react the forces. Therefore, mirror designs which minimize the volume of high conductivity copper while maintaining low resistivity reflecting surfaces have been developed. Because of the small volume of the metal, the thermal inertia of the movable mirror is relatively poor and particular attention must be paid to this component.

TRANSMISSION EFFICIENCY MEASUREMENT

The transmission efficiency measurement was performed using a low power rf source and diode detector. A schematic representation of the setup is shown in Fig. 1. The rf was generated by a 10 mW gun oscillator the output of which is converted from the $TE_{1,0}$ rectangular mode to the $HE_{1,1}$ hybrid mode that is used in the DIII–D ECH transmission line. The converted $HE_{1,1}$ wave passes through a phenolic mode filter, which reduces the undesired surface mode admixture during propagation. The filtered rf enters a corrugated wave guide section which can be a short piece of wave guide, ~50 cm long, for calibration or an actual transmission line or component for measurement. The receiver is a diode detector with another phenolic surface mode filter, an $HE_{1,1}$–$TE_{1,0}$ mode converter and an isolator in an arrangement symmetric to the transmitter. In this setup, some reflections at the receiver are unavoidable even with the use of an isolator before the detector. Therefore, standing waves can introduce errors in the measurement at fixed frequency. In order to minimize these errors, the frequency is swept. Figure 2 shows the frequency dependence of the detector signal for a short wave guide calibration test. The horizontal axis corresponds to a 0.2 GHz frequency sweep. The general trend toward larger received signal at higher frequency is primarily due to the frequency dependence of the source output power, but includes the uncalibrated frequency response of the detector.

In order to reduce errors from standing waves, the transmission efficiency was obtained by averaging the signal over the frequency sweep. The three step procedure was as follows: (1) Measure the frequency averaged signal with a short wave guide section (calibration). (2) Measure the frequency averaged signal with the component under test (measurement). 3) Return to the calibration setup and repeat the measurement. After verifying in (3) that there had been no change in generated power, the precision attenuator is adjusted to match the received signal from the "measurement". The attenuation required to match the signal in (2) then indicates the transmission efficiency. Twelve separate calibrations over a five day period showed ±1% scatter in the calibration results. Therefore, the error of the measured efficiency is ±2%.

Using this technique, efficiency measurements were made on complete transmission lines, segments of lines and combinations of components. In order to connect with the measurements performed in [1], the transmission efficiency of six miter bends connected series was measured. In contrast with the previous measurements wchich gave, 0.92% loss per miter, the recent measurements gave 1.5% loss per miter. Straight waveguide loss had been calculated at 5% per 100 m for the 31.75 mm diameter corrugated wave guide at 110 GHz. Polarizing miter bends were calculated to have 1.5% loss per miter but were not measured separately. Using the theoretical calculations plus measurements on the existing components and waveguide lines, Fig. 3 was constructed to compare the expected and measured efficiencies. The measured losses are seen to be at least twice the theoretical expectation and in some cases 3.5 times the expectation.

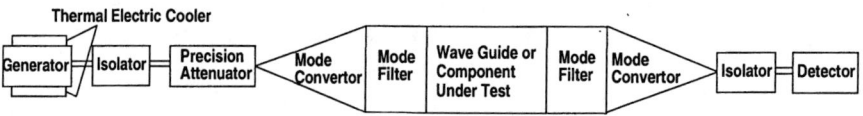

FIGURE 1. Schematic view of the set up.

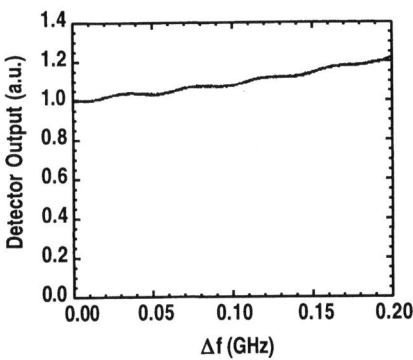

FIGURE 2. Frequency dependence of the calibration setup.

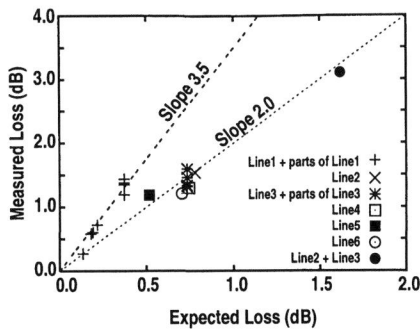

FIGURE 3. Transmission line efficiency. The horizontal axis is calculated using 1% loss per miter bend, 1.5% loss per polarizer and 5% W/G loss per 100 m.

Several additional tests were performed to attempt to identify problems with the measurements. One check was to fill the test waveguide with nitrogen to eliminate any moisture in the line. Another was to exchange the input and output mode converters, on the assumption that one converter could be generating appreciable non-$HE_{1,1}$ power. If one mode converter were faulty, the non-$HE_{1,1}$ mode will decay in the waveguide, and the interchanged converter setup will show a different transmission efficiency from the original. The angular distribution of rf which is radiated from the generator assembly (rf source + mode converter + mode filter) was also measured. In all checks, no evidence was found pointing to an error in the measurements. These investigations are continuing.

LAUNCHER TEMPERATURE MEASUREMENT

The temperatures of the launcher mirrors were inferred from measurements using Resistance Temperature Detectors (RTDs), which are attached to the back surfaces of the mirrors. In order to infer the peak surface temperature from the back surface measurements, it is necessary to simulate the thermal propagation of the mirror. The movable launcher mirrors, which have the poorest heat transfer to ambient, have a unique laminated design, which is shown in Fig. 4. A thin Glidcop reflecting surface is supported by the laminated structure of stainless steel and Glidcop. The stainless steel serves to reduce the eddy current forces at disruptions and to increase the mechanical strength. For such a complex structure, a three dimensional thermal simulation is required. The simulation, performed with the finite element code COSMOS, includes the temperature dependence of the electrical resistivity for copper. The resistance doubles at 800°C compared to room temperature.

A recent study has shown the necessity for calibration of the thermal impedance between the mirror and the RTD since the usual thermal grease cannot be used [2]. The thermal impedance is determined by comparing the simulation and a calibration in which a point on the mirror surface is held at 0°C and the RTD response is measured. Using the thermal impedance determined in this way, the surface temperature during rf injection can be estimated by comparing the simulation and the measurement. There was agreement between measurements and simulation, when the temperatures were compared 200 s after the pulse thus validating the model.

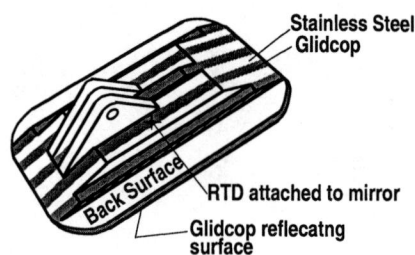

FIGURE 4. Schematic view of the movable mirror.

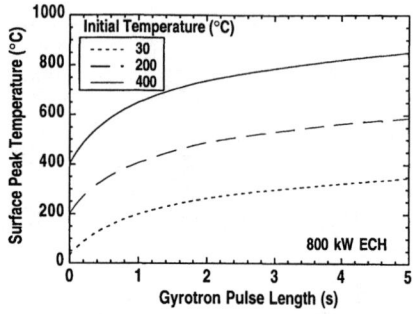

FIGURE 5. Peak surface temperature as a function of the gyrotron pulse length. The initial temperature was changed in the three cases.

Based on this result, the peak surface temperature of the mirror at the center of the rf beam can be calculated as a function of the pulse length with the initial temperature as a parameter (Fig. 5). The rf beam radius on the movable mirror was 5.5 cm and the rf power absorption fraction was assumed to be 0.21% with the angle incidence θ defined as the angle between the normal to the surface and the rf beam of 56°. The absorbed power scales as $1/\cos\theta$. As shown in Fig. 5, if the initial temperature is 400°C, the mirror surface temperature will exceed the maximum target value of 800°C in less than 4 s. Usually, the baseline temperature of the movable mirror increases by about 100°C at the end of a day of operation with many 2 s rf pulses and injected power of 700 kW. Therefore, the surface temperature will never exceed 600°C. However, there is a possibility of occasional arcs at the surface, which can cause an anomalously high baseline temperature. To prevent such events, an alarm with set point of 300°C as measured by RTDs has been installed.

SUMMARY

Transmission efficiency measurements using a low power rf source shows a lower transmission line efficiency for the DIII–D ECH system than predicted by theory. There is a possibility that high order modes have contaminated the measurements. This investigation continues. The launcher temperature simulation shows that even if the initial mirror temperature is 300°C, the peak temperature during an 800 kW, 5 s pulse is less than 800°C.

ACKNOWLEDGMENT

Work supported by U.S. Department of Energy under DE-FC02-04ER54698, DE-AC05-76OR00033. The launcher mirrors were produced by the Princeton Plasma Physics Laboratory.

REFERENCES

1. J. L. Doane, C. P. Moeller, Int. J. Electronics **77**, 489–509 (1994).
2. K. Kajiwara, C. B. Baxi, et al., Proc. 15th Top. Conf. on Radio Frequency Power in Plasmas, 325–330 (2003).

Long Pulse ECH Plasma in LHD

S. Kubo*, Y. Yoshimura*, T. Shimozuma*, H. Igami*, T. Notake*,
R. Kumazawa*, T. Seki*, K. Saito*, Y. Nakamura*, T. Mutoh*, K. Ohkubo*
and LHD Experimental Group*

*National Institute for Fusion Science, 322-6 Oroshi cho, Toki 509-5292, Japan

Abstract. Demonstration of a long pulse or a steady state operation of ECH and sustainment of non-collapsed plasma only by ECH is important in LHD from both the confinement device and the heating system engineering points of view. A gyrotron with a diamond output window is introduced and operated at the power level of 150 kW for more than 1 hour after modification of the cooling and evacuation system of the ECH transmission line. The power of about 110 kW injected into LHD is used to sustain the plasma with the electron density of 1.5×10^{18} m^{-3} and central temperature of more than 1.0 keV for 3900 sec. The gas puffing rate is carefully controlled so that the plasma density does not exceed the critical value above which the plasma collapsed for given injection power, magnetic configuration and wall condition of LHD. The results of gyrotron operation, transmission system modification for long pulse and optimizations of the magnetic field configuration of LHD and gas puffing for a given injection condition are discussed.

INTRODUCTION

Long pulse ECRH discharge using gyrotron at the power level of the order of 100 kW has been performed in the helical systems. The demonstration of long pulse or steady state operation has been one of the most important issues in the helical system, due to its non-disruptive nature. In early 90s, ATF in ORNL demonstrated 4667 s operation of 28 GHz 70 kW gyrotron sustaining electron density of 3×10^{18} m^{-3} but very low electron temperature. In LHD, 1200 s plasma sustainment is demonstrated at 1.0×10^{18} in 2000, by 84 GHz 50 kW injection under 95% duty operation of the gyrotron[1]. A CW gyrotron capable of delivering 200 kW at MOU is introduced in LHD. After upgrading transmission system in LHD, the operation of this CW gyrotron at 84 GHz up to 756 s with 72 kW injection power was successful and sustained the plasma at the density of 2.4×10^{17} m^{-3} and electron temperature of 240 eV in the beginning of 2004[2]. This paper reports further improvement of the transmission system and the results of extension of plasma duration up to 3900 s at the injection power level of 110 kW.

CW ELECTRON CYCLOTRON HEATING SYSTEM IN LHD

8-gyrotrons with 6-3.5 inch and 2-1.25 inch waveguide transmission system is routinely used in LHD, mainly for high electron temperature plasma production and sustainment[2]. CW gyrotron is connected to one of the 1.25 inch evacuated waveguide systems via waveguide switch.

Since the antenna that was originally used for the 1.25 inch waveguide system has stainless steel mirrors without active cooling inside LHD, the corrugated horn antenna placed near the mirror antenna is used for the CW experiment. Up to the 756 s operation, several components have been over heated and damaged. The temperature rise of the transmission components also made the excess out gassing from the waveguide wall that had limited the pulse duration of the injection.

In order to reduce the temperature rise in the waveguide system, almost all the straight waveguides are covered with copper plates with a water cooling pipe. The comparison of the temperature rise in several parts of the waveguide transmission line before and after the enforcement of the cooling are shown in Fig. 1 a). Due to a poor gas conductivity in the 1.25 inch waveguide, the excess out gas from the wall cause severe arcing problems in the waveguide. In order to keep the pressure in the waveguide less than 1.0 Pa, evacuation section is newly developed and nine of this type of pumping section are distributed along the 62 m waveguide system. The comparison of the pressure rise in several parts of the transmission line before and after these improvements are shown in Fig. 1 b).

The DC break with a small gap in the waveguide and vacuum sealed by a ceramic disk had been used. The leakage microwave have heated this disk locally and finally made a crack which lead the break in the vacuum. This ceramic is replaced by an aluminum disk with the same size but alumite coated for electric insulation.

Out gassing from the waveguide wall decreased due to the efficient conditioning by increased pumping rate and enforced cooling. Saturation pressure level after long pulse operation decreased shot by shot.

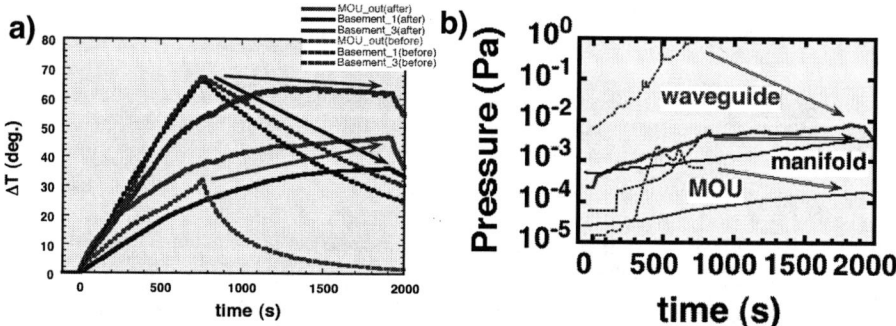

FIGURE 1. a) Time evolution of temperature rise in the transmission components during long pulse operation before enforcing the cooling (dotted lines) and after (solid lines). b) Time evolution of the pressure in the transmission line at MOU, mid-way on the waveguide line, and evacuation manifold before (dotting lines) and after (solid lines) improvements.

LONG PULSE DISCHARGE

After basic short pulse experience using the same configuration as CW experiment, magnetic field of 1.48 Tesla at the magnetic axis of 3.6 m is selected for the CW operation. This magnetic configuration corresponds to the second harmonic heating. Maximum available power was about 110 kW. Longest continuous power injection shot is shown in Fig. 2 a). The injection power is estimated from the measurement at the dummy load near the LHD. Previous long pulse experience show that the gas feed control is critical to keep as high density as for the power level of 100 kW without radiation collapse in LHD. The gas puffing rate is finely adjusted manually by mass flow controller. In the beginning 2000 s is devoted to keep constant density at 1.5×10^{18} m^{-3} that is about half of the critical density for available power. Electron temperature measured by ECE radiometer is kept more than 1.5 keV at the center. Although ECE is still optically grey in this low density regime, Thomson scattering data available in the beginning of this shot supports this measurement.

After the elapsed time of 2000 s, gas puffing rate is increased several times to try higher density. In Fig. 2 b) are shown expanded trace of this first trial from 2000 to 2300 s. Gas puffing rate is increased linearly from 2050 to 2100 s and kept at 0.006 Pa m^3 s^{-1} after that, but the density is kept increasing, indicating that the density exceeded the critical level for radiation collapse. Similar trails are performed several times within the same discharge. The shot is terminated manually by the stop of ECH injection due to just the limitation of the data acquisition, but not from any hardware troubles.

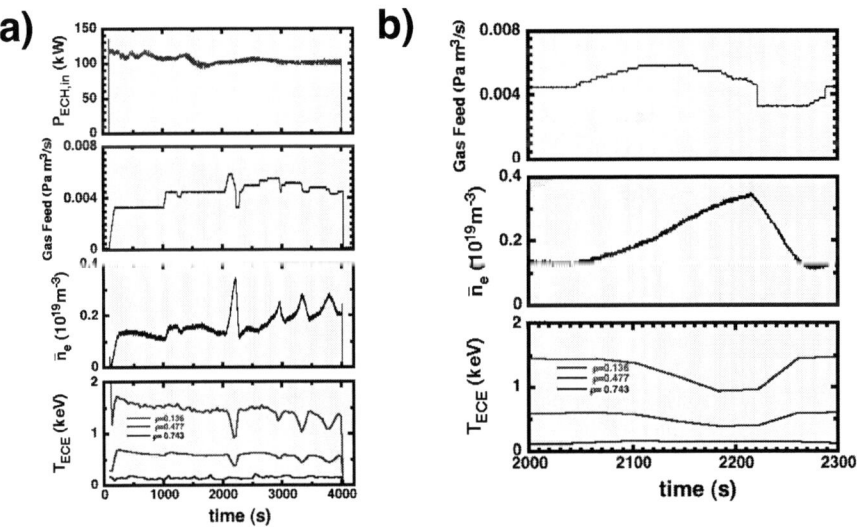

FIGURE 2. a) 65 minutes discharge. From top to bottom, time evolutions of injection power, gas feed, electron density, and electron temperature from ECE at $\rho = 0.136$, 0.477 and 0.743. b) Time expanded waveform from 2000 to 2300 s of a).

Gyrotron is stably operated for 3900 sec. As shown in Fig. 3 a), power supply for the beam, body voltages is well regulated and the beam current is maintained at 9 to 10 A by manually controlling the heater power. Almost all parameters of the gyrotron attained at saturation level as shown in Fig. 3 b), here, are shown the temperature differences in each cooling channel of the CW gyrotron components, saturation power level for each is also shown. The calibrated power at the output of MOU was 163 kW. The waveform of the power monitor set at second and last second miter bend is shown in Fig. 3 c). The gyrotron oscillation and the modes in the transmission line are stabilized after 1500 sec.

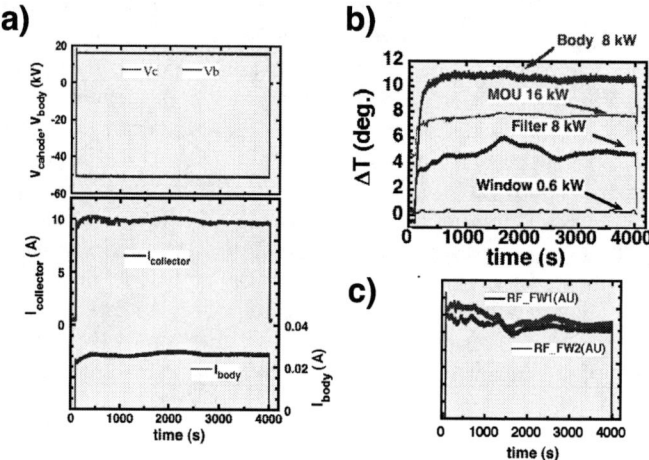

FIGURE 3. a) CW Gyrotron attains full steady state conditions at the output power of 160 kW at MOU. The heater power is manually controlled so as to keep the beam current at near 9 A. Time traces of body and cathode voltage (top), beam current and body current (bottom). b) Temperature differences and estimated power loss at each component of gyrotrons during 3900 s operation. c) Waveform of the power monitor set at gyrotron side (FW1) and LHD side (FW2).

CONCLUSION

After several improvements in the transmission line, full CW operation of the ECRH system at the power level of 110 kW became possible in LHD. This CW power injection successfully sustained the plasma of central electron temperature 1.5 keV, and density of 1.5×10^{18} m^{-3} for 65 minutes.

REFERENCES

1. T. Shimozuma, et al., *Fusion Engineering and Design*, **53**, 525–536 (2001).
2. S. Kubo, et al., *Plasma Physics and Controlled Fusion*, **47**, A81–A90 (2005).

Formation of Spherical Tokamak by ECH without Center Solenoid in the LATE Device

H. Tanaka*, M. Uchida*, T. Yoshinaga*, J. Yamada*, S. Yamaguchi[†] and T. Maekawa*

*Graduate School of Energy Science, Kyoto University, Kyoto, Japan
[†]Graduate School of Science, Kyoto University, Kyoto, Japan

Abstract.
In the LATE device, Spherical Tokamak (ST) plasmas are formed by electron cyclotron heating (ECH) without center solenoid. Two types of experiments are described: (1) slow formation of ST plasmas and (2) spontaneous formation of ST plasmas with rapid-current-rise. A ST formation scenario is discussed including the current-drive mechanism, particle confinement and MHD equilibrium.

Keywords: Electron Cyclotron Heating, Spherical Tokamak, Non-inductive Current Start-up
PACS: 52.50.Sw, 52.55.Fa, 52.55.Wq

INTRODUCTION

Removing the center solenoid from the fusion reactor based on Spherical Tokamak (ST) concept is desirable, because there is a limited central space for achieving the low aspect ratio [1, 2]. So providing a start-up scenario without inductive method is one of the important issues in ST research. Electron cyclotron heating (ECH) is one of candidates because breakdown and current initiation can be carried out simultaneously and the required equipment for microwave power injection is only a small launcher remote from the plasma.

In the LATE device, which has no center solenoid, it has been demonstrated that a ST plasma can be formed by using only a microwave power near the electron cyclotron range of frequency [3, 4]. In this report, we describe two types of experiments, (1) slow formation of ST plasmas and (2) spontaneous formation of ST plasmas with rapid-current-rise, and also discuss the current-drive mechanism in conjunction with particle confinement and MHD equilibrium.

SLOW FORMATION

The LATE device has a cylindrical vacuum chamber with an inner diameter of 1 m and a height of 1 m. The center pole has a diameter of 11.4 cm and encloses a 60 turn toroidal field coil. There are 3 pairs of vertical field coils and each coil current is controlled by pre-programming respectively. The plasma current is measured by flux loops and the current profile is obtained by least-square-error method with model current distribution. Figure 1 shows the time evolution of slow formation of a ST plasma. In this experiment,

2.45 GHz microwave power is fed from 3 magnetrons and injected obliquely to the toroidal field with the electric field vector on the equatorial plane from outboard side. The toroidal coil current is 60 kAT and the fundamental EC resonance layer locates at $R = 13.7$ cm. The breakdown occurs under a steady vertical field ($B_V = 12$ G) with a weak microwave power of ~ 0.5 kW. With natural decrease of the neutral gas pressure and by increasing the microwave power P_{inj}, plasma current increases up to 2 kA and the initial closed flux surfaces are formed. Then, B_V is increased slowly at a rate of ~ 47 G/s and P_{inj} is also increased up to 30 kW in total. The plasma current increases almost proportionally to B_V satisfying the MHD equilibrium condition. At the last stage, B_V is kept constant and steady plasma current of 7.2 kA is maintained till the end of the microwave pulse. The last closed flux surface has an aspect ratio of $R_0/a = 20.5$ cm $/14.8$ cm $= 1.4$, an elongation of $\kappa \simeq 1.6$ and $q_{edge} \simeq 34$. The current amounts 12 percent of the toroidal coil current of 60 kAT. Hard X-ray emissions with photon energy of ~ 30 keV increase strongly above $I_p \simeq 2$ kA, showing the production of large fraction of high energy tail electrons. The plasma current center locates near the 2nd harmonic EC resonance layer and the line-averaged electron density certainly exceeds the plasma cutoff density. These facts suggest that the 2nd harmonic ECH by the mode-converted electron Bernstein wave takes place. During the B_V ramp, the decay index of B_V is decreased from 0.1 to 0.02 to obtain large elongation of plasma cross section and large plasma current. A feedback control of vertical position is operated and it is essential to avoid the disruption by outgas due to direct loss of high energy tail electrons to the wall.

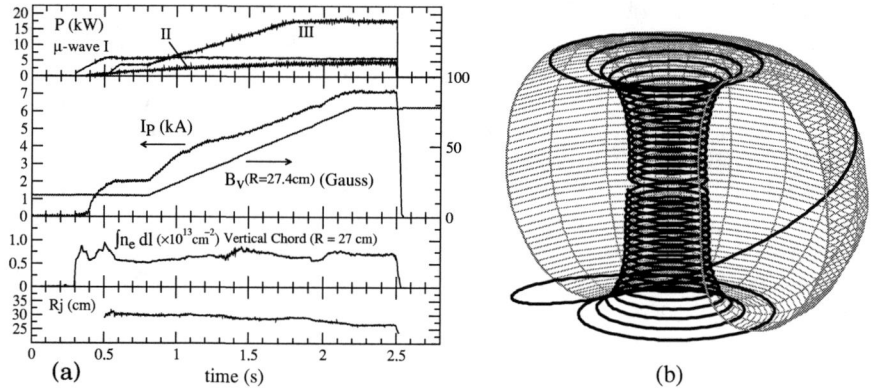

FIGURE 1. (a) Waveforms in slow formation with 2.45 GHz microwave. (b) A magnetic field line on the last closed flux surface at the last stage ($t = 2.3$ s, $I_p = 7.2$ kA)

SPONTANEOUS FORMATION

Under a steady vertical field, spontaneous formation of ST plasmas occurs. In Figure 2, the waveforms are shown for the injection of 5 GHz, 130 kW, 60 ms microwave pulse at $B_V = 85$ G. The plasma current begins to flow after the breakdown and increases slowly up to 2.1 kA. At this moment of $t = 89.5$ ms, the current profile is elongated vertically near the 2nd EC resonance layer and the field lines are helix (open field configuration).

Then the 1st rapid-current-rise occurs and I_p increases from 2.1 kA to 4.1 kA within 1.5 ms ($t = 91.7$ ms). The current profile expands to the high field side and a small closed flux surface attached to the center post appears. After the 2nd rapid-current-rise ($t = 94.7$ ms), I_p reaches 7.4 kA. Both the current profile and the closed flux surfaces expand to the low field side, while the center post limits the high field side. At the final steady stage ($t = 122.5$ ms), the current profile is detached from the center post and expands to the outboard chamber wall. At this time, 2/3 of the total current flows outside of the last closed flux surface. Weak hard X-ray emissions are observed even before the rapid-current-rise. At just before the rapid-current-rise, X-ray measurement by absorption method shows that the photon energy increases from ~ 20 keV to ~ 40 keV as B_V is increased from 14 G to 30 G. This fact suggests that high energy tail electrons are produced before the rapid-current-rise and their energy increases with B_V, which is consistent with the single particle confinement discussed in the later section.

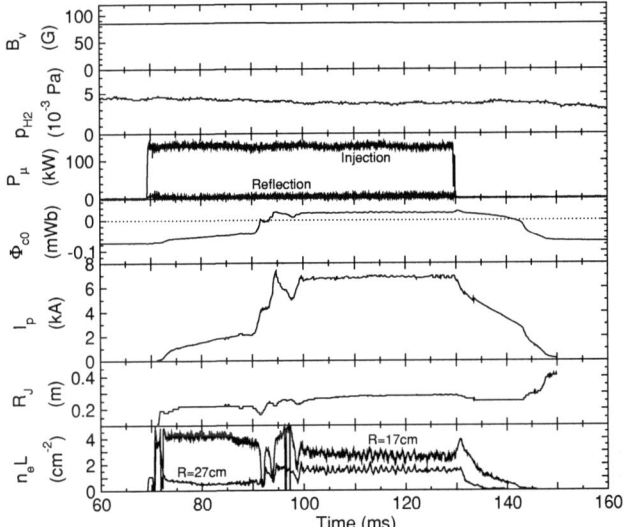

FIGURE 2. Waveforms in spontaneous formation with a 5 GHz, 130 kW, 60 ms microwave pulse

DISCUSSION

In the open field configuration, a toroidal current can be driven by plasma pressure such that the charge separation caused by the toroidal drift is canceled out and it is inversely proportional to B_V [5]. Assuming the current flows in the torus whose major and minor radii are R and a respectively, the relation between I_p and B_V can be written as $\bar{I}_p = 1/\bar{B}_V$, where $\bar{I}_p = \sqrt{\frac{\mu_0}{8} \frac{I_p}{\pi a \sqrt{<p>}}}$ and $\bar{B}_V = \sqrt{\frac{2}{\mu_0} \frac{(R/a) B_V}{\sqrt{<p>}}}$.

Early stage before the rapid-current-rise, observed I_p is proportional to the electron pressure measured by Langmuir probe. Then, in this stage, plasma current is driven by plasma pressure.

The time scale for particle and energy confinement and radial force balance is roughly estimated as several tens of microseconds, while time scale for increasing plasma current is several milliseconds. Then, the plasma is in equilibrium during the current rise and must be satisfy the radial force balance condition. The vertical field necessary for this force balance is expressed as $\bar{B}_V = A_f \bar{I}_p + 1/\bar{I}_p$, where A_f is a form factor determined by the plasma shape and internal inductance [6]. The first term in the right side corresponds to the vertical field which balances the current hoop force and is proportional to I_p. The second term corresponds to the vertical field which balances the pressure ballooning force and is inversely proportional to I_p.

When the plasma current is begun to flow, I_p is small and the second term is dominant (pressure-ballooning-force-dominant region). The pressure-driven current scales as $\bar{I}_p = 1/\bar{B}_V$ and satisfies the equilibrium. In the usual tokamak operation with closed flux surfaces, I_p is large and the first term is dominant (current-hoop-force-dominant region). Experimentally, the slow formation of ST plasmas shows that the equilibrium condition is in the current-hoop-force-dominant region. And the rapid-current-rise and spontaneous formation of closed flux surfaces overcomes the transition from the pressure-ballooning-force-dominant region to the current-hoop-force-dominant region. Therefore, there may be another current drive mechanism which cause the rapid-current-rise and bring the plasma to the current-hoop-force-dominant region.

A possible mechanism is anisotropic confinement of electrons in the velocity space [7]. In the open field configuration, the vertical motion of electrons is sum of the toroidal drift and the vertical component of the parallel motion along the field line. The electrons whose toroidal drift is cancelled by the vertical component of the parallel motion along the field line can be confined in toroidal direction and carry the forward current. This anisotropic confinement characteristics and quasi-linear diffusion by ECH will produce anisotropic velocity distribution and the toroidal plasma current will increase. Once the plasma current increases, the self poloidal field modify the field lines and the good confinement area becomes wider in the velocity space. The local magnetic mirror is produced and precession current by the trapped electrons may also drive the forward current. When the self poloidal field exceeds the external vertical field, a small closed flux surface appears in the high field side. Then the broadening of the current channel in the low field side occurs and the rapid-current-rise may take place.

REFERENCES

1. Y.-K. M. Peng, C. A. Neumeyer, P. J. Fogarty, et al., *Proc. 20th IAEA Fusion Energy Conf.*, IAEA-CN-116/FT/3-1Rb, Vilamoura, Portugal, 2004.
2. S. Nishio, K. Tobita, K. Tokimatsu, et al., *Proc. 20th IAEA Fusion Energy Conf.*, IAEA-CN-116/FT/P7-35, Vilamoura, Portugal, 2004.
3. M. Uchida, K. Higaki, T. Yoshinaga, et al., *J. Plasma Fusion Res.*, **80**, 83 (2004).
4. T. Maekawa, H. Tanaka, M. Uchida, et al., *Proc. 20th IAEA Fusion Energy Conf.*, IAEA-CN-116/EX/P4-27, Vilamoura, Portugal, 2004.
5. C. B. Forest, Y. S. Hwang, M. Ono, et al., *Phys. Plasmas*, **1**, 1568 (1994).
6. V. D. Shafranov, *Review of Plasma Physics*, Vol.2, p.117.
7. T. Shimozuma, J. Takahashi, H. Tanaka, et al., *J. Phys. Soc. Jpn.*, **54**, 1360 (1985).

RF PLASMA APPLICATIONS

Acceleration of Dense Flowing Plasmas using ICRF Power in the VASIMR Experiment

Jared P. Squire for the VASIMR Team

Muñiz Engineering, Inc., Advanced Space Propulsion Laboratory, NASA Johnson Space Center, Houston, TX 77058

Abstract. ICRF power in the Variable Specific Impulse Magnetoplasma Rocket (VASIMR) concept energizes ions (> 100 eV) in a diverging magnetic field to accelerate a dense (~ 10^{19} m^{-3}) flowing plasma to velocities useful for space propulsion (~100 km/s). Theory predicts that an ICRF slow wave launched from the high field side of the resonance will propagate in the magnetic beach to absorb nearly all of the power at the resonance, thus efficiently converting the RF power to ion kinetic energy. The plasma flows through the resonance only once, so the ions are accelerated in a single pass. This process has proven efficient (~ 70%) with an ICRF power level of 1.5 kW at about 3.6 MHz in the VASIMR experiment, VX-30, using deuterium plasma created by a helicon operating in flowing mode. We have measured ICRF plasma loading up to 2 ohms, consistent with computational predictions made using Oak Ridge National Laboratory's EMIR code. Recent helicon power upgrades (20 kW at 13.56 MHz) have enabled a 5 cm diameter target plasma for ICRF with an ion flux of over 3×10^{20} s^{-1} and a high degree of ionization. This paper summarizes our ICRF results and presents the latest helicon developments in VX-30.

Keywords: ICRF, single pass, magnetic beach, helicon, plasma propulsion, experiment
PACS: 52.50.Dg, 52.50.Qt, 52.75.Di

INTRODUCTION

The Space Shuttle Main Engines (SSME) are an incredible achievement of rocket development and engineering. They expel propellant at velocities approaching 5 km/s, about the best a chemical rocket engine can do. When performing trajectory analysis for interplanetary travel,[1] these rockets require huge amounts of propellant to perform missions, which limits the capabilities of our space program. To achieve higher velocities, we must develop a high power (MW) electrically driven plasma rocket. One such system is the Variable Specific Impulse Magnetoplasma Rocket (VASIMR).[2] In space we are power limited, so we must use the power most efficiently, which leads to optimum propellant velocities. Propellant velocities of about 30 to over 100 km/s enable efficient missions. This corresponds to ion kinetic energies of 100 to 1000 eV, depending on the propellant.

These electric plasma rockets must also be highly reliable and operate for possibly years. The VASIMR approach has adopted an rf-driven magnetized plasma for both the plasma source and acceleration. Coupling high power density (> 5 MW/m^2) to the plasma with antennas could minimize material erosion that limits most thrusters' life. A helicon[3,4] is our choice for the plasma source. Ion Cyclotron Resonant Frequency

(ICRF) acceleration of the plasma flow produces the necessary ion energies. This paper presents recent results that clearly demonstrate an efficient ICRF acceleration effect in a dense (10^{19} m^{-3}) plasma flow. Also, we present recent advances in the helicon plasma source performance.

EXPERIMENTAL CONFIGURATION

The data were taken in the VASIMR experiment (VX-30) that is shown in Fig. 1. There are three basic stages in the device, helicon plasma source, ICRF acceleration and then magnetic expansion. Figure 1 also shows a typical axial magnetic field profile. The magnetic field is generated by four cryogenic electromagnets that are integrated into the vacuum chamber and are driven independently. The peak magnetic field capability is 1.3 tesla. The helicon plasma source section has a magnetic field of about 0.15 T. The magnetic field peaks downstream and then plateaus at about 0.5 T where the ICRF antenna is located. The magnetic field then expands into a large vacuum chamber. A quartz tube with an inner diameter of 9 cm is sealed against the upstream end plate and passes through the helicon antenna. It stops just upstream of the ICRF antenna. A gas baffle is located at the peak of the magnetic field to help prevent neutral gas from escaping the plasma source and to act as a plasma limiter for the ICRF antenna. Deuterium gas is regulated through the end plate into the quartz tube.

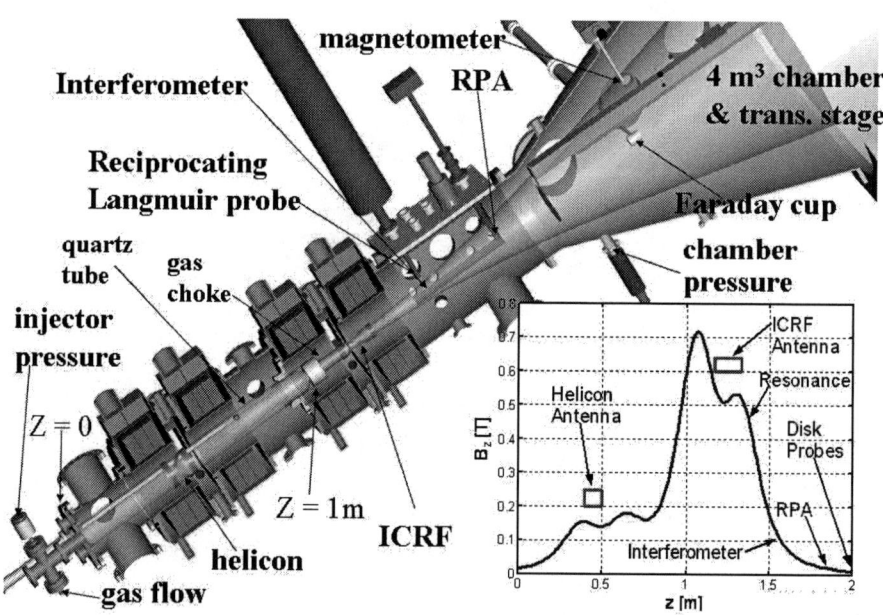

FIGURE 1. The VX-30 device and a typical axial magnetic field profile.

The total rf power capability of the system is approximately 30 kW. The helicon plasma source is driven by a 25 kW rf transmitter at 13.56 MHz. The ICRF system is driven by a 1.5 kW transmitter operating from 2 to 4 MHz; the drive frequency for ICRF powered experiments is 3.6 MHz. Plasma loading measurements were performed using a network analyzer and the high Q resonant impedance matching circuit.

The primary plasma diagnostics include a density interferometer located approximately 0.2 m downstream of the ICRF antenna, a reciprocating Langmuir probe at the same location, two independent retarding potential analyzers (RPA) for ion energy measurements, a swept rf compensated Langmuir probe to measure local plasma potential, and an array of six planar Langmuir probes made of tungsten disks facing into the plasma flow and biased in ion saturation. Neutral pressure is monitored at the gas injection and in the expansion chamber.

Figure 2 shows the temporal behavior of a typical discharge. The pulse length is limited to about one second, mostly because of the neutral pressure build up in the expansion chamber. Once the deuterium gas flow reaches the set point, the helicon rf power is applied. When the discharge is established, the pressure at the injection point increases by almost a factor of 10 to over 0.3 torr. We apply the ICRF power once the discharge appears to reach a steady state. The ICRF pulse lasts for typically 300 ms.

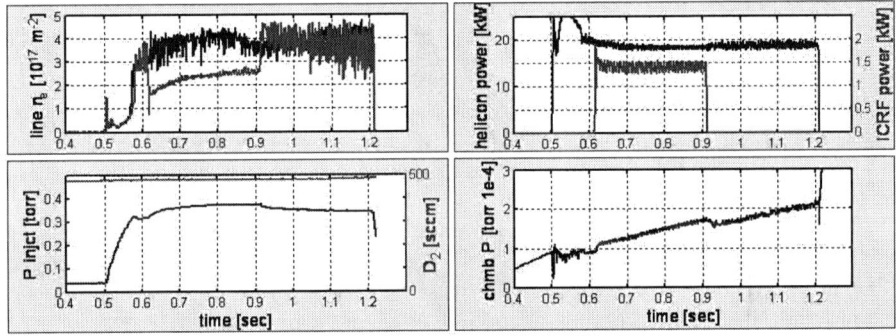

FIGURE 2. Temporal data for a typical discharge. Line integrated density, rf power, injector neutral flow and pressure, and chamber pressure.

SINGLE PASS ICRF ACCELERATION OF A DEUTERIUM PLASMA FLOW

The approach we take is a variation of the classic Stix[5] beach ICRH using the ion cyclotron wave (ICW) launched from the high field side of the resonance. There are, however, two distinct differences.[6] The first is that the plasma ions pass through the ICRF antenna and resonance only once. This is not a confinement device. The second is that the resonance is located less than one antenna length downstream from the end of the antenna. The wave hardly propagates before being absorbed at the resonance.

We have confirmed that we are indeed launching the ICW by examining the plasma loading behavior as a function of frequency near the resonant frequency. We performed a shot-to-shot frequency sweep of the plasma loading, with highly repeatable plasma conditions. We find a clear peak of almost 2 ohms when the resonance is located at the downstream end of the ICRF antenna. This is consistent with computational modeling using the EMIR code.[7,8] Figure 3a shows the plasma loading plotted as a function of the drive frequency normalized to the ion cyclotron frequency, f_{ci}, at the center of the antenna. Quantities of $f/f_{ci} < 1$ place the resonance downstream of the antenna center. The peak in the loading at $f/f_{ci} = 0.95$ indicates that the rf power is propagating downstream and taking axial distance to penetrate into the plasma. With a circuit resistance of 0.24 ohms, a plasma coupling efficiency (η_A) of over 90% is achieved (Fig. 3b). This is critical for an efficient rocket. We chose to run ICRF power experiments with the resonance about one antenna diameter downstream, putting the normalized frequency at $f/f_{ci} = 0.9$ and $f = 3.6$ MHz.

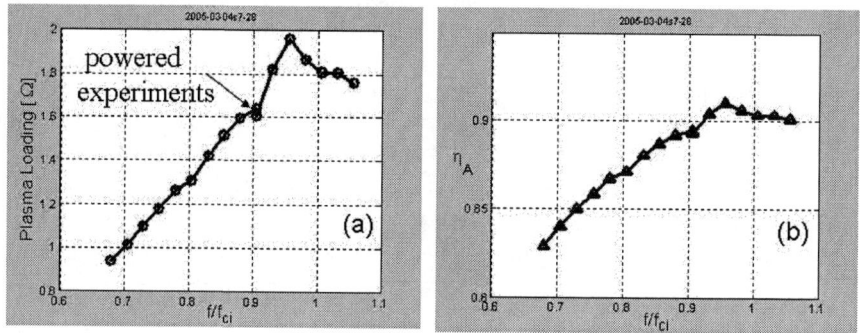

FIGURE 3. a) Plasma loading and b) Plasma coupling efficiency

The first obvious indication of plasma flow acceleration is the density drop observed by the interferometer when ICRF power is applied. Of course, this assumes that the ICRF neither causes particle loss nor creates further ionization. Figure 1 shows that the density drops to about 0.6 of the value without ICRF, which is consistent with the kinetic energy of the plasma flow tripling.

The RPA measurement is in very good agreement with the energy gain inferred by the density drop. Figure 4a contains the derivative of the ion current characteristic versus the ion repeller voltage, which indicates the energy of the bulk ions. The average energy of the ions increases from 10 to 27 eV with application of ICRF, or almost tripling of the energy. The energy boost is $E_{ICRF} = 17 \pm 2$ eV, which accounts for a large fraction of the input ICRF power, as we will discuss later. Additionally, one of our RPAs can rotate at an angle to the plasma flow from shot-to-shot. From a series of many shots, we obtain a 2d contour of the ion distribution function. Figure 4b shows the contour without ICRF and Fig. 4c with ICRF applied. Clearly the ion flow velocity nearly doubles and evidence of pitch angle scattering is observed in the distribution of ions, since the distribution spreads along a constant absolute velocity circle. This is likely due to the high neutral background pressure. Previous experiments[9] at lower gas flow rates observe an ion jet contained within a 10 degree velocity angle.

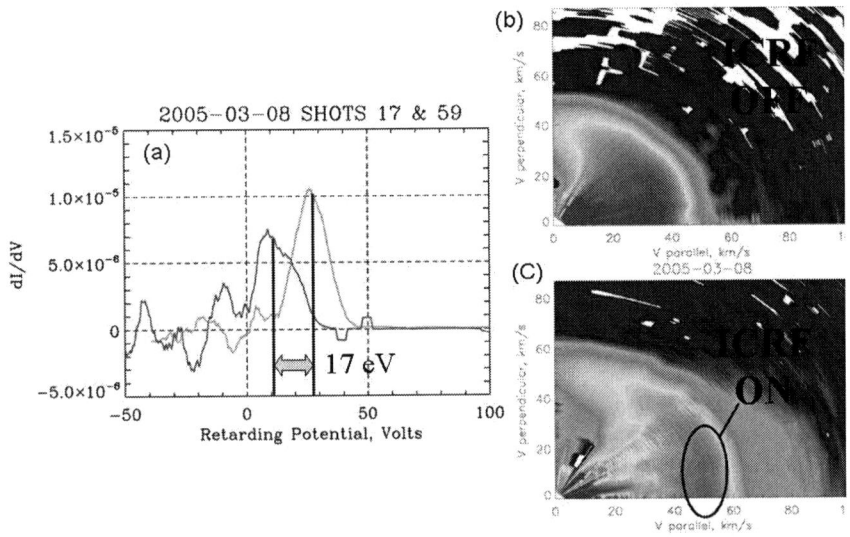

FIGURE 4. RPA data with and without ICRF. a) energy distribution, b) and c) 2d ion velocity distribution contours.

To perform a power balance, we measure the total ion current using the planar Langmuir probe array. Figure 5 shows the radial profile of the measured ion current density, with and without ICRF. The total plasma flux, $\Gamma_i = 3.1 \pm 0.6 \times 10^{20}$/sec, is conserved with application of ICRF, even though the profile shape becomes more peaked. The measurement is well downstream of the ion resonance where the applied magnetic field is substantially weaker, yet the plasma beta is only about 1%, so we expect a small variation from a flow along the vacuum magnetic field. We presently have no explanation for the radial profile shape change.

With the ion kinetic energy and total flux measured, we can compare the power in the ion flow to the input ICRF power. Considering the circuit efficiency, we measure coupled ICRF power to plasma as $P_{ICRF} = 1.25$ kW. Combining the RPA and probe array measurements, we determine that the ICRF ion kinetic power boost is $P_{ion} = 840 \pm 190$ W. This gives an ICRF acceleration efficiency of $\eta_A = P_{ion}/P_{ICRF} = 67 \pm 15\%$. A majority of the ICRF power is accounted for in the ion flow. Considering that the antenna is not highly directional (modeling indicates approximately 60 to 70 % directionality)[10] almost all of the launched ICW power is converted to

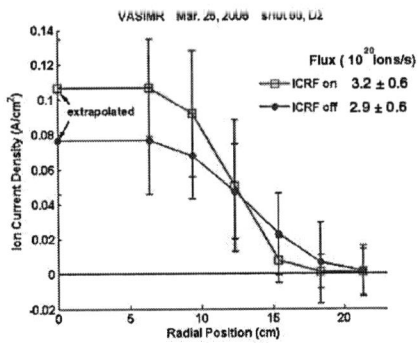

FIGURE 5. Ion current density radial profiles measured with and without ICRF power.

ion kinetic energy at the resonance as predicted.[6]

As further evidence that the plasma flow is accelerated significantly, we have measured the force on a target plate that we insert in the plasma exhaust. These data were taken at an earlier date when we had about one quarter of the plasma flux of the recent data. Nevertheless, we see that the force on the plate more than doubles with the application of about 1.5 kW of ICRF. Figure 6 shows the temporal response of the force on the plate.[11]

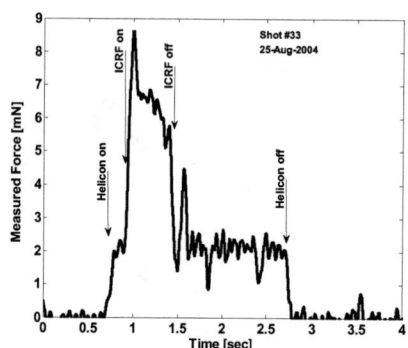

FIGURE 6. Force on the target plate. ICRF pulse for 0.5 second at about time = 1 sec.

Finally, there is an open question of the upper limit to the energy boost achievable by this ICRF method. In the present experiments, the ion energy boost is ICRF power limited. Previous experiments, in which the ICRF power to plasma flux ratio was much higher, achieved ion kinetic energy boosts of over 160 eV.[12] Also, the plasma load does not change significantly with application of 1.5 kW. The circuit is tuned to plasma using the network analyzer and the VSWR stays nearly unity under power. Given this, we expect that we can achieve an efficient energy boost of more than 200 eV on the present high plasma flux, with the application of over 10 kW of ICRF power.

DEUTERIUM HELICON OPERATION AT 20 KW

An efficient high flux helicon plasma source is critical for providing the desired target for ICRF acceleration. We have, therefore, put recent effort in increasing our helicon power and performance. The helicon performance parameter we would like to maximize is the ion current produced per input rf power (A/kW). For the data presented here we determine the ion current from the reciprocating Langmuir probe using the formula,

$$I_i = C_p \frac{\sqrt{e}}{L} \int \frac{I_{sat}}{a_s} r dr, \qquad (1)$$

where L is the length of the cylindrical probe, I_{sat} is the ion saturation current and a_s is the area of the plasma sheath around the probe. The formula is the classic Langmuir probe interpretation,[13] assuming sonic flow, with a correction factor, C_p, based on comparison to the density interferometer. Another very important figure of merit is the gas utilization, defined as,

$$\eta_g = \frac{I_i}{q_e \Gamma_g}, \qquad (2)$$

where Γ_g is the input atomic neutral flux. We want this quantity close to unity to assure a high degree of ionization at the ICRF antenna and for an efficient rocket.

Based on positive results from Oak Ridge National Laboratory,[14] where higher densities were achieved at magnetic fields well above the lower-hybrid resonance, we investigated the helicon performance as a function of applied magnetic field. We set the magnetic field fixed and tuned the rf system to the plasma load at about 20 kW. We then scanned the input gas flow rate from shot to shot and retuned the rf system, if necessary, to maintain the power approximately constant. Figure 7a summarizes the results, where we plot the source performance versus the input atomic neutral rate for four magnetic field profiles, Fig 7b. The solid curves represent the curves for $\eta_g = 1$. The error bar primarily reflects the uncertainty in the probe model, since shot reproducibility is very good. The helicon operates best at almost 5 times the lower hybrid resonant field.

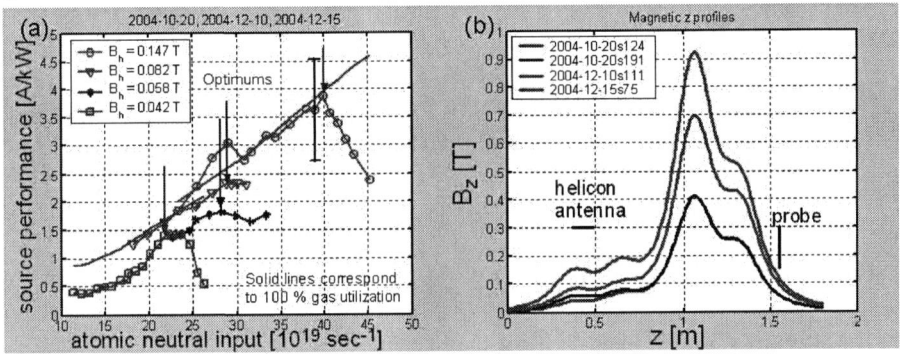

FIGURE 7. a) Gas flow scan for 4 magnetic field profiles. b) The 4 magnetic field profiles.

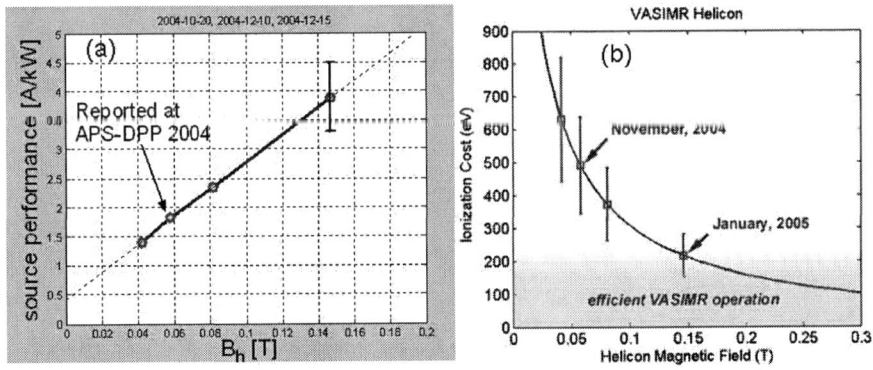

FIGURE 8. a) Optimal source performance versus helicon magnetic field. b) Ionization energy cost.

To evaluate the scaling of the source performance with applied magnetic field at the helicon antenna, we plot the peak of each curve in Fig. 7a versus that parameter. Figure 8a shows that the scaling is amazingly linear over a large range, nearly a factor of four. The higher magnetic fields were only limited by difficulty with plasma start up and arcing at the vacuum feedthrough. An important parameter for propulsion is the effective energy cost of ionization, essentially the inverse of the source performance and small factors considering circuit efficiency and ion kinetic energy. Figure 8b plots this parameter and shows that we are approaching a cost of 200 eV/ion. This is entering the range for an efficient rocket performance for deuterium ion exhaust speeds over 100 km/s.

CONCLUSION

ICRF acceleration of a dense ($>10^{19}$m^{-3}) plasma is clearly demonstrated and confirmed by multiple diagnostics. ICRF power is efficiently (90%) coupled to the plasma flow, consistent with launching the ion cyclotron wave. The applied ICRF power approximately triples the kinetic energy of the flowing helicon plasma source. The efficiency of the acceleration process is nearly 70%, in the present configuration. No signs of nonlinear limitations are observed, indicating that an efficient high flux (> 4×10^{20}) of ions exceeding 200 eV is achievable. A recent helicon power upgrade has enabled a deuterium source efficiency of nearly 4 A/kW. These combined results show great promise for an efficient rf driven plasma rocket system.

ACKNOWLEDGMENTS

This work was supported by the NASA, Lyndon B. Johnson Space Center.

REFERENCES

1. Stuhlinger, E., *Ion Propulsion for Space Flight*, New York, McGraw-Hill, 1964.
2. Chang Diaz, F.R., *Scientific American* 283 (2000) 72.
3. Boswell R.W. and Chen, F.F., *IEEE Transactions of Plasma Science* 25, 1229 (1997).
4. Chen, F.F. and Boswell, R.W., *IEEE Transactions of Plasma Science* 25, 1245 (1997).
5. Stix, T.H., Waves in Plasmas, American Institute of Physics, New-York NY, 1992.
6. Arefiev A.V., Breizman B.N., Theoretical components of the VASIMR plasma propulsion concept, *Phys of Plasmas*, 11 2942 (2004).
7. Carter, M.D., *private communication*.
8. Carter M.D., Baity F. W. Jr., Barber G. C., Goulding R. H., Mori Y., Sparks D. O., White K. F., Jaeger E. F., Chang Díaz F. R. Squire J. P. (2002) Comparing experiments with modeling for light ion helicon plasma sources, Physics of Plasmas, 9 5097 (2002).
9. Bering, E. A., III, F. R. Chang-Díaz, and Squire, J.P. "The use of RF waves in space propulsion systems." *Radio Science Bulletin*, No. 310 p. 92, (2004).
10. Ilin, A.V., Chang Diaz, F.R., et al., "Plasma Heating Simulation in the VASIMR System" 43nd AIAA Aerospace Sciences Meeting and Exhibit, Reno, NV, AIAA (2005).
11. Chavers, D.G., Chang-Díaz, F.R. Breizman, B.N., and Bengtson, R.D., "Momentum flux measurements using an impact thrust stand," Bulletin of the American Physical Society, Program of the 46th Annual Meeting of the Division of Plasma Physics, Savannah, GA, 49, 295 (2004).
12. Glover, T.W., F.R. Chang Diaz , et al. "Principal VASIMR Results and Present Objectives," Space Technology and Applications International Forum, Albuquerque, NM, 2005.
13. Hutchinson, I. H., *Principles of Plasma Diagnostics*. Cambridge, Cambridge University Press 1987.
14. Mori, Y., Nakashima, H., Baity, F.W., Goulding, R.H., Carter, M. D. and Sparks, D.O., *Plasma Sources Science and Technology* 13, 424 (2004).

Capacitive Systems for Dielectric Plasma Etch

Daniel Hoffman

Applied Materials, Inc., 974 E. Arques Avenue, Sunnyvale, CA 94086

Abstract. Two and three frequency capacitive systems are being used to generate weakly ionized plasma in $Ar/O/C_xF_y$ chemistries at the millitorr pressure range. One or two of the frequencies are generally used to accelerate ions (by sheath rectification) while the third is generally used to independently raise plasma density to levels sufficient for etching. The choice of frequencies is based on plasma impedances, which then yield rf voltages that can either consume power by creating a DC plasma sheath or consume power by creating plasma density (when sheath power is minimized). Within the two frequencies that create sheaths, the choice of ion energy spreads is determined by ion sheath transit time relative to an rf cycle. Technology challenges arising from the required plasma creation include significant intermodulation, very high cross talk between generators, and the avoidance of arcing through on any of the surfaces in contact with the plasma (including the gas injection system). The etch chamber is designed such that all generators are directly linked on a single coupling point or the plasma directly connects one launcher to another. We discuss and analyze different frequency ranges and their impact on chamber design.

Keywords: plasma sources, capacitive sources, ion energy, dielectric etch
PACS: 52.50.Dg, 52.77.Bn, 89.20.Bb

OVERVIEW

The design of a plasma etcher, particularly for dielectric applications has migrated toward middle densities ($1\text{-}10^{10}$ cm^{-3} vs. $>10^{11}$ cm^{-3}) and away from inductive systems to capacitive systems. While higher etch rate is associated with higher density, and inductive systems are efficient at producing high densities, the rate has been counterbalanced by the need to etch the mask at a substantially slower rate than the oxide itself. At very high densities, the mix of process gases commonly used today (usually containing $Ar/O/C_xF_y$) is capable of etching both oxide and mask at nearly equal rates. Thus, this sets the target density range of operation and the forms the basis of chamber design.

While density is the overall driver of chamber design, other features, such as the ion energy distribution function must be incorporated into the design. This paper attempts to explain the choices made in designing advanced dielectric etch tools. Table 1 summarizes the relevant parameters.

TABLE 1. Differences between new and previous generation dielectric etch tools

Parameter	Newer Tools	Previous Generation	Reason for Change
Type of Discharge	Capacitive	Inductive	Δn range, T_e
Plasma Density Range	$1\text{-}10^{10}$ cm^{-3}	10^{11} cm^{-3}	Photoresist selectivity
Working Pressure	1-1000 mT	<100 mT	Low pressure for dense materials, high for porous
Electron Energy	1-3 eV	1-5 eV	Photoresist selectivity
Minimum Pure Bias Pressure	< 30 mT	> 70 mT	Tool robustness
Chamber Materials	SiC, Si, Y$_2$O$_3$, Quartz	Anodized aluminum, Quartz	Chamber cleaning
Gas Flow	Over wafer, dual zone	Wafer edge, single zone	Alter local residence over wafer
Average Ion Energy	100-3000 eV	300-1000 eV	Sometimes need sparse energetic ions for hard materials
Ion Energy Distribution	Tunable from narrow to broad	Uncontrolled	Greater range of materials to be processed

DENSITY AND ELECTRON ENERGY CONSIDERATIONS

In creating an ideal chamber, it is preferred that the source create plasma density and a bias contribute only that level of energy needed to etch. The DC bias is created by rf power rectified by the plasma to have a large negative potential that accelerates ions into the wafer. Thus, power delivered to the plasma by the bias system is distributed between two major functions: sustaining a dc sheath and creating plasma density. The fraction of power allocated to these functions is frequency-dependent, because higher frequency systems generally have lower impedances (and, therefore, associated rf voltages and rectified DC sheaths).

From Figure 1, one can infer that generally bias frequencies are best below 20 MHz, while unnecessary sheath losses in a plasma density source can be avoided when the frequency exceeds 150 MHz. Also, given designed plasma volume and target density, the power required for low 10^{10} cm^{-3} density can range from many kW at frequencies below 60 MHz to a few kW for frequencies greater than 150 MHz.

Most inductive sources are effectively precluded if the desired density range is in the mid 10^{10} cm^{-3}. As the source power is increased from zero to some reasonable level, the discharge changes from capacitively coupled (by the high voltages on the coil) to inductive where the plasma images currents beneath the coil. For practical systems, this occurs right in the heart of the target design range: mid 10^{10} cm^{-3} density. Figure 2 data compare capacitive vs. typical inductive systems.

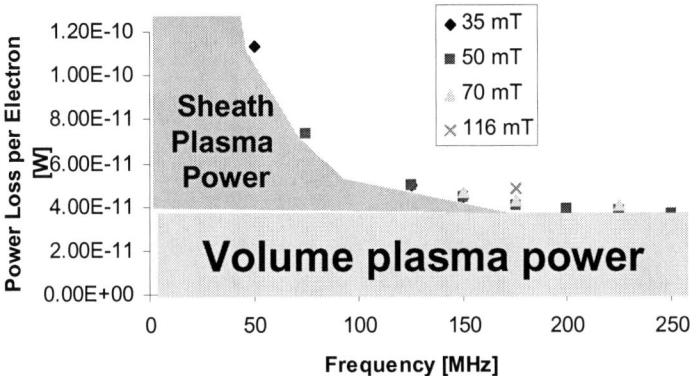

FIGURE 1. Power required for a given density vs. frequency. The pink area can be attributed to sustaining the sheath necessitated by rf voltages, while the green is associated with chamber density losses.[1]

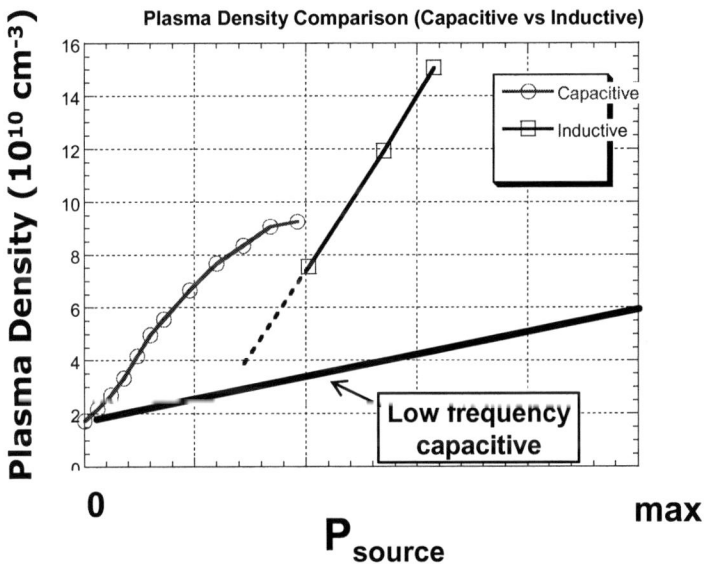

FIGURE 2. Density of capacitive and inductive sources

Electron energy must also be taken into consideration when selecting a source frequency. When a plasma is very warm, its energy can break down process gas molecules into smaller and smaller radicals until the molecule is fully decomposed (dissociated) into its elemental makeup. The optical emission from a typical plasma can show the density effect. From Figure 2, one sees that the density increases nearly linearly with source power. In Figure 3, the low-power optical emission spectrum shows CF$_2$ emission lines with little evidence of C emission. However, at the highest density, the C and Ar emission dominate. In this condition, the carbon has been detached from the C, and can effectively etch through most materials indiscriminately, thereby etching both mask and oxide. While not appropriate for the majority of etch processes, this high dissociation condition does have a use. After the wafer has been etched, the high-density plasma can be used to clean the chamber and is more effective than using low source.

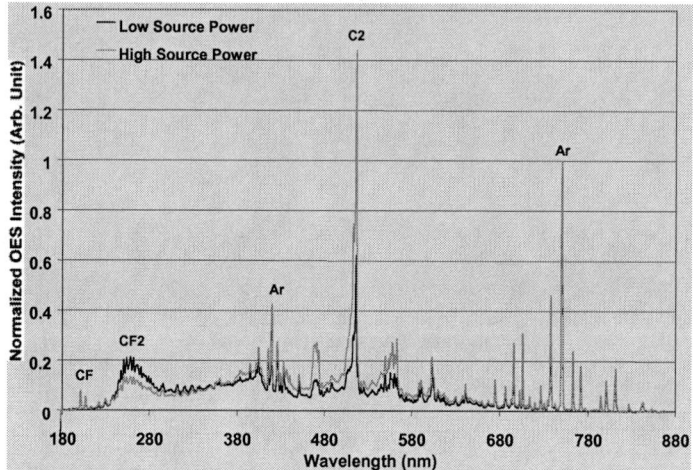

FIGURE 3. Optical emissions show the degree of dissociation occurring in a typical low density vs. high density plasma.

CHARGED SPECIES UNIFORMITY

While the use of high frequency is very beneficial for expanding the density range, low voltage at non-wafer surfaces, chamber cleaning, and extending the operating pressure window, a challenge arises when trying to preserve etch uniformity across the wafer. Various factors can amend uniformity: feed structure, receiver antennas, chamber shaping, or edge-of-source impedance design. In this case, we consider use of DC fields, named the Charged Species Tuning Unit (CSTU). By applying small magnetic fields, generated with solenoid coils, intrinsically center-high plasma distributions can be flattened. A simple solenoid coil generates both axial and radial field. Experimentally, it was determined that the axial field exerts the most influence on density transformation. Figure 3 shows the measurements of this effect.

ION ENERGY DISTRIBUTION

A single rf sine wave, in the presence of typical process plasma can generate a DC offset that accelerates ions into the wafer. If the average ion time to transit this DC sheath is sufficiently short, ions will not respond to the reversing rf field. However, if the transit time is long compared to an rf period, ions can enter the sheath when the phase of the rf electric field either constructively or destructively interferes with the

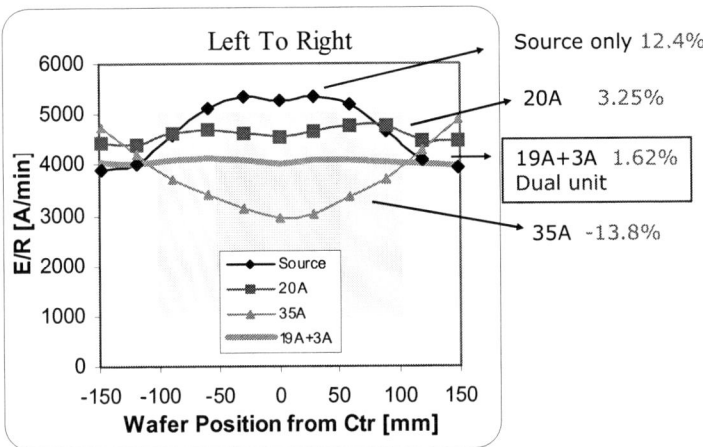

FIGURE 4. Plasma uniformity vs. radius as CSTU currents are adjusted. Uniformity can be transformed from center high to center low by adjusting solenoid current.

DC accelerating potential. The maximum constructive interference is approximately twice the rf amplitude, while the destructive interference is limited to zero potential. The choice of frequency determines these bounds for practical sheath thickness (~5mm) and potentials (~1000 V): 2 MHz can lead to potentials that range 0.3-1.8 times the input amplitude, while 13 MHz is sufficiently high frequency relative to ion transit time that the energy spread is ~0.8-1 times the input energy.

The need to broaden the ion energy range has led to examination of frequencies vs. ion energy spread.[2] Given this need, the choice of frequencies is based on the ion transit time relative to rf period and a second factor, density vs. sheath production per unit rf power. Analysis of very low density plasma (10^9 10^{10} cm-3) shows that bias power of dual frequencies (where one is low with wide ion energy spread and one is high with narrow spread) can yield the required control over the desired spread (Figure 5).

The impact of the source on the ion energies is two fold: if the frequency of the source is very high, the additional rf voltage is negligible and the source shifts average energy roughly proportional to 1/sqrt(density). However, if the source frequency is sufficiently low, the rf voltage must be incorporated into the evaluation of the source impact (Figure 6).

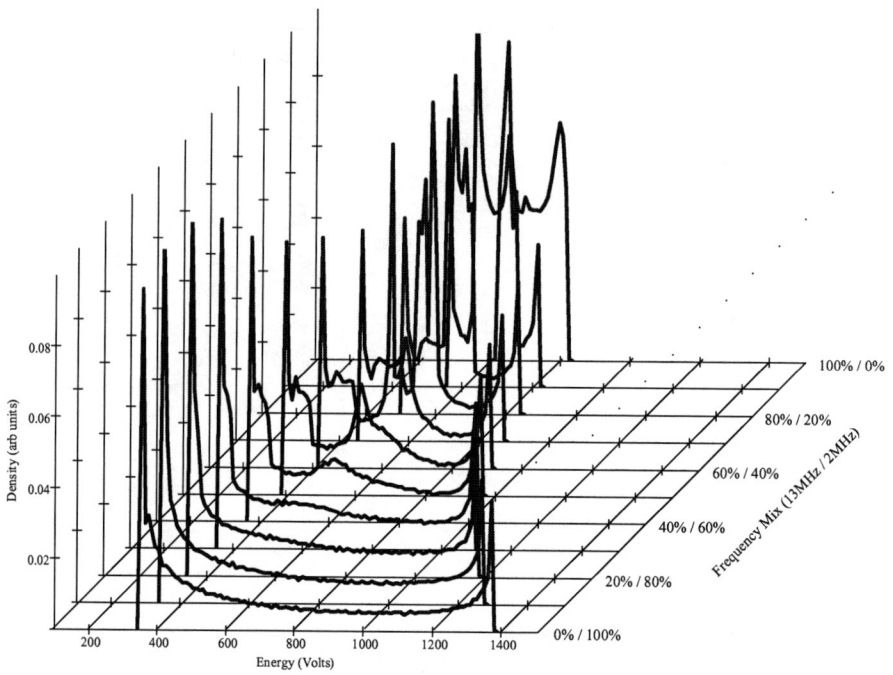

FIGURE 5. Ion energy distribution vs. fraction of power that is 2MHz.

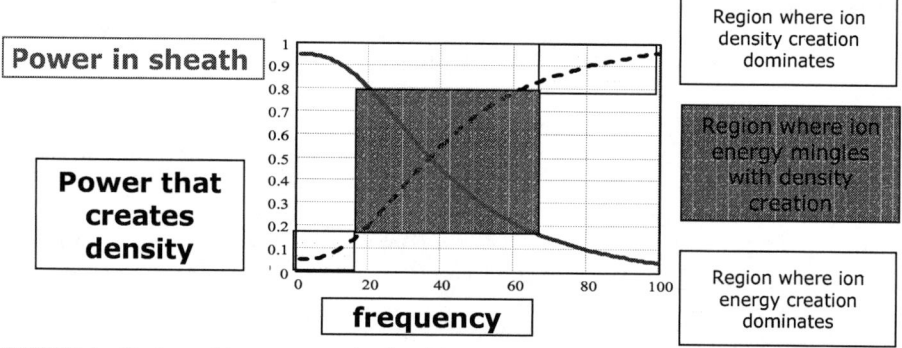

FIGURE 6. Regions of ion energy evaluation (30 mT). At higher pressures, the same curve applies, but the fraction of power that creates density at low frequency increases.

If the plasma space is transformed into a density/ion energy volume where the X axis is the minimum energy, Y axis is maximum energy, and the Z axis is plasma density, the operating volume of the etch chamber can be readily defined. Figure 7 shows how mixing 2 MHz and 13 MHz adjusts ion energy distribution with minimal impact on plasma density, while Figure 8 shows how the source raises density and lowers average energy in proportion to the square root (sqrt) of density. The goal in designing an etcher is to achieve the most orthogonal control possible of plasma parameters. Accordingly, in designing the etcher, frequencies selected from those shown in the figure were those that optimized orthogonality. If the frequencies chosen fall in the middle zone of Figure 6, the operating density/ion energy volume is substantially limited, which reduces the process capability of the tool.

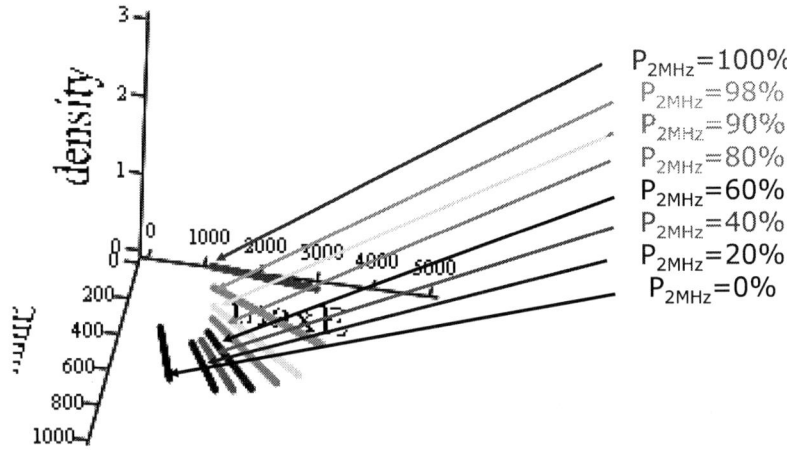

FIGURE 7. Ion density/plasma density volume scaled from prototype lab tool. X axis is minimum ion energy, Y axis is maximum, and Z axis is density. Bias power is swept from min power to max with differing mixes of 2 MHz and 13 MHz. For a fixed frequency spectrum, the ion density increase is much smaller than the 10X power sweep.

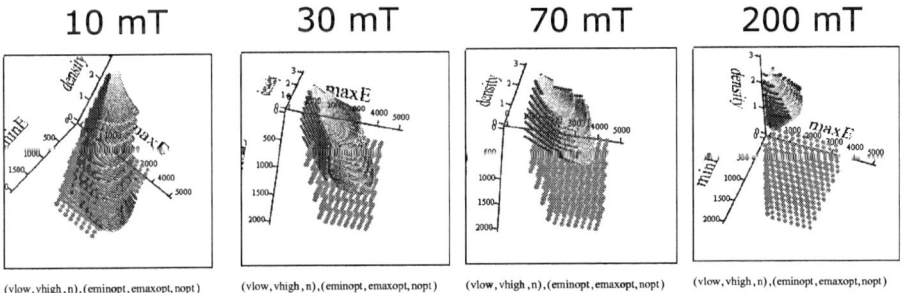

FIGURE 8. Ion density/plasma density volume scaled from prototype lab tool. X axis is minimum ion energy, Y axis is maximum, and Z axis is density. Source power is included in this measurement, and transforms the planes of Figure 7 to a volume by virtue of the density enhancement. Green indicates the target etch region for very deep high aspect ratio etch.

SUMMARY

The design of advanced dielectric etchers is strongly dependent on the need for good plasma density control, plasma uniformity, and ion energy distribution. The correct choice of frequencies can yield the most orthogonal system: source frequency optimized around density formation and bias optimized around ion energy spread.

ACKNOWLEDGMENTS

The author appreciates contributions to these studies by Jang Gyoo Yang, Doug Buchberger, Steve Shannon, and Doug Burns of the Applied Materials Dielectric Etch Division.

REFERENCES

1. Berra, K., "High Source Power Dual Damascene Trench Etch Process Development Using an Advanced Dielectric Etch Tool," Proceedings of the International Symposium on Dry Process, p. 197, 2003.
2. Shannon, S., "The Impact of Frequency Mixing on Sheath Properties – Ion Energy Distribution and V_{DC}/V_{RF} Interaction," Journal of Applied Physics, May 2005.

RF Power Coupling And Plasma Transport Effects In Magnetized Capacitive Discharges

P. M. Ryan[a], M. D. Carter[a], and D. J. Hoffman[b]

[a]*Oak Ridge National Laboratory, Oak Ridge, TN*
[b]*Applied Materials, Santa Clara, CA*

Abstract. Static magnetic fields have been used to expand the operational envelope, increase power efficiency, and control processing parameters in capacitively-coupled radio frequency plasma discharges. A simple physical model has been developed to investigate the roles of the plasma dielectric tensor and plasma transport in determining the ion flux spatial profile along a wafer surface over a range of plasma density, neutral pressure and magnetic field strength and orientation. The model has been incorporated into the MORRFIC code[1] and calculations have been made for a capacitively-coupled 300-mm etch tool operating at frequencies greater than 100 MHz. A Lieberman sheath model[2] show effects that can occur when the sheath voltages are made to be consistent with the driven RF fields in the collisional unmagnetized limit; other sheath models will also be considered. A two-dimensional transport model accounts for magnetized cross-field diffusion. Results isolate magnetic field effects that are caused by modification of the plasma dielectric from transport effects that are caused by the reduced electron mobility perpendicular to the magnetic field.

Keywords: Capacitively-coupled plasma sources, plasma etch tools
PACS: 52.50 Dg

INTRODUCTION

Control of the ion flux profile at the surface of the wafer is important in optimizing the etch rate uniformity of a capacitively-coupled plama processing tool. One such method of uniformity control uses solenoidal B-fields [3]. This paper introduces a computational model designed to evaluate the contributions from plasma transport and rf power deposition in the magnetized medium to the reactive ion flux distribution at the wafer surface.

THE MODEL

RF Power Deposition and Plasma Generation

The RF power deposition is calculated from Maxwell's equations in the frequency domain

$$\nabla \times \vec{E} = i\omega\vec{B} \qquad (1)$$

$$\nabla \times \vec{B} = \mu_0 \vec{J} - i\omega\mu_0\varepsilon_0 \overline{K} \cdot \vec{E} \qquad (2)$$

where $\overline{K}$ is the magnetized cold plasma dielectric tensor. The energy available for ionization is proportional to the RF power deposited inside a volume comparable to a mean-free path for inelastic electron-neutral collisions.

$$S(\vec{x}) = \frac{n_n(\vec{x})\int_{Vi} P_{rf} d^3x}{E_{ion}(\vec{x})\int_{Vi} n_n d^3x} \qquad (3)$$

For a magnetized plasma where $\rho_e < \lambda_{inelastic}$, the electron motion is constrained perpendicular to B. The model averages the RF power over a flux tube having length $\pm \lambda_{inelastic}$ along the field line and a radius of $\rho_{eip}\sigma_{en}/\sigma_{inelastic}$. Here ρ_{eip} is the electron gyroradius corresponding to the energy required for ionization of the gas and the cross section ratio is roughly the number of collisions before significant energy loss by the electrons. The average energy required for ionization, E_{ion}, can be represented in terms of σ_j, the inelastic collision cross section resulting in energy loss E_j, and σ_i, the ionization cross section, where $\langle\rangle$ represents an average over the electron distribution function.

$$E_{ion} \equiv \frac{n_e n_n \sum_j \langle \sigma_j v \rangle E_j}{n_e n_n \langle \sigma_i v \rangle} \qquad (4)$$

Plasma Transport

The model employs the 2D diffusion equation for plasma transport

$$S(\perp,\parallel) \approx -\frac{\partial}{\partial \parallel} D_\parallel \frac{\partial n(\perp,\parallel)}{\partial \parallel} - \frac{\partial}{\partial \perp} D_\perp \frac{\partial n(\perp,\parallel)}{\partial \perp} \qquad (5)$$

where S is the local source rate and $D_\parallel$, $D_\perp$ are the diffusion coefficients parallel and perpendicular to the magnetic field. We assume ambipolar diffusion along field lines,

$$D_\parallel \approx D_A \qquad (6)$$

and classical cross-field diffusion in strongly magnetized regions

$$D_\rho \approx \frac{n_n \sigma_{en} v_{te}}{2} \rho_e^2 \qquad (7)$$

where v_{te} and ρ_e are the electron thermal speed and gyroradius. Friction effects caused by ion thermal motion ($m_i \gg m_e$) allow the ambipolar diffusion coefficient to be simplified to the product of the acoustic velocity and the electrostatic potential gradient scale length,

$$D_A \approx C_s L_i = \frac{C_s}{n_n \sigma_{in}} \sqrt{\frac{T_e}{\gamma T_i}} \qquad (8)$$

Ambipolar diffusion at weak magnetic fields transitions to classical cross-field diffusion for strong magnetic fields according to

$$D_\perp \approx \frac{D_A D_\rho}{D_A + D_\rho} \qquad (9)$$

The boundary conditions at either end of a field line are governed by diffusion into the sheaths at the ion acoustic speed, C_s. For the magnetized plasma case where variations in the perpendicular direction are small, the density variation along a field line is parabolic. For the special case where each sheath boundary has the same D_A and C_s, the pre-sheath feeds plasma into the sheaths at the same rate. The sheath thicknesses must then be adjusted to make the plasma flow consistent with the applied voltages at each end of the field line.

A sheath model based on Lieberman's work was iterated in MORRFIC[4]. For low plasma densities or high powers, the iteration process converged but, for some parameters of experimental interest, the electric field in the sheath collapsed locally. The sheath thickness was fixed for the calculations presented here.

THE GEOMETRY

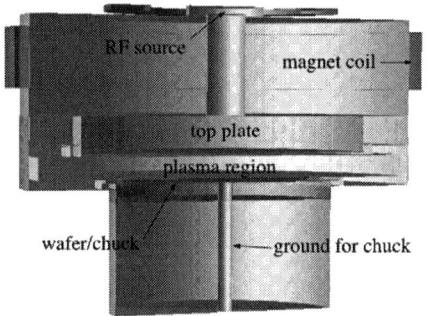

FIGURE 1. Geometry of capacitive discharge chamber showing the plasma region between the powered top plate and the chuck (grounded) and the dc solenoidal coil.

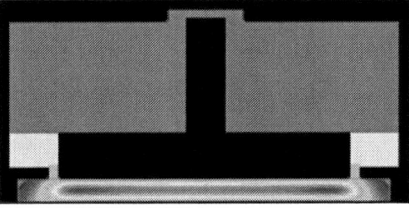

FIGURE 2. Example of the plasma density distribution in a magnetized discharge. The spatial uniformity of the ion flux to the wafer can be controlled with the applied magnetic field.

The geometry of the capacitive discharge chamber used in this study is shown in Figure 1. It is a simplified version of a commercial dielectric etch tool, with the aspect ratio altered sufficiently to avoid disclosing any proprietary information while revealing the fundamental properties of the discharge. High frequency source power (>150 MHz) is applied to the upper electrode to initiate and sustain the discharge. The lower electrode, or chuck, on which the wafer sits, is normally rf-biased with the application of one or more lower frequencies (2-12 MHz); for simplicity, the lower electrode remains grounded in these studies. One magnetic field coil is shown in Figure 1 and is used to produce a diverging B-field at the wafer surface; more such coils can easily be added to tailor the field shape. In these studies, an identical image coil (not shown) could also be positioned below the plane of the wafer to generate either pure mirror fields (B-radial = 0) or cusp fields (B-axial = 0) at the wafer surface.

INITIAL RESULTS

The effect of the radial component of the magnetic field on the plasma density profile at the wafer can be understood in terms of an enhanced confinement of ions at the edge. On axis the magnetic connection length L_c is equal to the gap spacing, L_G, while off-axis $L_c \approx L_G(1 + B_r^2/B_z^2)^{1/2}$. This contributes to an increased confinement time, as shown in Figure 3, that scales as

$$\bar{\tau} \approx \left[\left(1+\frac{B_r^2}{B_z^2}\right)^{1/2} + \frac{L_G}{6L_i}\left(1+\frac{B_r^2}{B_z^2}\right)\right]\left(1+\frac{L_G}{6L_i}\right)^{-1} \quad (10)$$

where L_i is an electrostatic potential gradient scale length. Figure 4 is a MORRFIC calculation showing the depletion of central density and increase in edge density as the magnetic coil current is increased.

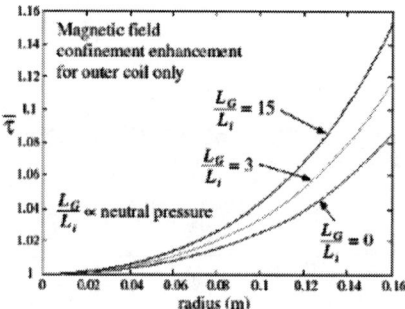

FIGURE 3. For magnetic field strengths and neutral collision parameters such that the electron transport is dominated by the magnetic field, the radial component of the static magnetic field can enhance the plasma confinement at the wafer edge.

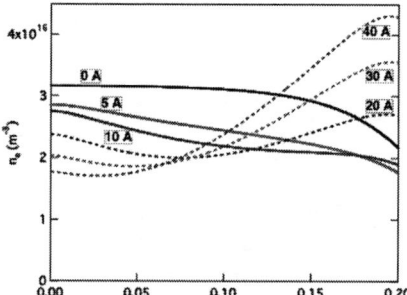

FIGURE 4. MORRIC calculations of plasma density at wafer surface as a function of solenoid coil currents for a magnetic cusp geometry. The source power is 400 W, $n_0 \sim 10^{20}$ m^{-3}, and the sheath thickness is fixed at 1.5 mm.

ACKNOWLEDGMENTS

The authors gratefully acknowledge Valery Godyak of Osram Sylvania for enlightening discussions on plasma sheath modeling. This work was supported in part by Applied Materials and by Oak Ridge National Laboratory, managed by UT-Battelle, LLC, for the USDOE under contract DE-AC05-00OR22725.

REFERENCES

1. Carter, M. D., 57th Gaseous Electronics Conference (Bunratty, Ireland, 2003)
2. Leiberman, M. A., IEEE Trans on Plasma Sci, **16** 638 (1988)
3. Hoffman, D. J., this conference.
4. Carter, M. D., D'Ippolito, D. A., Myra, J. R., and Russell, D. A., this conference.

A Simulation Approach for ICRF Plasma Thruster Antennas

G. Vecchi, L. Valitutti, V. Lancellotti, R. Maggiora, D. Milanesio

Dipartimento di Elettronica, Politecnico di Torino, Torino, Italy

Abstract. In the past twenty years plasma-based propulsion systems have found increasing aerospace interest; although they were initially conceived as rockets for interplanetary missions, more recent advances in plasma-based concepts have led to the identification of radio-frequency (RF) generation and acceleration systems as capable of providing not only continuous thrust, but also controllable exhaust velocities, as required in maneuvering applications. The most interesting such studies for plasma propulsion are those focused on the possibility of coupling radio frequency power to plasma, exploiting the possibility of having very efficient devices to generate and heat the plasma, magnetically confining it in a trap in the heating region, so that ion can escape the magnetic trap only when they are energetic enough to be converted into direct out-going flow which provides the thrust. The structure of this system is therefore based on of three stages where plasma is respectively generated, heated and expanded in a magnetic nozzle. The heating stage acts as an amplifier; here plasma is heated by the radio frequency waves by the process of ion cyclotron resonance. It has been developed and tested a numerical tool for the electromagnetic modeling of the ICRF antenna, of the RF booster unit of plasma thrusters, and of the RF-plasma interactions. The latter is studied in the critical ICRF acceleration region by setting up a convenient Electromagnetic (EM) analytical and numerical model based on the Moment-Method solution of a suitable set of integral equations. Solution of the relevant integral equation directly provides the electric surface current density induced on antenna conductors, but the ultimate quantity to be computed is the circuit characterization (e.g. admittance matrix) at the input ports.

ANTENNA MODELLING AND PROBLEM FORMULATION

The acceleration unit of a plasma thruster generally employs the ICRF technology to efficiently transfer large RF powers to magnetized plasmas; the actual configuration of this RF booster comprises two counter-driven loops antennas encircling the plasma column (see Fig. 1a). By developing a suitable model of these particular radiating elements it has been possible to reduce the complexity of the addressed radiation problem. Indeed a full geometric modelling of plasma-thrusters RF booster unit would require setting up a plasma description in terms of a circular cylinder, possibly non uniform in the radial direction, while plasma density and temperature are well approximated as azimuthally invariant. The electromagnetic model of the above situation is significantly complex and numerically heavy. On the other hand, studies carried out in the case of plasma-facing ICRH antennas in fusion experiments (e.g. [1, 2, 3]) have shown that the approximation of a flat plasma and conductors provides a reasonable approximation in the presence of a typical, dense plasma, provided that the current carrying conductors and the plasma surface remain parallel. Thus the idea is to carve the ICRF loop antenna and stretch it as to form a strap antenna, while the plasma column is modelled as a slab as well. In the original geometry the conductors encircle the plasma, which implies a 2π azimuthal

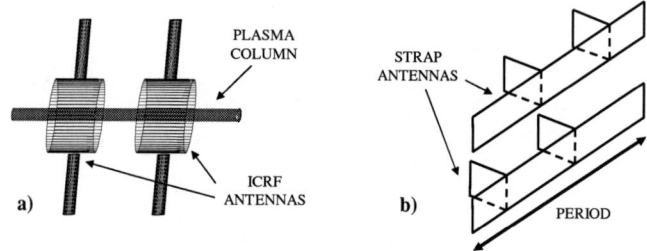

FIGURE 1. (a) Sketch of the ICRF antennas in the RF booster unit of electromagnetic plasma thrusters and (b) unit cell of the periodic strap model of ICRF antennas.

periodicity; therefore a better approximation ensues if the antenna is assumed as constituted by a periodic repetition of infinite replicas of straps electrically connected and laid down parallel to the plasma interface, as in Fig. 1b. A radial cross section of the unit cell of the periodic model of electromagnetic-plasma-thrusters acceleration stage is depicted in Fig. 2a. The analysis of the radiation by bodies periodically displaced in space can be effectively faced by means of Floquet's theorem; in particular the antenna system concerned here is periodic just along one spatial dimension, so that the 1D form of the Floquet's theorem is actually involved [4].

All the metallic parts of the antenna system are considered perfect electric conductors (PEC), since in practice the conductivity σ is high; on the other hand, a finite value of σ may be accounted for successively as a perturbing parameter in the solution process. The high speed plasma flow, present in the plasma thrusters, is supposed to rapidly and totally absorb the ion cyclotron waves launched by the magnetized electrode-less acceleration unit [5]; finally, since wave absorption in the resonance region is not considered self-consistently in the present plasma model, as a first approximation, the velocity of the plasma flow, whose main effect is to shift the ion cyclotron resonance frequency [6], is neglected.

The solving procedure [7] is based on the conceptual separation of the antenna from the plasma by means of the Equivalence Theorem (ET) [4] invoked twice. First the ET is applied to the infinitely thin volume V_A bounded by two plane surfaces S_{A-}, S_{A+}, which implies introducing unknown magnetic and electric current densities (see Fig. 2c). Notice that $S_{A\pm}$ can be placed arbitrarily anywhere between the antenna and the plasma interface. V_A is then filled with a PEC plate to accomplish antenna/plasma separation; due to the boundary condition at a PEC interface, only the equivalent magnetic current will contribute to the fields. In the antenna region now the ET is applied on a surface S_C enfolding all the replicas of the antenna straps and feeders, and the infinite (along $\hat{y}$; conversely finite along $\hat{z}$) conducting plate on which they are settled (see Fig. 2c). As a result, the electromagnetic field vanishes outside V_C, then this infinite volume is totally substituted with *free space*: the antennas and the bottom wall are removed as depicted in Fig. 2d. The unknown electric current density extends on the whole S_C surface specified by the replaced conducting elements, while the magnetic current is null thereon.

As a whole, the two-step application of the ET yields two unknown current densities, $\underline{M}$ and $\underline{J}_C$. Two coupled integral equations simply stem from enforcing the boundary and

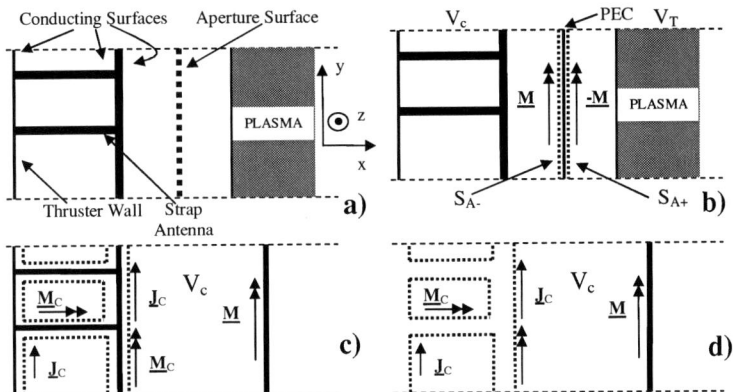

FIGURE 2. (a) Radial cross section of the unit cell of the periodic model of ICRF antennas, (b) first ET application: antennas and plasma separation, (c-d) second ET application: antenna conductors are removed.

continuity conditions the fields must obey on S_C and S_A. Since the radiating system is periodic along $\hat{y}$, the field radiated by $\underline{M}$, $\underline{J_C}$ have the form $\underline{E}(\rho,y) = \underline{E}_P(\rho,y)\exp(-j\beta_0 y)$, where "P" labels periodic functions and β_0 is imposed by the primary (source) field $\underline{E}^p$; a similar expression holds for the magnetic field. In particular to get the proper antenna phasing in this experiment, $\beta_0 = 0$ and $\underline{E}^p = \underline{E}_P^p$. The dependence of the fields on the currents is linear and represented by suitable surface integrals with appropriate kernels (i.e. Green's functions) [8]. More precisely, in the antenna region the free space Green's function is needed; by exploiting the properties of periodic functions, the one-dimensional periodicity is shifted to the Green's function, which becomes a slowly converging series. To improve convergence the Shanks' transform [9], whose parameters are selected by means of a *Montecarlo* optimization code [7], has been applied.

The integral equations are solved by the Moment Method (MoM) [10], both in spatial and spectral domain, the latter being the natural one wherein the plasma is easily described. In particular the unknown currents are expanded as a linear combination of a finite set of subdomain basis functions, which are defined over rectangles [1], in view of the adopted regular shape of the antennas geometry. The coefficients of this linear combination represent the actual unknowns of the problem, which in turn has been transformed from a system of integral equations into an algebraic system. The entries of the matrix representing the system are computed by the developed code.

NUMERICAL RESULTS

Once the current densities $\underline{M}$, $\underline{J_C}$ have been computed, the antenna input impedance (see e.g. [1, 3]) as well as other common parameters can be obtained. In particular, the coupling resistance $R_c = 2P_{rad}/|I|_{max}^2$ (P_{rad} identifies the absorbed RF power, I the current along the strap), is a key figure of merit that characterizes the efficiency of the ICRF antennas for thruster applications.

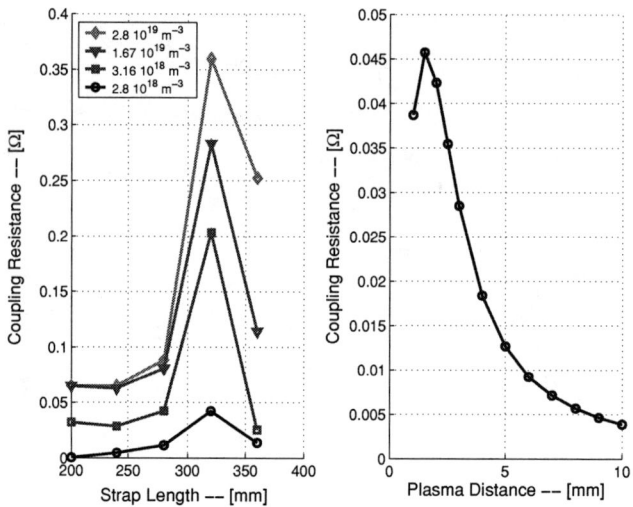

FIGURE 3. R_c as a function of the strap length (left) and the plasma/strap distance (right).

The code has been completely tested in vacuum, while preliminary results with plasma show good agreement with data available in literature. As an example of the capabilities of the code, in Fig. 3 plots of R_c versus the strap length for various plasma densities and R_c versus the plasma distance from the straps are reported. In the latter case the straps are 320-mm long, 25-mm wide and 5-mm apart from each other (see Fig. 1b), whereas the plasma comprises deuterium and has a typical density of $2.8 \cdot 10^{19}$ m^{-3} [6]. The axial magnetic field B_0 was assumed 0.2 T, which yields a corresponding cyclotron frequency $f_{cD} \approx 1.9$ MHz. Simulations show that an optimum strap length and an optimum plasma-strap distance exist for which the coupling is maximized. Other parametric studies can be easily obtained by the code, thus yielding useful information for the ICRF antenna design phase.

REFERENCES

1. R. Maggiora, V. Lancellotti, G. Vecchi, V. Kyrytsya 2004 *Nucl. Fusion* **44** 846
2. R. Maggiora *et al.* 2005 *Proc. 16th RFPP*, Park City, UT
3. V. Lancellotti *et al.* 2005 *Proc. 16th RFPP*, Park City, UT
4. A. F. Peterson, S. L. Ray, R. Mittra 1998 *Computational Methods for Electromagnetics* (New York: IEEE Press)
5. E. Bearing, F. Chang-Diaz, J. Squire, 2004 *The Radio Science Bulletin*. No-310
6. V. Ilin, R. Franklin, F. Chang-Diaz, J. Squire, D. Carter, 2005 *Proc. 43 AIAA*, No-0949, pp.1–12
7. L. Valitutti 2005 Radio-Frequency Heating in Plasma Thrusters *MS Thesis* University of Illinois at Chigaco and Politecnico di Torino
8. L. B. Felsen, N. Marcuvitz 1973 *Radiation and Scattering of Waves* (Englewood Cliffs: Prentice Hall)
9. D. Shanks, 1955, *Journal of Mathematical Physics*, No-34, pp.1–42
10. R. Harrington 1993 *Field Computation by Moment Methods* (New York: Oxford)

Studies of ICRF Discharge Conditioning (ICRF-DC) on ASDEX Upgrade, JET and TEXTOR

A. Lyssoivan[1], R. Koch[1], D. Van Eester[1], G. Van Wassenhove[1],
M. Vervier[1], R. Weynants[1], H.G. Esser[2], V. Philipps[2], G. Sergienko[2],
E. Gauthier[3], V. Bobkov[4], H.-U. Fahrbach[4], D.A. Hartmann[4],
J.-M. Noterdaeme[4,5], V. Rohde[4], W. Suttrop[4], I. Monakhov[6],
A. Walden[6], TEXTOR Team, ASDEX Upgrade Team
and JET EFDA Contributors*

[1] *LPP-ERM/KMS, Association EURATOM-BELGIAN STATE, 1000 Brussels, Belgium*♦
[2] *Institut für Plasmaphysik FZ Jülich, EURATOM Association, D-52425 Jülich, Germany*♦
[3] *Association EURATOM-CEA, DSM-DRFC, CEA Cadarache, 13108 St Paul lez Durance, France*
[4] *Max-Planck Institut für Plasmaphysik, EURATOM Association, D-85748 Garching, Germany*
[5] *Gent University, EESA Department, B-9000 Gent, Belgium*
[6] *Association EURATOM-UKAEA, Culham Science Centre, Abingdon OX14 3DB, UK*
See J. Paméla, Fusion Energy 2004 (Proc. 20th Int. Conf. Vilamoura 2004) IAEA Vienna, paper OV1/2
♦*Partners in the Trilateral Euregio Cluster (TEC)*

Abstract. The present paper reviews the recent results achieved in the ICRF-DC experiments performed in helium/hydrogen mixtures in the non-circular tokamaks ASDEX Upgrade and JET and first tests of the ICRF discharges in helium/oxygen mixtures in the circular tokamak TEXTOR. Special emphasis was given to study the physics of ICRF discharges. A new recipe for safe and reliable RF plasma production [$<n_e> \sim (3-5) \times 10^{17}$ m^{-3}, $T_e \sim (3-5)$ eV] with improved antenna coupling efficiency (by 1.5-3 times) and improved radial/poloidal homogeneity was proposed and successfully tested: coupling the RF power in the FW-IBW mode conversion scenario in plasmas with two ion species. The first results on ICRF wall conditioning in helium/hydrogen and in helium/oxygen mixtures are analyzed.

Keywords: ICRF antenna, ICRF discharge, ITER, Wall conditioning
PACS: 52.25.Jm, 52.35.Hr, 52.40.Fd, 52.40.Hf, 52.50.Qt

INTRODUCTION

To apply the wall conditioning procedure in the future superconducting fusion reactors like ITER in between shots, only discharges fully compatible with the presence of a permanent high magnetic field can be used. The alternative ICRF Discharge Conditioning (ICRF-DC) technique was proposed and developed on the circular tokamaks TEXTOR, TORE SUPRA and HT-7. The technique demonstrated complete compatibility with the toroidal magnetic field, high wall conditioning efficiency [1-4] and was proposed for between-pulse conditioning in ITER [5]. The encouraging efficiency of the wall conditioning achieved on limiter tokamaks stimulated next steps in the development of the ICRF-DC on divertor tokamaks [6]. In this paper, the physics of ICRF discharge and the recent results achieved on ASDEX Upgrade, JET and TEXTOR are discussed.

ICRF-DC CHARACTERIZATION

The ICRF-DC experiments on reviewed tokamaks have been performed using the standard ICRF systems without any modifications in hardware and under the conditions mentioned elsewhere [1-6].

Neutral Gas RF Breakdown

The initiation of ICRF discharge in a toroidal magnetic field B_T results from the absorption of RF energy mainly by electrons [7,8]. The RF $\tilde{E}_z$-field (parallel to the B_T-field) is thought to be responsible for this process. In the general case of a poloidal loop-type ICRF antenna with a tilted Faraday shield (FS), the RF $\tilde{E}_z$-field in vacuum can be induced *electrostatically* and *inductively* [8]. However, in the ICRF band for the present-day fusion devices (~10–100 MHz), for most of the antenna κ_z-spectrum, the RF waves cannot propagate in the vacuum vessel: $\kappa_\perp^2 = \omega^2/c^2 - \kappa_z^2 < 0$, where $\kappa_\perp$ is the perpendicular wave-vector. Hence, the neutral gas breakdown and initial ionization may only occur locally at the antenna-near $\tilde{E}_z$-field (evanescent in vacuum). To avoid deleterious effects of the neutral gas breakdown and arcing inside the antenna box, the frequency of RF generators and the RF voltage/power at antenna straps were reduced to minimal values, while still meeting the requirements for ICRF breakdown outside of the antenna box [8]. Figure 1 shows the transition from the RF breakdown phase to the ICRF discharge phase in JET. It is clearly seen that the gas breakdown occurs after some delay and shows up in a drop in the antenna RF voltage (averaged over four radiating straps) and in a burst in the H_α emission (measured toroidally at ~130° away from the antenna port).

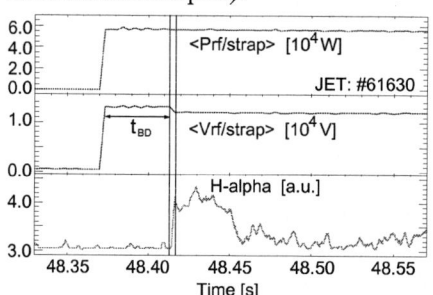

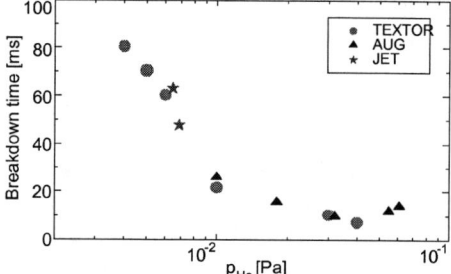

FIGURE 1. The transition from the neutral gas breakdown phase ($\omega \geq \omega_{pe}$) to the ICRF discharge phase ($\omega \leq \omega_{pe}$) in JET.

FIGURE 2. Pressure dependence of the RF breakdown time derived from the V_{RF} and H_α emission signals ($P_{RF/ant.strap} \approx 30-50$ kW, $f \approx 30$ MHz, $\omega = 4\omega_{cHe+} = 2\omega_{cD} = \omega_{cH}$), after [6].

From the point of view of ICRF system operation, such a correlation is the sign of RF discharge initiation outside of the antenna box and subsequent propagation of the initial low-density plasma ($\omega \geq \omega_{pe}$, $n_e(r) \leq 1.4 \times 10^{13}$ m^{-3} for $f=33.9$ MHz) along the magnetic field lines. The pressure dependence of the neutral gas breakdown time (associated with the RF voltage drop and the occurrence of the initial peak in the H_α emission) is plotted in Fig.2. Data from three tokamaks (TEXTOR, AUG and JET) were found to be in good agreement for the similar RF voltages (~8–13 kV), frequencies (~30–34 MHz) and the same pressures. It may be an indication that the antenna RF voltage (the antenna-near $\tilde{E}_z$ electric field) plays a fundamental role in the neutral gas breakdown and that the breakdown time is independent of the machines size.

ICRF Plasma Build-up

After the first (gas local breakdown) phase of the RF discharge, as ω_{pe} becomes of the order of ω, plasma waves can start propagating in a relay-race regime governed by the antenna κ_z-spectrum. The collisional absorption of the waves (T_e~3–5 eV during the ionization phase [3,8]) stimulates further space ionization of the neutral gas and plasma build-up in the torus (plasma phase). Such a non-resonant coupling allows RF plasma production at any B_T [8]. However, at plasma densities below a threshold for the FW propagation, coupling efficiency of the poloidal antennas is rather low in plasmas with a single ion species, $\eta = (R_{ant} - R_{loss})/R_{ant}$ ~20–30%. It results in the RF plasma build-up mainly at the machine low field side, LFS (antennas side) [6]. Both, antenna coupling and plasma homogeneity could be dramatically improved when a gas mixture of $H_2/(He+H_2)$~0.1–0.3 was used (Figs.3,4).

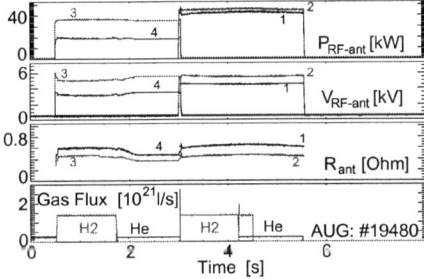

FIGURE 3. Effect of hydrogen/helium mixture on antenna coupling in AUG: $f_{ant3}=f_{ant4}=30.0$ MHz, $f_{ant1}=f_{ant2}=36.5$ MHz, $B_T=2.35$ T

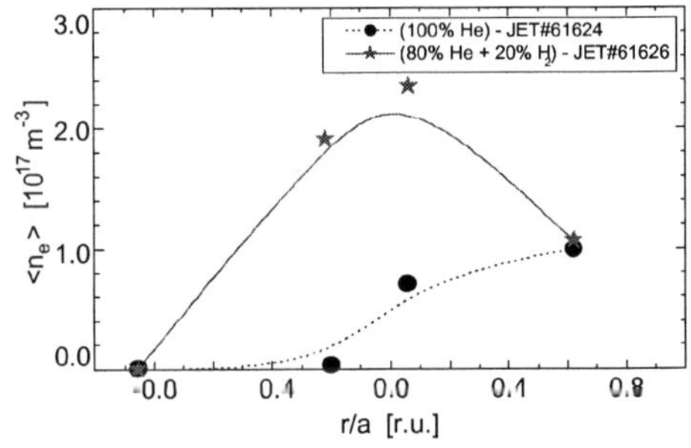

FIGURE 4. The line-averaged plasma density profiles for two ICRF discharges in JET: in pure helium (dashed) and in a gas mixture of $H_2/(He+H_2)$~0.2 (solid), $P_{RF-pl}\approx 70$ kW, $f=33.9$ MHz, $B_T=2.45$ T.

In the AUG case, the improved performance was achieved when the ICR layer $\omega=\omega_{cH}$ and the nearby FW-IBW mode conversion layer were shifted to the LFS. As a result, the antenna coupling efficiency increased up to 3 times at $f=30.0$ MHz (Fig.3). In the JET case, the radial extension of RF plasmas towards the HFS was clearly seen from the multi-channel FIR interferometer data (Fig.4). This effect has been predicted from the electron energy deposition profiles calculated with the 1D RF code [9] for helium RF plasmas with different H concentrations.

Analysis of the core atomic spectroscopy and the VUV spectroscopy data showed appearance of the H_α, D_α and HeI (neutral) lines during the JET ICRF discharges.

Assuming an equilibrium (coronal) ionization balance, $T_e\sim2-5$ eV was derived from the ionization stages observed for shots at the gas pressure $p_{tot}\approx(2-6)\times10^{-3}$ Pa.

All ICRF-DC experiments performed until now reported on the generation of high-energy fluxes of H (with energies up to 60 keV) and of D atoms (up to 25 keV), which were detected by a neutral particle analyzer (NPA) [3,6,8]. Clear evidence of tail formation in the distribution functions of H and D atoms was observed at higher ion cyclotron harmonics ($\omega=2\omega_{cH}=4\omega_{cD}$). This fact may be understood in terms of RF quasilinear diffusion: ion cyclotron harmonic heating tends to accelerate the faster particles more, with tail formation at higher energy than for fundamental heating [10].

ICRF Wall Conditioning Tests

One of the major issues in ITER is the retention of tritium in re-deposited carbon layers. The removal of hydrocarbon layers by oxygen is considered as the most promising method. Directly related pilot experiments have successfully been performed on TEXTOR, addressing ICRF discharge initiation and a-C:H-film removal in the O_2/He mixtures. The mass-spectrum analysis of the residual gas revealed that injected oxygen was converted into CO and CO_2 (Fig.5). The estimated C-erosion rate was ~0.13 nm/s assuming to be homogeneous over whole vessel area. In the AUG case, the

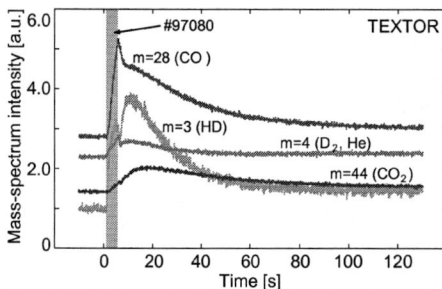
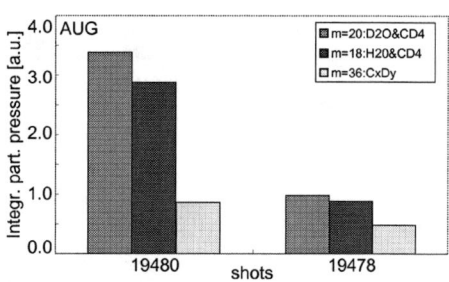

FIGURE 5. ICRF-DC in the O_2/He mixture in TEXTOR: $P_{RF-pl}\approx50$ kW, $f=29.0$ MHz, $B_T=2.3$ T, He-flow, O_2-puff (15.8 mbarl during RF pulse).

FIGURE 6. Effect of ICRF-DC in AUG in helium/hydrogen mixture (#19480, $P_{RF-pl}\approx50$ kW) and in pure helium (#19478, $P_{RF-pl}\approx30$ kW).

conditioning efficiency was found higher in the H_2/He mixtures compared with the pure helium gas (Fig.6) due to better both antenna coupling and plasma homogeneity.

ACKNOWLEDGEMENTS

This work has been performed under the European Fusion Development Agreement.

REFERENCES

1. H.G. Esser, et al., *Journal of Nuclear Materials* 241-243 (1997) 861-866.
2. A. Lyssoivan, et al., in *14th Topical Conf. on Radio Frequency Power in Plasmas, Oxnard 2001*, AIP Conf. Proceedings 595, New York 2001, p.146-149.
3. E. de la Cal and E. Gauthier, *Plasma Phys. Control. Fusion* 39 (1997) 1083-1099.
4. J.K. Xie, et al., *Journal of Nuclear Materials* 290-293 (2001) 1155-1159.
5. ITER Physics Guidelines, N19 FDR 1 01-07 13R0.1, (2001), pp.109-110.
6. A. Lyssoivan, et al., *Journal of Nuclear Materials* 337-339 (2005) 456-460.
7. M.D. Carter, et al., *Nuclear Fusion* 30 (1990), pp.723-730.
8. A. Lyssoivan, et al., *Final Report on ITER Design Task D350.2*, LPP-ERM/KMS 114 (1998).
9. D.Van Eester and R.Koch, *Plasma Phys. Control. Fusion* 40, 1949-1975 (1998).
10. R. Koch, Transactions of Fusion Sciens and Technology 45 (2004) 203-210.

Experimental Results of the Coaxial Multipactor Experiment (CMX)

Timothy P. Graves*, B. LaBombard*, S.J. Wukitch* and I.H. Hutchinson*

MIT-PSFC, Cambridge, MA 02139

Abstract. A multipactor discharge is a resonant condition for electrons in an alternating electric field. This discharge can be disruptive to RF circuits, cavities, and resonators by detuning the circuit and/or by seeding an arc with a partially developed multipactor discharge. The Coaxial Multipactor Experiment (CMX) investigates this discharge with goals of measuring the electron distribution, current, and absorbed power from the non-uniform RF field in coaxial transmission lines. CMX has a unique experimental setup which can support a multipactor discharge in a short section of continuous transmission line. A retarding potential analyzer with secondary electron suppression measures the multipactor electron distribution. Results depict a narrow, relatively high-energy distribution of electrons which exhibits energy dependence on frequency and not pressure below 1 mtorr. Each distribution has an energetic tail which extends to the maximum RF cavity voltage. Monte Carlo simulations reproduce the measured distributions and show dependence on the initial electron energy and phase distributions.

Keywords: Multipactor, Coaxial, RF breakdown, RF plasma heating
PACS: 52.80.Pi

INTRODUCTION

A multipactor discharge is a low voltage phenomena in which the electrons impact a surface in resonance with an alternating electric field [4, 6]. Electrons hit with sufficient energy to cause electron multiplication by secondary emission; i.e. the secondary emission coefficient, $\delta(E) \geq 1$. Because the discharge requires only electrons emitted from surfaces, multipactoring generally occurs under high vacuum conditions. Multipactor susceptibility spans a frequency range from MHz to tens of GHz [6]. Since its discovery, there has been little research on multipactoring because the majority of RF transmission lines are pressurized and therefore not susceptible to multipactor discharges. In the cases of fusion, accelerator, and space research, vacuum transmission lines are typically unavoidable and the frequencies and gap sizes are often in the range for multipactor discharges. These discharges can disrupt operation of the RF systems by detuning the circuit, dropping the Q factor, and limiting the circulating power. Multipactor discharges can lead to high voltage arcs or other catastrophic discharges in some cases.

For stripline or parallel plate geometries, the electric field is uniform in space and multipactor voltage is equal to the square of frequency and gap size, but for the coaxial transmission line, the electric field is non-uniform and inversely proportional to the radius. Further complications arise when a magnetic field is present, causing higher voltage limits and more complicated electron trajectories [5]. It has also been found that multipactor discharges can interact with travelling or mixed waves and move with the wave, possibly to a region of higher voltage, increasing the risk of arcing [7].

THE COAXIAL MULTIPACTOR EXPERIMENT (CMX)

CMX is a unique experiment specifically designed to study multipactor discharges in the coaxial and parallel plate geometry. A continuous transmission line with a high vacuum section makes CMX very similar to the situation in ICRH fusion experiments. The experiment is a high Q resonator with frequencies of 50 MHz to 150 MHz. The experimental goals of CMX are to determine the multipactor electron distribution functions, current density, and their dependence on frequency and pressure. Also, CMX investigates the spatial distribution and absorbed power of the discharge.

Experimental Setup

A 5W, RF source powers a high Q resonator (1500) which provides a voltage gain on the unmatched side of the RF circuit. A standard 10 cm (4") copper coaxial transmission line passes continuously through the multipactor vacuum chamber (MVC), in which 15 cm of the line is in UHV conditions ($10^{-8} torr$). The maximum voltage, typically 100-200V, is placed inside the MVC by means of a moveable shorted stub. RF measurements include three pairs of directional couplers: one on the matched side, and two on either side of the MVC. Figure 1 illustrates the experimental setup.

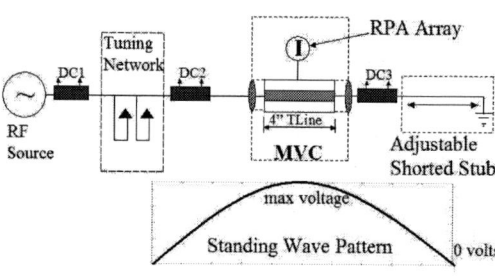

Figure 1: CMX Experimental Setup

An array of 12 retarding potential analyzers (RPA) with secondary electron suppression measure the electron distributions and current density of the multipactor discharge. Electrons enter the RPA through an entrance aperture of 3 mm and pass through a grounded and suppression grid before reaching the collector. The suppression grid is biased 20 volts negative relative to the collector to suppress all secondary electrons. The RPA voltage is swept up to 500V to measure the current as a function of voltage. A Keithley 602 electrometer and differential amplifier amplify the small nanoamp level current measured by the RPA. All RF directional coupler and RPA collector data are digitized by a National Instruments 1 MHz, 32 channel digitizer.

Experimental Results

Results from CMX's coaxial multipactor discharge depict non-maxwellian electron distribution functions shown in Figure 2. The figure illustrates the distributions at 4 different frequencies with 2 bulk groups of electrons: low energy, non-multipacting electrons and higher energy, multipacting/resonanting electrons. The -20V suppression voltage of the RPA prevents the collection of any electrons with energy less than 20eV.

For all frequencies, the high energy electron group is approximately at the same energy, but the distribution broadens for increasing frequency, and the tail cut-off extends to the corresponding maximum chamber voltage. Figure 2 illustrates the tail cut-off of each distribution in the inset section.

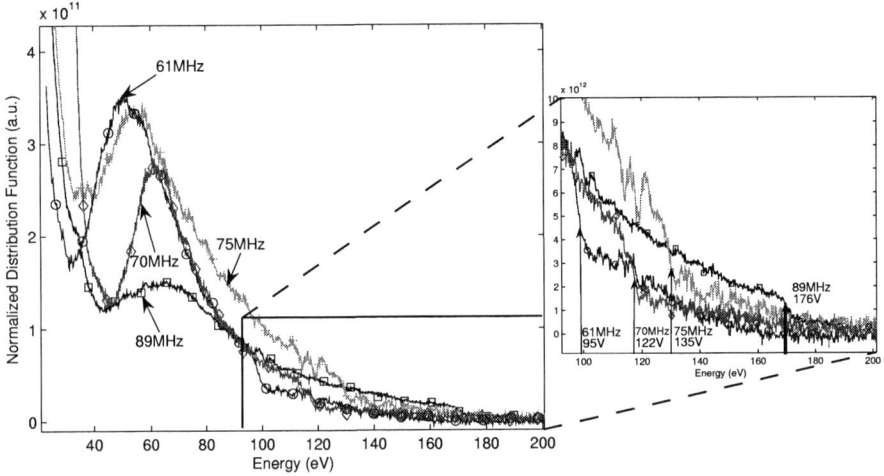

Figure 2: Electron distribution functions for 4 different frequencies. The inset depicts each distribution with a high energy cut-off roughly equal to the maximum MVC voltage.

The electron distributions show little dependence on pressure up to 1 mtorr as expected. The electron mean free path is just approaching the system size at 1 mtorr. Also, the discharge is uniformly distributed azimuthally around the coaxial region.

MONTE CARLO SIMULATION

In order to better understand the experimentally determined electron distribution functions, the steady-state multipactor discharge was modelled by a Monte Carlo simulation. The simulation is a 1-D, non-coupled electron code which solves the equations of motion in the non-uniform field by a Runga-Kutta solver. The randomly sampled Monte Carlo variables are the initial electron energy determined by an appropriate distribution function [2] and the relative phase to the RF field.

Figure 3a depicts distribution functions strikingly similar to the experimental distributions shown in Figure 2. The distributions broaden with frequency and extend out to the maximum line voltage as with the measured distributions. Figure 3b shows that the majority of the electrons impact the outer conductor with the proper phase to be resonant; i.e. they hit at 2π cycle time, just as the field changes sign. Figure 3c shows that electrons with sufficient energy to create and sustain the electron multiplication (E>100eV) hit at this resonance time.

For multipactor onset for 70MHz at 120V, simulations show $3-5\%$ of electrons in multipactoring distributions are above 100eV. Below the onset voltage no multipactor is seen experimentally, and no electrons with E>100eV are seen in simulations, implying the electrons with E>100eV sustain the discharge. E_{min} where $\delta = 1$ is unknown for the MVC surface, but it can be reduced down by 50% of $E_{min} = 200eV$ for pure copper because of surface impurities. [1].

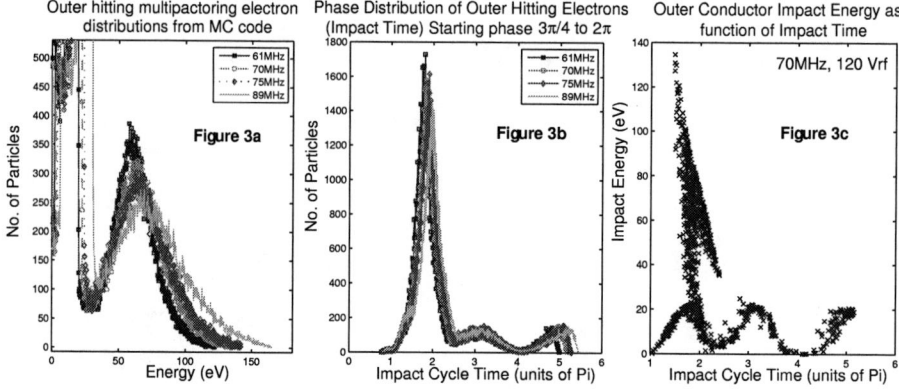

Figure 3: Monte Carlo Simulation results for outer conductor electron impacts. Figure 3a depicts the electron energy distribution of 10^5 particles. 3b depicts the electron impact time distribution, and 3c depicts the electron energy as a function of impact time.

CONCLUSIONS

CMX has repeatedly produced and studied a coaxial multipactor discharge that raises the MVC pressure signifying a desorption of surface gases. The measured electron distributions show 2 main electron groups: a low energy, non-multipacting group, and a higher energy (60ev) multipacting group with a tail energy up to the maximum voltage. Monte Carlo simulations give very similar distributions to the experimental findings. The distribution shape is due to initial electron energy and phase distributions. Only electrons which have both the appropriate initial energy and phase will be able to contribute to the growth and sustainment of the multipactor. By integrating the distributions, only $3-5\%$ of the electrons need to hit with energy above the assumed $\delta > 1$ threshold for multipactor onset.

REFERENCES

1. H. Bruining. *Physics and Applications of Secondary Electron Emission*. McGraw-Hill, NY, 1954.
2. M.S. Chung and T.E. Everhart. *J. Appl. Phys.*, 45(2), 1974.
3. R.A. Kishek, Y.Y. Lau, and D. Chernin. *Phys. Plasmas*, 4(3), March 1997.
4. R.A. Kishek, Y.Y. Lau, and et. al. *Phys. Plasmas*, 5(5), May 1998.
5. S. Riyopoulos, D. Chernin, and D. Dialetis. *Phys. Plasmas*, 2(8), August 1995.
6. J.M. Vaughan. *IEEE Trans. Electron Devices*, 35(7), July 1988.
7. P. Yla-Oijala. TESLA 97-21, November 1997.

AUTHOR INDEX

A

Akutsu, T., 15
Alberti, S., 387
Anderson, J. K., 341
Andrew, P., 273
Angioni, C., 379
Aniel, T., 307
Arnoux, G., 387
Artaud, J. F., 307
Ashikawa, N., 98

B

Bae, Y. D., 250
Baity, F. W., 74, 194
Basiuk, V., 150, 269, 307
Basse, N., 114, 327
Batchelor, D. B., 23, 46, 287
Baxi, C. B., 407
Beaumont, B., 182, 210
Beaumont, P., 122, 273
Beck, W., 327
Becker, W., 242
Behn, R., 399
Bell, R. E., 86
Bergkvist, T., 50, 273
Bernabei, S., 66, 82, 86, 327
Berry, L. A., 3, 23, 42, 46, 287
Bers, A., 269, 337, 353
Biewer, T. M., 66, 82, 86
Bigelow, T. S., 337
Bilato, R., 38, 134
Birus, D., 242, 246
Bobkov, V., 90, 273, 445
Bonoli, P. T., 23, 31, 46, 114, 118, 138, 287
Bosia, G., 150, 174, 178, 182
Boswell, C., 114
Bottino, A., 399
Boxer, A. C., 395
Brambilla, M., 38, 134, 287
Brand, P., 371
Braun, F., 238
Braune, H., 371
Brémond, S., 150, 174, 178, 182, 210

Brix, M., 273
Brizard, A. J., 146
Brzozowski, J., 273
Budny, R., 118
Burke, D. R., 319

C

Calabrò, G., 311
Callis, R. W., 407
Camenen, Y., 399
Cardinali, A., 295
Carter, M. D., 3, 23, 46, 218, 222, 287, 337, 437
Castaldo, C., 295
Caughman, J. B., 337
Cengher, M., 341
Cesario, R., 295
Chan, V. S., 15, 31
Chang, C. S., 118
Chantant, M., 150, 210
Chavan, R., 403
Chikaraishi, H., 98
Childs, R., 327
Chiu, S. C., 15
Choi, C., 106
Choi, M., 15, 31, 74
Clayton, D. J., 319
Coda, S., 357, 399
Colas, L., 150, 178, 182, 210, 214
Cooper, W. A., 62
Cox, W. A., 265, 341

D

Dammertz, G., 371
D'Azevedo, E., 23, 46, 287
Decker, J., 269, 337, 353, 361
deGrassie, J. S., 110
De La Luna, E., 122
De Vries, P., 122
Diem, S., 66, 337
D'Ippolito, D. A., 3, 23, 218, 222, 287
Dumont, R. J., 23, 257
Dumortier, P., 158, 190, 198, 202

Durodié, F., 158, 186, 194
Dux, R., 379

E

Edlund, E., 114
Efthimion, P. C., 337
Ejiri, A., 365
Ekedahl, A., 90, 210, 307, 315
Ellis, R., 327
Ellsworth, J. L., 395
Erckmann, V., 371
Eriksson, L.-G., 54, 90, 110, 273
Ershov, N. M., 46, 337
Esser, H. G., 445
Evrard, M., 158, 186

F

Fable, E., 399
Fadnek, A., 194
Fahrbach, H.-U., 38, 445
Faudot, E., 150, 214
Felton, R., 122
Fiore, C., 114, 118
Forest, C. B., 265, 341
Fredd, E., 74, 327, 337
Freudenberg, K. D., 194
Fujita, T., 279
Fukuyama, A., 15
Funaba, H., 98

G

Gantenbein, G., 371, 379
Garnier, D. T., 395
Gauthier, E., 445
Giroud, C., 273
Giruzzi, G., 257, 307
Goetz, J. A., 319, 323
Goniche, M., 150, 210, 307
Goodman, T. P., 399
Gorelenkov, N., 114
Gorelov, I. A., 407
Goto, M., 98
Goulding, R. H., 194
Gowers, C., 122

Graves, T. P., 449
Green, M. T., 407
Greenough, N., 327
Grimes, M., 327
Gunn, J. P., 150
Günter, S., 379
Gwinn, D., 327

H

Hanada, K., 365
Hansen, A. K., 395
Harling, J., 122
Hartmann, D. A., 242, 246, 445
Harvey, R. W., 23, 46, 74, 106, 287, 337, 357
Heidbrink, W. W., 74
Hellsten, T., 50, 54, 273
Henderson, M., 403
Henderson, M. A., 399
Heuraux, S., 150, 214
Higaki, H., 98
Hoang, G. T., 303
Hoffman, D. J., 429, 437
Hong, B. G., 250
Hosea, J., 66, 327, 337
Hosea, J. C., 74, 82, 86, 106, 194
Hosoyama, H., 279
Hubbard, A., 114
Hughes, J., 118
Hutchinson, I. H., 449
Huysmans, G., 269

I

Ichimura, M., 98
Ida, K., 98
Ide, S., 279, 365
Igami, H., 98, 345, 411
Ikeda, K., 98
Imbeaux, F., 307
Irby, J., 327
Isayama, A., 279
Isobe, M., 15

J

Jaeger, E. F., 3, 23, 42, 287
Jaeger, F., 46, 337
Jaun, A., 142

Jayakumar, R. J., 265
Joffrin, E., 273
Johnson, T., 50, 54, 273

K

Kajiwara, K., 407
Kaneko, O., 98
Karim, I., 395
Kasahara, H., 98
Kasilov, V., 15
Kasparek, W., 371
Kato, A., 98
Kaufman, A. N., 142, 146
Kaufman, M. C., 319, 323
Kawahata, K., 98
Kesner, J., 395
Kiptily, V., 122, 273
Kirov, K. K., 315
Koch, R., 445
Koert, P., 327, 331
Komori, A., 98
Kramer, G. J., 114
Kubo, S., 98, 411
Kumazawa, R., 15, 98, 411
Kung, C. C., 327
Kwak, J. G., 98
Kwak, J.-G., 250
Kyrytsya, V., 166, 230

L

LaBombard, B., 327, 449
Lamalle, P. U., 90, 122, 158, 186, 190, 198, 202, 234
Lancellotti, V., 166, 230, 441
Lao, L., 15
Laqua, H. P., 371
Lawson, K., 122
Laxåback, M., 50, 122, 273
LeBlanc, B. P., 66, 82, 86
Lerche, E. A., 122, 234
Leuterer, F., 379
Lin, L., 114
Lin, Y., 114, 118, 138, 206
Liptac, J., 327
Loesser, G. D., 194, 327
Lohr, J., 265, 407

Lomas, P., 122
Louche, F., 158, 190, 198
Luce, T. C., 265
Luo, Y., 74
Lyssoivan, A., 90, 445

M

Maekawa, T., 345, 415
Maggi, C., 38
Maggiora, R., 166, 230, 441
Mahar, S., 395
Mailloux, J., 90, 273, 307, 315
Makowski, M. A., 265
Manini, A., 379
Mantsinen, M., 90, 273
Mantsinen, M. J., 122
Maraschek, M., 379
Marlétaz, B., 387
Marmar, E., 114, 327
Marmillod, Ph., 387
Martin, Y., 387
Martynov, An., 399
Masuzaki, S., 98
Matthews, G., 273
Mauel, M., 395
Mayoral, M.-L., 90, 122, 273
Mazon, D., 307
McMahon, S. M., 341
Mellet, N., 62
Meo, F., 90, 122, 273
Messiaen, A. M., 158, 190, 198, 202
Michel, G., 371
Milanesio, D., 166, 230, 441
Mitarai, O., 365
Mitteau, R., 150
Miyazawa, J., 98
Monakhov, I., 90, 273, 445
Morisaki, T., 98
Morita, S., 98
Motojima, O., 98
Mück, A., 379
Muir, A., 407
Münich, M., 379
Murakami, M., 74
Murakami, S., 15
Mutoh, T., 15, 98, 411
Myra, J. R., 3, 23, 218, 222, 287

N

Nagaoka, K., 98
Nagayama, Y., 98
Nakajima, N., 15
Nakamura, Y., 98, 411
Narihara, K., 98
Nelson, B. E., 194
Nelson-Melby, E., 357
Neu, R., 379
Nguyen, F., 273
Nightingale, M., 194
Nikkola, P., 399, 403
Nishimura, K., 98
Noda, N., 98
Notake, T., 98, 411
Noterdaeme, J.-M., 90, 110, 122, 273, 445
Nowak, S., 387
Nunes, I., 122

O

Ogawa, H., 98
Ohkubo, K., 98, 411
Ohyabu, N., 98
Oikawa, T., 279
Oka, Y., 98
Okuda, H., 287
Oliva, S. P., 323
Ortiz, E. E., 395
Osakabe, M., 15, 98

P

Paoletti, F., 295
Parisot, A., 138, 166, 206
Parker, R., 138, 327, 331
Pavlo, P., 349
Peeters, A. G., 379
Pericoli Ridolfini, V., 311
Perkins, R., 118
Peterson, B. J., 98
Petty, C. C., 74, 106, 265, 273
Peysson, Y., 269, 307
Philipps, V., 445
Phillips, C. K., 23, 66, 82, 226, 287
Piazza, G., 122

Pinsker, R. I., 31, 74, 106, 341
Pochelon, A., 399
Ponce, D., 407
Popovich, P., 62
Porkolab, M., 74, 106, 114, 118, 138, 206
Porte, L., 387
Prater, R., 74, 106, 265
Preinhaelter, J., 337, 349

R

R., P. Moreau, 210
Rachlew, E., 273
Ram, A. K., 138, 269, 337, 357, 361
Ramponi, G., 403
Rantamaki, K., 90
Rasmussen, D. A., 194, 337
Rohde, V., 445
Russell, D. A., 3, 218, 222
Ryan, P., 66, 82
Ryan, P. M., 437
Ryter, F., 379

S

Saibene, G., 403
Saida, T., 15
Saito, K., 15, 98, 411
Sakamoto, M., 98
Sakata, S., 279
Salmi, A., 90
Sanchez, F., 403
Santala, M., 90, 122
Sartori, R., 273
Sasao, M., 15
Sato, K., 98
Sauter, O., 399, 403
Schilling, G., 114, 118, 206, 327
Seki, M., 279
Seki, T., 15, 98, 411
Sergienko, G., 445
Sharapov, S., 90
Shevchenko, V., 349
Shidara, H., 403
Shimozuma, T., 98, 411
Shimpo, F., 98
Shiraiwa, S., 365

Shoji, M., 98
Smirnov, A. P., 46, 74, 106, 337, 357
Smithe, D. N., 23, 46, 226, 287
Snipes, J., 114
Sparks, D. O., 194
Squire for, J. P., 421
Staebler, A., 273
Stutman, D., 82
Sudo, S., 98
Sueoka, M., 279
Suttrop, W., 38, 445
Suzuki, T., 279
Svidzinski, V., 341
Swain, D. W., 66, 82

T

Takahashi, C., 98
Takase, Y., 98, 365
Takeiri, Y., 98
Takeuchi, N., 98
Tala, T., 273
Tanaka, H., 345, 415
Tang, V., 31
Taylor, G., 337, 349
Tennfors, E., 273
Terry, D., 327, 331
Terry, J., 327
Terry, J. L., 118
Thomas, M. A., 319, 323
Thumm, M., 371
Tokuzawa, T., 98
Torii, Y., 98
Tracy, E. R., 142, 146
Tsumori, K., 98
Tuccillo, A., 273

U

Uchida, M., 415
Urban, J., 349
Ushigome, M., 365

V

Vahala, G., 349
Vahala, L., 349
Valitutti, L., 441
Valovic, M., 349

Van Eester, D., 58, 90, 122, 273, 445
Van Wassenhove, G., 445
Van Zeeland, M. A., 74
Vecchi, G., 166, 230, 441
Vervier, M., 158, 198, 202, 445
Vieira, R., 327
Villard, L., 62

W

Wagner, D., 379
Walden, A., 445
Walden, W., 273
Wallace, G., 327, 331
Walton, R., 194
Wang, S. J., 250
Watanabe, T., 98
Watari, T., 15, 98
Weitzner, H., 130
Wendorf, J., 246
Wesner, F., 246
Weynants, R., 158, 445
Wilgen, J. B., 337
Wilson, J. R., 66, 82, 86, 327, 337
Wright, J., 31
Wright, J. C., 23, 46, 138, 287
Wukitch, S. J., 114, 118, 138, 166, 206, 331, 449

Y

Yamada, H., 98
Yamada, J., 415
Yamaguchi, S., 415
Yokota, M., 98
Yoon, J. S., 250
Yoshimura, Y., 98, 411
Yoshinaga, T., 415

Z

Zaks, J., 327
Zastrow, K.-D., 273
Zhao, Y. P., 98
Zohm, H., 379, 403
Zucca, C., 399
Zweben, S. J., 118
Zwingman, W., 307